Unsteady Computational Fluid Dynamics in Aeronautics

FLUID MECHANICS AND ITS APPLICATIONS
Volume 104

Series Editor: Andre THESS
Department on Mechanical Engineering
Ilmenau University of Technology
98684 Ilmenau, Germany

Aims and Scope of the Series

The purpose of this series is to focus on subjects in which fluid mechanics plays a fundamental role.

As well as the more traditional applications of aeronautics, hydraulics, heat and mass transfer etc., books will be published dealing with topics which are currently in a state of rapid development, such as turbulence, suspensions and multiphase fluids, super and hypersonic flows and numerical modeling techniques.

It is a widely held view that it is the interdisciplinary subjects that will receive intense scientific attention, bringing them to the forefront of technological advancement. Fluids have the ability to transport matter and its properties as well as to transmit force, therefore fluid mechanics is a subject that is particularly open to cross fertilization with other sciences and disciplines of engineering. The subject of fluid mechanics will be highly relevant in domains such as chemical, metallurgical, biological and ecological engineering. This series is particularly open to such new multidisciplinary domains.

The median level of presentation is the first year graduate student. Some texts are monographs defining the current state of a field; others are accessible to final year undergraduates; but essentially the emphasis is on readability and clarity.

For further volumes:
www.springer.com/series/5980

P.G. Tucker

Unsteady Computational Fluid Dynamics in Aeronautics

Springer

P.G. Tucker
Department of Engineering
Whittle Laboratory
University of Cambridge
Cambridge, UK

ISSN 0926-5112 Fluid Mechanics and Its Applications
ISBN 978-94-007-7048-5 ISBN 978-94-007-7049-2 (eBook)
DOI 10.1007/978-94-007-7049-2
Springer Dordrecht Heidelberg New York London

Library of Congress Control Number: 2013945727

Printed on acid-free paper

Springer is part of Springer Science+Business Media (www.springer.com)

To my Family

Preface

In 2001 I published 'Computation of Unsteady Internal Flows'. This text was largely based around incompressible flow solver methods and hence typically lower speed flows. The key premise behind the original text was that, in some sense, most engineering flows are intrinsically unsteady (even if just due to turbulence). However, because of computational expense, this aspect is often ignored. Of course computing power continues to rise. The use of Graphical Processor Units for number processing is showing promise with rival technologies beginning to emerge.

Detached Eddy Simulation and related eddy resolving methods have added impetus to the use of unsteady Computational Fluid Dynamics. Simulations that potentially rival tremendously expensive rig/wind tunnel tests are now appearing. A notable shoot from this emerging era is work around 2007 at the US Airforce Laboratory, who performed DES for a F/A-18 fighter configuration. Tail buffet was explored and successful comparison made with real flight data (in terms of spectral shape of surface pressure data). This situation was not unforeseen. Around 1975, Chapman, Director of Aeronautics at NASA, proposed, using well founded scientific arguments,[1] that when computers reached 10^{14} flops, eddy resolving simulations that could rival aerodynamic tests would emerge. Modern high performance computing provision now exceeds Chapman's expectations, reaching Peta scale and beyond. Hence, now the ability to directly predict turbulence, for complex engineering systems, without recourse to accuracy reducing assumptions becomes ever closer—even if advances in solver technology have not been as extensive as perhaps expected by Chapman. The current text focuses on aerospace. Hence, unlike the former, it also includes discussion of compressible flow technology.

With the projected demand for air transport set to double the world aircraft fleet by 2020 it is becoming urgent to take steps to reduce environmental impact with respect to noise and other emissions. Hence, the current text, hopefully, will contribute, in some sense, to the quest to use computers to improve aircraft

[1] Note, Chapman's outer boundary layer scalings are optimistic but this aspect is less critical than the inner scalings.

and thus impact on this pressing environmental need. To make major technological breakthroughs, ultimately, extremely close airframe and engine integration will be needed. This gives the requirement for coupled engine-airframe simulations. Also, increasingly multi-physics simulations will be required. Such endeavors do not marry well with the obvious accuracy benefits provided by making turbulent eddy-resolving simulations. Hence, this text attempts to explore these tensions.

In preparing the text, great effort has been made to remove errors of a typographical nature. Apologies for the errors that are doubtless found.

I would like to express my gratitude to past Researchers who have helped run many of the simulations contained in this text. Especial thanks are due to my longest serving team members—Drs. R. Jefferson-Loveday and J. Tyacke.

The original text was prepared in WORD. Then Vadlamani Nagabhushana Rao lead an intrepid team who kindly converted the text to LaTeX, properly linking references figures and equations to the text. I am very grateful to the LaTeX team: Ahmed Al-Shabab; Zaib Ali, Jiahuan Cui, Mahak Mahak; James Page; Vadlamani Nagabhushana Rao, Robert Watson and Xiaoyu Yang.

Cambridge, UK
January 2013

Paul G. Tucker

Contents

Nomenclature[1]

Lowercase Roman

a_{ij}	Anisotropy tensor
c	Chord, speed of sound, wave velocity, turbulence modelling constant or aspect ratios
c'	Pseudo acoustic speed
c_p	Specific heat capacity at constant pressure
c_v	Specific heat capacity at constant volume
d	Normal wall distance
d_{ij}^{n-1}, d_{ij}^{n}	Edge lengths in spring analogy before and after movement, respectively
$\tilde{d}$	Turbulence length scale in DES
$\tilde{d}_P$	Wall distance from Poisson equation
f	Frequency
f_a	Frequency of 1st harmonic of feedback loop (see Chap. 4)
f_{sw}	Body force wake activation function
f_w, f_{v1}, f_{v2}, f_d	Functions in the Spalart-Allmaras turbulence model
$f(\tilde{d})$	Function in Hamilton-Jacobi equation
$\tilde{f}_p, \tilde{f}_b, \tilde{f}_{KH}, \tilde{f}_{BL}$	Blade wake/disturbance, shedding, Kelvin-Helmholtz and boundary layer dimensionless frequencies
$g(d)$	Function in Hamilton-Jacobi equation
h	Blade thickness or heat transfer coefficient
i, j, k	Array or grid point location identifiers
k	Thermal conductivity, turbulent kinetic energy, temporal weighting function component or variable to ensure that the acoustic wave speed is similar to the particle speed

[1]The nomenclature is setout as follows—First lowercase Roman letters are given followed by uppercase. Then Greek lowercase followed by uppercase symbols are given. Then superscripts and subscripts are setout. Over bars are then listed followed by special symbols and operators. Finally, the abbreviations used in the text are summarized.

$k_f = \omega c/2U_\infty$	Reduced frequency of aerofoil pitching
k_{ij}	Coefficient in spring analogy
k_P, k_I, k_D	Free parameters in proportional integral controller
l	Turbulence or correlation length scale or ratio of edge lengths
l_μ, l_ϵ	Turbulence model length scales
$\dot{m}$	Mass flow rate
n	Number of resolved Fourier coefficients, direction cosine or number of iterations
n_{cpl}	Number of global solution cycles
n_p	Pressure switch in body force model
p	Static pressure, number of coefficients used in scheme or number of stages
q	Heat flux or solution variable
r, θ, z	Cylindrical polar spatial coordinates
r_d	Shielding function in delayed DES
r_e, r_n	Exact and numerical wave numbers
rms_ϕ	Normalised root mean square change
s	Transition function, streamwise distance from trailing edge, surface or surface area
t	Time
t_{fp}	Time for a fluid particle to pass through a blade passage
t_i	Turbulent time scale
t_l	Time period needed to gain a satisfactory average
t_p	Periodic time scale
$t_r = t - \|x - y\|/c$	Retarded time
t_t	Transient time scale
u, v, w	Instantaneous x, y, z, velocity components
w	Wave number, velocity component or wake half width
w_c	Cut-off wave number
w'	Numerical wave number
x, y, z	Spatial coordinates

Uppercase Roman

A	Global representation of spatial discretization, central nodal coefficient, amplitude or Roe matrix
$\hat{A}$	Root mean square of quantity in correlation coefficient for zero space and time separations
A_μ, A_ϵ	Turbulence model constants
A_ω	Average cross sectional area normal to vorticity vector
B	Amplitude
C	Courant number ($u\Delta t/\Delta x$), constant or amplitude
C_s	Smagorinsky constant
C_l, C_μ, C_w	Turbulence model constants
C_t	Safety factor
$C_{\Delta t}$	Time domain objective
C_{ij}	Clark terms

C_{pt}	Total pressure loss coefficient
C_D	Drag coefficient
C_f	Skin friction coefficient
C_M	Moment coefficient
C_p	Surface pressure coefficient
D	Time step to diffusion time scale ratio or diameter scale
E	Solution error, flux term or energy in turbulence energy spectrum
F_p, F_n	Forces parallel and normal to blade passages, respectively
$[F_S]$	Force matrix
F	Flux term
F_{SST}	Delayed DES function in Menter SST framework
G	Flux, grid distortion term or filter kernel/operator
H	Representation of step height or closing disc extent
$H(x, \overline{\Delta}, \tau)$	Correlation between velocity and temperature located at x
I	Second moment of area tensor
I_o	Principal axis system
Id	Identity matrix
J	Jacobian or flux component
K_n, K_{vd}, K_w, K_l, K_p	Wake body force model calibration constants
K_0	Constant in Kolmogorov energy spectrum
$[K_s]$	Stiffness matrix
$[K_f]$	Fluid system matrix
L	Length scale, integration range, linear turbulent stress component or wave operator
L_{vK}	von Karman length scale
$\tilde{L}$	Length scale in Hamilton-Jacobi equation
L_{ij}	Leonard terms
m	Order of scheme
M	Mach number or capacitance or mass operator/matrix
N	Number of mesh points, realizations or blades
Nb	Number of blades
N_{CD}	Order of central difference scheme
Nf	Number of Fourier modes
$\overline{NL}$	Non-linear turbulent stress component
N_{NB}	Marker zone variable in zonal hybrid RANS-LES
$N_{P\phi}$	Number of mesh points for each solution variable
N_{UP}	Order of upwind scheme
$Nu = qL/k\Delta T$	Nusselt number
N_ϕ	Number of solution variables
P	Number of processors
P_k, P_{shear}	Production of turbulence energy
$Pr = \mu c_p/k$	Prandtl number
Q	Jacobian or under-relaxation function
R	Gas constant, radius scale or residual

Re	Reynolds number
$R(x, \overline{\Delta}, \tau)$	Correlation coefficient, for velocity, located at x
$[R]$	Coupling matrix
S	Source or strain term
S_{ij}	Mean strain rate tensor
St	Strouhal number
T	Temperature
$\mathbf{T}$	Transformation matrix
Te	Execution time
Ti	Turbulence intensity
T_{ij}	Lighthill stress tensor
TV	Total variation
U, U_θ	Blade and fluid tangential velocities, respectively
U, V, W	Instantaneous contravariant velocity components
$[U]$	Displacement matrix
U_c	Bulk or convection velocity
U_o	Bulk velocity
u_τ	Friction velocity
U_∞	Free stream velocity
V	Relative velocity in a blade passage, general velocity scale or volume
Vol	Cell volume
Wf	Weighting function
X, Y, Z	Dimensionless spatial coordinates

Lowercase Greek

α	Dimensionless weighting parameter, phase angle, latency parameter in LNS model, design variable, fraction of rotor speed at which stall modes rotate in compressor (Chap. 4), blade metal angle (Chap. 4), angle of attack
β	Optimization range, weighting control or compressibility parameter
$\gamma = c_p/c_v$	Ratio of specific heats, weighting parameter in compact scheme or intermittency
δ	Boundary layer thickness, step function or small number/perturbation
ϵ	Turbulence dissipation rate, small number, scaling parameter in HJ equation, (specified) error tolerance/level or flux limiter
ϵ_0	Positive relaxation parameter
ϵ_1	Numerical smoothing parameter
η	Parameter that defines time levels in discretized equations, transformed spatial variable or adiabatic film cooling effectiveness
θ	Momentum thickness or angle
κ	von Karman constant

λ	Temporal discretization control parameter, Eigen values, viscosity coefficient ($-2\mu/3$), wave speed (in LES filter definition) or length, or inverse velocity scale
μ	Dynamic viscosity
μ_t	Turbulent viscosity
ν	Kinematic viscosity
ν_t	Turbulent kinematic viscosity
ξ, η, ζ	General, transformed coordinates
ρ	Fluid density
σ	Diffusion Prandtl number or turbulence fluctuation scale
τ	Transformed temporal coordinate, shear stress, turbulence time scale or time shift
τ_s	Correlation time scale
ϕ	General variable
ψ	Difference between current value and a predefined dispersion level, ratio of mean flow and turbulence time scales
ω	Frequency (turbulence) or vorticity

Uppercase Greek

Γ	Diffusion coefficient or Jacobian matrix
Δ	Filter width
Δt	Time-step length
$\overline{\Delta}$	Space shift in turbulence correlation coefficients (Chap. 6)
Λ	Adjoint variable
Φ	General variable or shock switch parameter
Ω	Angular velocity

Superscripts

cen	Pertaining to a central difference
dis	Associated with dissipation
H	High order component
inv	Pertaining to inviscid components
l	Index
L	Low order component or lower boundary (Chap. 4)
n	Time level
new	Latest value
old	Previous value
U	Upper boundary (Chap. 4)
vis	Pertaining to viscous components
ΔT	Variable computed with a coarse time step
$'$	Perturbation or first derivate of variable
$''$	Second derivative of variable or coherent, low frequency, unsteadiness scale amplitude
$+, *$	Dimensionless distance in wall units or dimensionless variable

Subscripts

a	Pertaining to actual value
ave	Average value
c	Pertaining to chord or centroid
cl	Centerline value
$conv$	Convective flux component
ctr	Pertaining to central difference scheme
DB	Pertaining to database
DES	Pertaining to the DES model
exp	Pertaining to experimental value
$ERROR$	Pertaining to error
f	Unsteadiness associated with vacillation
fp	Relating to a particular moving fluid particle
g	Pertaining to grid
HJ	Pertaining to HJ equation
i, j, k	Array subscripts pertaining to the axial, radial and tangential directions, respectively
int	Interface value in hybrid RANS-LES
IGV	Pertaining to IGV
k	Pertaining to turbulence kinetic energy or midpoint of a face
$k–\omega$	Pertaining to $k–\omega$ turbulence model
l	Index
L	Pertaining to information on the left or large eddy scales
LES	Pertaining to LES model
max	Maximum value
min	Minimum value
num	Pertaining to numerical value
NB	Neighboring values
o	Reference value or pertaining to centre of gravity
P	Process linked to a particular variable
R	Pertaining to information on the right or rotor (Chap. 4)
$RANS$	Pertaining to RANS model
$smth$	Smoothing component
s	Pertaining to solid
S	Pertaining to small turbulent scales, stator (Chap. 4)
SCL	Pertaining to quantity obeying the Space Conservation Law
SGS	Pertaining to the subgrid scale
$target$	Target value
u, v, w	Pertaining to listed velocity components
x, y, z	Pertaining to the x, y and z directions, respectively
z, r, θ	Pertaining to the axial, radial and tangential directions respectively
ϵ	Pertaining to ϵ
θ	Pertaining to temperature fluctuations
ϕ	Pertaining to the variable ϕ
Δt	Variable represented on a finer temporal 'grid' of Δt

Overbars

$\tilde{}$	Variable expressed in the frequency domain or that is dimensionless
$\bar{}$	Averaged or filtered value
$\rightarrow$	Coarser filtered value
$\hat{}$	Dimensionless variable

Special Symbols/Operators

int	Round up value to the nearest integer
$\Im$	Imaginary part
max	Operator to take maximum value
$N(a, b)$	Normally distributed random number operator with mean a and standard deviation b
$NS(\phi)$, $NS^s(\phi)$	Navier-Stokes and steady Navier-Stokes operator
$\Re$	Real part
$URANS(\phi)$	Unsteady RANS operator
$\delta(\varphi)$	Dirac delta function
δ_{ij}	Kronecker delta ($\delta_{ij} = 1$ if $i = j$ and $\delta_{ij} = 0$ if $i \neq j$)
ε_{ijk}	Alternating symbol
$\langle\rangle$	Phase averaged variable

Abbreviations

ADI	Alternating Direct Implicit
ALE	Arbitrary Eulerian-Lagrangian
APE	Acoustic Perturbation Equations
AUSM	Advection Upstream Splitting Method
BEM	Boundary Element Method
BD	Backwards Difference
Bi-CGSTAB	BiConjugate Gradient STABilized method
CAA	Computational AeroAcoustics
CABARET	Compact Accurately Boundary-Adjusting high-REsolution Technique
CD	Central difference scheme
CEN2	Second order central difference scheme
CFD	Computational Fluid Dynamics
CN	Crank-Nicolson scheme
COM6	Compact 6th Order Scheme
CPR	Correction Procedure via Reconstruction
DES	Detached Eddy Simulation
DG	Discontinuous Galerkin
DNS	Direct Numerical Simulation
DRP	Dispersion Relation Preserving
DSM	Deterministic Stress Model
EE	Euler Equation
ENO	Essentially Non-Oscillator
FCT	Flux Corrected Transport
FE	Finite Element

FRAM	Filtering Remedy And Methodology
FWH	Ffowcs-Williams and Hawkings
GMRES	Generalized Minimum RESidual
HJ	Hamilton-Jacobi
HPT	High-Pressure Turbine
HPTR	HPT-Rotor
IBA	Interpolation Based Algorithm
ILES	Implicit Large Eddy Simulation
IPT	Intermediate Pressure Turbine
IPTS	IPT-Stator
LDA	Laser Doppler Anemometry
LDDRK	Low Dissipation and Dispersion Runge-Kutta
LEE	Linearized Euler Equation
LES	Large Eddy Simulation
LNS	Limited Numerical Scales
LNSE	Linearized Navier-Stokes Equations
LPT	Low-Pressure Turbine
MDICE	MultiDisciplinary Computing Environment
MILES	Monotone Integrated Large Eddy Simulation
MUSCL	Monotone Upstream-centred; Schemes for Conservation Laws
MST	Mean Source Terms
NLAS	Non-linear Acoustics Solver
NLDE	Nonlinear Disturbance Equation
NLES	Numerical Large Eddy Simulation
NSS	Nearest Surface Search
NURBS	Non-Uniform Rational B-Splines
OASPL	Overall Sound Pressure Level
ODE	Ordinary Differential Equation
OGV	Outlet Guide Vane
PANS	Partially Averaged Navier-Stokes
PI	Proportional Integral
PIV	Particle Image Velocimetry
PPW	Points Per Wave
PSD	Power Spectral Density
RANS	Reynolds Averaged Navier-Stokes
RK	Runge-Kutta scheme
RO	Reduced Order
ROM	Reduced Order Model
RPM	Random Particle Mesh
SAS	Scale Adaptive-Simulation
SCL	Space Conservation Law
SD	Spectral Difference
SHASTA	SHarp And Smooth Transport Algorithm
SST	Shear Stress Transport

SV	Spectral Volume
T-S	Tollmien-Schlichting
TVD	Total Variation Diminishing
UP	Upwind Scheme
URANS	Unsteady Reynolds Averaged Navier-Stokes
VLES	Very Large Eddy Simulation
WALE	Wall Adapting Local Eddy-viscosity
WENO	Weighted Essentially Non-Oscillatory

Chapter 1
Introduction

1.1 Aerospace Challenges

With the projected demand for air transport set to double the world aircraft fleet by 2030 it is becoming urgent to take steps to reduce the environmental impact of aircraft in terms of CO_2, NOx, acoustic and other harmful emissions. In order to enable sustained growth whilst limiting the environmental impact of air transport the Advisory Council for Aeronautics Research in Europe (ACRE) has set challenging targets for emission levels from gas turbines. For a typical modern engine as shown at the centre of Fig. 1.1 (surrounded by some of the CFD simulation zones that will be considered here), a turbine entry temperature increase of 50 °C equates to a reduction of 836 tonnes of CO_2 per aircraft per annum. Hence, predictive technologies that provide greater understanding of the hostile environment around the combustor-turbine interface and therefore the use of elevated combustor outlet temperatures are of tremendous environmental importance. Also, computational models that allow the design of low drag airframes are key.

With regards to noise, in the worst case it can be more than just annoying, potentially being a contributory factor towards illnesses such as hypertension (Rosenlund et al. 2001). Hence, again (ACRE) has set the target of reducing perceived noise levels by 50 % by 2020 [European Commission, 2001]. Noise is frequently very much controlled by turbulent structures. Hence, being able to use high fidelity LES (Large Eddy Simulation) and DNS (Direct Numerical Simulation)—see later—to understand and explore ways of modifying turbulence structures is important. These techniques are discussed in detail in Chap. 3.

At take off, as well as the propulsive jet, engine fans are another key noise source. Recently CFD (Computational Fluid Dynamics) has played a great role in reducing tonal noise from fans. However, making inroads into broadband noise has been tough and LES may prove a useful tool for thus purpose. During the low engine power settings at landing it becomes important to consider airframe noise.

The ACARE targets are extremely challenging. Perhaps the only way of plausibly meeting them is to take a system level approach and explore the engine and airframe design more holistically. For example, ultra-high bypass ratio engines offer

P.G. Tucker, *Unsteady Computational Fluid Dynamics in Aeronautics*,
Fluid Mechanics and Its Applications 104, DOI 10.1007/978-94-007-7049-2_1,

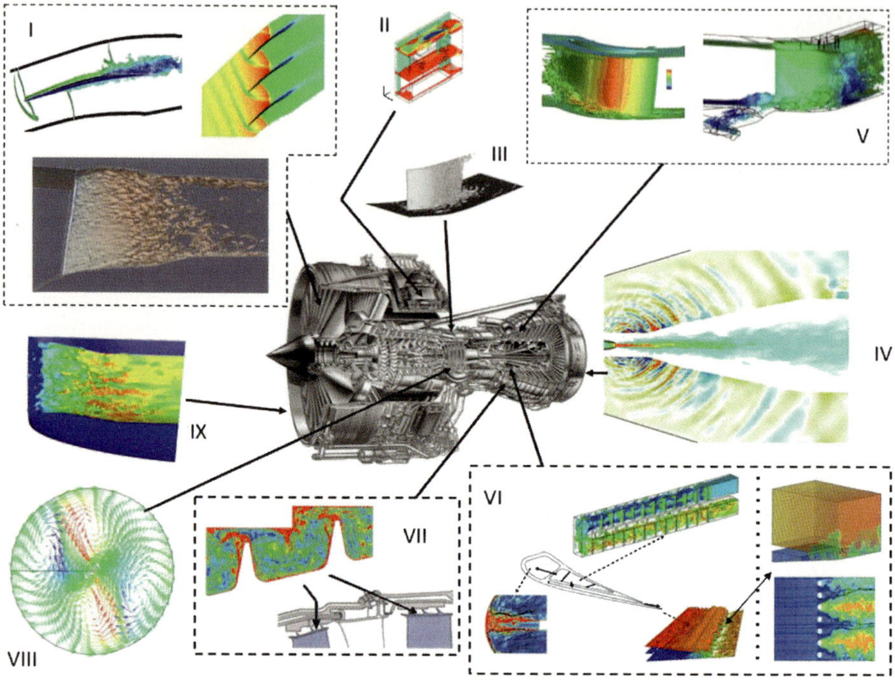

Fig. 1.1 Modern gas turbine aero engine and simulation zones considered here (Tucker 2011b). Published with kind permission of Elsevier

tremendous potential to reduce noise and improve propulsive efficiency. However, the closer proximity of engines to the airframes creates great aerodynamic challenges. Indeed, the influence of airframe proximity on jet noise poses many questions that are as yet not fully understood. Clearly, the use of high accuracy unsteady techniques at the component level will play an important role. However, when it is born in mind the considerably higher cost of using unsteady techniques, the need for more holistic modelling creates a tension. A key question is how will higher fidelity unsteady CFD fit into the need for more multi-component and multi-objective holistic design?

For example, when designing a fan system for a gas turbine aero engine, the fan-OGV (Outlet Guide Vane) interaction will produce upstream propagating tonal and broadband noise. At engine part speeds, where the axial Mach numbers are low, noise can relatively easily radiate forwards. The use of fewer blades will give less broadband noise. However, the tonal noise could then rise. The OGVs can be sculpted to reduce acoustic radiation. However, of course aerodynamic efficiency is also key and there are strong mechanical constraints. Acoustic liners can be utilized but these add another layer of modelling complexity in making large unsteady fan simulations, which in themselves present a serious computational challenge. Therefore, it can be seen that the optimization of the fan-OGV is a complex multi-objective, multidisciplinary problem.

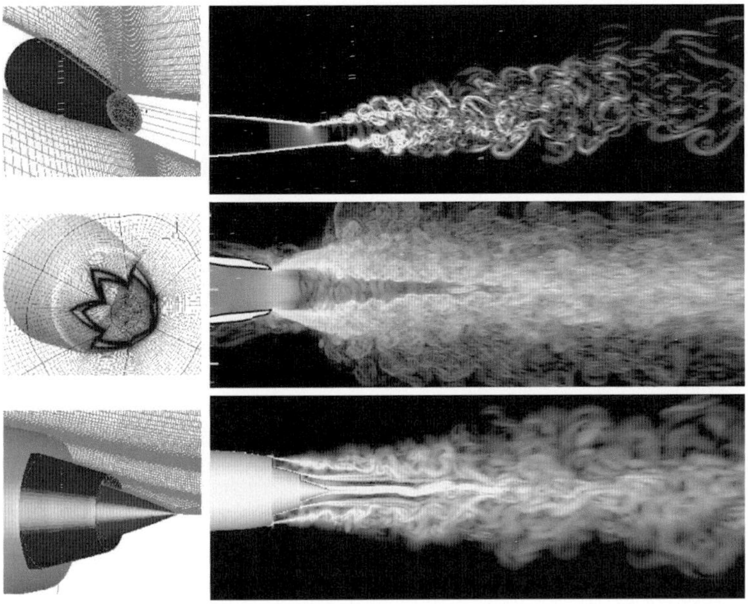

Fig. 1.2 Geometries, meshes and instantaneous flows for different complex nozzle geometries

To meet the highly challenging, longer term, ACARE targets there are, for example the following options: (a) use open rotor engines, (b) have engines mounted over wings—the wings providing noise shielding or (c) use blended wing body aircraft. The open rotor engines are a promising low energy propulsion option but present both global system and local flow physics problems. The global issues are where such a propulsion unit should be mounted and if the unit should be configured with the rotors at the back or at the front of the engine. The interaction of the blade wakes and tip vortices between the upstream and downstream rotors are key concerns. Such engines are well known to be potentially extremely noisy.

If it is wished to use a blended wing-body aircraft with boundary layer ingestion (to increase propulsive thrust through the engine absorbing lower momentum boundary layer fluid) then the engine airframe integration is again critical. The engine fan blades will pass through relatively low and high momentum fluid zones. This will result in temporal variations in blade incidence. The low momentum zone will propagate into the engine and can trigger compression system instabilities. Also, again, acoustics and mechanical constraints need to be factored into the analysis process. Therefore, again we have a complex multi-objective design optimization problem. Clearly, the development of future vehicles will require new levels of collaboration between engine and airframe manufacturers and also multi-component simulations. Chevron nozzles are an effective noise reduction technology. However, they involve complex unsteady flow physics and so the way they work is not well understood. As noted earlier, with ultra high bypass ratio engines, where the propulsive jet is in close proximity to the airframe understanding diminishes even further. Figure 1.2 shows LES related flow fields for different nozzle geometries. Each has

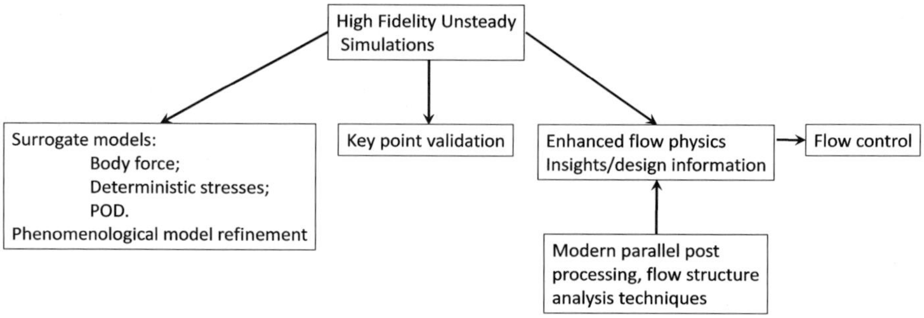

Fig. 1.3 Potential uses of high fidelity unsteady flow simulations and inputs

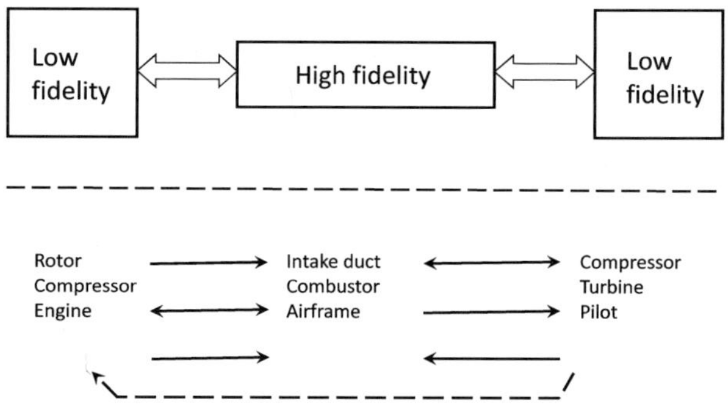

Fig. 1.4 The use of mixed fidelity simulations in aircraft propulsion

substantially different flow physics that needs ultimately to be considered in a system level context.

1.2 Large Scale Simulations

Figure 1.3 gives a schematic of potential uses of high fidelity simulations. Notably, a key use of such simulations is for exploring flow control and unlocking the complex flow physics processes that are needed to produce for example, ideally, laminar boundary layers.

High fidelity CFD can be used to calibrate lower order surrogate models and also as part of large-scale simulations that are of multi-fidelity. In the Stanford whole engine simulations (Schluter et al. 2005) the combustor is treated in LES mode with the surrounding engine in modelled as URANS (Unsteady Reynolds Averaged Navier-Stokes). In the open rotor intakes work of Loiodice et al. (2010), the unsteady rotor wakes are treated using body forces and the duct with URANS. The duct is then also connected to a body force representation of the compressor. Chapter 7 discusses multi-fidelity and coupled simulations. Clearly the tension between using large scale system level calculations and high fidelity techniques needs to be managed if the ultimate vision of Spalart and Bogue (2003) is to be reached, where simulations couple the engine airframe and even pilot interactions. Figure 1.4 shows

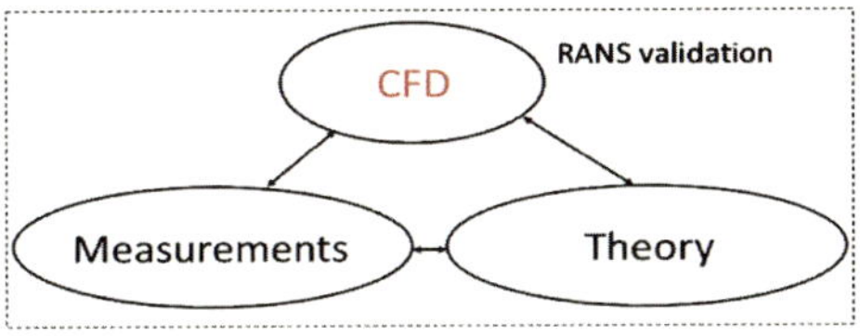

Fig. 1.5 Traditional analysis triad

potential mixed fidelity modelling scenarios for aircraft engines. The arrows indicate potential primary information flow paths.

Traditionally CFD has been best utilized a part of the analysis triad shown in Fig. 1.5. However, eddy resolving techniques such as LES are beginning to challenge this. Large scale, high-fidelity simulations with modern parallel pre and post-processing methods allows both high fidelity and relatively readily accessible flow insights. Hence, it is certainly now tenable to argue that to really understand many fluid flows one needs to do simulations in preference to experiments. A key area that needs further development is completely parallel processing from mesh generation though to post processing. Although these elements are available they are not, as yet, widely used. Figure 1.6, taken from Xia et al. (2010) shows the traditional hybrid serial-parallel solution path (see Frame (a)) and the ideal fully parallel process (Frame (b)).

As shown in Fig. 1.7 there a wide range of elements involved in making large scale unsteady simulations. These encompass computational science, geometry handling, mesh generation, high fidelity numerics and modelling. The high cost of large scale unsteady simulations, that leads to the need for efficient parallel processing based solution strategies are now outlined.

1.3 Computational Cost

1.3.1 Turbulence

A limiting item in the numerical modelling of aerodynamics and aeroacoustics is the modelling of turbulence. This strongly controls skin friction, heat transfer and noise sources. Turbulence involves a wide range of length and time scales. These can either be fully resolved on extremely fine grids—DNS. Alternatively all the turbulent scales can be fully modelled as with RANS (Reynolds Averaged Navier-Stokes). The extremes in computational cost between these limits have given rise to mixed approaches. For example LES, where there is an approximately 10 % modelling contribution for just the smaller scales. The larger are resolved on the grid. Generally, the most challenging regions for LES is the near wall regions. These are populated by numerous fine, flow aligned, and streak like structures. For example, for a Trent 1000 fan at cruise, estimates of Tucker (2011b) suggest there are 10^7 streaks to resolve on the grid. For a Boeing 777 at cruise Pope (2004) estimates 10^8 streaks (these estimates probably do not include the effects of pressure gradient and sweep).

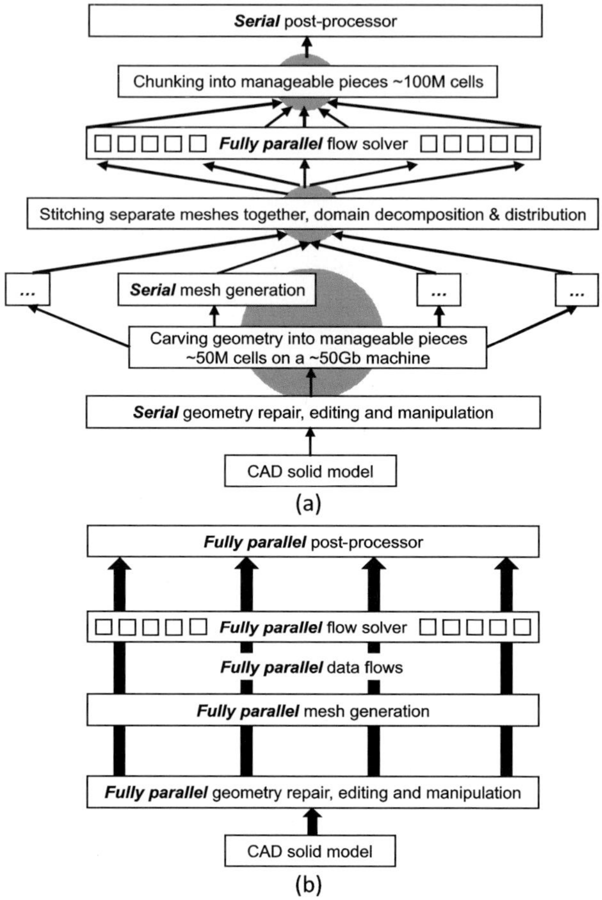

Fig. 1.6 Current and future paradigm of simulations. (**a**) Typical process chain. (**b**) End-to-end fully parallel process chain (Xia et al. 2010). Published with kind permission from Elsevier

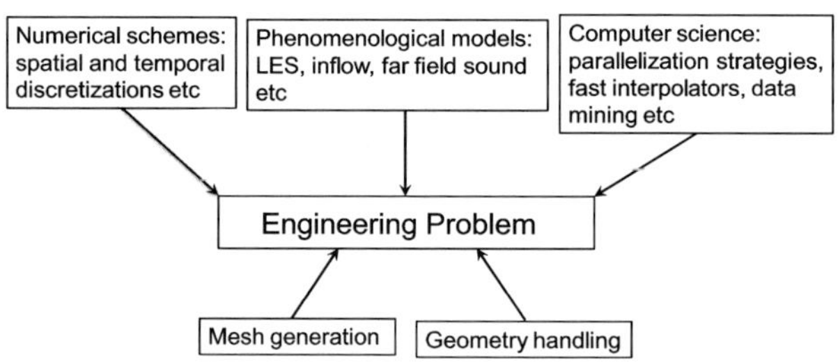

Fig. 1.7 Multiple elements needed for making large-scale simulations

Similar stark statistics relating to spatial and temporal scales, but this time for the fuselage of an Airbus A340-300, are brought out by Carpenter et al. (2007). According to their estimates for the Airbus 340-300, there are 2000 streaks m^{-1} and approximately 20×10^8 for the whole fuselage.

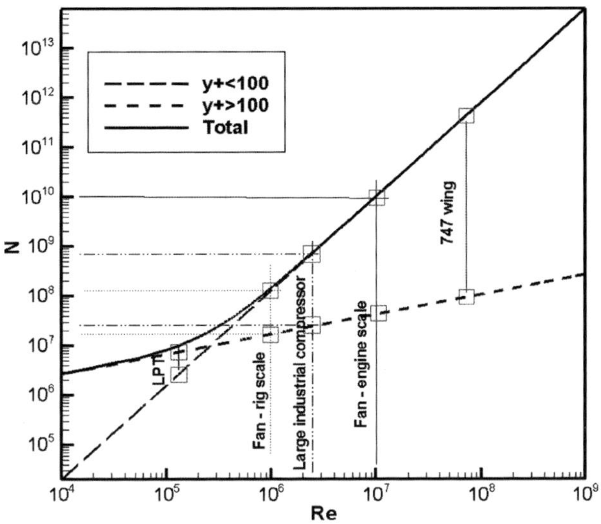

Fig. 1.8 Grid requirements for LES and hybrid LES-RANS (Tucker 2011b). Published with kind permission from Elsevier

1.3.2 Cost

Chapman et al. (1975) explored the outlook for application of LES and a range of other fidelity levels for aerodynamic modelling in an engineering design environment. The accuracy of Chapman's inner layer grid requirement estimates were recently reconfirmed by Choi and Moin (2012). Chapman concluded that in 10–20 years from the date of publication, that computers would become 1000 times faster reaching 10^{14} flops. At this point, LES for realistic engineering problems (needing of the order of 10^9 cells for correct resolution) would be feasible. The Los Alamos Roadrunner is a petaflop machine. So it seems that we have reached the point where more of Chapman's vision can be realized. However, in his estimates Chapman expected substantial improvements in numerical algorithms and these seem to have, to an extent, stagnated perhaps not reaching Chapman's expectations. Although the speed of processors has reached a nearly constant level, the computational power has continued to increase with parallel architectures. We are now over ten years ahead of the proposed date of cost effectiveness, which has provided an equivalent increase in computing power. However, a key issue is parallel scaling. Another aspect is that Chapman's scalings underestimated the outer layer requirements and are based on highly optimized near wall grid structures to resolve turbulence.

Even with such grids Chapman shows that computational cost for pure LES will scale with roughly $Re^{5/2}$. The use of a RANS layer, to just cover these streaks, which extend to about $y^+ = 100$ (the superscript indentifies the use of wall units), results in an approximate Re^1 scaling (Piomelli 2008; Choi and Moin 2012). This approach is called hybrid RANS-LES and different approaches are discussed in Chap. 3.

Figure 1.8 is similar to that shown in Piomelli and Balaras (2002) and Leschziner et al. (2009). Here is adapted with more turbomachinery in mind. For airframes the

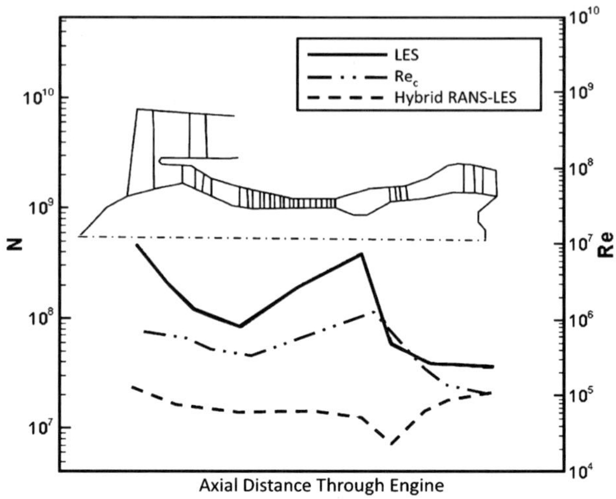

Fig. 1.9 Hybrid RANS-LES and LES grid requirements in different zones for a medium-sized gas turbine (Tucker 2011a). Published with kind permission from Elsevier

Reynolds numbers are an order of magnitude higher and hence making the points to be made here even starker. Figure 1.8 essentially plots the number of boundary layer grid points, N, against Re_c—the chord based Reynolds number. It is assumed that span considered is equal to chord, c. The short dashed line gives N for $\delta^+ > y^+ \geq 100$. The boundary layer thickness is δ. The long dashed, sloping line, is for $y^+ < 100$. The full details of the rationale for the estimates are given in Tucker (2011b). It should be stressed the estimates are only order of magnitude. They relate to individual aerofoils. If cylindrical endwalls (or a fuselage in the airframe context) are to be accounted for, N must be substantially scaled. For the outer region, the grid requirements are not strongly independent of Re. This is a key aspect. Here, the outer layer grid count is estimated from looking at successful LES for a range of workers. On average, these take N in an approximate δ^3 outer volume as around 1500.[1] The full line in Fig. 1.8 is the sum of the inner and outer zone grid requirements. For a small low-pressure turbine $Re_c \approx 1 \times 10^5$. At this Re, the $y^+ < 100$ grid requirements are modest. Hence, LES is practical—assuming the inertial subrange is of sufficient extent. For a fan, where the Re is at least an order of magnitude higher (see dotted vertical line for a rig scale fan, and full line for a full scale fan) resolving the $y^+ < 100$ zone adds around two to three orders of magnitude to the grid requirements. The Re_c for a large industrial compressor (Köller et al. 2000) is around 2.5×10^6 (see vertical dash dot line). Therefore, again resolution of the $y^+ < 100$ zone is high. Note, the coloured square symbols are used to emphasise the disparity in computational cost for LES and hybrid approaches.

This disparity in computational cost between LES and hybrid RANS-LES is further illustrated in Fig. 1.9, adapted from Tucker (2011a). This figure shows hybrid RANS-LES and LES grid requirements in different engine zones. The estimates are

[1]Estimates of other workers are higher but the current figure fits in with a range of successful LES.

for a medium-sized gas turbine. The Reynolds number distribution data through the engine is taken from Mayle (1991). It can be clearly seen that except for the low-pressure turbine zone, where the Reynolds number is low, the use of hybrid LES-RANS produces around an order of magnitude reduction in grid requirements relative to RANS. Nonetheless, the cost of a hybrid approach, although cheap, relative to LES, comes at a substantial cost relative to RANS. Indeed, as a rough estimate, an LES simulation would need around 5000 times the computational time of RANS (Gourdain et al. 2012). Again, it should be stressed that the grid requirements plotted are for blades where the span to chord ratio is unity. Hence, the numbers need scaling up for high aspect ratio blades, the incorporation of endwalls and the modelling of wakes. Hence, it can be seen that the modelling of turbulence presents some key computational challenges and hence computing cost.

Nonetheless, the decision to design and develop a new turbofan engine is an expensive one with development costs easily exceeding £0.5 billion (Cumpsty 2003). In light of such investment, there is a significant incentive to reduce development costs. The cost of a typical rig test is around £0.75 million (Place 1997). Hence, clearly the development of a computational tool giving insight into complex flow physics at a lower cost is desirable. For compressor and turbine hybrid LES-RANS simulations a small circa 100-core cluster can be used costing approximately £50,000 (Tucker 2011a). Assuming a lifetime of 3 years the cost of one simulation is approximately £100 (Klostermeier 2008). Whilst such low Reynolds number ($Re = 5 \times 10^5$) simulations are trivial compared with a full rig test, the figures show a potential economic advantage in pursuing LES. This is particularly so given the level of physical insight they can potentially give and the emergence of cheap commodity hardware. For jet nozzle cases, containing 12×10^6 cell meshes, and shown in Chap. 6, one simulation costs approximately £2,000 on the same cluster. Experience suggests that there can sometimes to be a variation of up to 10 in the time taken to run a simulation for the different, structured and unstructured solvers. Hence the solver used and parameters of the simulation (such as whether the code is run in explicit or implicit mode, the time step and total number of time steps required) can have a significant impact on the total cost. As well as the financial cost, the elapsed time taken to perform simulations is important. It is easily feasible to gain almost linear speed up to 1,000 processors with standard CFD technology (Hills 2007). Hence an $N = 12 \times 10^6$ jet case, for example, could be completed in 4 days. As scientific computing moves towards many core architectures, with cluster facilities becoming cheaper (a 750 core cluster can be bought for around £100,000) such highly parallelized codes are attractive. Graphical Processing Units (GPU's) also show promise for improving computing power v cost (Brandvik and Pullan 2008). The peak rate of floating point operations is an order of magnitude greater than the maximum achievable on a contemporary CPU (Brandvik and Pullan 2008). This is made possible by constraining the functionality of the GPU and employing a highly parallelized architecture. Notably, with a GPU, the power of the 112-core cluster noted earlier can be gained at about 1/10th the cost. This makes the exploration of industrial LES warrant serious attention. Nonetheless, it should be noted the above costs neglect electricity and this is substantial. Indeed, as pointed out by

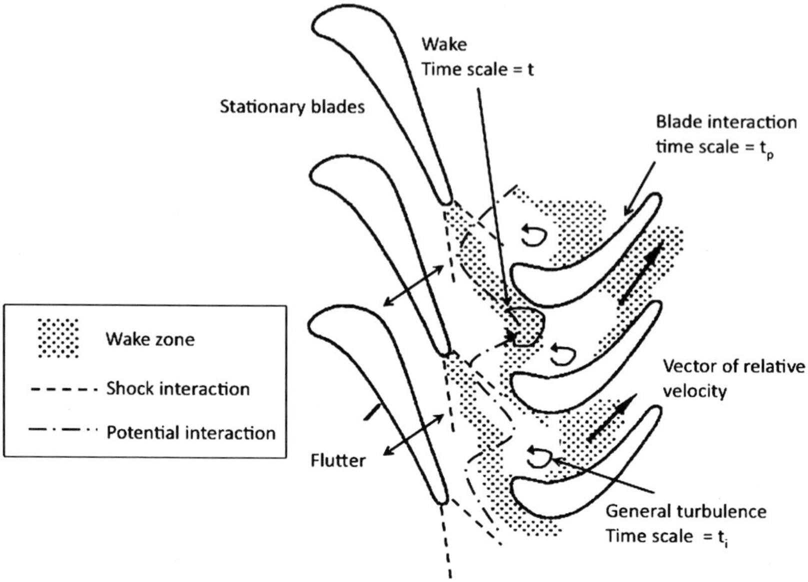

Fig. 1.10 Schematic of wake interaction from an upstream stator with a downstream rotor (Tucker 2011a). Published with kind permission from Elsevier

Hield (2012), the energy consumption in electricity used in running high fidelity simulations can be much higher than that of a compressor rig on a per revolution basis. Next, the areas where unsteady CFD in aerospace is important are outlined. First turbomachinery flows are discussed and then airframe.

1.4 Unsteady Flow Sources

1.4.1 Turbomachinery

Prediction of Efficiency

Turbomachinery flows, like may others, are inherently unsteady. A key aspect is that turbomachinery flows are characterized by multiple rows of adjacent stationary and rotating blades (stages). These interact, giving rise to multiple harmonics. Blade rows can be seen in Fig. 1.1. The wakes from upstream blade rows and shock systems from trailing edges will interact in a complex unsteady fashion with downstream (end even upstream) blades rows creating unsteadiness. Figure 1.10 gives a schematic indicating these interactions.

As noted by Meneveau and Katz (2002) and Rhie et al. (1995), the local unsteady internal fluid stresses resulting from the movement of the wakes can be typically of a similar or higher magnitude to the turbulent stresses. For a compressor Suryavamshi et al. (1997) observe the root-mean-square of total pressure unsteadiness to be nearly 10 %. Similarly, Bardoux et al. (1999) attribute a 70 % change in wake loss to unsteady influences.

Early turbine stages will be subject to substantial circumferential temperature non-uniformities, from the combustors. Hence, the rotors will periodically pass through hot zones. Also, there will be related migration influences. These are discussed later in relation to blade life.

Even though the potential field decays much faster than the wake field, the interaction of the potential field from the adjacent blades will cause unsteadiness and acoustic waves. This acoustic energy creates boundary layer unsteadiness and hence minor additional loss. Fritsch and Giles (1992) estimate this to be just under 0.1 % in isentropic efficiency. Leboeuf and Charbonnier (2003), when modelling a transonic turbine, observed numerous pressure reflections between the stator and rotor blade rows. This resulted in amplification of radial mixing and strong unsteadiness correlations. The interaction of wakes and shocks with boundary layers, as well as giving direct loss, will also give rise to a time varying circulation losses. This is a result of viscous dissipation of vorticity shed due to the temporal variations in circulation. This can alter predicted isentropic efficiencies by up to 0.3 % (Fritsch and Giles 1992). Also, the unsteady wakes can strongly control transition (Coull 2009). This, in turn, controls the reattachment and separation of flow resulting in (in the low-pressure turbine) a complex cyclic unsteady process (Michelassi et al. 2003). This can cause in a profound global loss influence. Notably, as indicated by Ning and He (2001), unsteady vortex shedding from blades significantly influence the base pressures and hence profile loss. With regards to this, List et al. (2008) show that, when blade rows are in close proximity, bow shocks from downstream blades can enhance upstream vortex shedding. Furthermore, Raverdy et al. (2003) and Matsuura and Kato (2006) demonstrate that the vortex shedding from blade trailing edges, can couple with suction surface separation bubbles enhancing turbulence. Dufour et al. (2009) observe (using LES) a similar effect for a high-pressure turbine. Montomoli et al. (2011) find that wake unsteadiness in upstream blade rows can substantially modify losses near end walls. For low aspect ratio blades the losses in these zones can be substantial. Evidently, oscillations in (rotor blade) circulation, resulting from rotor-stator interactions in fans and compressors, can result in circumferential variations in total temperature in the stationary frame. This temporal behaviour clouds the measuring of efficiency by 1.5 % discrepancies—(Giles 1992; Shang et al. 1989). Sims-Williams et al. (2012) observe that for a turbine blade, at blade extremities adjacent to stationary surfaces the temporal total pressure variation can be 70 % of the dynamic head of the inflow.

'Clocking' is known to, in a complex fashion, influence turbine performance. With this, the circumferential alignment of the stator blades on different stages can be altered. This will subtly alter the upstream stator-rotor wake/turbulent flow field interactions with the adjacent downstream stator. The study of this interaction is only possible through unsteady CFD. Notably, a key focus of clocking studies is the low-pressure turbine zone. The low Reynolds numbers in this area mean that wake induced transition is a key process. Such complex transitional flow physics is hard to capture using URANS. Hence LES/DNS appears a key tool for clocking studies. Finding an optimal clocking position is complex. As ever, there is much contradictory evidence with regards to optimal positions.

As shown by Baines (2010) and Winterbone et al. (1991) turbochargers are subjected to pulsating inflow. Baines shows, this can alter efficiency by nearly 30 %. As shown by Guelich and Bolleter (1992), for radial compressors blade wakes, as well as resulting in an aerodynamic loss can cause fatigue and also substantial acoustic related problems. Bousquet et al. (2012) demonstrate that as a result of these unsteady sources, there is a need to account for unsteady influences when accurately exploring the stability limit of centrifugal compressors.

All the above listed unsteadiness sources are unlikely to be recovered as useful energy. Instead they will give loss through viscous dissipation. The total of numerous small and also substantial losses (or predictive errors of loss) noted above will have a key design influence. Fractions of a percent improvement in efficiency are of considerable worth over the operating life of an engine.

Operability, Life and Fatigue

Blades immediately downstream of combustors will be subjected to extreme levels of unsteadiness from large-scale structures. These have proved hopeless for steady CFD models to adequately characterize. Hence, as shown by Dufour et al. (2009), steady approaches can give rise to 50 % errors in predicted heat transfer. Accurate heat transfer prediction is important for modelling blade life. Also, as shown by Butler et al. (1989), in turbines, hot streaks, from the combustor migrate to the turbine rotor pressure surface. Hence circumferential variations in inlet temperature will result in unsteady thermal loading and potential fatigue. Note, the hot streak temperature can be twice that of the surrounding fluid. For compressors, as observed by Kerrebrock and Mikolajczak (1970), there is a similar migration mechanism. This again gives rise to the need to consider unsteadiness.

Unsteady flow modelling is also needed when studying the influence of inlet distortions on compressor stability. Under these circumstances the fan and the downstream rotors will move in and out of low total pressure zones. The exposure time of rotors to this distorted zone and its extent are key interrelated stability factors. Figure 1.11 shows intake distortion arising from unsteady separated flow during aircraft climb. This type of separated flow distortion will also occur when an aircraft is making relatively sharp turns. Also, as shown by Cousins (2004), the intake flow can be distorted by structural elements such as struts. For aircraft with supersonic inlets complex, potentially unsteady, intake shock structures can arise (Cousins 2004). These can produce unsteady intake distortion. It is worth noting that even for essentially basic boundary layer flows unsteady shock movement can occur (Hadjadj 2012).

Ground ingested vortices can also be highly unsteady forms of intake distortion. Vortices can change their location at the engine face rapidly with time. During ground based operation, in cross-winds, separation can also occur on the nacelle as

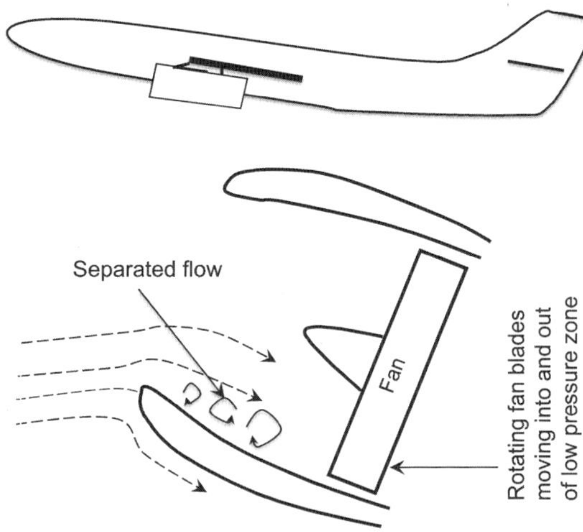

Fig. 1.11 Intake distortion arising from unsteady separated flow during aircraft climb (adapted from Cousins 2004)

shown in Fig. 1.12. When the engine in ingesting relatively low levels of mass flow a large separation can occur, as show in Fig. 1.12a. At higher mass flows the extent of the unsteady separated zone is much lower. This is shown in Fig. 1.12b. However, again, this intrinsically unsteady separated zone will also have the additional unsteady flow component related to the fan blades passing through it. Figure 1.12c shows the variation of distortion level against the intake mass flow.

Figure 1.13 gives contour plots of entropy function (an energy loss measure) through a three stage fan at the following planes: (a) Upstream of inlet guide vane and downstream of (b) Stator 1, (c) Stator 2, and (d) Stator 3. An intake total pressure loss has been represented over a small sector to represent some form of idealized engine intake distortion. The purpose of this type of simulation is to explore the transfer of distortions through a compressor and its impact on performance. A key aspect of these simulations is that they are performed on GPUs. Hence, such simulations, involving multiple hundreds of millions of cells can nowadays be performed with a modest financial outlay.

In blended wing body and military engines the fan stages can be downstream of complex inlet ducts. Traditionally the flow distortion from these has been treated as 'steady'. However, naturally the flow field seen by the blades is unsteady even if the flow field distortion generated by the intake is steady. However, as observed by Burcham and Hughes (1970), Van Deusen and Mardoc (1972), and Hercock and Williams (1974) for military engine intakes the unsteady (dynamic) distortion levels can be as high as the steady.

With regards to unsteady distortion, historical work (Cousins 2004) has suggested that distortion patterns need to persist through the period of one revolution to be of significance. The reduced frequency, expressed in terms of the residence time of the fluid in the blade passage to the time period that the blade is inside a disturbance can be used to understand blade dynamic distortion resilience (Cousins 2004). Clearly we need to ascertain the response time of the blade. Cousins (2004) characterizes this as proportional to the time that it takes a fluid particle to reach the

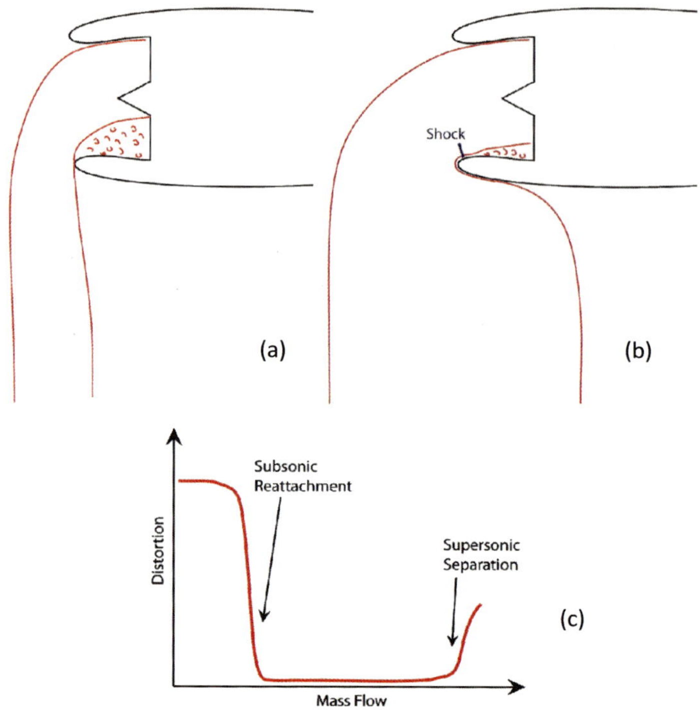

Fig. 1.12 Flow separation on an intake in crosswind: (**a**) lower intake mass flow; (**b**) higher intake mass flow and (**c**) distortion level with mass flow (Judge 2012)

throat of an axial compressor from the leading edge. This idea is shown schematically in Fig. 1.14a. The rationale for this response time estimate is that beyond the throat the major diffusion process starts. Hence, the disturbance reaching this point will have greatest impact on the delicate diffusion process that takes place in a compressor blade passage. The period of a fluid particle reaching the throat from the leading edge is likely to substantially exceed the boundary layer eddy turnover periods. This helpfully ensures a spectral gap for turbulence modelling purposes. Such issues are discussed in Chap. 3. As noted by Cousins (2004), the unsteady blade loading caused by distortion is a key source of high cycle fatigue failure.

In open rotor engines, the unsteady wakes from upstream rotors can strongly influence the intake duct flow. The latter feeds air into the compressor. Hence, controlling the distortion seen at the compressor face (Loiodice et al. 2010). This rotor-duct-compressor system in itself gives rise to a large-scale unsteady coupled system. Surge also is a large-scale coupled system process. With this, the working fluid surges back and forth through all the compressor stages. However, surge is preceded by stall. As shown in Frame (b) of Fig. 1.15, this involves large-scale unsteady separations. As shown by Emmons et al. (1955), these, if the process continues, are followed by the precession of low reverse velocity, long length scale (in

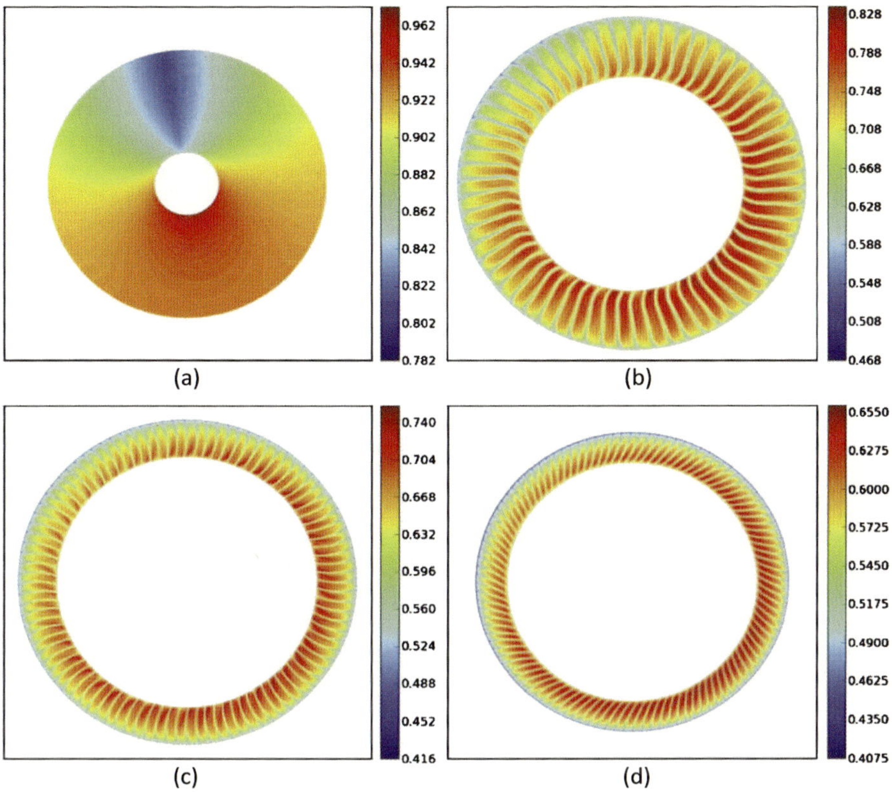

Fig. 1.13 Contour plots of the entropy function through a three stage fan at planes (**a**) Upstream of inlet guide vane and downstream of (**b**) Stator 1, (**c**) Stator 2, and (**d**) Stator 3 (Tucker 2011a). Published with kind permission from Elsevier

the azimuthal direction) zone(s) around the annulus. Frame (c) of Fig. 1.15 gives a schematic of this process.

The process is as follows. Passage (ii) in Frame (b) of Fig. 1.15 has stall. This diverts the flow and increases the incidence in passage (iii) and potentially other passages to the left of Passage (iii). This produces an extended blocked flow region shown in Frame (c) of Fig. 1.15. This blocked region also gives a flow component to the right. This helps to decrease incidence and hence recover stall. The potential for recovery on the right hand side and stall growth on the left means that the stall cell zone effectively rotates around the annulus in the same direction as the blades (when viewed in the stationary frame). For part-span stall (i.e. where the stall cell does not occupy the blades full span) the cell(s) rotate at around between 50 and 80 % of the blade speed. For full span it rotates at between 30 and 50 %.

Returning to stall, there are two key known routes to stall. Both are unsteady flow processes. They are termed modal and spike inception (Camp and Day 1998). As shown in Fig. 1.15a, the modal relates to, two-dimensional, long-wavelength

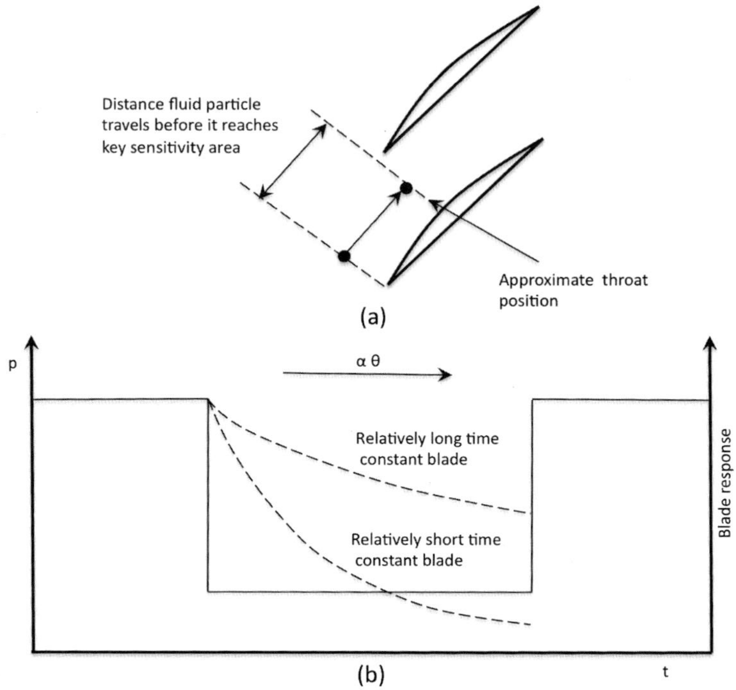

Fig. 1.14 Relationship between blade time response and system input: (**a**) blade time response and (**b**) system input and time response (adapted from Cousins 2004)

azimuthal disturbances. They exist over the full blade span. The modes have a relatively slow development (and can trigger spikes). In the absolute frame, modes tend to travel at about 40 % of the rotor speed. This is a bit slower than fully developed stall cells. Spike inception length scales are much smaller. As a result of flow separation they generally initiate at the blade tip. They relate too much more three-dimensional disturbances. Spikes move faster, at around 70 % of the rotor speed. Indeed, there is evidence to suggest that they move at the rotor speed for a short period of time, as they form, but this is not concrete (Young 2012). The occurrence of these two different modes is strongly related to the blade tip clearance. The modal disturbance occurs for larger tip clearances. However, as with many instability growth processes, in complex machines, there is contradictory evidence relating to the precise nature of the process and what triggers it. A key means of attempting to reduce the prevalence of stall is the use of tip treatments. With these, the engine casing adjacent to the blade tips have complex grooves in varying arrangements. Eddy resolving CFD would appear an ideal tool for gaining a greater understanding of blade tip physics, especially when there are casing treatments.

As noted above, the stall process can be followed by surge. This is a large-scale system unsteadiness. To replicate, simulations need domains with large upstream and downstream extents. The downstream domain needs to somehow replicate the tremendous energy/storage capacity of the combustor system. This energy source

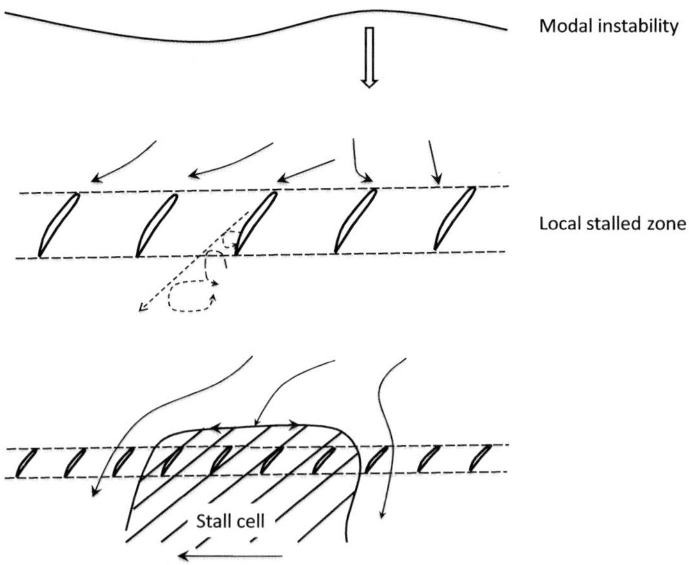

Fig. 1.15 Stall cell inception and propagation

can force the flow to reverse and for flames to erupt from the front of an engine. In the surge process, this flow then reverses again, flowing back towards to combustor. The flow direction reversal process repeats through time, until an engine shutdown is initiated.

Small Scale Geometrical Features

Figure 1.16 illustrates a few complex geometrical features found around turbine blades. Frames (a) and (b) show seals at the blade extremities. The Frame (a) labyrinth type seal is often called a shroud seal. Frame (c) shows an inter-blade seal. The misalignment that can occur between blades is evident. Frame (d) gives a highly simplified section through a turbine blade. The section allows some of the internal cooling features to be seen. Cooling fluid is expelled from the trailing edge. The induced shear layer can produce large scale Kelvin-Helmholtz vortices.

Nowadays, necessary blade forms for good compressor and turbine performance are relatively well understood. Hence, engine manufacturers are focusing hopes for performance gains on more secondary flows. These include leakage and sealant flow systems. Typically they involve small geometrical features that can promote unsteadiness. The potential range of turbine blade tip geometrical features is extensive—see for example Saxena et al. (2004). The situation is similar for compressors, where, for example, as noted above, casing treatments can be used.

Cardwell (2005) shows a range of leakage gaps that can occur between blades in gas turbines. As shown by Reid et al. (2007), these small gaps can have a substantial influence on the highly non-linear flow physics and hence performance.

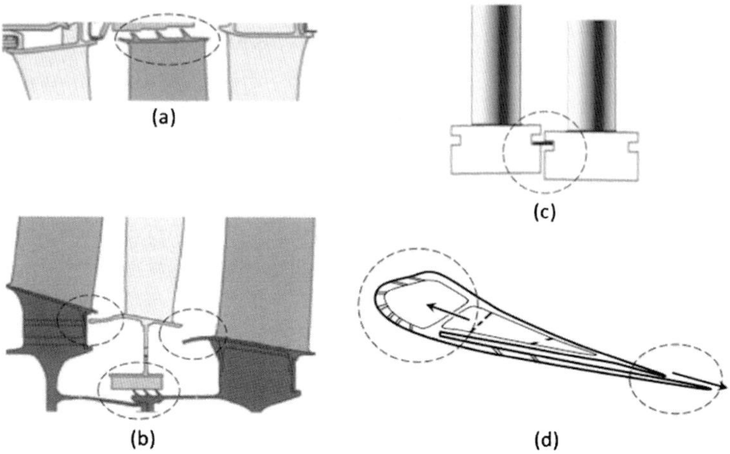

Fig. 1.16 Complex geometrical features around turbine blades: (**a**), (**b**) blade extremity seals, (**c**) inter-blade seals and (**d**) blade cooling flows

Small-scale features, such as fences Moon and Koh (2001) & Camci and Rizzo (2002) have also been explored to reduce secondary flow losses in the turbine passages themselves. Again these present a small geometrical scale flow modelling challenge. The importance of tip leakage flows was recognised over 50 year ago (Herzig et al. 1954; Allen and Kofsky 1955). Wallis et al. (2000) recognised the potential for blade tip performance gains, placing mini-blade like turning devices in a shroud seal zone. These were intended to extract work from the flow passing through the tip and reduce aerodynamic losses. Based on measurements, Pfau et al. (2001) concluded that the shroud cavity flow is likely to be a source of high losses in turbines. Pfau et al. (2001) extend their study, suggesting potential 0.2–0.5 % improvements in overall efficiency. Rehder and Dannhauer (2007) consider tip leakage flow control for shrouded rotors finding a substantial impact on flow and heat transfer. O'Mahoney et al. (2011) show that inter blade seal (in the blade hub zone) flows, between rotating and stationary blades, can be highly unsteady. They involve complex three-dimensional shear layers. These can potentially impinge on downstream surfaces, thus generating an acoustic feedback loop. The blade passing may also potentially excite shear layer cavity interaction resonances. As observed by Bladh et al. (2012), for compressors, the axial inter-blade type gaps, shown in Fig. 1.16c, can produce extreme dynamic excitation of downstream blades—5 times the excitation level caused by wakes. Leichtfuss et al. (2012) indentify the unsteady flow impact of gaps at the tips of variable angle inlet guide vanes and rotors in compressors.

The turbine blade internal cooling involves turbulence promoting ribbed passages, flow impingement zones and, as noted above, can involve the generation of trailing edge shear layers. Figure 1.17 shows typical flow fields in these turbine zones. They are based on LES type computations taken from Tucker (2011b). Notably, for turbines small-scale coolant jets are also used for heat shielding purposes.

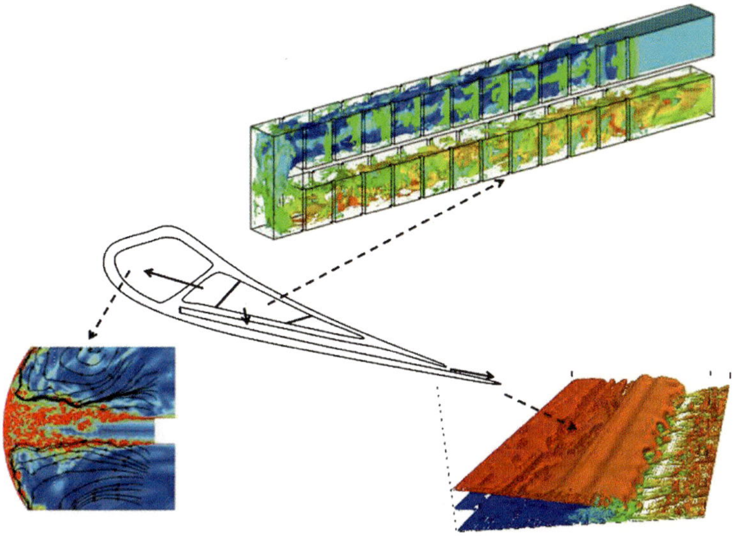

Fig. 1.17 Flow fields related to turbine cooling (Tucker 2011b)

These can also promote unsteady mixing. Hence, the modelling of such unsteady flow promoting small-scale features seems a key area of importance.

Small-scale features are also used for flow control in airframes and are present for structural reasons. Hence, this is another key modelling area for airframes.

Internal Systems

The blades of gas turbines are mounted to rotating and stationary disk assemblies. Rotor-stator systems (Owen and Rogers 1989) occur where stationary disks are adjacent to rotating. These can generate substantial coherent unsteady flow structures. These three-dimensional structures rotate at a velocity around half that of the disk speed. They appear as unsteadiness in the main annulus (Jakoby et al. 2004). The flow in the cavities of the high-pressure compressor drums, to which the compressor blades are mounted, is intrinsically unstable and hence unsteady. Unsteady vortex breakdown regimes can occur and also large-scale cyclonic and anticyclonic vortices will be present (Farthing et al. 1992a,b). Frame (VIII) of Fig. 1.1 shows these being captured in an LES type computation. These, vortices can rotate slower than or faster than the cavity depending on the temperature gradient (Tucker 2001). The flow can also vacillate between different distinct flow regimes with vortex breakdown also occurring.

Moving, even further radially inwards in the engine there are bearing and oil systems. The former, for land-based turbines, can be hydrodynamic journal bearings. The unsteady orbiting of these can cause turbine-bearing failure (De Jongh et al. 1996). Notably the internal and external air systems are coupled, ideally needing large-scale, coupled, unsteady simulations to ensure total design confidence. Notably the internal air systems transient heat transfer behaviour controls disk thermal expansion and hence blade-tip clearance and hence potentially stall and surge.

Acoustics

It is not just efficiency that has a strong economic and environmental impact. The increasing population densities around airports means that noise is a key problem. It is the 2nd major aerospace priority in the United States and also Europe. To fully calculate aeroacoustics, turbulence temporal information is frequently essential. This characterizes the acoustic source. Jet noise is still a key take off noise component. Frame (IV) of Fig. 1.1 shows the turbulent acoustic source zone and wave fronts emanating from this for an LES type computation of jet noise. The unsteady interaction of the propulsive jet with the wing and flaps is a key emerging noise source for landing. Through the development of high bypass ratio engines the jet noise has substantially reduced. Consequently, Tyler (1962) started to explore compressor/fan noise arising from rotating pressure fields spinning with the compressor rotors. Fan noise has both tonal and broadband components (Ashcroft and Nurnberger 2001) (Frame (I), Fig. 1.1). These arise from unsteady/rotating pressure patterns (as noted), shock structures but also the unsteady interaction of the rotor with flow disturbances such as atmospheric turbulence or inlet distortions. The latter potentially arises from airframe boundary layers. Also, the unsteady rotor-stator interactions will produce sound, notably the interaction of the rotor wakes and tip vortices with downstream stators (Zhao et al. 2012). Acoustic resonances can occur in combustion chambers, bleed valves and engine intakes. Source mechanisms are discussed by Bartels and Sayma (2007).

As noted earlier, rotor-stator interactions will also give rise to aerodynamic loss. Hence, there is a key link between aeroacoustics and aerodynamics. This is particularly so when it is noted that acoustic feedback can substantially modify the mean flow. Also, as noted by Moore (1977), the aero engine internal noise can modify the propulsive jet structure. Hence, although noise is a byproduct of turbulence it can also influence the flow structure itself. Indeed, Epstein et al. (1986) explored to use noise to control compressor stability. Williams and Hawkings (1969) also outlines the potential of noise to both reduce/cancel noise and also enhance aerodynamic performance. The latter are not unchartered waters and pose great research challenges. Nonetheless, the rich flow physics and acoustics insights provided by LES, along with the potential to precisely control excitations might enable advances in the area of using sound to reduce sound and also aerodynamic losses.

Combustion noise is emerging as an important source of external noise from aero engines (Flemming et al. 2005). This is partly because the other engine noise sources (such as from the propulsive jet and fan) have been reduced through improved understanding and advanced technologies. However, it is also because the novel lean premixed combustors, which are necessary to reduce pollutants, burn more unsteadily. There are two sources of combustion noise. Direct sound is due to the pressure fluctuations generated by the unsteady combustion. These acoustic waves are then modified as they propagate through the turbine—again an unsteady process. In-direct noise is due to the convection through the turbine of entropy waves or local hot spots produced by the unsteady combustion. As these hot-spots accelerate through the turbine they generate noise. Combustion noise has both broadband and tonal elements relating to instabilities excited by combustion. The tonal modes can be powerful and have catastrophic consequences. It is the broadband that is of interest for radiated noise. Specifically, as flow passes through the turbine there are noise sources due to the following interactions: (a) Viscous interactions between blade rows; (b) Potential field interactions and (c) Interaction of temperature gradients (entropic source) with blade rows.

Shock Buffet

Shock buffet can occur inside gas-turbine blade passages and their intakes. It involves large scale, relatively low frequency (relative to the turbulence scales) traversing of a shock over a surface. For aerofoils this takes place over suction side. The shock movement can give rise to strong lift variations. Trapier et al. (2008), explore shock buffet in a supersonic intake. They identify shock buffet as arising from an acoustic feedback mechanisms. In airframes shock buffet can cause structural deflections and give rise to fatigue failure. It can also limit the cruising speed of aircraft (Deck et al. 2005). Some flow unsteady sources in airframes are discussed next.

1.4.2 Unsteady Flow and Airframes

Figure 1.18 shows some of the aircraft simulation zones considered here. Notably, frames (III-V) explore aircraft-engine integration. Frame (III) considers the downwash from the rotors on a tilt rotor aircraft. Frames (IV–V) look at the integration of a bypass engine with a wing and idealized flap. Frame (VI) considers the use of a wing to shield ground-based observers from engine noise. Frame (II) is a simulation for a high lift configuration. It involves massively separated flow, as does Frame (I) for an aerofoil at a high angle of attack. Some unsteady flow aircraft aspects are discussed next.

Dynamic Stall and Spin

Plunging and pitching aerofoils give rise to dynamic stall. This term refers to the delayed separation arising for a rapidly changing angle of attack. The study of dynamic stall is important for fully characterizing aircraft performance (Visbal 2011) and helicopter aerodynamics. Such studies are also relevant, at low Reynolds number, to

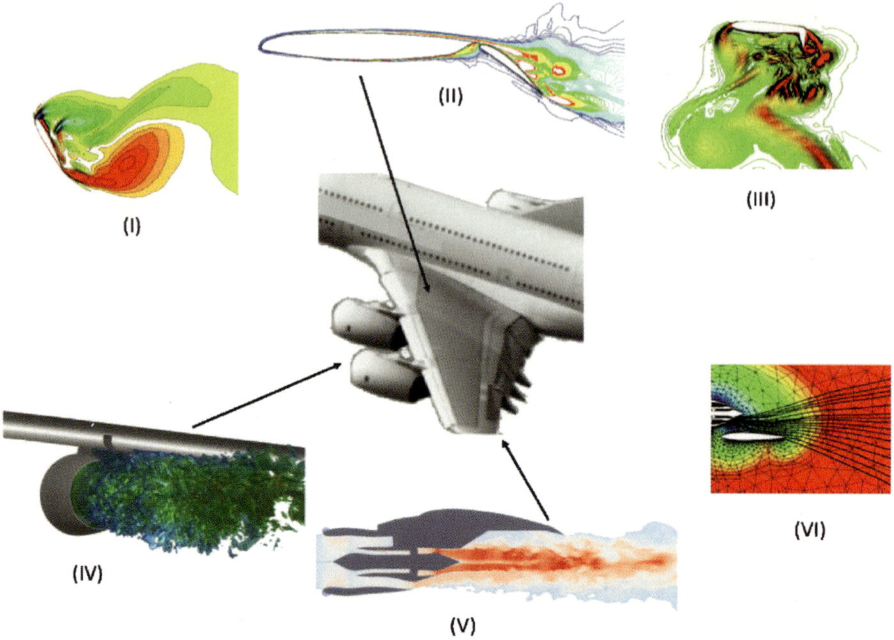

Fig. 1.18 Aircraft and some simulation zones considered

the understanding natural fliers. These are of currently of topical interest. With helicopters the blade pitch is changed. This is because, as the craft speed increases, the retreating blade speed reduces. To match the lift of the forward advancing blade, the angle of attack on the retreating blade must be increased. The resulting high variations in pitching moment can cause structural fatigue (Haase et al. 2009). As noted by Forsythe et al. (2004), aircraft spin is a key area where unsteady CFD can be of used. The exploration of this, through flight tests is hard. Also, unsteady CFD could be used to explore the engine intake distortion state during this process.

Delta Wings

Gursul (2005) outline numerous unsteady flow mechanisms for delta wings. For these, the leading edge vortex produces an intense, curved shear layer. Hence, there is Kelvin-Helmholtz instability. Unsteady vortex wandering can occur. This has also been observed for wing tip vortices (Baker et al. 1974; Devenport et al. 1996; Green and Acosta 1991). The mechanisms for the unsteady, vortex wandering process, is evidently not fully understood. However, a potential element is Biot-Savart interaction with smaller surrounding vortices. There is also the well-known unsteady vortex breakdown mechanism. Oscillation (can be quasi-periodic) in the vortex breakdown location can take place. This is followed by an unsteady helical mode instability, that breaks down to turbulence. The spectrum of dimensionless (in terms of chord

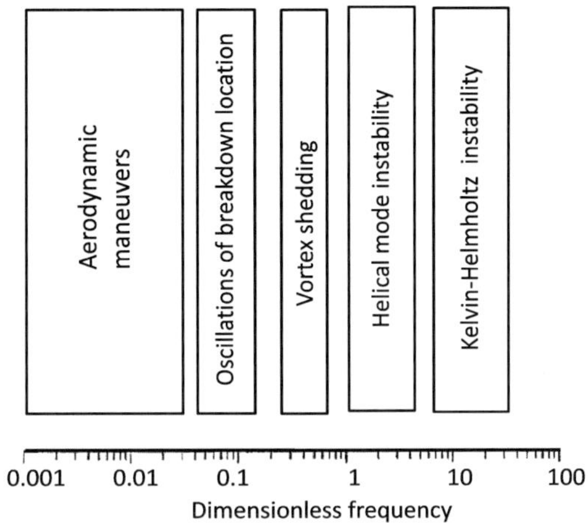

Fig. 1.19 Unsteady frequency range over delta wings (Gursul 2005)

and free stream velocity) unsteady frequencies for delta wings is shown in Fig. 1.19. With regards to eddy resolving numerical simulations a key aspect to observe is the wide range of time scales. On top of the several unsteady flow mechanisms discussed above, is the interaction of these through lag and hysteresis mechanisms for wings exhibiting unsteady motion.

High-Lift Devices

Increasing improvements in engine noise levels are bringing airframe noise to the fore. The high-lift devices (slats and flaps) deployed during landing produce large-scale unsteady flow structures and hence substantial noise. The slat has multiple shear layers producing large-scale unsteady flow structures with impingement (Paschal et al. 2000; Takeda et al. 2001). Since the slat is active for the full wingspan it is a particularly critical noise source. Like Gursul, Cummings et al. (2004) also note a range of unsteadiness components when making eddy-resolving simulations for a high lift configuration.

Other Aspects

As noted by Fu et al. (2007), the horseshoe vortex in wing body junctions can exhibit large-scale unsteadiness resulting in noise. Some relatively weak, minor unsteady vortices can also be observed in this zone. These are important to characterize since they control separation.

Loads control is key to allowing safe and lighter, more energy efficient aircraft. Spoiler type devices can potentially be used to reduce excessive wing loads. Spoilers, as normally used, result in large-scale unsteady separated flows.

A key area where the modelling of unsteadiness is needed is aeroelasticity (Bartels and Sayma 2007). This involves the interaction of the aerodynamics and the

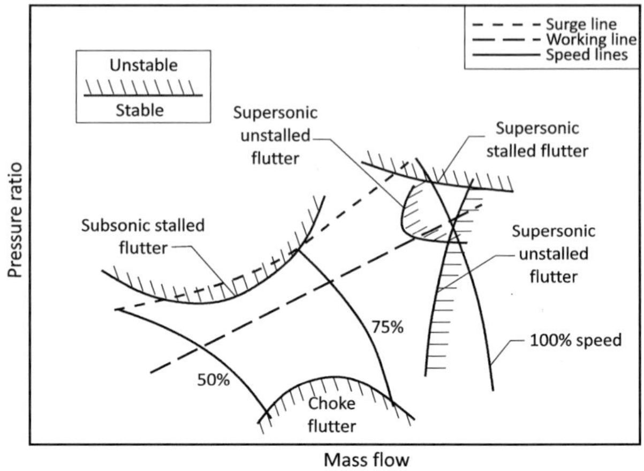

Fig. 1.20 Schematic of the typical flutter boundaries for an axial compressor taken from May (2012)

structural forces (inertia, elastic with damping). Flutter is the key subset of aeroelasticity in which aerodynamics forces couple with a system's natural inertial modes of vibration to produce periodic oscillations. The hope is that, unlike with the well-known Tacoma Narrow bridge disaster, the disturbances do not grow and at worst are observed as a mild buzzing in the aircraft's structure. Figure 1.20 gives a schematic of the typical flutter boundaries for an axial compressor taken from May (2012). Clearly the compressor designer needs care in avoiding the numerous flutter zones as well as the basic aerodynamic limitations such as stall.

A further aeroelasticity subset is dynamic or forced response. This involves the study of the response of a wing to say a sudden gust or a fan to an intake distortion. The other related area is buffet—noted earlier. Non-synchronous blade vibration is a specific kind of aeroelastic behaviour. With this, a rotating instability, like the stall cells (discussed above) cause structural excitation (Im and Zha 2012). Control surface free-play-induced oscillations can occur (Bartels and Sayma 2007). Also, acoustic resonance oscillations can take place where an aerodynamic unsteadiness mode couples to a structural mode (Bartels and Sayma 2007). The aircraft control systems also influence aeroelastic response (aeroservoelasticity).

The operability of modern engines and aircraft is strongly controlled by computers (see Frame II, Fig. 1.1). Flows in these systems frequently involve bluff objects and hence large-scale unsteady separated flow (Tucker 2001).

Most turbomachinery design calculations are performed in steady mode. However, the use of unsteady CFD offers the potential for more radically different designs. For, example in low-pressure turbines, consideration of the impact of unsteady wakes enabled dramatic reductions in the numbers of blades required and hence weight.

From the above it can be seen that unsteadiness exists in a vast array of areas in aircraft and gas turbines. It can involve a wide range of time scales and spectral content. Indeed, for gas turbines the more stages, the higher the number of important harmonics that need to be resolved by the chosen numerical scheme. Also, on top of the core harmonic content there is the potential for distortion interactions in the compressor. In turbines there will be hot spots and migratory components related to these. There will also be acoustic feedback processes modifying the actual flows. This all places increasing demands on both the temporal and spatial schemes. The unsteadiness sources frequently link to environmental noise and some times structural integrity. Hence, problems are intrinsically multi-objective in nature. A key, pacing item, in all of the above, is the modelling of turbulence. This is currently most widely achieved through solving the RANS equations with a model for turbulence. Hence, this aspect is discussed next.

1.5 Predictive Accuracy of RANS

Gas turbine aero engines are well evolved. Hence, as noted earlier, potential efficiency gains as generally hard won. As discussed much of the focus for these gains is through controlling the complex, small-scale feature related, flows and the interaction of these with the main, more conventionally considered flow areas. Also, utilizing real roughness or mitigating its influence is a key research area. The combustor has temperatures approaching 2000 K. The gasses from this are in close proximity to coolant flows, creating strong temperature gradients. These gradients are subjected to a strong centripetal acceleration field—they are in a rotating system. This all can result in highly unstable flow (reminiscent of the classic Rayleigh Bernard instability) zones and additional turbulence production. As clearly shown by Hanjalinc (2005), even the most advanced RANS models totally fail when modelling such flow physics. There are zones of extreme streamline curvature and rotating surfaces in close proximity to stationary. This can all give rise to the extremes of laminarized or highly turbulent flows and transitional states. Many separated flows, generated by small scale features are hard to reliably deal with using RANS. For example, ribbed passage flows are not amenable to low order modelling (Iacovides and Raisee 1999).

Eddies passing through sealant gaps will be exposed to a highly strained and potentially accelerated mean flow fields. Flow fields in intakes or cross-winds have both high acceleration and streamline curvature. This all gives rise to laminarized flow, then separation followed by transition. Turbine blades immediately downstream of a combustor are exposed to large scale eddies from the combustor. Rapid distortion theory is applicable to the turbulence (Tucker and Liu 2007) in this zone and specialized RANS turbulence model treatments needed.

As noted earlier, the internal air systems of gas turbine engines have a range of rotating disc configurations. There can be substantial temperature gradients giving rise to buoyancy forces. In practice the cavity geometries can be complex with small-scale features such as bolts. In a rotating system such features can give rise to what initially is surprising flow behaviour. Indeed, rotating disc flows can be highly complex, involving laminar, transitional and turbulent regions. The turbulent structures can be highly anisotropic and potentially, there can be, large-scale three-dimensional external flow structures (Launder et al. 2010). Unsteady, vortex breakdown can occur (Owen and Pincombe 1979). Even for the most basic case of a rotating disk in a quiescent environment, the boundary layers themselves are intrinsically three-dimensional. Also, the three-dimensional structures that are external to them can impose further complex three-dimensionalities. This all challenges the Boussinesq approximation used in RANS models (Elena and Schiestel 1996). With complexities arising from rotational and buoyancy forces RANS models can perform poorly. Notably, inviscid cores can arise. Turbulence models with diffusion terms will erroneously diffuse high near wall turbulence into these cores thus corrupting heat transfer predictions. Hence, LES offers substantial improvements in predictive accuracy.

Combustors have swirl which is challenging for most eddy viscosity models. They also have large scale unsteady structures (promoted though shear layers and jet interactions) intended to promote mixing. Hence, these flows are poorly suited to RANS predictions. Notably, the large unsteady structures strongly control the flame propagation (Boileau et al. 2008). As shown by Sloan et al. (1986), even for a simple isothermal combustor the predicted centre line velocity distributions, for different RANS models, can vary by hundreds of percent. Also, the trend towards lean, premixed combustion and the transient oscillations that this produces is likely to advance the case for using LES (McGuirk and Taylor 2002). It is worth noting that accurately predicting the unsteady phase between pressure and heat release is critical to accurately predicting thermo-acoustic instabilities and predicting the performance of pulsed gain combustors.

Jets in cross flow have highly anisotropic, unsteady, large-scale structures. These are challenging for RANS models. However, predicting jets in cross flow is important for understanding the dynamics of vertical take off and landing aircraft (when transitioning between flight regimes), nacelle deicing and fuel mixing in combustors. They are also key components turbine film cooling.

There is much interest, nowadays, at looking at off-design performance for airframes. Key areas, are, for example, flaps deployed at high angles of attack and landing gear aerodynamics. Also, hopes for performance gain at design are being placed in utilizing turbulence modifying flow control methods. Mitigating environmental noise is becoming increasingly important.

It seems unlikely, that RANS can adequately model much of the flow physics discussed above without very careful calibration. Time averaging is intrinsically present in the RANS process. For unsteady flows, the averaging for this must be distinct. It must not average out any unsteady process that it is wished to study (unless some satisfactory phenomological model is incorporated to account for the unsteady process and the fluid forces involved in this). Ideally unsteady LES related techniques are necessary. Hence, much of this text focuses on this approach. When using LES and also exploring acoustics the accuracy of numerical schemes becomes more critical. Indeed, for unsteady flows, in general, the accuracy of the spatial discretization becomes more important as does computational speed. Hence, Chap. 2 focuses on numerical methods. Chapter 3 looks at the key area of turbulence and it's modelling. Chapters 4 and 5, look at unsteady aerodynamic modelling. Chapter 6, then moves onto aeroacoustics. Chapter 7 looks at coupled problems. These include those with solid and fluid zones and those with distinctly different fluid systems. The major focus of this text is the gas-turbine aero engine and ultimately how this interacts with the airframe. Hence, as part of this unsteady modelling of airframes is included.

References

H.W. Allen, M.G. Kofsky, Visualization study of secondary flows in turbine tip regions. NASA Technical Note 3519 (1955)

G. Ashcroft, D. Nurnberger, A computational investigation of broadband noise generation in a low-speed axial fan. AIAA Paper No. AIAA-2001-0512 (2001)

N.C. Baines, Turbocharger turbine pulse flow performance and modelling—25 years on, in *9th International Conference on Turbochargers and Turbocharging, Institution of Mechanical Engineers*, London, 19–20 May 2010

G.R. Baker, S.J. Barker, K.K. Bofah, P.G. Saffman, Laser anemometer measurements of trailing vortices in water. J. Fluid Mech. **65**(2), 325–336 (1974)

F. Bardoux, F. Leboeuf, C. Dano, C. Toussaint, Characterization of deterministic correlations for turbine stage. Part 1: time average flow analysis, in *Proceedings of the ASME Paper 99-GT-100*, Indianapolis (1999), pp. 1–9

R.E. Bartels, A.I. Sayma, Computational aeroelastic modelling of airframes and turbomachinery: progress and challenges. Philos. Trans. R. Soc. A, Math. Phys. Eng. Sci. **365**(1859), 2469–2499 (2007)

T.R. Bladh, Q. Zhuang, J. Hu, J. Hammar, Leakage-induced compressor blade excitation due to inter-segment gaps, in *Proceedings of ASME Turbo Expo 2012* (2012). GT–2012–70040

M. Boileau, G. Staffelbach, B. Cuenot, T. Poinsot, C. Bérat, LES of an ignition sequence in a gas turbine engine. Combust. Flame **154**(1), 2–22 (2008)

Y. Bousquet, X. Carbonneau, I. Trébinjac, Assessment of steady and unsteady predictions for a subsonic centrifugal compressor stage, in *Proceedings of ASME Turbo Expo 2012 GT2012*, Copenhagen, Denmark, 11–15 June 2012. ASME Paper No. GT2012-68567

T. Brandvik, G. Pullan, Acceleration of a 3D Euler solver using commodity graphics hardware, in *46th AIAA Aerospace Sciences Meeting and Exhibit*, Reno, Nevada, 7–10 January 2008. No. AIAA-2008-605

F.W. Burcham Jr., D.L. Hughes, Analysis of in-flight pressure fluctuations leading to engine compressor surge in an F-111A airplane for Mach numbers to 2.17. AIAA Paper No. AIAA-1970-624 (1970)

T.L. Butler, O.P. Sharma, H.D. Joslyn, R.P. Dring, Redistribution of an inlet temperature distortion in an axial flow turbine stage. J. Propuls. Power **5**(1), 64–71 (1989)

C. Camci, D.H. Rizzo, Secondary flow and forced convection heat transfer near endwall boundary layer fences in a 90 turning duct. Int. J. Heat Mass Transf. **45**(4), 831–843 (2002)

T.R. Camp, I.J. Day, A study of spike and modal stall phenomena in a low-speed axial compressor. J. Turbomach. **120**, 393–401 (1998)

N.D. Cardwell, Effects of realistic first-stage turbine endwall features. MSc Thesis, Virginia Polytechnic Institute, 2005

P.W. Carpenter, K.L. Kudar, R. Ali, P.K. Sen, C. Davies, A deterministic model for the sublayer streaks in turbulent boundary layers for application to flow control. Philos. Trans. R. Soc. A, Math. Phys. Eng. Sci. **365**(1859), 2419–2441 (2007)

D.R. Chapman, H. Mark, M.W. Pirtle, Computers vs. wind tunnels for aerodynamic flow simulations. Astronaut. Aeronaut. **13**, 12–35 (1975)

H. Choi, P. Moin, Grid-point requirements for large eddy simulation: Chapman's estimates revisited. Phys. Fluids **24**, 011702 (2012)

J.D. Coull, Wake induced transition in low pressure turbines. PhD thesis, Department of Engineering, The University of Cambridge (2009)

W.T. Cousins, History, philosophy, physics, and future directions of aircraft propulsion system/inlet integration, in *Proceedings of ASME Turbo Expo 2004 Power for Land, Sea, and Air*, Vienna, Austria, 14–17 June 2004. ASME Paper No. GT2004-54210

R.M. Cummings, S.A. Morton, J.R. Forsythe, Detached-eddy simulation of slat and flap aerodynamics for a high-lift wing, in *42nd Aerospace Sciences Meeting and Exhibit* (2004), p. 1

N. Cumpsty, *Jet Propulsion: A Simple Guide to the Aerodynamic and Thermodynamic Design and Performance of Jet Engines* (Cambridge University Press, Cambridge, 2003)

F.M. De Jongh, P.G. Morton, R. Holmes, The synchronous instability of a compressor rotor due to bearing journal differential heating. discussion. J. Eng. Gas Turbines Power **118**(4), 816–824 (1996)

S. Deck, P. Spalart, P. Sagaut, J. Forsythe, K.D. Squires, Numerical simulation of transonic buffet over a supercritical airfoil. AIAA J. **43**(7), 1556–1566 (2005)

W.J. Devenport, M.C. Rife, S.I. Liapis, G.J. Follin, The structure and development of a wing-tip vortex. J. Fluid Mech. **312**(67), 106 (1996)

G. Dufour, N. Gourdain, F. Duchaine, O. Vermorel, L.Y.M. Gicquel, J.F. Boussuge, T. Poinsot, *Large Eddy Simulation Applications*. VKI Lecture Series Numerical Investigations in Turbomachinery: The State of Art, Rhodes-St-Genese, Belgique, 21–25 Sept. 2009

L. Elena, R. Schiestel, Turbulence modeling of rotating confined flows. Int. J. Heat Fluid Flow **17**(3), 283–289 (1996)

H.W. Emmons, C.E. Pearson, H.P. Grant, Compressor surge and stall propagation. Trans. Am. Soc. Mech. Eng. **77**(3), 455–469 (1955)

A.H. Epstein, F. Williams, E.M. Greitzer, Active suppression of aerodynamic instabilities in turbomachines. J. Propuls. Power **5**(2), 204–211 (1986)

P.R. Farthing, C.A. Long, J.M. Owen, Rotating cavity with axial throughflow of cooling air: flow structure. J. Turbomach. **114**(237) (1992a)

P.R. Farthing, C.A. Long, J.M. Owen, Rotating cavity with axial throughflow of cooling air: heat transfer. J. Turbomach. **114**(229) (1992b)

F. Flemming, C. Olbricht, B. Wegner, A. Sadiki, J. Janicka, F. Bake, U. Michel, B. Lehmann, I. Röhle, Analysis of unsteady motion with respect to noise sources in a gas turbine combustor: isothermal flow case. Flow Turbul. Combust. **75**(1), 3–27 (2005)

J.R. Forsythe, K.D. Squires, K.E. Wurtzler, P.R. Spalart, Detached-eddy simulation of the F-15E at high alpha. J. Aircr. **41**(2), 193–200 (2004)

G. Fritsch, M. Giles, Second-order effects of unsteadiness on the performance of turbomachines, in *37th International Gas Turbine and Aeroengine Congress and Exposition*, vol. 1 (1992)

S. Fu, Z. Xiao, H. Chen, Y. Zhang, J. Huang, Simulation of wing-body junction flows with hybrid RANS/LES methods. Int. J. Heat Fluid Flow **28**(6), 1379–1390 (2007)

M.B. Giles, An approach for multi-stage calculations incorporating unsteadiness. ASME Paper No. 92-GT-282 (1992)

N. Gourdain, L.Y.M. Gicquel, E. Collado, RANS and LES for the heat transfer prediction in turbine guide vane. J. Propuls. Power **28**(2), 423–433 (2012). doi:10.2514/1.57717

S.I. Green, A.J. Acosta, Unsteady flow in trailing vortices. J. Fluid Mech. **227**(1), 107–134 (1991)

J.F. Guelich, U. Bolleter, Pressure pulsations in centrifugal pumps. J. Vib. Acoust. **114**(2), 272–279 (1992)

I. Gursul, Review of unsteady vortex flows over slender delta wings. J. Aircr. **42**(2), 299–319 (2005)

W. Haase, M. Braza, A. Revell, *DESider—A European Effort on Hybrid RANS-LES Modelling: Results of the European-Union Funded Project, 2004–2007*, vol. 103 (Springer, Berlin, 2009)

A. Hadjadj, Large-eddy simulation of shock/boundary-layer interaction. AIAA J. **50**(12), 2919–2927 (2012)

K. Hanjalinc, Invited talk, in *9th UK Heat Transfer Conference*, 5–7 September 2005 (Manchester University Press, Manchester, 2005)

R.G. Hercock, D.D. Williams, Distortion-induced engine instability. Aerodynamic response. AGARD, LS72-Paper (3) (1974)

H.Z. Herzig, A.G. Hansen, G.R. Costello, A visualization study of secondary flows in cascades. NACA Rep. 1163 (Supersedes NACA TN 2947) (1954)

P. Hield, Personal communication (2012)

N. Hills, Achieving high parallel performance for an unstructured unsteady turbomachinery CFD code. Aeronaut. J. **111**(1117), 185–193 (2007)

H. Iacovides, M. Raisee, Recent progress in the computation of flow and heat transfer in internal cooling passages of turbine blades. Int. J. Heat Fluid Flow **20**(3), 320–328 (1999)

H.S. Im, G.C. Zha, Simulation of nonsynchronous blade vibration of an axial compressor using a fully coupled fluid/structure interaction. ASME GT2012-68150 (2012)

R. Jakoby, T. Zierer, K. Lindblad, J. Larsson, D.E. Bohn, J. Funcke, A. Decker et al., Numerical simulation of the unsteady flow field in an axial gas turbine rim seal configuration. GT2004-53829, ASME Expo (2004)

M. Judge, Personal communication (2012)

J.L. Kerrebrock, A.A. Mikolajczak, Intra-stator transport of rotor wakes and its effect on compressor performance. ASME J. Eng. Power **92**(4), 359–368 (1970)

C. Klostermeier, Investigation into the capability of large eddy simulation for turbomachinery design. PhD thesis, University of Cambridge (2008)

U. Köller, R. Mönig, B. Kösters, H.A. Schreiber, Development of advanced compressor airfoils for heavy-duty gas turbines—Part I: design and optimization. J. Turbomach. **122**(3), 397–405 (2000)

B. Launder, S. Poncet, E. Serre, Laminar, transitional, and turbulent flows in rotor-stator cavities. Annu. Rev. Fluid Mech. **42**, 229–248 (2010)

F. Leboeuf, D. Charbonnier, Development of a model for the deterministic stresses in a transonic turbine stage, in *Proceedings of the Fifth European Conference on Turbomachinery*, Prague (2003)

S. Leichtfuss, C. Biela, H.P. Schiffer, F. Heinichen, Influence of inlet guide vane wakes on the passage flow in a transonic axial compressor, in *Proceedings of ASME Turbo Expo 2012 GT2012*, Copenhagen, Denmark, 11–15 June 2012. ASME Paper No. GT2012-69485

M. Leschziner, N. Li, F. Tessicini, M. Leschziner, N. Li, F. Tessicini, Simulating flow separation from continuous surfaces: routes to overcoming the Reynolds number barrier. Philos. Trans. R. Soc. A, Math. Phys. Eng. Sci. **367**(1899), 2885–2903 (2009)

M.G. List, S.E. Gorrell, M.G. Turner, Investigation of loss generation in an embedded transonic fan at several gaps using high-fidelity, time accurate CFD, in *Proceedings of the ASME Turbo Expo Power for Land, Sea and Air GT 2008*, Berlin, Germany (2008). Paper No. GT2008-51220

S. Loiodice, P.G. Tucker, J. Watson, Coupled open rotor engine intake simulations, in *Proceedings of the 48th AIAA Aerospace Sciences Meeting and Exhibit*, Orlando, Florida, Jan. 2010. Paper No. AIAA–2010–840

K. Matsuura, C. Kato, Large-eddy simulation of compressible transitional cascade flows with and without incoming free-stream turbulence. JSME Int. J. Ser. B **49**(3), 660–669 (2006)

M. May, Reduced order modeling for the flutter stability analysis of a highly loaded transonic fan, in *Proceedings of ASME Turbo Expo 2012 GT2012*, Copenhagen, Denmark, 11–15 June 2012. ASME Paper No. GT2012-69775

R. Mayle, The role of laminar-turbulent transition in gas turbine engines. ASME Trans. J. Turbomach. **113**, 509–537 (1991). *ASME, International Gas Turbine and Aeroengine Congress and Exposition, 36th*, Orlando, FL, 3–6 June 1991

J.J. McGuirk, A. Taylor, Numerical simulation of gas turbine combustion processes-present status and future trends. J. Gas Turbine Soc. Jpn. **30**(5), 347–361 (2002)

C. Meneveau, J. Katz, A deterministic stress model for rotor-stator interactions in simulations of average-passage flow. J. Fluids Eng. **124**(2), 550–554 (2002)

V. Michelassi, J.G. Wissink, W. Rodi, Direct numerical simulation, large eddy simulation and unsteady Reynolds-averaged Navier–Stokes simulations of periodic unsteady flow in a low-pressure turbine cascade: a comparison. Proc. Inst. Mech. Eng. A, J. Power Energy **217**(4), 403–411 (2003)

F. Montomoli, H.P. Hodson, L. Lapworth, RANS–URANS in axial compressor, a design methodology. Proc. Inst. Mech. Eng. A, J. Power Energy **225**(3), 363–374 (2011)

Y.J. Moon, S.R. Koh, Counter-rotating streamwise vortex formation in the turbine cascade with endwall fence. Comput. Fluids **30**(4), 473–490 (2001)

C.J. Moore, The role of shear-layer instability waves in jet exhaust noise. J. Fluid Mech. **80**(02), 321–367 (1977)

W.E.I. Ning, L.I. He, Some modeling issues on trailing-edge vortex shedding. AIAA J. **39**(5), 787–793 (2001)

T.S.D. O'Mahoney, N.J. Hills, J.W. Chew, T. Scanlon, Large-eddy simulation of rim seal ingestion. Proc. Inst. Mech. Eng., Part C, J. Mech. Eng. Sci. **225**(12), 2881–2891 (2011)

J.M. Owen, J.R. Pincombe, Vortex breakdown in a rotating cylindrical cavity. J. Fluid Mech. **90**(01), 109–127 (1979)

J.M. Owen, R.H. Rogers, *Flow and Heat Transfer in Rotating-Disc Systems*. Rotor–Stator Systems, vol. 1 (Research Studies Press Ltd., Baldock, 1989)

K. Paschal, L. Jenkins, C. Yao, Unsteady slat-wake characteristics of a high-lift configuration. AIAA Paper No. AIAA-2000-0139 (2000)

A. Pfau, M. Treiber, M. Sell, G. Gyarmathy, Flow interaction from the exit cavity of an axial turbine blade row labyrinth seal. J. Turbomach. **123**(2), 342–352 (2001)

U. Piomelli, Wall-layer models for large-eddy simulations, in *Proceedings of the 46th AIAA Aerospace Sciences Meeting and Exhibit*, Reno, Nevada, 7–10 January 2008. AIAA Paper No. AIAA-2008-0603

U. Piomelli, E. Balaras, Wall-layer models for large-eddy simulations. Annu. Rev. Fluid Mech. **34**(1), 349–374 (2002)

J.M.M. Place, Three-dimensional flow in core compressors. PhD thesis, University of Cambridge (1997)

S.B. Pope, Ten questions concerning the large-eddy simulation of turbulent flows. New J. Phys. **6**(1), 35 (2004)

B. Raverdy, I. Mary, P. Sagaut, N. Liamis, High-resolution large-eddy simulation of flow around low-pressure turbine blade. AIAA J. **41**(3), 390–397 (2003)

H.J. Rehder, A. Dannhauer, Experimental investigation of turbine leakage flows on the three-dimensional flow field and endwall heat transfer. J. Turbomach. **129**(3), 608–618 (2007)

K. Reid, J. Denton, G. Pullan, E. Curtis, J. Longley, The interaction of turbine inter-platform leakage flow with the mainstream flow. J. Turbomach. **129**(2), 303–310 (2007)

C.M. Rhie, A.J. Gleixner, D.A. Spear, C.J. Fischberg, R.M. Zacharias, Development and application of a multistage Navier–Stokes solver. Part I: multistage modeling using bodyforces and deterministic stresses. ASME Trans. J. Turbomach. **120**, 205–214 (1995)

M. Rosenlund, N. Berglind, G. Pershagen, L. Järup, G. Bluhm, Increased prevalence of hypertension in a population exposed to aircraft noise. Occup. Environ. Med. **58**(12), 769–773 (2001)

V. Saxena, H. Nasir, S.V. Ekkad, Effect of blade tip geometry on tip flow and heat transfer for a blade in a low-speed cascade. J. Turbomach. **126**(1), 130–138 (2004)

J.U. Schluter, X. Wu, S. Kim, S. Shankaran, J.J. Alonso, H. Pitsch, A framework for coupling Reynolds-averaged with large-eddy simulations for gas turbine applications. Trans. Am. Soc. Mech. Eng. J. Fluids Eng. **127**(4), 806 (2005)

T. Shang, A.H. Epstein, M.B. Giles, Blade row interaction effects on compressor measurements, in *Proceedings of the AGARD 74th Specialist Meeting on Unsteady Aerodynamic Phenomena in Turbomachines*, Luxembourg (1989). AGARD Conference Proceedings No. 468

D. Sims-Williams, F. Forster, G. Ingram, Reconstruction of the unsteady pressure field in a low speed linear cascade, in *Proceedings of ASME Turbo Expo 2012 GT2012*, Copenhagen, Denmark, 11–15 June 2012. ASME Paper No. GT2012-69156

D.G. Sloan, P.J. Smith, L.D. Smoot, Modeling of swirl in turbulent flow systems. Prog. Energy Combust. Sci. **12**(3), 163–250 (1986)

P.R. Spalart, D.R. Bogue, The role of CFD in aerodynamics, off-design. Aeronaut. J. **107**(1072), 323–329 (2003)
N. Suryavamshi, B. Lakshminarayana, J. Prato, J.R. Fagan, Unsteady total pressure field downstream of an embedded stator in a multistage axial flow compressor. J. Fluids Eng. **119**(4), 985–994 (1997)
K. Takeda, G.B. Ashcroft, X. Zhang, P.A. Nelson, Unsteady aerodynamics of slat cove flow in a high-lift device configuration. AIAA paper **706**, 2001 (2001)
S. Trapier, S. Deck, P. Duveau, Delayed detached-eddy simulation and analysis of supersonic inlet buzz. AIAA J. **46**(1), 118–131 (2008)
P.G. Tucker, *Computation of Unsteady Internal Flows* (Kluwer Academic, Norwell, 2001)
P.G. Tucker, Computation of unsteady turbomachinery flows: Part 1—hierarchy and example simulations. ASME J. Turbomach. **134**(2), 021023 (2011a), 10 pp.
P.G. Tucker, Computation of unsteady turbomachinery flows: Part 2—LES and hybrids. Prog. Aerosp. Sci. **47**, 546–569 (2011b)
P. Tucker, Y. Liu, Turbulence modelling for flows around convex features giving rapid eddy distortion. Int. J. Heat Fluid Flow **28**(5), 1073–1091 (2007)
J.M. Tyler, Axial flow compressor noise studies. SAE Transact. **70**, 309–332 (1962)
E.A. Van Deusen, V.R. Mardoc, Distortion and turbulence interaction—a method of evaluating engine inlet compatibility. J. Aircr. **9**(1), 16–22 (1972)
M.R. Visbal, Numerical investigation of deep dynamic stall of a plunging airfoil. AIAA J. **49**(10), 2152–2170 (2011)
A.M. Wallis, J.D. Denton, A. Demargne The control of shroud leakage flows to reduce aerodynamic losses in a low aspect ratio, shrouded axial flow turbine. ASME Turbo. Expo. 2000, Munich, Germany (2000)
J.E.F. Williams, D.L. Hawkings, Sound generation by turbulence and surfaces in arbitrary motion. Philos. Trans. R. Soc. Lond. Ser. A, Math. Phys. Sci. **264**(1151), 321–342 (1969)
D.E. Winterbone, B. Nikpour, H. Frost, A contribution to the understanding of turbocharger turbine performance in pulsating flow, in *International Conference on Internal Combustion Research, Paper No. C433/011* (1991), pp. 19–28
H. Xia, P.G. Tucker, W.N. Dawes, Level sets for CFD in aerospace engineering. Prog. Aerosp. Sci. **46**(7), 274–283 (2010)
A. Young, Personal communication (April 2012)
L. Zhao, W.-Y. Qiao, Z.-Q. Mu, P.P. Chen, Tip clearance effects on unsteady behavior of the tip leakage vortex and stator/rotor interaction tonal noise, in *Proceedings of ASME Turbo Expo 2012 GT2012*, Copenhagen, Denmark, 11–15 June 2012. ASME Paper No. GT2012-68549

Chapter 2
Computational Methods for Unsteady Flows

2.1 Introduction

In this chapter various temporal discretization schemes will first be discussed. Then the spatial discretization is considered. Also, peripheral solution approaches will be addressed, such as prediction of the pressure field along with simultaneous equation solvers. Finally, the impact of grid topology on the accuracy of solutions is considered. There is a strong emphasis on LES.

2.2 Overview of Temporal Discretizations

The major element of any CFD procedure is the solution of convection and diffusion transport equations. Like the spatial, the temporal discretization process needs to incorporate assumptions of how variables will change with time and also how this time variation varies over each cell. Generally, the process involves two key initial stages. First, just the spatial terms are discretized (again profile assumptions are involved). These terms are generally represented here by the global symbol A. The temporal derivative remains, giving a semi-discrete Ordinary Differential Equation ($\partial\phi/\partial t = A$).

This process of reducing the full governing equations to semi-discrete Ordinary Differential Equations (ODEs) is called the *Method of Lines*. The second stage involves discretization of the ODEs. There are numerous discretization assumptions and methods originating from general solution procedures for ODEs. The most common procedures, which have found application in CFD, will be described here. Most attention will be focused on schemes widely used in practical CFD. The following generic approaches are presented: two-level, three-level, predictor-corrector and splitting.

P.G. Tucker, *Unsteady Computational Fluid Dynamics in Aeronautics*, Fluid Mechanics and Its Applications 104, DOI 10.1007/978-94-007-7049-2_2,

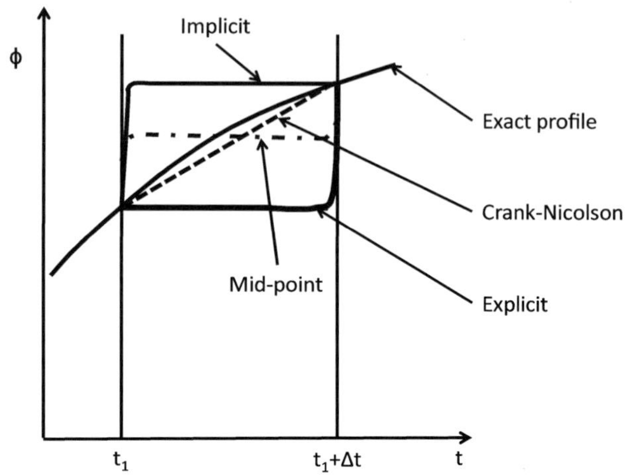

Fig. 2.1 Temporal profile assumptions

2.3 Temporal Profile Assumptions for Variables

2.3.1 Dependent Variable Changes with Time

When discretizing the unsteady flow equations an assumption must be made of how dependent variables change with time. Basic profile assumptions are given in Fig. 2.1. These can mathematically be represented as

$$\phi = Wf^{n+1}\phi^{n+1} + Wf^{n}\phi^{n} + Wf^{n-1}\phi^{n-1} \tag{2.1}$$

The precise numerical values, for different weighting coefficients, is given later. The discretized transport equations can be written using the method of lines, noted earlier, as below

$$(1+\eta)\frac{\Delta[\rho_{\mathrm{i}}\phi_{\mathrm{i}}]^{n}}{\Delta t} - \eta\frac{\Delta[\rho_{\mathrm{i}}\phi_{\mathrm{i}}]^{n-1}}{\Delta t} = A(\phi) \tag{2.2}$$

where A represents the discretized spatial derivatives and source terms. Also,

$$\Delta[\rho_{\mathrm{i}}\phi_{\mathrm{i}}]^{n}/\Delta t = \left[(\rho\phi_{\mathrm{i}})^{n} - (\rho_{\mathrm{i}}\phi_{\mathrm{i}})^{n-1}\right]/\Delta t,$$

$$\Delta[\rho_{\mathrm{i}}\phi_{\mathrm{i}}]^{n-1}/\Delta t = \left[(\rho\phi_{\mathrm{i}})^{n-1} - (\rho_{\mathrm{i}}\phi_{\mathrm{i}})^{n-2}\right]/\Delta t$$

Equation (2.2) offers the possibility of both two and three-level (involving $\phi^{n}, \phi^{n-1}, \phi^{n-2}$) schemes. To implement these, values of ϕ *not involved in time derivatives* $(\Delta(\rho_{\mathrm{i}}\phi_{\mathrm{i}})/\Delta t)$ are replaced with Eq. (2.1). The possible number of temporal discretizations available to CFD is extensive. We can define

$$Wf^{n} = \lambda, \qquad Wf^{n-1} = 1-\lambda-k, \qquad Wf^{n-2} = k \tag{2.3}$$

Table 2.1 Values of η, λ and k for commonly occurring time schemes

Scheme	η	λ	k	Stability	Order	Other features
Explicit (Euler)	0	0	0	$0 \leq C^2 \leq 2D \leq 1$	1	–
Implicit (Backwards Euler)	0	1	0	Unconditionally stable	1	–
Crank-Nicolson (one-step trapezoidal)	0	1/2	0	Unconditionally stable	2	–
Leapfrog	−1/2	0	0	Unconditionally unstable	2	–
Adams-Moulton	0	5/12	1/12	–	3	–
Galerkin	0	3/5	0	Unconditionally stable	1	–
Exponential	0	$1/(1 - e^{A_{i\phi}\Delta t}) - 1/A_{i\phi}\Delta t$	0	Unconditionally stable	1	–
Lax-Wendroff	0	0	0	$0 \leq C^2 \leq 2D \leq 1$	–	Γ augmented by $(\Delta t u^2)/2$
DuFort-Frankel	0	0	0	$C < 1$, for accuracy $C^2 \leq 1$	2	$\phi_i^{n-1} = \frac{\phi_i^n + \phi_i^{n-2}}{2}$ in diffusion terms

Then, some values of η, λ and k, for frequently occurring time schemes are given in Table 2.1. Note, in this table, $C = u\Delta t/\Delta x$ is the Courant number. For compressible flows, this needs to be re-expressed as $C = (|u| + c)\Delta t/\Delta x$ where c is the speed of sound. Hence, for compressible flow the velocity scale is replaced by the maximum acoustic wave speed. The more severe stability constraint is the diffusive. This involves the parameter, $D = \Gamma\Delta t/\rho(\Delta x)^2$where Γ is a diffusion coefficient. Notice that for the Lax-Wendroff Scheme, Γ is augmented by $(\Delta t u^2)/2$.

2.3.2 Spatial Variation of the Time Derivative

The time derivative, in the governing equations can either be assumed constant over a cell or to vary in some fashion *consistent* with underlying space discretization. With finite volume/difference methods, as used here, the former *lumped approach*, indicated for one-dimension in Fig. 2.2, is adopted. For finite elements, the later *consistent formulation* is common. With this, the time derivative is multiplied by what is called a mass or capacitance matrix. Effectively, this spreads the time derivative over several nodes. For a one-dimensional uniform mesh case, involving linear shape functions/elements the capacitance or mass operators are such that $M_{i-1} = 1/6$, $M_i = 2/3$, $M_{i+1} = 1/6$. This results in the piecewise-linear $\partial(\rho\phi)/\partial t$ profile indicated in Fig. 2.2. The time derivative spreading will introduce extra computational cost. However, there are accuracy benefits (Fletcher 1997). It is possible

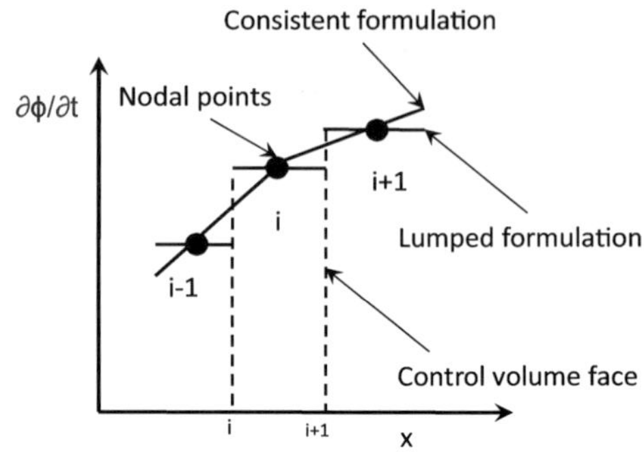

Fig. 2.2 Spatial variation of time derivative

to combine differing discretization techniques, such as having a finite element space discretization with finite difference temporal modelling (Segerlind 1976).

2.4 Two-Level Schemes

A key two-level scheme, finding popularity in more academic LES is the neutrally dissipative Crank-Nicolson (CN) scheme. As with many schemes, it finds different implementations aimed at promoting stability (Giles 2004; Hujeirat and Rannacher 1998) and efficiency (Beam and Warming 1976; Briley and McDonald 1975; Lacor 1999; Jameson 1991). It is possible to use hybridized approaches. For example, in the LES work of Talha (2012) and Manoha et al. (2000) (trailing edge noise) the explicit Adams-Bashforth scheme (see later) is used for the convective terms and CN for diffusive. With regards to CN implementation variants, both one and two-legged discretizations are possible. Here the distinction is being made whether the underlying discretization function is applied to the variable being solved for or some function of it. For LES, all these choices can have a significant impact. For example, as shown in Fig. 2.3, there is an over a 10 % variation in Reynolds stresses, between different CN scheme implementations. The profiles shown are for a free shear flow. The symbols are for two different sets of measurements. The labels, CN1 and CN2 identify the two Crank-Nicolson implementations. CN1 is a stabilized form. Full details for this case can be found in Tucker (2008). Note, for $0.5 \leq \lambda \leq 1$, two-level schemes are unconditionally stable. This does not mean that for $\lambda < 1$ solutions will be physically realistic but instead that errors will reduce with time.

2.4.1 General Explicit Schemes

When $\lambda = 0$ (with $k = \eta = 0$) the explicit forward difference Euler scheme is gained. This first order scheme is stable for $C^2 \leq 2D \leq 1$. The restriction $2D \leq 1$ implies that when the mesh spacing is halved the time-step must be made four times smaller. This is a severe time-step restriction. Therefore, the explicit Euler scheme

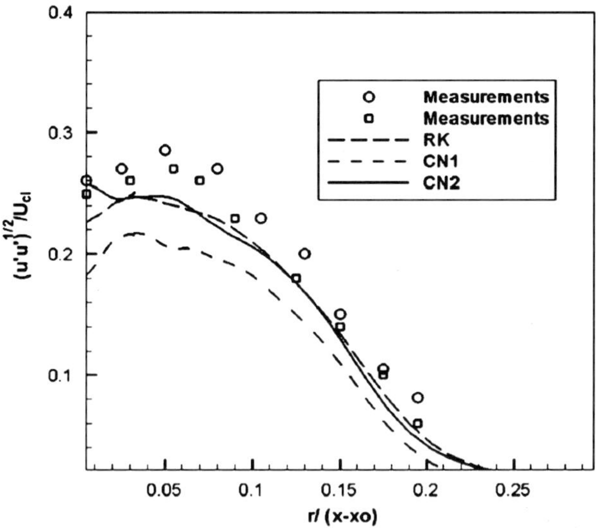

Fig. 2.3 Radial variation of streamwise Reynolds stress in self-similar region of a round jet for different temporal schemes (from Tucker 2008). Published with kind permission from Elsevier

is only suitable for problems where high temporal resolution is required and carefully tailored grids can be constructed. The latter requirement avoids needless small cells which place an excessive global time step restriction. For problems where ϕ is changing slowly with time, the time-step restriction makes the scheme inefficient and implicit schemes should be used.

Higher order explicit schemes can be implemented. However, generally the higher the order, the more severe the stability restriction. Explicit methods do not require simultaneous equation solvers. Therefore, they lend themselves to vector and parallel processing.

2.5 Higher-Level Schemes

As termed here, higher-level methods use data stored at several (> 2) levels/points. Generally for industrial CFD applications, no more than 3–4 levels are implemented. The most common multipoint methods are the Adams types. Using polynomial temporal fits both explicit and implicit Adams type schemes can be constructed. The former are referred to as Adams-Bashforth and the latter Adams-Moulton methods. The Adams-Bashforth scheme can be expressed as

$$\phi^{n+1} - \phi^n \simeq \Delta t \sum_{l=0}^{p} Wf_l \left(\frac{\partial \phi}{\partial t} \right)^{n-l} \tag{2.4}$$

In the above p defines the number of coefficients that the scheme uses. The application of the Adams-Bashforth method (with CN) to trailing edge noise is given by Manoha et al. (2000). In conjunction with Eqs. (2.2)–(2.3) various Adams multipoint methods can be gained by selecting different values of η, λ and k. Parameter

Table 2.2 Weighting functions for backwards difference schemes

N_{BD}	Wf^{n+1}	Wf^{n}	Wf^{n-1}	Wf^{n-2}	Wf^{n-3}	Wf^{n-4}
1	1	−1	–	–	–	–
2	3/2	−4/2	1/2	–	–	–
3	11/6	−18/6	9/6	−2/6	–	–
4	25/12	−48/12	36/12	−16/12	3/12	–
5	137/60	−300/60	300/60	−200/60	75/60	−12/60

values, for the implicit 3rd order Adams-Moulton method are given in Table 2.1. Values for other schemes are given by Beam and Warming (1982). Notably, Tam and Webb (1993) optimize the Adams-Bashforth scheme for acoustics problems to minimize dispersion/frequency errors.

The leapfrog (midpoint) scheme, approximates the time derivative using a second order central difference. With stability modifications the scheme is used for meteorological (Hignett et al. 1985) and oceanographic predictions. The DuFort-Frankel is also a three-level, multipoint, second order scheme. It addresses the leapfrog scheme's lack of stability. However, for accuracy there is the severe restriction $C^2 \ll 1$ which limits the scheme's practical use. These schemes find little aerospace application.

2.5.1 Gear Schemes

Gear schemes are implicit backwards difference methods. Hence, for a discretized equation right hand side, A, at a time level $n+1$ we have the following temporal discretization

$$\left[\frac{\partial \phi}{\partial t}\right]_{BD} = \sum_{n-(N_{BD}-1)}^{n+1} \frac{Wf^n}{\Delta t} = -A^{n+1} \tag{2.5}$$

The weighting functions, Wf, for the different order backwards difference (N_{BD}) schemes are given in Table 2.2.

As can be seen from Appendix A, even for a 5th order Gear scheme the numerical error can be higher than for the CN scheme. Hence, the latter lower storage scheme seems preferable and accuracy grounds. However, stability is lower and, especially in an industrial context, this is important. Higher level schemes can need more careful attention to starting solutions—more data is needed.

2.6 Other Temporal Discretization Methods

Predictor-corrector methods are another key class of temporal scheme. Probably the most widely used temporal integration schemes in aerospace engineering, especially

Table 2.3 Some Runge-Kutta scheme coefficients

Coefficient	SRK4	LDDRK5-BB	LDDRK6-BB	LDDRK4-H	LDDRK5-H	LDDRK6-H
Wf_1	1	1	1	1	1	1
Wf_2	1/2	1/2	1/2	1/2	1/2	1/2
Wf_3	1/6	0.16525035366	0.16591977136	0.162997	0.166558	1/3!
Wf_4	1/24	0.03937258598	0.04091973204	0.040757	0.0395041	1/4!
Wf_5	–	0.00714909644	0.00755570439	–	0.00781071	0.0078100
Wf_6	–	–	0.00089142126	–	–	0.0013214

for acoustics, are explicit Runge-Kutta (RK) schemes. These are used for both unsteady flows and integrating to steady state solutions. Essentially, the RK technique can be considered as using a carefully selected blend of lower order methods. These are intended to give a desired higher order of accuracy or a larger stability 'foot print' in a stability diagram. The blend of lower order methods gives rise to multiple stages of solutions with different weighting coefficients—Wf. Explicit Runge-Kutta methods are most common. However, there are implicit variants that allow larger C. Explicit RK schemes offer the potential for high order with low storage. Most RK schemes are tuned to have an enlarged stability foot print. However, optimizing for stability potentially makes them unsuited to computational acoustics. For this, small amplitude, long length waves generally need to be accurately modeled over considerable distances. Also, as will be discussed later (when dealing with spatial dicretizations), a scheme can have a high formal order but not be that accurate at practical time step resolutions. Hence, especially for acoustics, potentially high order schemes can be further optimized to reduce dissipation and dispersion to form what are known has LDDRK (Low Dissipation and Dispersion RK) schemes. For a p stage RK scheme, that advances a solution from the nth to the $n+1$ time step the following can be written

$$\phi^0 = \phi^n \tag{2.6a}$$

$$\phi^l = \phi^n + Wf_l \Delta t A\left(\phi^{l-1}\right) \quad \text{for } l = 1, \dots, p \tag{2.6b}$$

$$\phi^{n+1} = \phi^p \tag{2.6c}$$

where Wf_l are the coefficients of the algorithm. Table 2.3 shows the coefficients for a standard 4-stage scheme (SRK4). This is 4th order for linear systems and 2nd for nonlinear. Also given in the table are 5 (LDDRK5-BB) and 6 (LDDRK6-BB) stage schemes from Bogey and Bailly (2004). The table also includes schemes of Hu et al. (1996) for different numbers of stages (LDDRK*-H). These schemes have been optimized for low dissipation and dispersion. They are likely to show clear benefits for aeroacoustics problems or where such problems are combined with LES.

To look at the numerical error in an RK scheme the amplification factor is needed. This is the ratio, in the frequency domain, of predicted values at the current and previous time steps. In what follows, the operation in the frequency domain is identified using a tilda as an 'overbar'. In this domain, the RK scheme looks like (Hu et al. 1996)

$$\tilde{\phi}^{n+1} = \tilde{\phi}^{n}\left(1 + \sum_{l=1}^{p} Wf_l(-iuw'\Delta t)^l\right) \tag{2.7}$$

In the above u is the convection speed in the following equation $\partial\phi/\partial t + u\partial\phi/\partial x = 0$ and w' is the numerical wave number. The numerical amplification, r_n, factor can be expressed as

$$r_n = \frac{\tilde{\phi}^{n+1}}{\tilde{\phi}^{n}} = 1 + \sum_{l=1}^{p} Wf_l(-i\sigma)^l \tag{2.8}$$

where $\sigma = uw'\Delta t$. As noted by Hu et al. the exact amplification factor is $r_e = e^{-i\sigma}$. The numerical factor, as can be seen, is some polynomial approximation to the analytical amplification factor equation. Indeed, the order of the polynomial used to characterize the numerical amplification reflects the scheme's potential order. Hence, the maximum order of a p stage scheme is p (Hu et al. 1996). Hu et al. optimize $|r_n - r_e|^2$as a function of $uw'\Delta t$. This optimizes the sum of the dissipation and dispersion errors. This optimization involves the minimization of the following integral

$$\int_0^{\beta} \left|1 + \sum_{l=1}^{p} Wf_l(-i\sigma)^l - e^{i\sigma}\right|^2 d\sigma \tag{2.9}$$

In the above β defines the range of σ for which the optimization applies. Notably, when used in conjunction with high order compact schemes, Kim and Lee (1996) found rather non-monotonic behavior with regards to accuracy benefits and order of RK schemes. Morris et al. (1997) apply the LDDRK scheme of Hu et al. (1996) to the solution of nonlinear disturbance equations (see Chap. 4) when looking at jet noise.

Tam and Webb (1993) optimise the Adams-Basforth scheme with respect to dissipation and dispersion. When giving a 36 % weighting to dispersion error and 64 % to dissipation they come up with the following Eq. (2.4) weighting factors: $Wf_0 = 2.30255809$; $Wf_1 = -2.49100760$; $Wf_2 = 1.57434093$; $Wf_3 = -0.38589142$.

More precise details of this weighting process between dissipation and dispersion is discussed later when dealing with Dispersion Relation Preserving (DRP) spatial schemes.

Splitting methods are a further key class of temporal integration scheme. A disadvantage of implicit methods is the expense of solving large simultaneous equation sets. For explicit methods this is not necessary, but instead there are greater stability restrictions. With splitting methods, implicit and explicit schemes are combined. In certain circumstances improvements in computational performance are made. The most widely known splitting method is the Alternating Direction Implicit (ADI) scheme of Peaceman and Rachford (1955). With this, in two-dimensions, one full time-step Δt, essentially consists of two $\Delta t/2$ half steps. For the first half time-step,

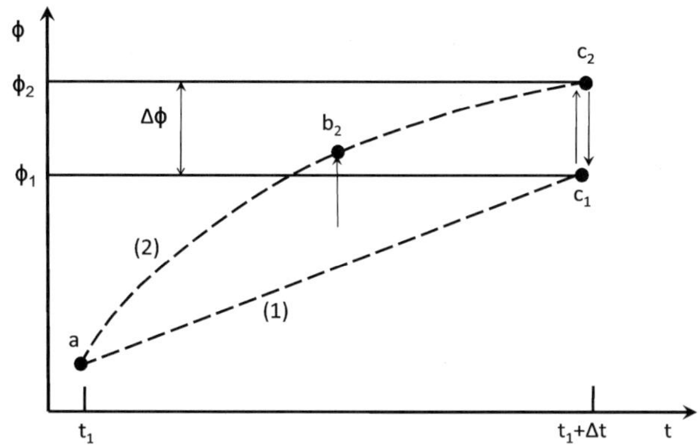

Fig. 2.4 Schematic of basic time step adaptation approaches

spatial derivatives in say an x coordinate direction are treated implicitly while those in the y direction are treated explicitly. For the second half step, this procedure is reversed. Douglas and Gunn (1964) presents a second order splitting scheme which in three-dimensions is unconditionally stable.

2.7 Elementary Solution Adapted Time-Step Approaches

As noted, through the method of lines, the discretized flow equations can be cast as temporal ODEs. Then, in principle, any standard time adaptive ODE approach (Press et al. 1989; Ascher and Petzold 1998; Brenan et al. 1996) can be applied to the solution of the unsteady flow equations. Possible standard approaches have many similarities to spatial adaptations. Temporal truncation error estimates are here initially grouped into the following four classes:

(I) Altered Scheme—solutions for schemes of the same order but different natures are made. By analysis of truncated Taylor series terms solution errors can be inferred;
(II) Altered Step—solutions with one long and two short time-steps are compared and
(III) Altered Order—solutions of different orders are compared.

Figure 2.4 gives a schematic of two potential comparative solutions (1) and (2). These can be used in assessing the temporal truncation error, where $\Delta t \propto \phi_2 - \phi_1$. Solution Element (1) ($a$–$c_1$) could involve lower accuracy. Element (2) (a–b_2–c_2) either involves two reduced steps or one high order step. Alternatively, elements (1) and (2) can have schemes of the same order but different natures.

Gresho et al. (1984), when using a 1st order forward/backward Euler procedure show from Taylor's series, for a scheme of order $n-1$, that

$$\Delta t^{new} = C_t \Delta t^{old} \left| \frac{\varepsilon}{E^{old}} \right|^{1/n} \tag{2.10}$$

where the superscripts 'new' and 'old' refer to solutions for different step sizes, 'new' being the latest. The parameter, ϵ, is a pre-set normalised error input value. This enables the maximum temporal solution error to be specified ($\epsilon = 0.001$ corresponds to about a 0.1 % solution error). Also, C_t, is a safety factor. The time step must be spatially constant for all the solution variables (N_ϕ). Hence, $E^{old} \propto \phi_2 - \phi_1$ (the difference in the solution for the two schemes or step lengths), needs to be a spatially averaged (or more conservatively a maximum) normalised (to be consistent with the ϵ definition) value. Gresho et al. propose the following average

$$E^{old} = \left[\frac{1}{N_\phi} \sum_{i\_=1}^{N_\phi} \left(\frac{1}{N_{P\phi}} \sum_{j=1}^{N_{P\phi}} \left(\frac{E_{P\phi}^{old}}{|\phi_P^{old}| + \phi_o} \right)^2 \right) \right]^{1/2} \tag{2.11}$$

where $N_{P\phi}$ is the number of nodal points for each variable (this is generally the same for all variables) and ϕ_o a reference scale. Evaluation of Eq. (2.11) requires storage of two full solutions. Hence, for a three-dimensional system it gives a significant storage burden. Also, it would seem a more refined volume weighting would be better. Some approaches to overcome this storage burden are outlined in Tucker (2002a).

Equation (2.10) is also arrived at when considering schemes of different order (Press et al. 1989; Skelboe 1977) or solutions of different step length. When using different step lengths three solutions are required, two of these being used in the error estimate. Whether the benefits of being able to adapt time-steps outweigh the significant computational overhead is likely to be problem dependent. As noted by Ascher and Petzold (1998), step-doubling approaches, although easy to implement, can be relatively expensive. When using higher order schemes with adaptive step strategies, instabilities and hence oscillations can occur. To help with this, the order of schemes can be limited (Brenan et al. 1996).

2.7.1 Relating Error Estimate to New Time-Steps

The error estimate is related to Δt through Eq. (2.10). Again, there are many ways in which Eq. (2.10) can be used. For stability reasons (Skelboe 1977), ideally, the maximum *swing* $((\Delta t^{new}/\Delta t^{old})_{\max})$ should be restricted. According to Ascher and Petzold (1998), Δt^{new} should also be limited to avoid round-off error. Tucker (2001) notes various time-step update protocols used by different workers.

The inferred computed error estimate can be used to effect a Richardson type extrapolation. If the time-steps are sufficiently small, this should yield an improvement to the solution. However, the radius of convergence of the Taylor (power) series is small. Hence, attempting this is probably best avoided. Tucker (2002a) explores the use of subcycling with adaptive time stepping (Gresho et al. 1984). With this, selected data is interpolated from the 1st to the 2nd comparative step. The objective being to make the 2nd step computationally cheaper.

2.7.2 Alternative Techniques

Predictor-corrector methods naturally lend themselves to time adaptation (Ascher and Petzold 1998). Gresho et al. (1980) successfully apply this, to a wake flow, in FIDAP (a commercial CFD program), with a second order Adams-Bashforth predictor step. Further adaptive time step studies for different applications are given by Tu et al. (1992), Reindl et al. (1991) and Hujeirat and Rannacher (1998). Crudely, the commercial CFX4 (Glanfield 2000) program adapts time-steps with respect to convergence rates. Muramatsu and Ninokata (1992) apply a novel fuzzy logic based Δt adaptation procedure. Bell and Surana (1994) use temporal adaptation in a space-time finite-element procedure. Based on integral error estimates, element sizes in the time domain are adapted (this is only done after spatial accuracy refinements fail to reduce errors below a specified level). Im (2000) successfully applies an adaptive time integration based around a variable-order, variable-step backwards differentiation formula to the modelling of a transient flame flow.

2.8 Unsteady Adjoint and Time Step Adaptation

The adjoint process has the accuracy of an objective function in mind. Then using a Taylor series we can write an estimate for an exact time domain objective $C_{\Delta t}(\phi_{\Delta t})$ about an estimate $C_{\Delta t}(\phi_{\Delta t}^{\Delta T})$ as

$$C_{\Delta t}(\phi_{\Delta t}) \approx C_{\Delta t}(\phi_{\Delta t}^{\Delta T}) + \left[\frac{\partial C}{\partial \phi}\right]_{\phi_{\Delta t}^{\Delta T}} (\phi_{\Delta t} - \phi_{\Delta t}^{\Delta T}) \tag{2.12}$$

In the above, $\phi_{\Delta t}^{\Delta T}$ represents a variable computed with a coarse time step of ΔT and represented on a finer temporal 'grid' of Δt. Similar, to Eq. (2.12) the residual vector, A, can be expressed as

$$A_{\Delta t}(\phi_{\Delta t}) \approx A_{\Delta t}(\phi_{\Delta t}^{\Delta T}) + \left[\frac{\partial A}{\partial \phi}\right]_{\phi_{\Delta t}^{\Delta T}} (\phi_{\Delta t} - \phi_{\Delta t}^{\Delta T}) \tag{2.13}$$

The above can be rearranged as

$$\phi_{\Delta t} - \phi_{\Delta t}^{\Delta T} = \left[A_{\Delta t}(\phi_{\Delta t}) - A_{\Delta t}(\phi_{\Delta t}^{\Delta T})\right]\left[\frac{\partial A}{\partial \phi}\right]_{\phi_{\Delta t}^{\Delta T}}^{-1} \tag{2.14}$$

Note, since

$$A_{\Delta t}(\phi_{\Delta t}) = 0 \tag{2.15}$$

Equations (2.12) and (2.14) can be combined to give

$$C_{\Delta t}(\phi_{\Delta t}) = C_{\Delta t}(\phi_{\Delta t}^{\Delta T}) - \left[\frac{\partial C}{\partial \phi}\right]_{\phi_{\Delta t}^{\Delta T}} \left[\frac{\partial A}{\partial \phi}\right]_{\phi_{\Delta t}^{\Delta T}}^{-1} A_{\Delta t}(\phi_{\Delta t}^{\Delta T}) \tag{2.16}$$

The furthest right hand group of terms in Eq. (2.16) are an estimate of the solution error

$$E = -\left[\frac{\partial C}{\partial \phi}\right]_{\phi_{\Delta t}^{\Delta T}} \left[\frac{\partial A}{\partial \phi}\right]_{\phi_{\Delta t}^{\Delta T}}^{-1} A_{\Delta t}(\phi_{\Delta t}^{\Delta T}) \tag{2.17}$$

Alternatively, the above could be viewed as a solution correction. For computational efficiency and convenience the following adjoint variable is defined

$$\Lambda_{\phi\Delta t}^{T}(\phi_{\Delta t}^{\Delta T}) = -\left[\frac{\partial C}{\partial \phi}\right]_{\phi_{\Delta t}^{\Delta T}} \left[\frac{\partial A}{\partial \phi}\right]_{\phi_{\Delta t}^{\Delta T}}^{-1} \tag{2.18}$$

Or, rearranging

$$\left[\frac{\partial A_{\Delta t}}{\partial \phi_{\Delta t}}\right]_{\phi_{\Delta t}^{\Delta T}}^{T} \Lambda_{\phi\Delta t}(\phi_{\Delta t}^{\Delta T}) = -\left[\frac{\partial C}{\partial \phi}\right]_{\phi_{\Delta t}^{\Delta T}}^{T} \tag{2.19}$$

However, Eq. (2.19) needs solution on the fine time domain and this we seek to avoid. Hence, it is rewritten as

$$\left[\frac{\partial A}{\partial \phi}\right]_{\phi_{\Delta T}}^{T} \Lambda_{\phi\Delta T} = -\left[\frac{\partial C}{\partial \phi}\right]_{\phi_{\Delta T}}^{T} \tag{2.20}$$

The above is a simultaneous equation set. It needs to be solved by backwards integration in time. The coarse time domain adjoint is interpolated onto the fine time domain. Of note is the matrix structure in Eq. (2.20). This is given below

$$\begin{bmatrix} - & - & & \\ & [\frac{\partial A^{n-2}}{\partial \phi^{n-2}}] & [\frac{\partial A^{n-1}}{\partial \phi^{n-2}}] & \\ & & [\frac{\partial A^{n-1}}{\partial \phi^{n-1}}] & [\frac{\partial A^{n}}{\partial \phi^{n-1}}]^T \\ & & & [\frac{\partial A^{n}}{\partial \phi^{n}}]^T \end{bmatrix} \begin{bmatrix} - \\ \Lambda_{\phi,\Delta T}^{n-2} \\ \Lambda_{\phi,\Delta T}^{n-1} \\ \Lambda_{\phi,\Delta T}^{n} \end{bmatrix} = - \begin{bmatrix} - \\ [\frac{\partial C}{\partial \phi^{n-2}}]^T \\ [\frac{\partial C}{\partial \phi^{n-1}}]^T \\ [\frac{\partial C}{\partial \phi^{n}}]^T \end{bmatrix} \tag{2.21}$$

Once $\Lambda_{\phi,\Delta T}$ has been computed, this field is interpolated onto the fine time 'mesh'. The error, E, is computed from

$$E = \Lambda_{\phi\frac{\Delta T}{\Delta t}}^{T} A_{\Delta t}(\phi_{\Delta t}^{\Delta T}) \tag{2.22}$$

The backwards integration, needed for the adjoint variable, means that an entire time solution is needed prior to the integration of this equation. Hence, this presents storage challenges (Mani and Mavriplis 2007). Despite this, Rumpfkeil and Zingg (2007) apply the unsteady approach that has just been noted, to the control of a range of unsteady flows. Ad hoc procedures are used to reduce the computational cost. These include storing/using data every other time step.

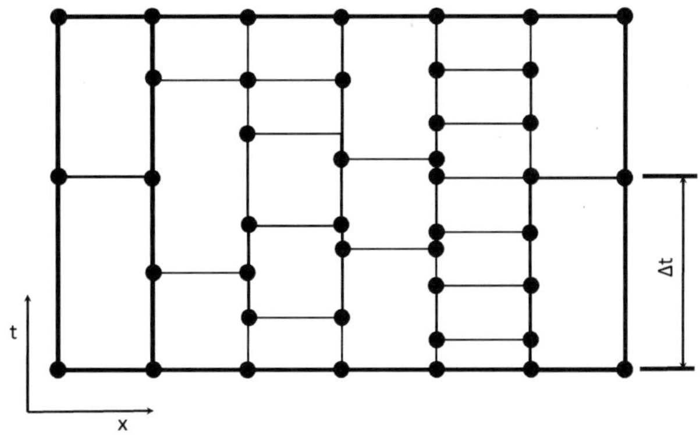

Fig. 2.5 A typical mesh, with a single spatial dimension

2.8.1 Adjoint Methods for Unsteady Flow Design Optimization

Adjoint methods enjoy most popularity for design optimization. However, the need to store adjoint time histories precludes practical use for unsteady flows. Despite this, Rumpfkeil and Zingg (2007) apply the unsteady approach to the control of a range of unsteady flows. Helpfully, turbomachinery calculations, for example, can involve flows with a discrete unsteadiness frequency(s), ω. Then, harmonic methods can be used where the flow variables, ϕ, are assumed to vary as

$$\phi(x,t) = \overline{\phi}(x) + A(x)cos(\omega t) + B(x)sin(\omega t) \tag{2.23}$$

Hence, if ω is specified, the unsteady flow time history is described if $\overline{\phi}(x)$, $A(x)$ and $B(x)$ are defined. Hence, the optimization problem is reduced to one involving *steady* equations. This makes the application of adjoints considerably simpler. The approach, (see Chap. 4) is extendable to multiple harmonics. Using the above harmonic approach, He and Wang (2011) carry out concurrent aerodynamic and aeroelastic adjoint based design optimization for a compressor fan.

2.9 Temporal Adaptation Using Space-Time Elements/Volumes

With space-time methods a mesh that extends into the temporal domain is used. Based on the temporal truncation error, this mesh can be adapted in time (Mani and Mavriplis 2010) or in space also. A key attraction of this approach is that regions of high unsteadiness activity can be treated with smaller time steps than the rest of the domain. The governing equations, in say a finite volume method, are integrated around temporal as well as spatial edges. A typical mesh, with a single spatial dimension, is shown in Fig. 2.5. With time varying, spatial mesh deformation the space-time domain could look as shown in Fig. 2.6.

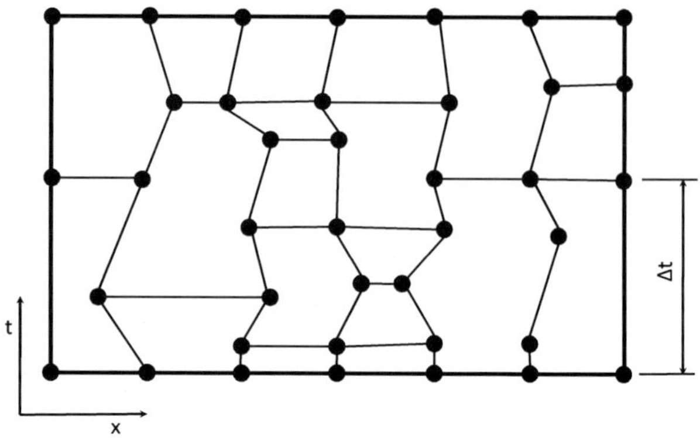

Fig. 2.6 Mesh, with a single, time-adapting, spatial dimension

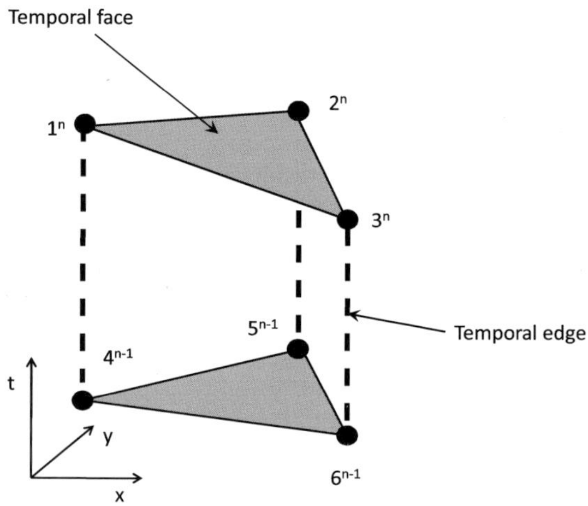

Fig. 2.7 A space-time element at two time locations (shaded zones are the areas of the temporal faces)

For the temporal discretization, the solution variable, ϕ, is assumed to vary in a piece-wise constant fashion. The time derivative is expressed as

$$\int_T \frac{dA\phi}{dt}dt = \frac{dA\phi}{dt}\Delta t \tag{2.24}$$

In the above, Δt is the edge length in the time-domain. Also, the A in the above is the face area normal to the time coordinate. This can be seen in Fig. 2.7. The time derivative is then discretized as the finite difference below

$$\frac{dA\phi}{dt}\Delta t = A^n\phi^n - A^{n-1}\phi^{n-1} \tag{2.25}$$

The n superscripts indicate different time levels. The time domain can essentially be assumed to be treated with upwind type differences. The spatial fluxes are integrated over space-time faces e.g. face 2–3–6–5 in Fig. 2.7. A space-time finite element procedure is outlined by Bell and Surana (1994).

With the space-time approach, where there are moving wake or shock zones, for example, the extent of the element faces in time can be reduced. It can be extended in the calmer (more steady) flow zones.

2.10 Convective Schemes for Unsteady Flow

In unsteady flows, particularly high gradients of variables can arise. This places greater demands on convective differencing schemes. As shown by Orkwis et al. (2002) and also noted by Fritsch and Giles (1992) unsteady flow related numerical artifacts can account for a significant proportion of predicted energy losses in turbomachinery blade rows. Also, (see Fritsch and Giles) numerical smoothing can erroneously drain unsteady flow energy from the main turbine passage zone (outside boundary layers). Therefore, for unsteady flows the use of specialized spatial schemes can be important.

It is also important to understand the spatial schemes interaction with the temporal (Kim and Lee 1996). For example, the accuracy of QUICK (a quadratic upstream biased interpolation—Leonard 1979) with first order forward Euler time differencing is essentially no better than when first order spatial upwinding is used (Nasser and Leschziner 1985). Therefore, it is best adapted, in a similar fashion to Leith's (1964) central difference based scheme, with extra terms that extend control volume face interpolations to include forward in time (linear in the case of Leith's scheme) estimates. This is the essence of QUICKEST (QUICK with Estimated Streaming Terms (EST)). Davis and Moore (1982) present a multi-dimensional version of QUICKEST, showing that a third order temporal integration is effectively gained from the lower order temporal scheme. Glass and Rodi (1982), Roache (1992), and Wallis and Manson (1997) present schemes which to varying degrees (the former loosely) are related to the EST approach. Glass and Rodi make use of a cubic Hermitian interpolation. These are also made use of in compact schemes (see later) by Tang and Baeder (1998).

The above EST related methods are also members of a family called semi-Lagrangian methods. This name arises because they can be partly derived by considering the Lagrangian trajectories of scalars on an Eulerian grid. Staniforth and Côté (1991) give a thorough review of semi-Lagrangian approaches, noting that, in one-dimension, the Eulerian Lax-Wendroff scheme can be considered as a special case of the semi-Lagrangian schemes.

An unusual, semi-Lagrangian related approach—ICED-ALE—is presented by Hirt et al. (1974). A similar technique called LINC is given by Butler (1971). These schemes are extensions to an earlier ALE (Arbitrary-Lagrangian-Eulerian) method devised by Hirt, in which grid nodes can be held fixed (Eulerian), move with the fluid (Lagrangian) or in any prescribed manner. Clearly, moving the grid with the flow, reduces the magnitude of the problematic convective terms. However, severe grid distortion can occur. It seems appropriate to mention here the fully Lagrangian vortex method (Khatir 2000). With this, vortex blobs or elements are convected with the flow in a Lagrangian fashion. Such approaches remove the convective term modelling problem but accounting for diffusion then becomes problematic.

Gresho et al. (1984) presents the Balancing Tensor Diffusivity (BTD) scheme (previously described by Dukowicz and Ramshaw 1979 and Crowley 1967). This involves adding a contribution to the physical diffusion term. This cancels the negative false contribution arising from the forward Euler scheme. For three-dimensional flows, to apply BTD, cross-stream diffusion terms are necessary and 19 stencil points are needed. The aforementioned schemes see little aerospace use being more of historical and academic interest.

For especially high spatial accuracy, *spectral* methods (generally involving a Fourier series based discretization, the order of which increases with the number of grid points) can be used. However, for complex geometries, these methods are of limited use. To extent these geometrical restrictions are overcome using *spectral element* techniques (Patera 1984) or compact difference schemes (Lele 1992) that have spectral like behaviour.

An interesting class of schemes is the kinetic energy (Jameson 2008b) and energy (Jameson 2008a) conserving. Kinetic energy conservation properties are ideal for LES and DNS. In practical terms, the approach seems to offer substantial gains in stability—even for flows with shocks (Allaneau and Jameson 2010a). The kinetic energy conserving scheme is simple to implement. It just needs a localized change in the way variables are averaged at the control volume interface straddled by the adjacent nodes. Hence, if L and R are nodes of cells adjacent to face LR, for kinetic energy conservation the momentum flux should satisfy the following condition

$$[\rho u_i u_j]_{LR} = \frac{1}{2}(\rho u_i)_{LR}(u_{j,L} + u_{j,R}) \tag{2.26}$$

The boundary conditions also need, ideally, to be reformulated to ensure global energy conservation and care needed with the pressure and other stress terms. The potential formulations for $(\rho u_i)_{LR}$ are discussed by Allaneau and Jameson (2010a). Modification (2.26) essentially means that the skew-symmetric form (see later) of the convective term is being solved for. This, on fine enough grids, results in the need for little smoothing—if any (giving a useful computational time saving). Later, tests are presented, showing the impact of unstructured cell topology for homogeneous decaying turbulence. Although results are not shown here, for these cases, the kinetic energy conserving scheme dramatically reduces the sensitivity of results to cell topology. The use of the approach in a DNS study for a plunging aerofoil is given by Allaneau and Jameson (2010b). Note, the use of staggered grids also ensures energy conservation (Kim and Moin 1985). However, this approach is considerably more complex to implement even in a basic structured solver.

2.11 Classical High-Order Approaches

2.11.1 Compact Schemes

Compact (or Pade) schemes are attractive in that they allow use of a relatively small finite difference stencil to gain high order accuracy. Broadly speaking they allow better resolution at higher waves numbers and offer the potential of spectral type accuracy but with greater geometrical flexibility. With regards to modern parallel computing, the smaller stencil of the compact scheme, has the attraction that the overlap at halo cells, at the grid interfaces, is smaller. Hence, there is less data flow. On the other hand, if applied in an implicit framework a tri-diagonal matrix inversion is required in every coordinate direction, for every variable and derivative for every iteration. In explicit mode the matrix inversions just need to be carried out every time step. Hence, even then this is costly. For a function, ϕ, a compact scheme, for a first derivative, on a uniform grid, can be expressed as

$$Wf_{i-1}\phi'_{i-1} + Wf_i\phi'_i + Wf_{i+1}\phi'_{i+1} = \alpha\frac{\phi_{i+1} - \phi_{i-1}}{2\Delta x} + \beta\frac{\phi_{i+2} - \phi_{i-2}}{2\Delta x} \tag{2.27}$$

Hence, it is not just the nodal values that are unknowns but also the derivatives at these nodal points. The coefficients in the above can be evaluated from Taylor series expansion about node i and some manipulation. Equation (2.27) can be expressed symbolically as

$$\left(A\phi'\right)_i = (S\phi)_i \tag{2.28}$$

On a uniform grid, the compact scheme will have better resolution (Note, resolution of a scheme is often defined as the minimum number of grid points per wavelength that a scheme needs to resolve a wave to reasonable accuracy and will be discussed further later) than the a finite difference scheme of equivalent order. For 2nd derivatives, the following equation can be used

$$\begin{aligned} &Wf_{i-1}\phi''_{i-1} + Wf_i\phi''_i + Wf_{i+1}\phi''_{i+1} \\ &\quad = \alpha\frac{\phi_{i+1} - 2\phi_i + \phi_{i-1}}{\Delta x^2} + \beta\frac{\phi_{i+2} - 2\phi_i + \phi_{i-2}}{4\Delta x^2} + \gamma\frac{\phi_{i+3} - 2\phi_i + \phi_{i-3}}{9\Delta x^2} \end{aligned} \tag{2.29}$$

Table 2.4 gives the coefficients for different 4th and 6th order accuracy compact schemes. Note, the final table elements are for a 6th order compact scheme. This is designed to be dissipative at higher waves numbers (Laizet and Lamballais 2009). The dissipation is intended to reduce aliasing errors. Hence, as with DRP schemes, the resolution of a scheme can be improved at the potential sacrifice of order. For example, Lele (1992) imposes the constraint that, at selected wave numbers, the scheme gives an exact correspondence between the actual and computed wave number. Kim and Lee (1996) used a more refined optimization approach connected to the concepts used for DRP. They also explore the complex relationship between solution accuracy and RK temporal integration order—finding this to be non-linear. Tang and Baeder (1998) use Hermetian polynomial and trigonometric series based approaches to perform further optimizations of compact schemes. To reduce the potential stencil size, in the matrix inversion embodied in Eq. (2.28), Hixon (2000) proposes a pre-factorization stage. Hence, a pentadiagonal system can be replaced, for example, by the product of tridiagonal systems. This approach, evidently, also makes the enforcement of boundary conditions more straightforward and is tested for a range of benchmark acoustic problems.

Table 2.4 Coefficients for different compact schemes

Derivative	Order	Wf_{i-1}	Wf_i	Wf_{i+1}	α	β	γ
ϕ_i'	4	–	1	–	3/2	0	0
ϕ_i''	4	1/10	1	1/10	6/5	0	0
ϕ_i'	6	1/3	1	1/3	14/9	1/9	0
ϕ_i''	6	2/11	1	2/11	12/11	3/11	0
ϕ_i''	6	0.479598	1	0.479598	0.42090	1.70207	−0.1637

Compact schemes can also be used for filtering (Lele 1992). Joo and Durbin (2009) use them, to gain stability when making hybrid RANS-LES (see Chap. 4) type computations of an idealized turbine blade.

2.11.2 Discontinuous Galerkin Scheme

Another means of getting high order on a compact stencil is through use of the Discontinuous Galerkin (DG) method. The method is similar to the Galerkin finite element method in that a shape function is used. The discontinuous shape function fields are (weakly) connected at boundaries though boundary conditions. The discontinuous shape function fields are shown in Fig. 2.8. At the discontinuous interfaces concepts from approximate Riemann solvers are used. The cells have multiple degrees a freedom. Hence, in LES/DNS grid node count terms, it is not straightforward to make direct comparisons when trying to explore computational performances. Interestingly, Broeckhoven et al. (2007), point out that unstructured triangular grids have better dissipation and dispersion properties for this approach than quadrilateral cells. This aspect is very different to when conventional finite volume methods are used. However, Broeckhoven et al. (2007) note that increased computational cost is to be expected relative to more conventional approaches. For the discontinuous Galerkin method, the mass or capacitance matrix is no longer global and evidently this is computationally convenient (Broeckhoven et al. 2007). An obvious key advantage of discontinuous Galerkin methods is that the small stencil will ensure low communication time when performing distributed parallel processing.

de Wiart et al. (2012) apply the DG method to the DNS for a low-pressure turbine. Wave number against modified wave number plots (real and imaginary) are presented in this work. These suggest the scheme has accurate dispersion properties. However, the dissipative, at high wave numbers, are larger then conventional high order finite difference schemes.

Atkins and Lockard (1999) apply the discontinuous Galerkin method to noise scattering from a two-dimensional slat and a three-dimensional generic wing-body configuration. The linearized Euler equations are solved. In 5th order mode the approach is found to be able to model the propagation of a wave to 100 wavelengths with just 3 % error and just 2 points per wavelength. The approach is

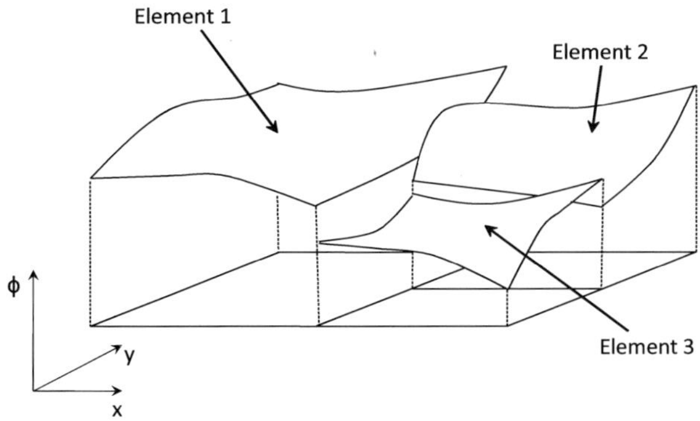

Fig. 2.8 Broken functions of high order polynomials used for the shape functions in the discontinuous Galerkin method (Adapted from de Wiart et al. 2012)

also found to give good parallel speedup and accuracy to be resilient to grid expansion. Stanescu et al. (2002) apply the discontinuous Galerkin method to sound propagation from an engine with a wing and fuselage. The non-linear Euler equations are solved. Birkefeld and Munz (2012) observe that without careful formulation, the discontinuous Galerkin method can be sensitive to grid quality and instabilities can arise after long integration periods. When solving the linearized Euler equations they utilize a hybrid approach. With this, near complex geometrical surfaces the unstructured discontinuous Galerkin method is used. This is linked to high order structured Cartesian grid zones away from surfaces. These have greater computational economy. The complete system is solved to 4th order in space and time.

2.11.3 Spectral Difference, Volume and CPR Methods

The spectral difference (SD) method is intended to be a simpler, more computationally economical alternative to DG (Wang et al. 2007), Spectral Volume (SV), and other high order methods for unstructured grids. The SD and SV (Liu et al. 2006) methods are similar to DG making use of piece-wise discontinuous polynomials. The DG, SV and SD methods are analogous to Galerkin finite element, finite volume and finite difference (Wang et al. 2007) techniques. The SD method of Wang et al. stores discrete variables within each cell in a rigid structured fashion. Thus it hybridizes some of the advantages of structured and unstructured approaches. The SV has a similar sub-cell structuring. Liu et al. demonstrate the application of SV to wave propagation. The correction procedure via reconstruction (CPR) scheme evidently unifies the above along with DG (Haga et al. 2011). The preliminary application of this approach to an ILES of an SD7003 aerofoil with transitional flow is given by Vermeire et al. (2013).

2.11.4 ENO/WENO

The ENO (Essentially Non Oscillatory—see Harten et al. 1987) and WENO (Weighted Essentially Non-Oscillatory—see Liu et al. 1994) schemes are an important class of method. They can be extended to arbitrary order. With them, a combination of stencils/polynomials of equal order and hence equal number of stencil points are used. However, the stencil points used have different levels of upwind and downwind bias. In smooth regions, a weighted combination of all stencils is used. For ENO, where oscillations are detected, one member of the family of stencils is used. The least oscillatory is selected. For WENO, at oscillations, the full family is used but with altered weightings. Typically, a smoothness indicator—based on pressure gradients—is used to activate the stencil/blending control. As noted by Hadjadj (2012), in LES terms the experiences of WENO performance are mixed. For example, when exploring shock/boundary layer interactions, Hadjadj just uses the WENO scheme around the shock. A standard centered scheme is used elsewhere.

2.12 High Resolution Spatial Schemes

2.12.1 DRP Schemes

Typically with numerical methods we first think of the order of a scheme. This can be established from use of the Taylor series. However, we also need to consider the resolution of a method. This relates to considering the dissipation and dispersion errors of schemes (both spatial and temporal). The can be ascertained through Fourier analysis. The DRP schemes look at both the accuracy and resolution of schemes. They seek a best compromise between these two aspects to yield the potentially most efficient scheme. Hence, we might approximate a first order derivate as below

$$\left.\frac{\partial\phi}{\partial x}\right|_i = \frac{1}{\Delta x}\sum_{k=-N}^{N} Wf_k\phi_{i+k} \tag{2.30}$$

With DRP (Tam and Webb 1993) the coefficients in the above are compromised from what they would be for a standard high order scheme so that the scheme's resolution is improved. Since the scheme of Tam and Webb is a 4th order centered/symmetric, just the dispersion error is optimized through considering the integral below

$$E = \int_{-\beta}^{\beta} \left|w\Delta x - w'\Delta x\right|^2 d(w\Delta x) \tag{2.31}$$

where w is the true wave number, w' the computed and β the optimization range. Tam and Webb optimize for wavelengths greater than $4\Delta x$. Lockard et al. (1995)

Table 2.5 DRP scheme coefficients from Broeckhoven et al. (2007)

Source	Wf_0	$Wf_1 = Wf_{-1}$	$Wf_2 = Wf_{-2}$	$Wf_3 = Wf_{-3}$
Tam and Webb (1993)	0	0.79926643	−0.18941314	0.02651995
Tam and Shen (1993)	0	0.77088238	−0.16670590	0.20843142

optimize for wavelength greater than $7\Delta x$. They consider the value of Tam and Webb too restrictive a constraint, corrupting accuracy in the low wave number range. Table 2.5 gives some DRP scheme coefficients, for Eq. (2.30) as summarized in Broeckhoven et al. (2007). Because the DRP scheme is centered and non-dissipative some smoothing can be required for stability. Another way of looking at this is that although the DRP schemes will precisely model long wave components they can also support short waves. Hence, point-wise oscillations can be found, as with standard centered finite difference schemes. These waves can propagate at extremely high velocities relative to the short wave components. Indeed, as noted by Tam et al. (1993) for the wave velocity, c, the following can be written

$$c \approx \frac{dw'}{dw} \tag{2.32}$$

Although Fig. 2.12a relates to standard compact and finite difference schemes it has a typical shape. The application of Eq. (2.32) to this curve will show that the numerical wave speed will become high at high wave numbers. The differentiation of the Fig. 2.12a curves will give a zone were $c \approx 1$ corresponding to long and well-resolved waves. There is also a zone where $0 < c < 1$ corresponding to dispersive waves. Finally, there is a zone where $c < 0$. This corresponds to what Tam et al. call 'parasite waves'. These travel at high speed and relate to the point-wise oscillations often observed in CFD with centred schemes. Tam et al. observe that these waves can be generated by non-smooth initial data.

To deal with these contaminating waves, Lockard et al. (1995) develop an asymmetric DRP scheme i.e. it has an offset or upwind type stencil. The dissipation means that w' has an imaginary part. Hence, when optimizing the scheme both the real and imaginary parts need to be considered and also the sign of the imaginary part. Hence, a weighting must be assigned to these two components in the optimization

$$E = \int_{-\beta}^{\beta} \left[Wf\left[\Re\left(w\Delta x - w'\Delta x\right)\right]^2 + (Wf - 1)\left(\Im\left(w'\Delta x\right)\right)^2 \right] d(w\Delta x) \tag{2.33}$$

In the above, Wf is a weighting function to change the emphasis between controlling dissipation and dispersion. Also, $\Re$ and $\Im$ identify the real and imaginary parts, respectively. However, the imaginary part can have differing signs and the negative part corresponds to dissipation. To ensure a negative imaginary part Lockard et al.

(1995) modify the equation above to

$$E = \int_{-\beta}^{\beta} \big[Wf\big[\Re(w\Delta x - w'\Delta x)\big]^2 + (Wf - 1)\big(\Im(w'\Delta x) - Wf_1 \sin^{Wf_2}(w\Delta x/2)\big)^2\big] d(w\Delta x) \tag{2.34}$$

where Wf_1 and Wf_2 are further control parameters. With Wf_1 negative (or zero) and Wf_2 a positive integer, dissipation (negative w') can be enforced.

The alternative to the approach of Lockard et al. is to specifically add a smoothing term. Again, the order of this can be controlled but also the wave number components that are damped. This strategy is similar to the well-known approach adopted by Jameson et al. (1981) who use 2nd and 4th order derivatives. Tam et al. (1993) produce a selective damping that is only active at high wave numbers. Also, crude explicit filtering can be used.

Notably, Morris et al. (1997) apply Tam and Webb's 4th order DRP scheme to the solution of nonlinear disturbance equations (see Chap. 4) when looking at jet noise. Agarwal and Morris (2000) use the approach to explore acoustic scattering from a ROBIN rotorcraft fuselage. Again, the 4th order DRP scheme is used but this time it is applied to the linearized Euler equations.

2.12.2 CABARET

CABARET (Compact Accurately Boundary-Adjusting high-REsolution Technique) is a distinctive, explicit, upwind, leapfrog based, second-order method (Karabasov and Goloviznin 2007, 2009). It extends the original work of Iserles (1986) and others. CABARET uses dual staggered variables in a compact space-time computational stencil. The scheme has low dispersion and dissipation error. The dispersion error is as low as that of the fourth order optimized—six order central schemes (Colonius and Lele 2004), for a wide range of wave and CFL numbers. Use is made of the CABARET scheme in Chap. 6. The data structure/stencil is shown in Fig. 2.9. The solid symbols give the locations of variables ϕ. The open symbols identify the location of $f(\phi)$. Unlike standard practice, here the whole indices identify faces and the half the cells. Considering the scalar conservation law $\partial\phi/\partial t + \partial f(\phi)/\partial x = 0$ and the Fig. 2.9 nomenclature the following can be written

$$\frac{\phi_C - \phi_E}{0.5\Delta t^{n+1/2}} + \frac{f_4 - f_5}{\Delta x_{i+1/2}} = 0 \tag{2.35}$$

where $\Delta t^{n+1/2} = t^{n+1} - t^n$ and $\Delta x_{i+1/2} = x_{i+1} - x_i$. The above assumes that all variables are known at the nth time level. Hence, the first stage is a space centred explicit in time predictor step. The time step is a half-step. The next stage involves another half step. However, this is a corrector step. Again it is space centred but this

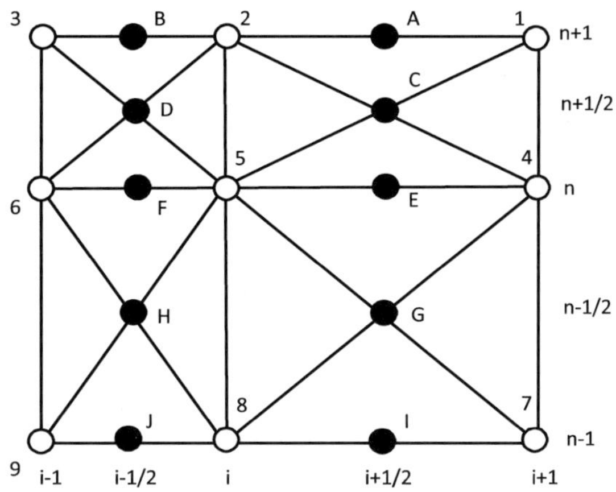

Fig. 2.9 CABARET data structure for a one-dimensional spatial domain (the solid circles represent the conservation variables and the open flux variables)

time backwards difference in time as given below

$$\frac{\phi_A - \phi_C}{0.5\Delta t^{n+1/2}} + \frac{f_1 - f_2}{\Delta x_{i+1/2}} = 0 \tag{2.36}$$

However, the fluxes f_1 and f_2 need to be determined. These are estimated using a linear extrapolation. Hence, for $f_1 = f_1(\phi)$

$$\phi_1 = 2\phi_c - \phi_5 \tag{2.37}$$

2.13 Convective Schemes for Density Based Solvers and Related Aspects

The convective schemes used in density based solvers can differ quite substantially to those used in pressure based solvers. This is because as well as resolving 'waves' convecting at the fluid speed (vortical and entropy) they also capture acoustic waves. For best efficiency there is the need to account for these differing wave speeds. The schemes in density-based codes also generally include some capability to deal with shocks.

Of note is the two-stage Flux Corrected Transport (FCT) SHASTA (Sharp And Smooth Transport Algorithm) of Boris and Book (1973) extended by Book et al. (1975). The first stage involves integrating the governing equations using a strongly diffusive method (incidentally, for this stage, Boris and Book's scheme has a semi-Lagrangian element), which gives, rise to a monotone solution. The second involves correcting this solution (by correcting fluxes), reducing the diffusive error by introducing an anti-diffusive element. Ideally, the anti-diffusive component is made smaller than the initial diffusive component, leaving, if possible, a residual diffusive component large enough in principal to cancel any dispersive error. The method easily extends to three-dimensions. Unlike SHASTA, to damp shock-induced oscillations, traditional methods can introduce levels of artificial viscosity, which based on diffusive stability limits, produce severe time-step restrictions. SHASTA has the key disadvantage that the initial diffusive stage can damp out maxima that will not be recovered in the anti-diffusion stage, thus giving rise to erroneous solution clipping. Motivated by this, Chapman (1981) presents the explicit FRAM (Filtering Remedy and Methodology) monotone solution approach. This has three stages. In the first, a higher order scheme is used to give a provisional time advanced solution. During the second stage, based on a local Lagrangian solution, bounds are calculated for the provisional solution. Where the initial solution is outside these bounds sufficient local dissipation is introduced to yield a monotone solution. For the first stage, several predictions presented by Chapman use the second order scheme of Crowley (1967) noted earlier.

By far the most popular density based solver convective scheme is the so-called flux-difference splitting scheme of Roe (1981). This is a Godunov-type scheme. It is based on the solution of a Riemann type problem. A key characteristic is that the control volume face flux interpolation is based on upstream and downstream interpolations to the face. The average of these two interpolations is taken. Also, the level of smoothing is controlled by a matrix of Eigen values. These are based on the difference between the fluid speed and the speed of sound. Notably, Roe's scheme sees wide spread use in LES. However, it was never designed with this purpose in mind. A related approach is flux vector splitting (Steger and Warming 1981). Another, relatively popular density based solver scheme is the Advection Upstream Splitting Method (AUSM)—see Liou and Steffen (1993).

2.13.1 The MUSCL Scheme

The MUSCL scheme (Monotone Upstream-centred Schemes for Conservation Laws) extends the Roe scheme to higher orders (Van Leer 1979). This is achieved through using higher order interpolations both upstream and downstream of the control volume face. This scheme is discussed in more detail later. Since, quite a few of the later results are MUSCL based further details are given below. An inviscid convective flux, E^{inv}, at a control volume face can be approximated as below

$$E_l^{inv} = \frac{1}{2}\left[E^{inv}(\phi_L) + E^{inv}(\phi_R) - |A|(\phi_R - \phi_L)\right] \tag{2.38}$$

The subscripts L and R represent left and right control volume face values. These values reside immediately to the left and right of a control volume face. Also, $A = \partial E/\partial \phi$ and ϕ represents primitive variables. The left and right interpolations can be expressed as

$$\begin{aligned}\phi_L = \phi_{i-1} + \big[&Wf_{i+2}\Delta_{i+2}^{\phi} l_{i+2} + Wf_{i+1}\Delta_{i+1}^{\phi} l_{i+1} + Wf_i \Delta_i^{\phi} l_i \\ &+ Wf_{i-1}\Delta_{i-1}^{\phi} l_{i-1} + Wf_{i-2}\Delta_{i-2}^{\phi} l_{i-2} + Wf_{i-3}\Delta_{i-3}^{\phi} l_{i-3}\big]\end{aligned} \tag{2.39}$$

Table 2.6 Coefficients of ϕ_L for different schemes

ϕ_L	Wf_{i+3}	Wf_{i+2}	Wf_{i+1}	Wf_i	Wf_{i-1}	Wf_{i-2}	Wf_{i-3}
$m=1$	0	0	0	0	0	0	0
$m=3$	0	0	0	2/6	1/6	0	0
$m=5$	0	0	−3/60	24/60	11/60	−2/60	0
$m=7$	3/420	−22/420	79/420	180/420	−34/420	4/420	0

Table 2.7 Coefficients of ϕ_R for different schemes

ϕ_L	Wf_{i+3}	Wf_{i+2}	Wf_{i+1}	Wf_i	Wf_{i-1}	Wf_{i-2}	Wf_{i-3}
$m=1$	0	0	0	0	0	0	0
$m=3$	0	0	0	2/6	1/6	0	0
$m=5$	0	0	−3/60	24/60	11/60	−2/60	0
$m=7$	3/420	−22/420	79/420	180/420	−34/420	4/420	0

and

$$\phi_R = \phi_i + \big[Wf_{i+3}\Delta^{\phi}_{i+3} l_{i+3} + Wf_{i+2}\Delta^{\phi}_{i+2} l_{i+2} + Wf_{i+1}\Delta^{\phi}_{i+1} l_{i+1} \\ + Wf_i \Delta^{\phi}_i l_i + Wf_{i-1}\Delta^{\phi}_{i-1} l_{i-1} + Wf_{i-2}\Delta^{\phi}_{i-2} l_{i-2}\big] \tag{2.40}$$

In the above

$$\Delta^{\sigma}_{i+n} = (\sigma_{i+n} - \sigma_{i+n-1}) \tag{2.41}$$

and

$$l_{i+n} = \frac{L_i}{L_{i+n}}, \quad L_{i+n} = \sqrt{\left(\Delta^x_{i+n}\right)^2 + \left(\Delta^y_{i+n}\right)^2 + \left(\Delta^z_{i+n}\right)^2} \tag{2.42}$$

The values of Wf_{i+n} for various upwind schemes are given in Tables 2.6 and 2.7, where m gives the scheme's order. Note, i is the index in the coordinate direction being considered.

For LES type simulations to control smoothing, Eq. (2.38) is modified to

$$E^{inv}_l = \frac{1}{2}\big[\underbrace{E^{inv}(\phi^{cen}_L) + E^{inv}(\phi^{cen}_R)}_{\text{central}} - \underbrace{\epsilon_1 |A| (\phi^{dis}_R - \phi^{dis}_L)}_{\text{dissipation}}\big] \tag{2.43}$$

where ϵ_1 is a variable parameter (discussed later) and a value of zero would give a pure central difference. Values above this introduce dissipation. To provide a second-order ($n=2$) centred term ϕ^{cen}_L and ϕ^{cen}_R use $m=1$ weighting coefficients. For a 6th order central term the $m=5$ coefficients are needed. Smoothing terms use $m=3$ coefficients for the $n=2$ scheme, and $m=7$ coefficients for the $n=6$ scheme. This leads to smoothing to the 4th and 8th power of the derivative being considered for $n=2$ and $n=6$ schemes, respectively.

2.13.2 Monotonicity

A desirable numerical scheme property is that it is Total Variation Diminishing (TVD). TVD's basis can be appreciated by considering a one-dimensional convection equation—$\partial\phi/\partial t + u\partial\phi/\partial x = 0$. For a solution, over a domain of length L, a variable called the *total variation*

$$TV = \int_0^L \left|\frac{\partial\phi}{\partial x}\right| dx \tag{2.44}$$

should ideally not increase with time. Based on this correct physical behaviour, TVD schemes are constructed so that

$$\int_0^L \left|\frac{\partial\phi^{n+1}}{\partial x}\right| dx \leq \int_0^L \left|\frac{\partial\phi^{n}}{\partial x}\right| dx \tag{2.45}$$

Or, if the above differentials are expressed as central differences, the following can be written

$$\sum_{x=0,L} |\phi_{i+1}^{n+1} - \phi_i^{n+1}| = \sum_{x=0,L} |\phi_{i+1}^{n} - \phi_i^{n}| \tag{2.46}$$

Enforcing the TVD constraint (introduced by Harten 1983) removes erroneous solution oscillations. TVD schemes are monotonicity preserving. This means that they do not produce any new extrema in a domain of L. Also, all local minima and maxima do not decrease or increase, respectively. To enforce monotonicity, for high order schemes flux or slope limiters are used. These limiters only become active in non-smooth flow regions. Basically, they allow blending of a low-order, high-resolution scheme with a high order but lower resolution scheme. There are numerous limiters. Relatively well known ones are the minmod, superbee (Roe 1986; Van Albada et al. 1982; Van Leer 1977). Limiters are necessary for the proper implementation of MILES schemes. For example, a 3rd order limiter (Drikakis and Rider 2004) results in accurate turbulence energy spectrums.

2.14 Preconditioning

Is seems worth stressing here that when using a compressible flow solver, at low Mach numbers the smoothing can become excessive and thus damp out flow unsteadiness. As already noted, for compressible flow solvers, as well as being scaled by ε_1, essentially, the smoothing term is also scaled by a matrix of Eigen values. These have the form

$$\lambda = (u, u, u \pm c) \tag{2.47}$$

The latter Eigen value scales pressure. When, the Mach number is low the magnitude of this Eigen value can become large thus smoothing the pressure field. In

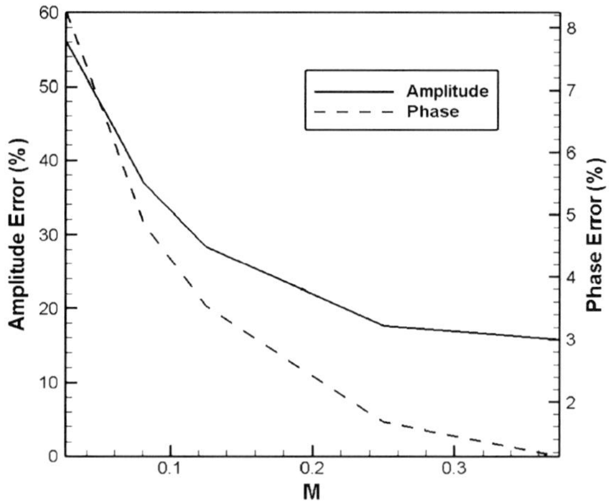

Fig. 2.10 Amplitude and phase errors for different Mach numbers. Published with kind permission from the ASME

practical terms this seems to have a knock on effect for the rest of the flow variables. This is shown in Fig. 2.10 which gives the phase and amplitude errors for a subcritical Tollmien-Schlichting wave. The case setup is discussed in Appendix A. We can see substantial errors with decreasing Mach number.

The essential form of the compressible flow equations can be expressed as

$$\frac{\partial Q}{\partial t} + \frac{\partial E}{\partial x} + \frac{\partial F}{\partial y} = \text{Viscous Terms} \tag{2.48}$$

where $Q = [\rho, \rho u, \rho v, e]^T$, $E = [\rho u, \rho uu + p, \rho uv, eu + pu]^T$, $F = [\rho v, \rho uv, \rho vv + p, ev + pv]^T$. In the forgoing, p is pressure, ρ density, u and v the velocity components and e total energy per unit volume. Using the Jacobian matrix below

$$\Gamma = \frac{\partial Q}{\partial Q_R} \tag{2.49}$$

Equation (2.48) can be re-expressed as

$$\Gamma \frac{\partial Q_R}{\partial t} + \frac{\partial E}{\partial x} + \frac{\partial F}{\partial y} = \text{Viscous Terms} \tag{2.50}$$

where now $Q = [p, u, v, T]^T$ and following Choi and Merkle (1991), after some rearrangement the following can be gained

$$\Gamma = \begin{pmatrix} 1/\beta M^2 & 0 & 0 & 0 \\ u/\beta M^2 & \rho & 0 & 0 \\ v/\beta M^2 & 0 & \rho & 0 \\ \frac{e+p}{\rho \beta M^2} - 1 & \rho u & \rho v & -\frac{\gamma p R}{\gamma - 1} \end{pmatrix} \tag{2.51}$$

In the above, R is the specific gas constant and T temperature, $\beta = \gamma RT$ and γ is the ratio of specific heats. The Eigen values for the above are now

$$\lambda = \left(u, u, \frac{u(1 + \frac{\beta M^2}{\gamma RT}) \pm c'}{2} \right) \tag{2.52}$$

However, in the above a key step is the introduction of the pseudo acoustic speed, c'. This is given below

$$c'^2 = u^2 \left(1 - \frac{\beta M^2}{\gamma RT} \right)^2 + 4\beta M^2 \tag{2.53}$$

where now $\beta = k\gamma RT$ and k is chosen to ensure that the acoustic wave speed is similar to the particle speed. Hence, in the above we have changed the scaling of the time derivative in the compressible solver through introduction of the Jacobian. The latter, with c' maintains well conditioned Eigen values at low speeds. However, the time accuracy has been lost.

The most popular form of preconditioning is that of Weiss and Smith (1995). For unsteady flows the scheme must be used in an implicit temporal framework with pseudo inner time step iterations. Otherwise, as with the Choi and Merkel approach above, if used explicitly time accuracy is lost. In an implict solution framework (where the preconditioning is applied to a pseudo time term) although the preconditioning cures the excessive damping the problem is shifted. The number of inner iterations becomes large. Also, most preconditioning strategies seem to neglect that in a three-dimensional flow there are three key Mach number components. Amalgamating these into one resolved component neglects the tensorial nature of the discretized equations. Hence, for highly anisotropic Mach number fields preconditioned results can be inaccurate. Then it seems better to switch, if possible, to pressure based solver technology. A specialist DNS scheme—flux entropy splitting—for compressible flow solvers is outlined by Sandham et al. (2002). A key point is that near walls the Mach number will always tend to zero. This is highly problematic for LES. The increased dissipation will kill the structures in the key turbulent production zone (at around 20 wall units from surfaces). Then a potential option is to use a near wall RANS layer. Such hybrid RANS-LES approaches are discussed in the next chapter. For example, for typical gas turbines beyond 60 wall units the Mach numbers are sufficiently high for a compressible flow solver to be effective.

2.15 Spatial Order and Solution Accuracy

An open question, especially with LES is the influence of numerical order on results. There are many successful LES and DNS that make use of schemes with 2nd order accuracy. Clearly the numerical order/scheme will influence the grid requirements.

It is straightforward, using Fourier based modified wave number analysis to study the accuracy of numerical differentiation. The property of interest can be expressed as $\phi(x) = e^{iwx}$ where w is the wave number. On differentiation this becomes

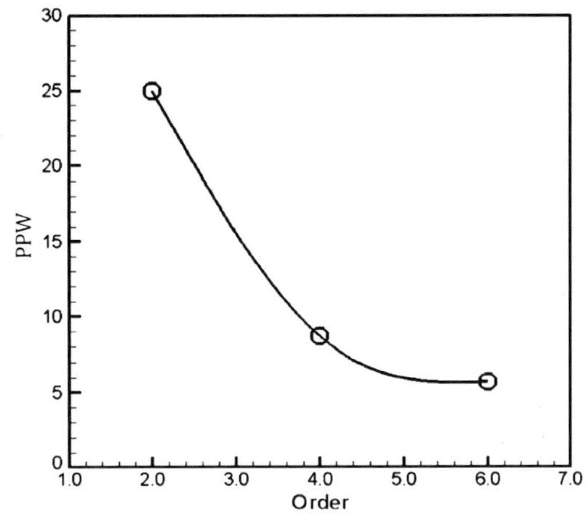

Fig. 2.11 PPW against scheme order for a wave number error of 1 %

$d\phi(x)/dx = iwe^{iwx} = iw\phi(x)$. The discrete scheme will give $iw'\phi(x)$ here w' is the, less accurate, modified wave number. We could quantify the error in the wave number as

$$\text{Error} = \frac{(w' - w)}{w} \tag{2.54}$$

For a spectral scheme the error in the above equation will be zero. Using the above noted analysis it is possible to estimate the ratio of say an eddy scale, l, (or wave length) to the grid spacing Δx. Hence, the points needed per wave length (PPW) for a specified error can be calculated. Figure 2.11, takes tabulated data from Lele (1992). However, it is plotted as the PPW for an error of 1 % based on Eq. (2.54) for different scheme orders. The lines are curve fits to Lele's data the symbols being the actual data. Substantial benefits appear for schemes of up to 4th order. It does not seem worth going beyond 6th order. Note, that the results discussed are for, uniform grid, and central difference schemes. Hence, there is no dissipative error (unless the grids are stretched) to consider. From Fig. 2.11, the use of high order schemes in LES/DNS might look promising. However, dissipation is more of a critical issue for LES and DNS. The dispersion can assist in yielding resolved scales and hence in an erroneous sense can perhaps be useful. However, for LES there is a lot of resolved energy at the interface between the resolved and modelled scales (this is especially so at high Reynolds numbers). Relative to DNS, the grid is coarse at this interface and then the theoretical justification for a high order scheme is compelling. As shown by Ghosal (1996), even for a scheme of order eight, the numerical contribution will be higher than the LES model contribution. The alternative to high order scheme use, to correct this theoretical problem would be to use a finer grid embedded under an explicit LES filter. However, this approach comes at an extreme cost. A compromise, according to Ghosal is to have a 4th order scheme with grid spacings under the filter of half the filter width. Then the LES model contribution will outweigh that of the numerical scheme. However, even this compromise gives almost a factor of ten increase in computational cost. The most concrete example of where high order is shown to be of benefit for LES is perhaps the work of Spyropoulos and Blaisdell (1998). In this LES of a spatially evolving supersonic boundary layer is performed. Strong theoretical justification is given by Chow and Moin (2003) for high order schemes in LES. Potential benefits for combustion modelling, where characterization of small-scale behaviour is more critical, are identified by Pitsch (2006). The benefits of a high order spectral scheme for modelling a basic boundary layer using DNS are identified by Laizet and Lamballais (2009). However, the substantial improvements are only evident in the fluctuating vorticity. Of note, it appears that more serious work on LES model development makes use of higher order schemes or larger filters with grid resolution under the filter (Geurts and Holm 2003).

2.15.1 Grid Stretching

Lele's analysis is for uniform grids. However, for practical industrial applications non-uniform grids are essential. Chung and Tucker (2003) repeat Lele's analysis for non-uniform hyperbolic sine and tangent grids. The ratio of largest to the smallest grid spacings is 100. This is consistent with typical DNS (where $\Delta y_{\text{min}} = 0.1$ and $\Delta y_{\text{max}} = 10$) and also for LES where typically $\Delta y_{\text{min}} = 1$ and $\Delta y_{\text{max}} = 100$). Hyperbolic sine grids are relatively popular for free shear layers and tanh boundary layers. Figure 2.12 plots dispersion and dissipation errors for compact and finite difference schemes on tanh grids with $N = 32$. The dotted line gives the exact result that a spectral scheme would provide. The other lines are for different levels of grid stretching and are as identified in the figure caption. Figure 2.12(a) shows how the wave number overshoots for a compact 6th order scheme when the grid becomes more stretched.

However, the key thing that happens for a non-uniform grid is that there is a finite imaginary component to w' representing dissipation. This is zero for uniform grids. As can be seen from Fig. 2.12(b), as the grid is stretched, for a compact scheme, the dissipation error (negative w' zone) becomes large. However, as Fig. 2.12(c) shows, for the centred scheme (2nd order) this error is much smaller. Hence, since finite difference schemes are more economical there is an attraction in staying with them for industrial flows.

This is further illustrated in Fig. 2.13. This plots the PPW for a 10 % dispersion error. The results are for a stretched sinh grid and a 4th order standard finite difference scheme (explicit) and compact scheme (implicit). The largest grid stretching, in the plot, is consistent with typical expansion ratios for LES/DNS. Figure 2.13 shows that for practical problems, where the grid needs stretching, the compact and standard finite difference schemes have similar accuracy. However, the latter needs less computational effort.

Figure 2.14 plots contours of instantaneous axial velocity for a round jet. The plot is intended to show the shear layer development for high and low order centred schemes (with higher order smoothers). Frame (a) is the solution for a 2nd order

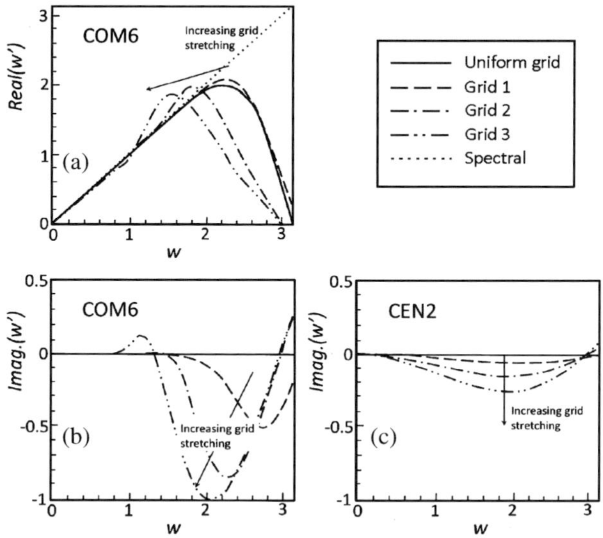

Fig. 2.12 Dispersion and dissipation errors for compact and finite difference schemes on tanh grids: (**a**) dispersion error for 6th order compact scheme; (**b**) dissipation error for 6th order compact scheme and (**c**) dissipation error for 2nd order central difference scheme (from Chung and Tucker 2003)

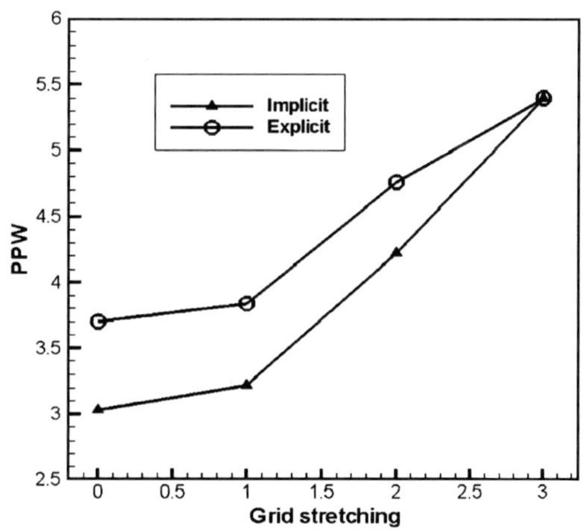

Fig. 2.13 Points per wave length (PPW) (for a 10 % dispersion error) with grid stretching for a 4th order standard finite difference scheme (explicit) and compact scheme (implicit)

scheme and (b) for a 6th order scheme. The solution is made on a modest 5 million cell grid for a jet Reynolds number of around 2×10^5. The high order scheme seems to show the development of more plausible structures in the near nozzle region. Notably, Ozyoruk and Long (1997) use 4th order central difference scheme solutions to predict fan noise emanating from an engine intake. They gain encouraging agreement with far field noise data.

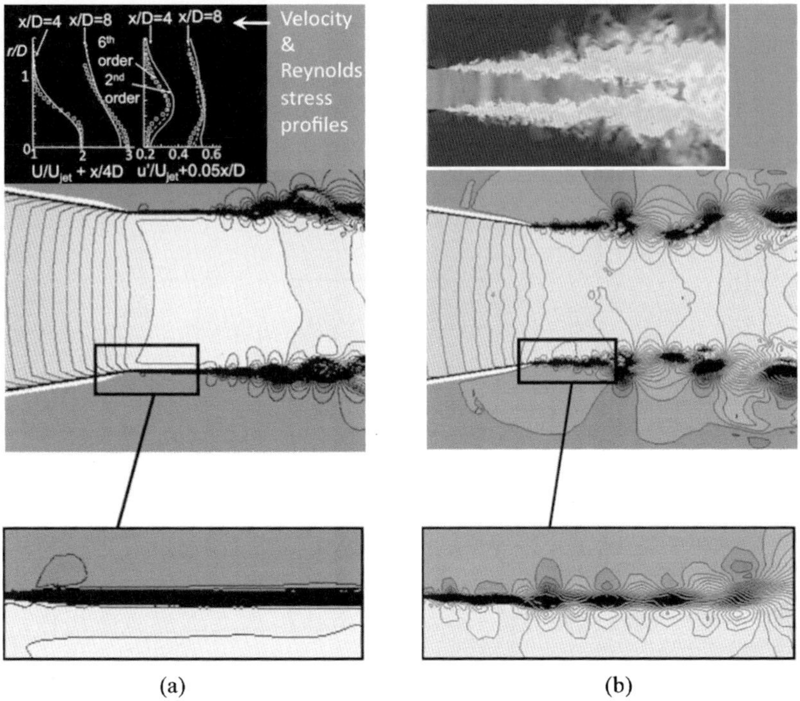

Fig. 2.14 Shear layer development in a jet for high and low order centred schemes: (**a**) 2nd order scheme and (**b**) 6th order scheme

2.15.2 High Order Upwinding

The stability of high order upwinding schemes makes them appear attractive for LES. However, as shown in the Fig. 2.15, the energy spectrum, even with 7th order upwinding results in a substantial damping of energy at the higher frequencies. Indeed for the Fig. 2.14 plot—with 5th order upwinding (Tucker 2004) the shear layer development is even more delayed (in fact it is totally suppressed) than for the 2nd order scheme.

2.15.3 Aliasing and Numerical Order

The shortest wavelength that can be resolved by a grid is twice the grid spacing. From Fourier analysis, flows can be seen to consist of a range of wave components. Because of the non-linear convective terms, without significant dissipation this range, with time, will grow. Energy from wavelengths that cannot be resolved by the grid (sub-grid scale wavelengths) can erroneously become combined with longer wavelengths. This is called aliasing.

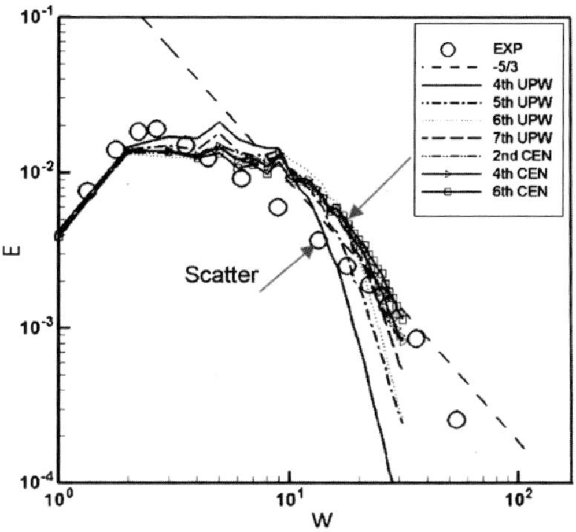

Fig. 2.15 Predicted energy spectrum for homogeneous decaying turbulence with various order upwind and central difference schemes. Published with kind permission from the ASME

Table 2.8 Various forms of convective term in the Navier-Stokes equations

Title	Tensor form	Source
Non-conservation	$u_j \frac{\partial u_i}{\partial x_j}$	–
Conservation	$\frac{\partial (u_j u_j)}{\partial x_i}$	–
Skew-symmetric	$\frac{1}{2}(\frac{\partial (u_j u_j)}{\partial x_i} + u_j \frac{\partial u_i}{\partial x_j})$	Arakawa (1966)
Rotational form	$u_j(\frac{\partial u_i}{\partial x_j} - \frac{\partial u_j}{\partial x_i}) + \frac{1}{2}\frac{\partial (u_j u_j)}{\partial x_i}$	Deardorff (1970)

According to Arakawa (1966), aliasing divergence is characterized by a smooth fluid flow structure degenerating into eddies. Evidently, once formed, the eddies intensify without limit. As a result, there is an explosive growth of kinetic energy. This results in the solution diverging. The convective term in the Navier-Stokes equations can be expressed in a range of forms. Some well-known forms are given in Table 2.8.

It is widely reported that discretization of skew-symmetric form results in substantially less aliasing error (see, for example, Blaisdell et al. 1996). On the other hand, the rotational form is observed to show substantial dissipation of turbulence. Horiuti and Itami (1998) attribute this to the discretization error when using low order schemes. From the chain rule of differential calculus, given below

$$\frac{\partial (u_j u_j)}{\partial x_i} = u_j \frac{\partial u_j}{\partial x_i} + u_j \frac{\partial u_j}{\partial x_i} \tag{2.55}$$

it can be seen that the last two terms in the rotational form should cancel. However, because of numerical truncation error in low order schemes they will not (Horiuti and Itami 1998). Ducros et al. (2000) show that in finite volume form the 2nd order central difference discretization naturally provides a skew-symmetric form. Hence, this could reflect why the 2nd order central difference scheme seems to perform well for LES. Joo and Durbin (2009) use the skew-symmetric form of the convective term

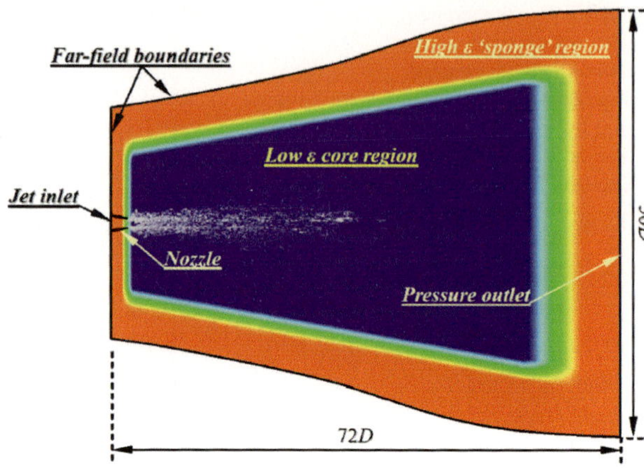

Fig. 2.16 Contours of ε_1 for a jet flow LES (*blue zone* corresponds to low ε_1). Published with kind permission from Elsevier

to gain stability when making hybrid RANS-LES type simulations for the cutback film cooling flow on a turbine blade representation.

When using spectral methods aliasing is prevented through use of the so-called 2/3 rule. With this, for $w > 2n/3$, where n is the number of resolved Fourier components, the modes are set to zero. For more dissipative schemes aliasing becomes less of a problem. The damping from upwinding suppresses the growth of the high wave number components. The substantial density variations accompanying compressible flow gives rise for further non-linearity but little is know regarding the implications of this for aliasing errors.

2.16 Smoothing Control

For high-speed flows, unless the grids are extremely fine it is necessary to introduce some numerical smoothing. Hence, typically centred fluxes, J_{ctr}, might be used with an additional smoother, J_{smth}. Generally, with 2nd order centred codes and shock free flow, this smoother would involve fourth order derivates.

$$J_{conv} = J_{ctr} + \varepsilon_1 J_{smth} \tag{2.56}$$

The smoother is typically pre-multiplied by a scaling parameter, ε_1. This can be reduced around a target zone of interest (Shur et al. 2003). To assist in damping reflections, it can become higher, in a smooth fashion, near boundaries. This is shown in Fig. 2.16 for a jet LES. Note, in the work of Shur et al. the blending is made between higher order upwinding and central differences but the two approaches are mathematically equivalent.

An alternative approach developed by Mary and Sagaut (2002) is to have a ε_1 field, which is flow dependent. Where there are excessive point-wise oscillations (dispersion error) ε_1 is increased. Where no dispersion error is detected it is decreased.

The approach works by taking a stencil of four nodes for an arbitrary control-volume interface. Wiggles in the primitive variables (p, u, v, w) are checked for by looking for the coexistence of a minimum and a maximum along the stencil. If a wiggle is detected, the local value of ε_1 is increased; otherwise it is decreased according to Eq. (2.57) below

$$\varepsilon_1^{new} = \begin{cases} \min[(\varepsilon_1^{old} + \Delta_\varepsilon), \varepsilon_1^{\max}] & \text{if wiggle is detected} \\ \max[(\varepsilon_1^{old} - \Delta_\varepsilon), \varepsilon_1^{\min}] & \text{if wiggle is not detected} \end{cases} \tag{2.57}$$

Here Δ_ε, $\varepsilon_1^{\max}$ and $\varepsilon_1^{\min}$ correspond to the increment in ε_1, the maximum allowable ε_1 and the minimum allowable ε_1. A parameter also controls the number of updates per time step when inner iterations are used. An investigation of this scheme can be found in Jefferson-Loveday (2008). Its performance for decaying homogeneous isotropic turbulence is explored with and without a standard Smagorinsky model. The scheme is found to be effective in keep solutions running with the lowest amount of dissipation in a average sense.

However, the scheme does lack certain sophistication in that the smoothing variable is both increased and decreased in a stepwise fashion. This can be improved through the use of a PI (Proportional and Integral) controller. The difference between the current level of dispersion and that which is desired (some low level) can be defined ψ. A smoothing increment, $\Delta\varepsilon_1$ can then be expressed using a PI controller as

$$\Delta\varepsilon_1(t) = k_p\psi(t) + k_I \int_0^t \psi(t)d\tau + k_D\frac{d\psi}{dt} \tag{2.58}$$

In the above k_p, k_I and k_D are tuneable parameters. The integral term avoids a steady state offset developing, thus avoiding the target value being reached. The final term avoids overshoots. Figure 2.17 gives a schematic showing the expected variation of ψ and ε_1 for the original scheme of Mary and Sagaut (Frame (a)) and the one using a PI controller (Frame (b)). As can be seen, from Frame (b), a more smooth variation of ε_1 is expected.

Figure 2.18 shows the performance of the PI scheme when predicting homogeneous decaying turbulence. Details of the case-setup can be found in Jefferson-Loveday (2008). The symbols are measurements. The chain dashed lines are results for a fixed ε_1 scheme. The full line is for the PI controller based scheme. There is evidence of much less dissipation with this scheme.

Strelets (2001) presents a flow physics based dispersion/dissipation control. With this

$$\varepsilon_1 = \varepsilon_{\max}\tanh\left(A^{c_1}\right) \tag{2.59}$$

and

$$A = c_2 \max\left\{\left[\left(\frac{C_{DES}\Delta}{l}\right)/g - 0.5\right], 0\right\} \tag{2.60}$$

In the above, Δ, is the LES filter width (see Chap. 3) and l is a turbulence length scale. This has a complex functional form that is not reported here. However, $l =$

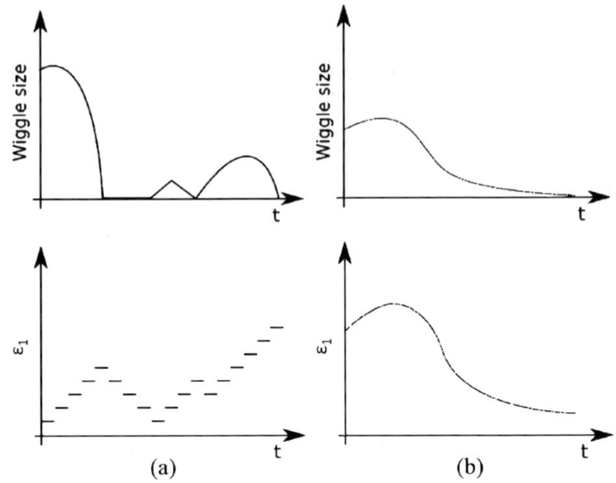

Fig. 2.17 Schematic of the expected variation of ψ and ε_1 with time: (**a**) original scheme of Mary and Sagaut (2002) and (**b**) PI controller based scheme

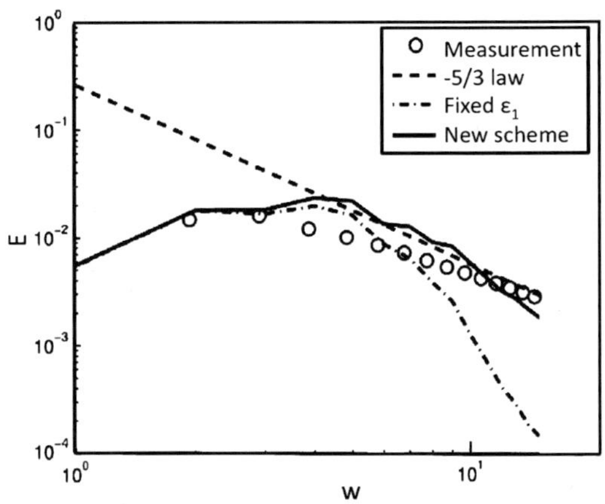

Fig. 2.18 Turbulence energy spectrum for homogeneous decaying turbulence with PI control scheme

$f(\mu_T, S, \Omega, \tau)$ where μ_T is turbulent viscosity, S strain rate, Ω vorticity and τ a convective time scale. The function g is given below

$$g = \tanh\left\{\frac{c_3\Omega_{\max}[S, \Omega]}{\max([S^2 + \Omega^2]/2, \varepsilon)}\right\} \tag{2.61}$$

The function g is intended to introduce more smoothing/upwind behaviour in irrotational but disturbed flow zones where $\Omega \ll 1$ and $S > 0$. Clearly, also low smoothing is sought in rotational flow zones. In the above, $\varepsilon \to 0$, $\varepsilon_{\max} = 1.0$, $c_1 = 3.0$, $c_2 = 1$ and $c_3 = 2$.

2.16.1 Shocks and LES

Around shocks a 2nd order derivative smoother can be activated via a pressure based shock detection switch. This is the most standard CFD procedure. The switch basically detects high pressure gradients. The difficulty when performing LES is that small scale eddies can be misinterpreted as shocks and hence the 2nd order smoother can become excessively active.

The switch of Ducros et al. (1999), given below, assists in alleviating this problem. It attempts to only make the shock smoothing active around shocks and not under-resolved turbulence.

$$\Phi = \frac{(\nabla.\mathbf{u})^2}{(\nabla.\mathbf{u})^2 + \omega^2 + \epsilon} \tag{2.62}$$

In the above ϵ is a small number to prevent division through by zero and ω is the resolved vorticity ($\nabla \times \mathbf{u}$). This function Φ, ranges between zero and unity—tending to unity around shocks.

Bisek et al. (2013) when making compact scheme LES use the Swanson and Turkel (1992) pressure gradient detector.

$$\Phi = \frac{|p_{i+1} - 2p_i + p_{i+1}|}{(1-\omega)(|p_{i+1} - p_{i-1}|) + \omega(p_{i+1} - 2p_i + p_{i+1})},$$
$$\begin{cases} \Phi > 0.05, & \text{Roe scheme} \\ \Phi \leq 0.05, & \text{Compact scheme} \end{cases} \tag{2.63}$$

In shock zones a flux limited (van Albada) Roe scheme is used, the high order compact scheme being used elsewhere.

2.17 Mesh Related Techniques

Unstructured moving meshes can be used in an ALE formulation. More details on such meshes can be found in Chap. 7. Yang et al. (1997) give a Cartesian, trimmed cell method for moving body problems. Where the geometry is especially complex, the immersed boundary method provides a more flexible alternative for modelling complex moving body cases (Pinelli et al. 2010). However, both the trimmed cell and immersed boundary methods are unsuitable for modelling flows where the prediction of accurate skin friction is important. Vortex methods, that generally do not require the use of grids, (Khatir 2000) naturally lend themselves to moving boundary problems. The mesh-less smooth particle hydrodynamics method (Marongiu et al. 2007, 2010) are also suitable. The smooth particle hydrodynamics methods were originally developed for astrophysics modelling such as modelling the formation of stars. Like most mesh-less methods it has been available for many decades. As originally formulated it is intended for highly compressible problems. Marongiu et al. apply the approach to a free surface Pelton turbine flow.

2.17.1 Body Fitted Grids

For finite volume schemes it is challenging to maintain formally high order numerical discretizations. This is much more straightforward with finite difference and compact schemes. However, then geometrical flexibility is limited. A way round this, to an extent, is to work in transformed coordinates. With this, the governing equations in an (x, y, z, t) coordinate system can be transformed into a (ξ, η, ζ, τ) system using the chain rule of differential calculus. The transformed computational grid is normally uniform and Cartesian. Hence, the application of higher order schemes is made, in a sense, relatively easy. The transformation (Rayner 1993) simply involves making the following substitutions in the governing (x, y, z, t) equations, where $\xi = \xi(x, y, z, t)$, $\eta = \eta(x, y, z, t)$, $\zeta = \zeta(x, y, z, t)$ and $\tau = \tau(t)$

$$\frac{\partial}{\partial x} = \left(\frac{\partial \xi}{\partial x}\right)\frac{\partial}{\partial \xi} + \left(\frac{\partial \eta}{\partial x}\right)\frac{\partial}{\partial \eta} + \left(\frac{\partial \zeta}{\partial x}\right)\frac{\partial}{\partial \zeta} + \left(\frac{\partial \tau}{\partial x}\right)\frac{\partial}{\partial \tau} \tag{2.64}$$

$$\frac{\partial}{\partial y} = \left(\frac{\partial \xi}{\partial y}\right)\frac{\partial}{\partial \xi} + \left(\frac{\partial \eta}{\partial y}\right)\frac{\partial}{\partial \eta} + \left(\frac{\partial \zeta}{\partial y}\right)\frac{\partial}{\partial \zeta} + \left(\frac{\partial \tau}{\partial y}\right)\frac{\partial}{\partial \tau} \tag{2.65}$$

$$\frac{\partial}{\partial z} = \left(\frac{\partial \xi}{\partial z}\right)\frac{\partial}{\partial \xi} + \left(\frac{\partial \eta}{\partial z}\right)\frac{\partial}{\partial \eta} + \left(\frac{\partial \zeta}{\partial z}\right)\frac{\partial}{\partial \zeta} + \left(\frac{\partial \tau}{\partial z}\right)\frac{\partial}{\partial \tau} \tag{2.66}$$

If the grid is moving, the chain rule also needs to be applied to the time derivative

$$\frac{\partial}{\partial t} = \left(\frac{\partial \xi}{\partial t}\right)\frac{\partial}{\partial \xi} + \left(\frac{\partial \eta}{\partial t}\right)\frac{\partial}{\partial \eta} + \left(\frac{\partial \zeta}{\partial t}\right)\frac{\partial}{\partial \zeta} + \left(\frac{\partial \tau}{\partial t}\right)\frac{\partial}{\partial \tau} \tag{2.67}$$

The bracketed derivatives above are called metrics. Variables treated as fixed in the partial differentiations contained in Eqs. (2.64) to (2.67) are indicated using subscripts below

$$\left(\frac{\partial}{\partial x}\right)_{yzt} \left(\frac{\partial}{\partial y}\right)_{xzt} \left(\frac{\partial}{\partial z}\right)_{xyt} \left(\frac{\partial}{\partial t}\right)_{xyz} \left(\frac{\partial}{\partial \xi}\right)_{\eta\zeta\tau} \left(\frac{\partial}{\partial \eta}\right)_{\xi\zeta\tau} \left(\frac{\partial}{\partial \zeta}\right)_{\xi\eta\tau} \left(\frac{\partial}{\partial \tau}\right)_{\xi\eta\zeta}$$

Clearly $\partial\tau/\partial x = \partial\tau/\partial y = \partial\tau/\partial z = 0$ and so grid distortion from boundary movement does not affect transformations (2.64) to (2.66).

Results presented later make use of the artificial compressibility method of Rogers and Kwak (1990). Hence, it seems useful here to outline the transformed equations in combination with this approach. With the artificial compressibility method the following artificial equation of state is used, $p = \beta^{-1}\rho$. In this relationship, β is an artificial compressibility parameter (see later) and τ is a pseudo time. Since the flow is assumed incompressible, the pseudo time derivative is integrated to a steady state. Application of the chain rule above to the Reynolds averaged Navier-Stokes equations, in their strong conservation form, gives (Jefferson-Loveday 2008)

$$\partial_\tau \mathbf{Q} + \partial_\xi \mathbf{E}^{inv} + \partial_\eta \mathbf{F}^{inv} + \partial_\zeta \mathbf{G}^{inv} = \left[\partial_\xi \mathbf{E}^{vis} + \partial_\eta \mathbf{F}^{vis} + \partial_\zeta \mathbf{G}^{vis}\right] \tag{2.68}$$

where

$$\mathbf{Q} = J^{-1}\begin{bmatrix} \overline{p} \\ \overline{u} \\ \overline{v} \\ \overline{w} \end{bmatrix}, \qquad \mathbf{E}^{inv} = J^{-1}\begin{bmatrix} \beta\overline{U} \\ \overline{u}\overline{U} + \xi_x\overline{p} \\ \overline{v}\overline{U} + \xi_y\overline{p} \\ \overline{w}\overline{U} + \xi_z\overline{p} \end{bmatrix}$$

$$\mathbf{F}^{inv} = J^{-1}\begin{bmatrix} \beta\overline{U} \\ \overline{u}\overline{V} + \eta_x\overline{p} \\ \overline{v}\overline{V} + \eta_y\overline{p} \\ \overline{w}\overline{V} + \eta_z\overline{p} \end{bmatrix}, \qquad \mathbf{G}^{inv} = J^{-1}\begin{bmatrix} \beta\overline{U} \\ \overline{u}\overline{W} + \zeta_x\overline{p} \\ \overline{v}\overline{W} + \zeta_y\overline{p} \\ \overline{w}\overline{W} + \zeta_z\overline{p} \end{bmatrix}$$

$$\overline{U} = \xi_x\overline{u} + \xi_y\overline{v} + \xi_z\overline{w}, \qquad \overline{V} = \eta_x\overline{u} + \eta_y\overline{v} + \eta_z\overline{w}, \qquad \overline{W} = \zeta_x\overline{u} + \zeta_y\overline{v} + \zeta_z\overline{w} \tag{2.69}$$

Note, the overbars, see Chap. 3, represent Reynolds averaging. The viscous flux terms are given by

$$\begin{aligned}
\mathbf{E}^{vis} &= J^{-1}(\xi_x\mathbf{E}^{vis} + \xi_y\mathbf{F}^{vis} + \xi_z\mathbf{G}^{vis}) \\
\mathbf{F}^{vis} &= J^{-1}(\eta_x\mathbf{E}^{vis} + \eta_y\mathbf{F}^{vis} + \eta_z\mathbf{G}^{vis}) \\
\mathbf{G}^{vis} &= J^{-1}(\zeta_x\mathbf{E}^{vis} + \zeta_y\mathbf{F}^{vis} + \zeta_z\mathbf{G}^{vis})
\end{aligned} \tag{2.70}$$

where

$$\mathbf{E}^{vis} = \begin{bmatrix} 0 \\ \tau_{xx} \\ \tau_{xy} \\ \tau_{xz} \end{bmatrix}, \qquad \mathbf{F}^{vis} = \begin{bmatrix} 0 \\ \tau_{yx} \\ \tau_{yy} \\ \tau_{yz} \end{bmatrix}, \qquad \mathbf{G}^{vis} = \begin{bmatrix} 0 \\ \tau_{zx} \\ \tau_{zy} \\ \tau_{zz} \end{bmatrix} \tag{2.71}$$

and

$$\begin{aligned}
\tau_{xx} &= (\mu + \mu_t)(4(\xi_x\overline{u}_\xi + \eta_x\overline{u}_\eta + \zeta_x\overline{u}_\zeta) \\
&\quad - 2(\xi_y\overline{v}_\xi + \eta_y\overline{v}_\eta + \zeta_y\overline{v}_\zeta + \xi_z\overline{w}_\xi + \eta_z\overline{w}_\eta + \zeta_z\overline{w}_\zeta))/3 \\
\tau_{yy} &= (\mu + \mu_t)(4(\xi_y\overline{v}_\xi + \eta_y\overline{v}_\eta + \zeta_y\overline{v}_\zeta) \\
&\quad - 2(\xi_x\overline{u}_\xi + \eta_x\overline{u}_\eta + \zeta_x\overline{u}_\zeta + \xi_z\overline{w}_\xi + \eta_z\overline{w}_\eta + \zeta_z\overline{w}_\zeta))/3 \\
\tau_{zz} &= (\mu + \mu_t)(4(\xi_z\overline{w}_\xi + \eta_z\overline{w}_\eta + \zeta_z\overline{w}_\zeta) \\
&\quad - 2(\xi_x\overline{u}_\xi + \eta_x\overline{u}_\eta + \zeta_x\overline{u}_\zeta + \xi_y\overline{v}_\xi + \eta_y\overline{v}_\eta + \zeta_y\overline{v}_\zeta))/3 \\
\tau_{xy} &= (\mu + \mu_t)(\xi_x\overline{u}_\xi + \eta_x\overline{u}_\eta + \zeta_x\overline{u}_\zeta + \xi_y\overline{v}_\xi + \eta_y\overline{v}_\eta + \zeta_y\overline{v}_\zeta) \\
\tau_{yz} &= (\mu + \mu_t)(\xi_y\overline{v}_\xi + \eta_y\overline{v}_\eta + \zeta_y\overline{v}_\zeta + \xi_z\overline{w}_\xi + \eta_z\overline{w}_\eta + \zeta_z\overline{w}_\zeta) \\
\tau_{zx} &= (\mu + \mu_t)(\xi_z\overline{w}_\xi + \eta_z\overline{w}_\eta + \zeta_z\overline{w}_\zeta + \xi_x\overline{u}_\xi + \eta_x\overline{u}_\eta + \zeta_x\overline{u}_\zeta)
\end{aligned} \tag{2.72}$$

Note, μ_t is eddy viscosity. Although the transformed differential equations are much larger than the original, they can be solved on a uniform grid computational plane. Hence, high order schemes can be readily applied. A key issue is the formation of the metric terms. Ideally, these must be also expressed at an order consistent with that of the base finite difference scheme. For highly non-orthogonal grids the formulation of the metric terms can have a critical accuracy impact (Visbal and Gaitonde 2002) as will non-smooth grid stretching (Gamet et al. 1999). Hence, care is needed. Visbal and Gaitonde (2002) contrast standard

$$\begin{aligned}\hat{\xi}_x &= y_\eta z_\zeta - y_\zeta z_\eta \\ \hat{\eta}_x &= y_\zeta z_\xi - y_\xi z_\zeta \\ \hat{\zeta}_x &= y_\xi z_\eta - y_\eta z_\xi\end{aligned} \tag{2.73}$$

and the conservation metric form of Thomas and Lombard (1979)

$$\begin{aligned}\hat{\xi}_x &= (y_\eta z)_\zeta - (y_\zeta z)_\eta \\ \hat{\eta}_x &= (y_\zeta z)_\xi - (y_\xi z)_\zeta \\ \hat{\zeta}_x &= (y_\xi z)_\eta - (y_\eta z)_\xi\end{aligned} \tag{2.74}$$

On skewed grids, they found that the latter held free stream conditions more accurately. Hence, Daude et al. (2012) also adopt this approach in their high-order, finite difference simulations.

There is an extensive body of literature on the use of high order schemes on curvilinear grids for acoustics. For example, Uzun and Hussaini (2009) use 6th high order compact schemes for complex chevron nozzle jet noise LES. In a related fashion, Lu et al. (2012) use a high order unstructured flux reconstruction method. The unstructured element shape is mapped into a locally regular transformed ξ, η, ζ system—a standard high-order procedure. The method is applied to transonic flow around a VKI-LS59 turbine blade.

2.17.2 Overset Grids

The Chimera (a Chimera is a mythological creature with a goats body, snakes tail and lions head) or overset grid method, described by Chesshire and Henshaw (1990), has seen popularity for modelling moving grid problems. With this, for example, a Cartesian grid often maps the flow domain and the body modelled by moving curvilinear overset grids.

Figure 2.19 shows a simple overset mesh and the isosurfaces of vorticity from a hybrid large eddy type simulation, for a static geometry case. Notably, Paliath et al. (2011) also use a compact overset code for complex chevron geometry nozzle LES. Spalart et al. (2003) use 5th order finite differences on curvilinear overset meshes. They explore flow control over an idealized V22 tilt rotor wing using detached eddy simulation (see Chaps. 3 and 5). Hedges et al. (2002) use the approach for landing gear aerodynamic studies. Rizzetta et al. (2008) apply a 4th order compact scheme (with 6th order smoothing) in an overset grid solver, studying flow control over turbine blades.

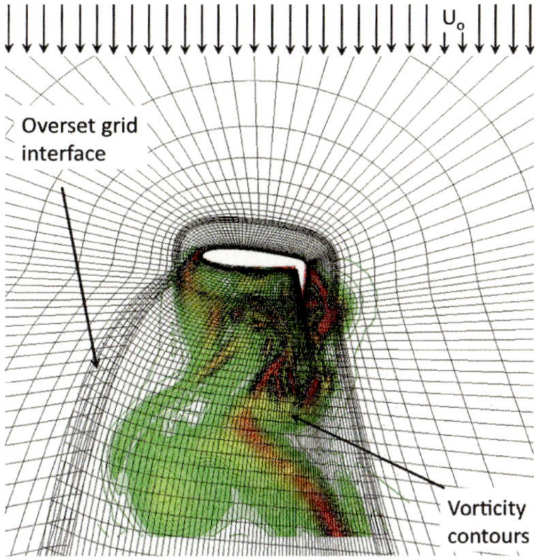

Fig. 2.19 Overset mesh and vorticity contours showing the flow field for a generic tilt-rotor wing geometry (from Tucker 2006). Published with kind permission from Wiley and Sons

However, with overset meshes, the key issue is how to produce a high order and sufficiently conservative interpolation between the different overlapping grids. This is compounded by the potential for several overlapping grids at the same point in space. Maintaining order is explored in Lee et al. (2011). Unstructured overset meshes, with relative movement can be a powerful tool for especially complex, moving grid problems (Nakahashi and Togashi 2000). The moving overset component allows a reduction in cell distortion. For distorting meshes, control volume face velocities need to be reformulated to obey space conservation (Demirdžić and Perić 1988). This aspect is discussed further in Chap. 7.

2.18 The Substantial Derivative

When computing unsteady flows, concepts relating to the substantial derivative can be helpful to consider. These can have implications for the numerical scheme. The substantial derivative, $(\partial\phi/\partial t)_{fp}$, is more usually expressed as $D\phi/Dt$. However, here the subscript fp is used. This enforces its meaning. This being the time rate of change in ϕ when a particular fluid particle is followed. The derivative is given below

$$\left(\frac{\partial\phi}{\partial t}\right)_{fp} = \underbrace{\left(\frac{\partial\phi}{\partial t}\right)_g + \left(\frac{\partial\phi}{\partial t}\right)_f}_{\partial\phi/\partial t} + (\mathbf{U} - \mathbf{U}_g)\,.\,\nabla\phi + G \tag{2.75}$$

The terms $(\partial\phi/\partial t)_g$ and $(\partial\phi/\partial t)_f$ together are those that would normally be related to unsteadiness i.e. $(\partial\phi/\partial t)$. However, here they are split out with $(\partial\phi/\partial t)_g$

being the component that can be reduced by minimizing $|\mathbf{U} - \mathbf{U}_g|$ where $\mathbf{U}_g$ is the grid/coordinate system velocity. For a rotating system $\mathbf{U}_g$ can be proportional to Ω_g the grid angular velocity. The $(\partial\phi/\partial t)_f$ represents any unsteadiness would potentially remain.

For certain systems (classified as exhibiting unsteady flow), the fluid structure may stay exactly or much the same, but simply translate (at a velocity $\tilde{\mathbf{U}}$) or rotate (at an angular velocity Ω) in space, relative to the coordinate system/grid. The mechanism for this translation or rotation could be fluid dynamic in origin or alternatively as the result of boundary movement. Examples of the former are a wave train moving around a cylindrical domain (Tucker 2002b) with an angular velocity, $\boldsymbol{\Omega}$, about a rotation axis or vortex wakes propagating downstream from aircraft wings (Steger and Kutler 1977). Translational movement of effectively a constant geometrical distortion when viewed in a frame of reference moving with the distortion, will give steady flow $((\partial\phi/\partial t)_g + (\partial\phi/\partial t)_f = 0)$. For some practical flows it is possible to make unsteadiness smaller by minimizing $(\partial\phi/\partial t)_g$ through appropriate choice of coordinate system. Also, reducing the magnitude of temporal terms should bring down some temporal solution errors. This could be achieved through the ALE process. Furthermore, moving coordinate systems can significantly lower convective term magnitudes. These terms, as indicated above, are generally challenging in terms of unsteady flow modelling accuracy. Tucker (1997) gives examples of solution accuracy gains by the use of rotating coordinate systems. The G term, in Eq. (2.75), relates to grid distortion, being applicable to moving boundary cases. For a three-dimensional transformed grid (ξ, η, ζ) system

$$G = \left(\frac{\partial \xi}{\partial t}\right)_g \left(\frac{\partial \phi}{\partial \xi}\right) + \left(\frac{\partial \eta}{\partial t}\right)_g \left(\frac{\partial \phi}{\partial \eta}\right) + \left(\frac{\partial \zeta}{\partial t}\right)_g \left(\frac{\partial \phi}{\partial \zeta}\right) \tag{2.76}$$

However, as implied earlier, the coordinate system can also be chosen to minimize G. Hence, inaccuracies arising from modelling these extra terms, in whatever form they might appear in the chosen discretization framework, will also be minimized.

To illustrate the alteration of grid velocity to minimize unsteadiness the case of a rotating cavity with an axial through flow is considered. This is representative of the flow in the high-pressure compressor drum of an aero engine. The case is fully reported in Tucker (2002b). Two regions of radial outflow occur for the conditions considers. These emanate from an inner axial through flow of coolant. Figure 2.20, Frame (a) gives cavity mid-axial plane velocity vectors. Frame (b) gives cavity mid-axial contours of $\partial u/\partial t$ (where u is axial velocity). These are for grid angular velocities Ω_g = 17.0 (left hand image) and 16.8 rad/s (right hand image). The larger Ω_g corresponds to the cavity angular velocity. The more uniform, Frame (b), right hand, Ω_g = 16.8 rad/s contours signify greater flow steadiness.

Sliding grid planes can be used to connect rotating to stationary blade zones. These have the advantage that in the rotating frame the relative velocities solved for are smaller and hence Peclet/Reynolds numbers smaller. The latter enables the use of centred schemes or schemes that are more centred in nature. These are less prone

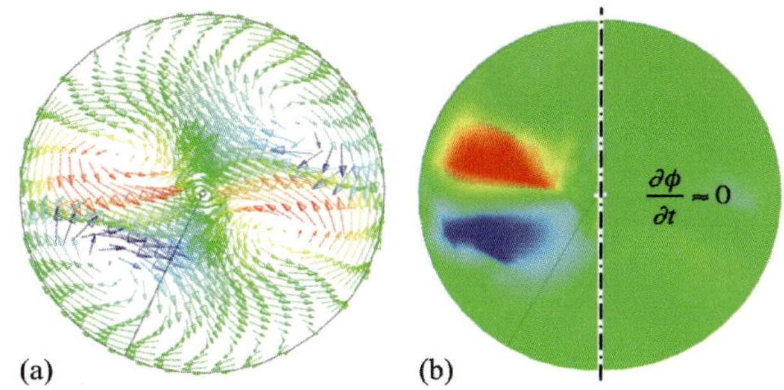

Fig. 2.20 Mid-axial velocity vectors and contours of for rotating cavity with an axial through flow of air: (**a**) Velocity vectors and (**b**) Contours for $\Omega_g = 17.0$ and 16.8 rad/s (adapted from Tucker 2002b). Published with kind permission from Elsevier

to dissipation and hence allow, for example, better wake preservation. The sliding planes allow appropriate minimization of unsteadiness. Of note, for the rotating wake body force modelling of Loiodice et al. (2010) the blade viscous body forces are cast in a rotating frame to enhance wake preservation. Vorticity confinement (Steinhoff and Underhill 1994) injects energy lost from dissipation back into flows. This approach is potentially useful for maintaining the strength of, for example tip, vortices on under-resolved grids.

2.19 Simultaneous Equation Solution

A discussion on the extensive field of advanced simultaneous equation solver technology is not attempted here. This is partly because for unsteady flows the flow equations become more parabolic and hence this issue is less critical. Hence, except for the Poisson based equations used in pressure-based solvers, efficient simultaneous equation solvers become less critical. Explicit temporal schemes do not require a simultaneous equation solver and can be easily parallelized. However, they have more severe time-step stability restrictions. Consequently they are only suitable when high temporal resolution is required. Implicit techniques essentially require a simultaneous equation solver. This is generally a major solution overhead. Approximate factorization or splitting methods successfully combine explicit and implicit techniques only requiring a one-dimensional solver. ADI is a member of this family. A relatively popular method is the Bi-CGSTAB (biconjugate gradient stabilized method). This is applied to LES of trailing edge noise by Manoha et al. (2000). Strong solvers such as GMRES (Campobasso and Giles 2003) and Recursive Projection Methods (Campobasso and Giles 2004) can force steady solutions where none exist. In an engineering design context, where fast solutions are needed, this can appear attractive. However, as shown by Krakos and Darmofal (2010) such solutions may not correspond to the time average of the unsteady solution.

Multigrid convergence acceleration is an especially efficient procedure and sees widespread use. Brandt (1980) describes a multilevel-based strategy tailored for modelling flows evolving with time. The basis of the method is Brandt's assumption that for evolution processes the high and low frequency (in a spatial sense) components can be very different. During early time-steps, the high frequency components change rapidly but then settle down. Then, the lower frequency components only need be resolved. The evolution of these can be modelled using coarser grids, the higher frequency component being accounted for by forcing terms originating from the finer grid. Therefore, after an initial period, the solution can mostly be integrated through time on a coarser grid. The fine grid is periodically revisited to update the relationship between the newer low frequency and older high frequency components.

When using Newton methods to compute unsteady flows the Jacobian re-evaluation is expensive. To circumvent this, Gresho et al. (1980) and Engelman and Sani (1986) explore the following two approaches: (a) Use adaptive time-stepping, with a small user defined time truncation error and (b) Use a quasi-Newton approach. With this the Jacobian is only re-evaluated every two or three time-steps. Option (a) results in the need for only one Newton iteration and hence Jacobian up-date per time-step.

To improve solution speed a parallel processing domain decomposition in time is evidently possible (Seidl et al. 1995; Burmeister and Horton 1991). Such approaches are not covered here and as yet do not see widespread use/acceptance.

2.20 Evaluation of the Pressure Field

The evaluation of the pressure field for density and pressure-based solvers is dramatically different. For pressure-based solvers with unsteady flow the PISO (Issa 1986) and fractional step (Kim and Moin 1985) approaches are regarded as the most ideal alternatives to the multitude of SIMPLE variants. However, in the author's experience these schemes do not offer any great performance benefits.

The AVPI scheme described by Jones and Marquis (1985) and Henkes (1990) is of note for unsteady flows. It reduces core memory requirements by avoiding the need to store additional coefficients required in popular projection methods. With AVPI, these coefficients become approximately equal to $\Delta t/\rho_i^{n-1}$ and so can be calculated on the fly. For the 1st order backwards Euler scheme and a constant density system, AVPI gives the following relatively simple Poisson equation for a correction, p', to the pressure field

$$\nabla^2 p'^{n+1} = \frac{\rho}{\Delta t} \nabla \cdot u^* \tag{2.77}$$

The star superscripted quantities are iteratively approximate. An overview of various techniques for computing p in pressure-based solvers is given in Tucker (2001).

2.20.1 *Pressure Subcycling*

As previously noted, explicit scheme stability requirements can dictate that smaller time-steps must be used than needed to resolve actual physical processes. In this

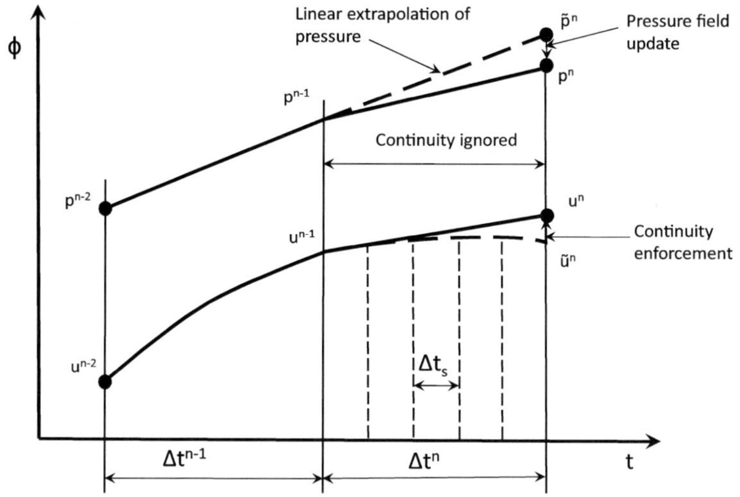

Fig. 2.21 Illustration of the subcycling procedure

situation, to lessen computational expense a procedure called subcycling, described by Gresho et al. (1984) can be used. The premise for the procedure is that, for the momentum equations (and other transport equations), explicit scheme stability is not linked to pressure and continuity constraints. The subcycling procedure, illustrated in Fig. 2.21, is as follows:

(I) Over a time period Δt^n, solve the momentum equations explicitly using minor steps, Δt_s, governed by stability limits. During this period use extrapolated pressure values.
(II) At the end of the period Δt^n, by correcting the velocity field, re-enforce continuity.
(III) Finally, use the continuity satisfying velocity field to update pressures.

By not requiring a pressure update for every minor time-step the above procedure saves computational effort.

2.20.2 Pressure-Velocity Coupling

To avoid the prediction of nonphysical oscillatory pressure fields, for incompressible flow solvers the pressure-velocity coupling has been a considerable area of research. For many years, a popular technique (Harlow and Welch 1965) has been to arrange the grids such that the relevant pressure nodes are coincident with the control volume faces of the momentum variable being solved for. To achieve this, staggered multiple control volumes must be used. Due to programming simplicity, collocated variable arrangements are preferred. The most well known collocated variable technique, to avoid unrealistic pressure field oscillations, is due to Rhie and Chow (1983). Essentially, this approach involves the introduction of pressure field smoothing. The smoothing properties of this approach, which applies dissipation to the pressure field, depend on time-step and under-relaxation values (Barton and Kirby 2000). The alternative Pressure Weighted Interpolation Method—Corrected (PWIMC) method of Kobayashi and Pereira (1991) is less dependent on under-relaxation parameter values. Barton and Kirby (2000) also present a method where performance is de-sensitized to time-step and under-relaxation values. Clearly these aspects are particularly important for LES/DNS.

2.20.3 Compressible Flow Solvers and Pressure Recovery

A key point to note is that for density based solvers, at low Mach numbers, the density variations in the time derivative of the continuity equation becomes extremely small. It is through these that the pressure field is recovered. This, when combined with the pre-conditioning issues noted earlier creates serious solution accuracy problems. Hence, compressible flow solvers are best used for high Mach number flows, where there are substantial density variations that can be used to recover pressure.

If density variations can be completely neglected the artificial compressibility method of Rogers and Kwak (1990) (based on the approach of Chorin 1967) can be effective. With this, the continuity equation is re-expressed, through a simplified equation of state, as below.

$$\frac{\partial p}{\partial \tau} + \beta \left(\frac{\partial u_j}{\partial x_j} \right) = 0 \tag{2.78}$$

The parameter β is tunable but has a fixed theoretical range (Rogers et al. 1986). The approach is intended to be used in a compressible solver framework. When used in this framework, like all other equations, the above pressure equation has a smoothing term. If β is too large, in steep pressure gradient areas, this will act as a substantial mass source term. However, with reasonable care the method is effective. A simple, alternative, expedient can be to scale down the geometry and increase velocity to preserve the Reynolds number. The higher velocity ensures a higher Mach number. If this is sufficiently high, then the compressible flow solver will have reasonable computational performance. The Mach number required for reasonable compressible solver performance can vary dramatically between different codes. Hence, there is a need to understand the CFD program's traits. However, for many flows this Mach number scaling strategy will give rise to excessive density variations is localized regions. Also, the Mach number tends to zero at walls.

2.21 Boundary Conditions

For unsteady flows the use of both non-reflecting inflow and outflow boundary conditions becomes especially important. If not, the multiple reflections of waves will ultimately lead to solution divergence. The simplest remedy is to do the following:

(I) Keep the boundaries well away from the key area of unsteadiness activity;
(II) Coarsen grids towards these boundaries and
(III) Increase the numerical dissipation towards the boundaries.

The latter two things ensures that reflected waves cannot be supported by the grid. However, clearly such strategies are computationally wasteful. Hence, if possible, more refined methods should be used. The literature on non-reflecting boundary conditions and potential variants is extensive. Hence, just some of the key approaches, that are used in results presented here, are discussed. The one-dimensional, characteristics based non-reflecting boundary condition of Giles (1990) has seen extensive use for compressible flows. However, the one-dimensional nature of this scheme limits its applicability. The convective boundary condition of Pauley et al. (1990) is simple to implement and given below

$$\frac{\partial \phi}{\partial t} + U_c \frac{\partial \phi}{\partial x} = 0 \tag{2.79}$$

In the above U_c is taken as the bulk mean velocity and ϕ would be the velocity components and other solution variables. The damping approach of Ashcroft and Zhang (2001) is also easy to implement. With this the equation below is used

$$\phi(l) = (1 - s)\phi_{\text{target}}(l) + s\phi(l) \tag{2.80}$$

ϕ_{target} being a target field and ϕ a computed field that is gradually, via a transition zone, pinned to the target field. A potential transition function is given below

$$s = \left|1 - (l/L)^{\beta}\right| \tag{2.81}$$

In the above, L is the width of the sponge/damped layer and $0 < l < L$. Also, $l = 0$ at the domain boundaries and $l = L$ is inside the domain. Following the approach of Shur et al. (2003), and as noted earlier, ideally, for LES etc, a centered scheme, is used in the target zone. This can then be blended with dissipative upwind differences at boundaries or the magnitude of smoothing terms increased. With this strategy, following Shur et al., when centered schemes are being blended with upwind differences, to assist with damping numerical reflections, the order of the upwinding scheme, N_{UP}, can also be reduced towards flow boundaries. This can be achieved through the following equation

$$N_{UP} = \max\left\langle int\left[\frac{N_{UP,MAX}\,l}{L}\right], 1\right\rangle \tag{2.82}$$

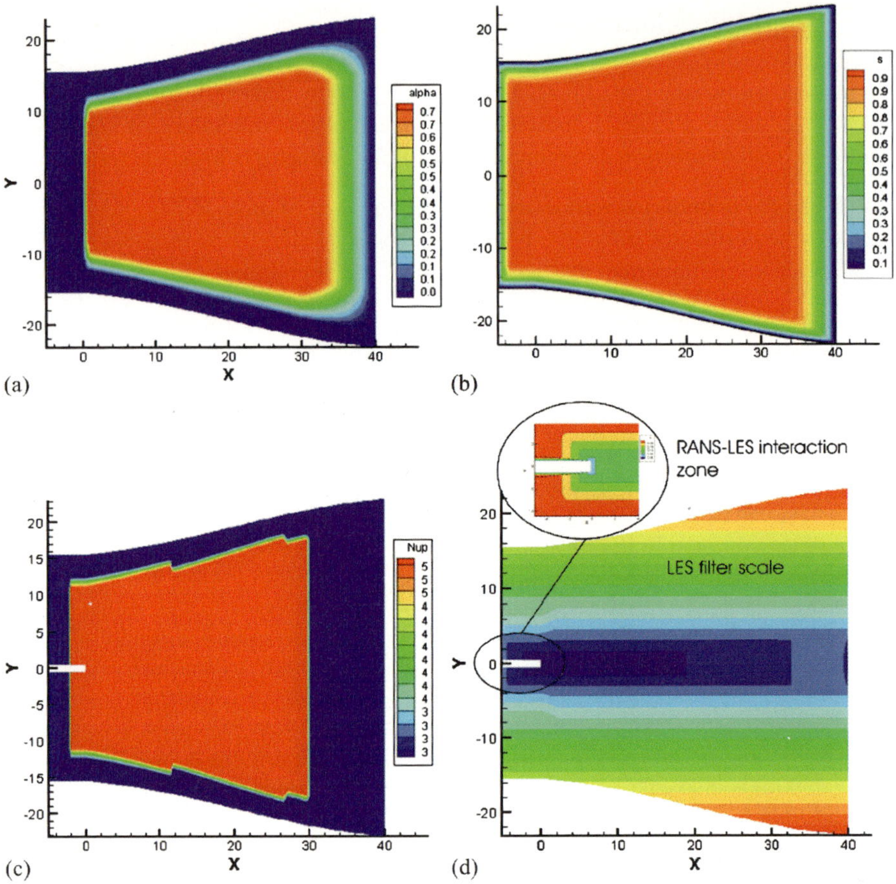

Fig. 2.22 Level set based weighting function distributions used for jet acoustic simulations: (**a**) upwind/centered scheme blending function (ϵ_1); (**b**) blending function (s) for sponges; (**c**) N_{UP} distribution and (**d**) modelled turbulence scales

where $N_{UP,MAX}$ is the maximum order of the upwind solution component (5 in the simulations of Shur et al.). Also, *int* represents the rounding up of the bracketed quantity to the nearest integer. The operator max takes the maximum value.

Figure 2.22 relates to jet LES computations. Level set fields, generated by differential Hamilton-Jacobi (HJ) equations (Sethian 1999) are presented, some after being subjected to further computational operations. For example, Frame (a) gives a blending field that allows a convective flux, J_{conv}, to be expressed as a weighted average of centred, J_{CD}, and upwind fluxes, J_{UP}. This is achieved via a blending parameter ε_1 where $J_{conv} = \varepsilon_1 J_{CD} + (1 - \varepsilon_1) J_{UP}$. The Frame (a) contours are equivalent to those in Fig. 2.16. Similarly, Frame (b) shows HJ computed values for s used in Eq. (2.80). Frame (c) gives the result of the application of Eq. (2.82) to a HJ level set field. Finally, Frame (d) gives length scale fields to model turbulence. This consists of a near wall distance field propagated to the LES filter scale with the HJ equation solved in its eikonal form. The combined figures show how the HJ equation can be used to yield a more unified hybrid RANS-LES (see Chap. 4) solution approach. This is through providing both non-reflecting boundary properties and also modelled turbulence length scales. Notably, HJ type equations are amenable to parallel solution on unstructured grids. Hence, they offer greater potential for large-scale computations where serial based surface distance search procedures can be slow.

The radiation boundary condition is based on the idea that for acoustic problems asymptotic solutions, at the far field, can be derived. These can be expressed as expansions, typically written as powers of $1/r$ where r is the distance from the sound source to the boundary. Differential equations are derived that satisfy the expansion. These differential equations are imposed as boundary conditions to the Euler equation. The differential equations only have outwards pointing characteristics. Hence, they do not need information at the domain boundary and can be discretized by offset differences. As a basic example, the far field boundary condition of Tam (1998), for a two-dimensional acoustic problem, is given below.

$$\frac{1}{U(\theta)}\frac{\partial \boldsymbol{u}}{\partial \tau} + \frac{\partial \boldsymbol{u}}{\partial \rho} + \frac{\boldsymbol{u}}{\rho} = 0 \tag{2.83}$$

In the above:

$$U(\theta) = M\cos(\theta) + \sqrt{1 - M^2\cos^2\theta} \tag{2.84}$$

where M is the far field Mach number and θ' the angle of the node point relative to the mean flow. The above can be extended to three-dimensions.

With the Perfectly Matched Layer (PML) approach Hu (2001) a perfect mathematical matching is enforced between the core solution zone and the extended grid sponge zone. The extended sponge zone is intended to perfectly damp all acoustic waves at all angles. The approach allows use of a more compact sponge zone. The formulation of Hu (2001) is for the linearized Euler equations.

Aero engine intakes can have acoustic liners. These need to be modelled using impedance boundary conditions. If it is assumed the unsteadiness varies as $e^{-i\omega t}$ then the impedance, $Z = R - iX$, where R is the resistance and X the reactance (imaginary component of acoustic impendence). An impedance boundary condition is used by Muhlbauer et al. (2008), but this time as an outflow condition when exploring pulsed gain combustion engines. A brief overview of impedance boundary conditions is given by Tam (1998).

2.22 Impact of Grid Topology on Solution Accuracy

To further explore the performances of some of the schemes discussed examples are considered for acoustic, vortical and Tollmien-Schlichting (T-S) waves in Appendix A. Here, in relation to grid topology influence, some key points will be sum-

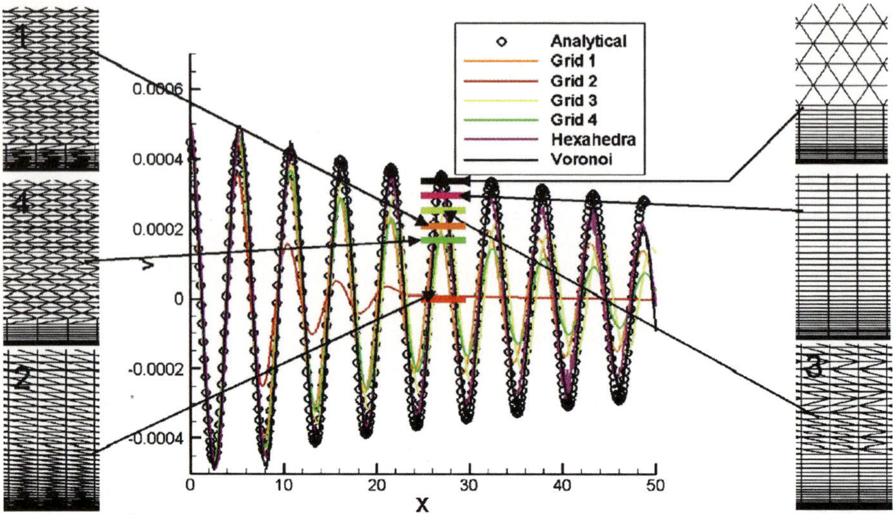

Fig. 2.23 Plot of T-S wave decay with streamwise distance for different grid topologies (from Tucker 2008). Published with kind permission from Elsevier

marized for T-S waves and also homogeneous decaying turbulence. Comparisons are made with an analytical solution for a subcritical T-S wave (Orszag 1971).

As noted above, the most popular options for dealing with complex aerodynamic geometries is either through use of overset curvilinear grids, unstructured grids or for highly complex problems a combination of the two. The former option is considered in Appendix A. It is shown that the overset interface can create a slight discontinuity in the mean flow field. Fortunately, this does not corrupt the T-S wave amplitude relative to the mean field. Unstructured grids allow more severe cell topology changes. These have greater potential solution impact and hence are dealt with here.

The unstructured grids shown in Fig. 2.23 are considered. Just grid snapshots for a lower channel half, that contains the T-S wave, are shown. Essentially, for these more triangulated grids, where possible, the nodal locations are the same as used for the hexahedral grid shown in the figure. However, due to the inclusion of triangulation the number of edges typically increases by around 30 %. Table 2.9 summarizes the number of edges for the different Fig. 2.23 grids.

The program considered here is an edge-based solver hence computational demands rise with the increasing number of edges. The grid referred to as Voronoi is shown in the top right hand corner of Fig. 2.23. As can be seen, this grid is in fact hybrid. The construction of a pure Voronoi grid, that can capture the sharp near wall velocity gradients, without radically altering the streamwise grid spacing, being impossible. A potential quasi-Voronoi grid, illustrating the aspect ratio problem, is shown in Fig. 2.24.

Table 2.9 Number of nodes and edges for the different unstructured grids

Grid	No. of nodes	No. of edges
Hex	34706	86113
1	35876	124284
2	34706	120169
3	37386	106717
4	35876	114044
Voronoi	24794	65381

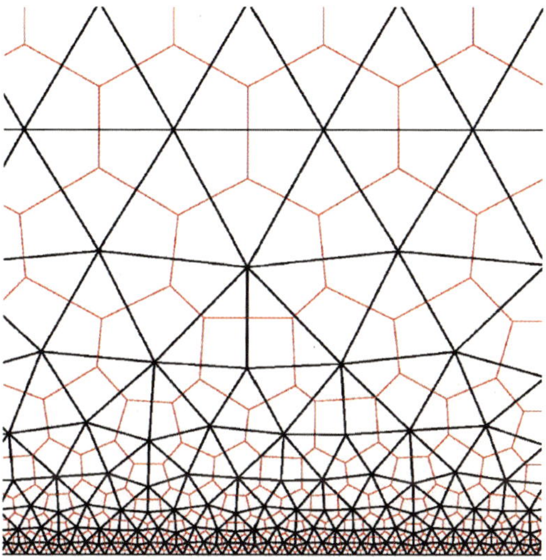

Fig. 2.24 Quasi-Voronoi grid refined in the near wall region (from Tucker 2008). Published with kind permission from Elsevier

The graph at the centre of Fig. 2.23 shows the decay of v' (perturbation in vertical velocity) for the T-S wave. The analytic solution to the Orr-Sommerfeld equation is represented by symbols. From Fig. 2.23 it is clear that Grid 2 (lower left hand corner) gives the worst accuracy.

As would be expected, in Fig. 2.23, the modest hexahedral grid shows reasonable agreement with the analytic solution. On the other hand the hybrid-Voronoi grid (even though the linking of the different grid topologies is far from smooth) does well. In fact, it has the highest accuracy. It is important to stress that this grid has around 40 % less nodes and over 30 % less edges than the hexahedral grid.

Figure 2.25 perhaps sheds some light on the computational performance issues. The hexahedral and Voronoi grids do well probably partly because the control volume face is orthogonal to the line connecting the nodes that straddle it. According to Moulinec et al. (2005), this property gives better energy conservation. Also, for the different grids there are different levels of mean flow alignment (and symmetry). Compared to grids 1 and 4, the Grid 2 triangulation appears to be least mean flow aligned. It exhibits the most control volume to node connecting line non-orthogonality. The randomized grid's nature ensures more faces exhibit the desir-

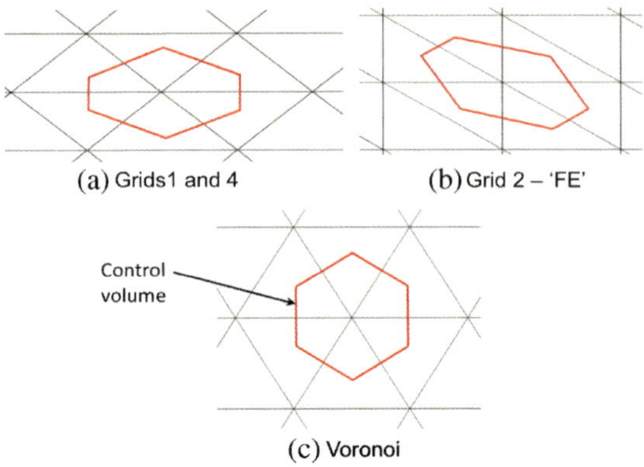

Fig. 2.25 Control volume shapes studied. Published with kind permission from Elsevier

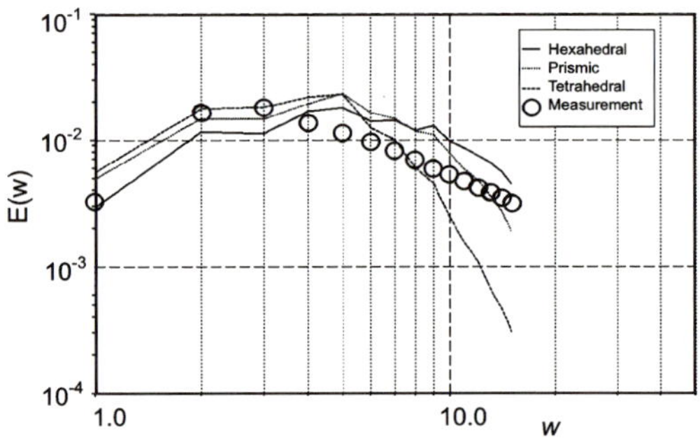

Fig. 2.26 Energy spectrum for homogeneous decaying turbulence computed on different grid topologies

able orthogonal condition i.e. at least one zone of the control volume will be akin to grids 1 & 4. Also, it should be stressed the randomized grids have a bigger accuracy enhancing hexahedral cell zone.

From this discussion it is clear that for LES/DNS the cell shape can potentially strongly influence results. There seems room for research into optimal grid topologies for LES. Based on the above findings, for DNS/LES type simulations it seems sensible to, where possible, use hexahedral cells having as isotopic nature as possible. As discussed by Broeckhoven et al. (2007) it can be critical to have smooth changes in grid spacing at internal boundaries. This is especially so for aeroacoustics, where a mismatch in the vorticity field, between an LES zone and acoustic propagation zone, can act as an artificial sound source. The hybrid Voronoi grid could well give rise to this issue. To avoid this problem, workers have explored carefully decaying the acoustic source term in the vicinity of this interface zone (Broeckhoven et al. 2007).

Figure 2.26 gives energy spectrums for LES type computations of the homogeneous decaying turbulence case explored earlier. The hexahedral grid locations are triangulated in differing ways. Again, the hexahedral grid does well. The prismatic which is like an extruded Voronoi also shows encouraging performance. However, the tetrahedral mesh, that lacks the orthogonality property shows strong numerical dissipation levels. It is worth noting that the kinetic energy scheme, discussed earlier remedies much of the grid sensitivity shown in Fig. 2.26.

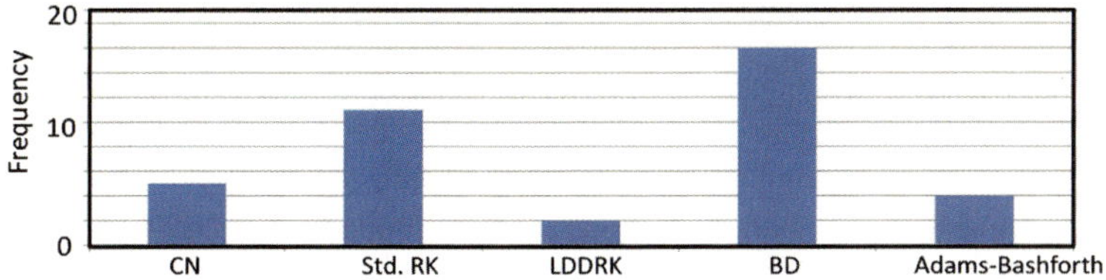

Fig. 2.27 Frequency of use of different temporal schemes

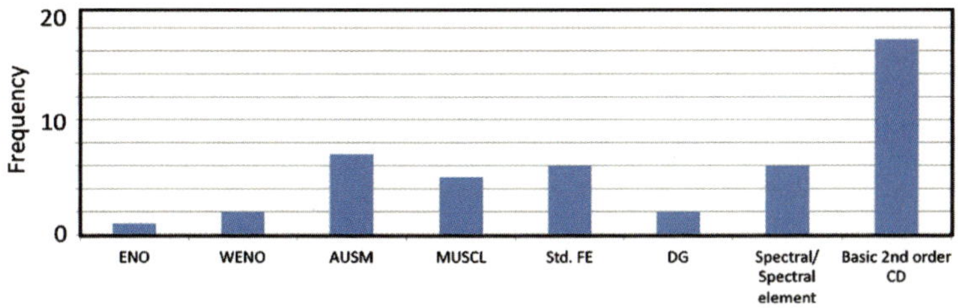

Fig. 2.28 Frequency of use of different spatial schemes

2.23 Frequency of Use of Different Numerical Approaches

The frequency of use, of the different numerical schemes, discussed above, are considered in this section. The statistics are based on the later chapters. In these, the application of such schemes to eddy resolving simulations is mostly overviewed. Figure 2.27 gives the frequency of use of different temporal schemes. The most popular appears to be the implicit backwards BD scheme. This is followed the RK and then the CN. Figure 2.28 gives the frequency of use of different spatial schemes. Like the temporal, it seems that the most popular scheme(s) are the more traditional. The basic 2nd order central difference (CD) scheme is by far the most popular. Although not tabulated, the later chapters shown the finite volume approach is considerably more popular than finite element method, as would be expected. Structured and unstructured grid methods have similar frequencies of use. Chimera grids have a low frequency of use (5 % of simulations). Also, the Roe scheme sees much more application than the Jameson—nearly an order of magnitude in frequency.

2.24 Conclusions

As can be seen from the brief overview of methods given in this chapter there are a wide range of schemes. Just small subsets of these find practical use. In the subsequent chapters further case customized (for different application areas) numerical approaches will emerge. Grid quality can have a strong solution impact as does the numerical scheme. Hence, particularly when making LES computations, attention should be paid to the numerical scheme and grid quality and how these two elements interact. The choice of solution frame of reference can also have a profound accuracy impact. Assessing when to use high order schemes is complex and again is linked to mesh quality. Increases in computational power allow the use of finer grids and time steps. Hence this helps alleviate accuracy deficiencies in schemes. However, schemes need sufficient order for grid resolution increases to make a worthwhile accuracy impact. Effective non-reflecting boundary conditions help remove the need for excessive numbers of grid nodes.

References

A. Agarwal, P.J. Morris, Direct simulation of acoustic scattering on a rotorcraft fuselage, in *Proceedings of 6th AIAA/CEAS Aeroacoustics Conference*, Lahaina, Hawaii, 12–14 June 2000. AIAA 2000–2030

Y. Allaneau, A. Jameson, Direct numerical simulations of a two-dimensional viscous flow in a shocktube using a kinetic energy preserving scheme, in *Proceedings of 19th AIAA Computational Fluid Dynamics*, San Antonio, Texas, 22–25 June 2010a. AIAA 2009–3797

Y. Allaneau, A. Jameson, Direct numerical simulations of plunging airfoils, in *Proceedings of 48th AIAA Aerospace Sciences Meeting Including the New Horizons Forum and Aerospace Exposition*, Orlando, Florida, 4–7 January 2010b. AIAA 2010–728

A. Arakawa, Computational design for long-term numerical integration of the equations of fluid motion: two-dimensional incompressible flow. Part I. J. Comput. Phys. **1**(1), 119–143 (1966)

U.M. Ascher, L.R. Petzold, *Computer Methods for Ordinary Differential Equations and Differential-Algebraic Equations*, vol. 61 (SIAM, Philadelphia, 1998)

G. Ashcroft, X. Zhang, A computational investigation of the noise radiated by flow-induced cavity oscillations. AIAA Pap. **512** (2001)

H.L. Atkins, D.P. Lockard, A high-order method using unstructured grids for the aeroacoustic analysis of realistic aircraft configurations. AIAA Pap. 99–1945 (1999)

I.E. Barton, R. Kirby, Finite difference scheme for the solution of fluid flow problems on non-staggered grids. Int. J. Numer. Methods Fluids **33**(7), 939–959 (2000)

R.M. Beam, R.F. Warming, An implicit finite-difference algorithm for hyperbolic systems in conservation-law form. J. Comput. Phys. **22**(1), 87–110 (1976)

R.M. Beam, R.F. Warming, Implicit numerical methods for the compressible Navier-Stokes and Euler equations, in *Computational Fluid Dynamics*. Lecture Series, vol. 1, Von Karman Institute for Fluid Dynamics, Belgium, 29 March–2 April 1982

B.C. Bell, K.S. Surana, A space–time coupled p-version least-squares finite element formulation for unsteady fluid dynamics problems. Int. J. Numer. Methods Eng. **37**(20), 3545–3569 (1994)

A. Birkefeld, C.D. Munz, Simulations of airfoil noise with the discontinuous Galerkin solver NoisSol. ERCOFTAC Bull. **90**, 28–33 (2012)

N.J. Bisek, D.P. Rizzetta, J. Poggie, Plasma control of a turbulent shock boundary-layer interaction. AIAA J. (2013). doi:10.2514/1.J052248

G.A. Blaisdell, E.T. Spyropoulos, J.H. Qin, The effect of the formulation of nonlinear terms on aliasing errors in spectral methods. Appl. Numer. Math. **21**(3), 207–219 (1996)

C. Bogey, C. Bailly, A family of low dispersive and low dissipative explicit schemes for flow and noise computations. J. Comput. Phys. **194**(1), 194–214 (2004)

D.L. Book, J.P. Boris, K. Hain, Flux-corrected transport II: generalizations of the method. J. Comput. Phys. **18**(3), 248–283 (1975)

J.P. Boris, D.L. Book, Flux-corrected transport. I. SHASTA, a fluid transport algorithm that works. J. Comput. Phys. **11**(1), 38–69 (1973)

A. Brandt, Multilevel adaptive computations in fluid dynamics. AIAA J. **18**(10), 1165–1172 (1980)

K.E. Brenan, S.L.V. Campbell, L.R. Petzold, *Numerical Solution of Initial-Value Problems in Differential-Algebraic Equations* (SIAM, Philadelphia, 1996)

W. Briley, H. McDonald, Solution of the three-dimensional compressible Navier-Stokes equations by an implicit technique, in *Proceedings of the 4th International Conference on Numerical Methods in Fluid Dynamics*. Lecture Notes in Physics (Springer, Berlin, 1975), pp. 105–110

T. Broeckhoven, J. Ramboer, S. Smirnov, C. Lacor, Numerical methods, in *Large-Eddy Simulation for Acoustics*, ed. by C. Wagner, T. Huttl, P. Sagaut (Cambridge University Press, Cambridge, 2007)

J. Burmeister, G. Horton, Time-parallel multigrid solution of the Navier-Stokes equations. Multigrid Methods III **98**, 155–166 (1991)

T.D. Butler, LINC method extensions, in *Proceedings of the Second International Conference on Numerical Methods in Fluid Dynamics* (Springer, Berlin, 1971), pp. 435–440

M.S. Campobasso, M.B. Giles, Effects of flow instabilities on the linear analysis of turbomachinery aeroelasticity. J. Propuls. Power **19**(2), 250–259 (2003)

M.S. Campobasso, M.B. Giles, Stabilization of linear flow solver for turbomachinery aeroelasticity using recursive projection method. AIAA J. **42**(9), 1765–1774 (2004)

M. Chapman, FRAM—nonlinear damping algorithms for the continuity equation. J. Comput. Phys. **44**(1), 84–103 (1981)

G. Chesshire, W.D. Henshaw, Composite overlapping meshes for the solution of partial differential equations. J. Comput. Phys. **90**(1), 1–64 (1990)

Y. Choi, C. Merkle, Time-derivative preconditioning for viscous flows, in *22nd AIAA, Fluid Dynamics, Plasma Dynamics and Lasers Conference* (1991). AIAA–91–1652

A.J. Chorin, A numerical method for solving incompressible viscous flow problems. J. Comput. Phys. **2**(1), 12–26 (1967)

F.K. Chow, P. Moin, A further study of numerical errors in large-eddy simulations. J. Comput. Phys. **184**(2), 366–380 (2003)

Y.M. Chung, P.G. Tucker, Accuracy of higher-order finite difference schemes on nonuniform grids. AIAA J. **41**(8), 1609–1611 (2003)

T. Colonius, S.K. Lele, Computational aeroacoustics: progress on nonlinear problems of sound generation. Prog. Aerosp. Sci. **40**(6), 345–416 (2004)

W.P. Crowley, Second-order numerical advection. J. Comput. Phys. **1**(4), 471–484 (1967)

F. Daude, J. Berland, T. Emmert, P. Lafon, F. Crouzet, C. Bailly, A high-order finite-difference algorithm for direct computation of aerodynamic sound. Comput. Fluids **61**, 46–63 (2012)

R.W. Davis, E.F. Moore, A numerical study of vortex shedding from rectangles. J. Fluid Mech. **116**(3), 475–506 (1982)

C.C. de Wiart, K. Hillewaert, P. Geuzaine, DNS of a low pressure turbine blade computed with the discontinuous Galerkin method, in *Proceedings of the ASME Turbo Expo*, Copenhagen, Denmark, 11–15 June 2012. GT2012–68900

J.W. Deardorff, A numerical study of three-dimensional turbulent channel flow at large Reynolds numbers. J. Fluid Mech. **41**(02), 453–480 (1970)

I. Demirdžić, M. Perić, Space conservation law in finite volume calculations of fluid flow. Int. J. Numer. Methods Fluids **8**(9), 1037–1050 (1988)

J. Douglas, J.E. Gunn, A general formulation of alternating direction methods. Numer. Math. **6**(1), 428–453 (1964)

D. Drikakis, W. Rider, *High-Resolution Methods for Incompressible and Low-Speed Flows* (Springer, Berlin, 2004)

F. Ducros, V. Ferrand, F. Nicoud, C. Weber, D. Darracq, C. Gacherieu, T. Poinsot, Large-eddy simulation of the shock/turbulence interaction. J. Comput. Phys. **152**(2), 517–549 (1999)

F. Ducros, F. Laporte, T. Souleres, V. Guinot, P. Moinat, B. Caruelle, High-order fluxes for conservative skew-symmetric-like schemes in structured meshes: application to compressible flows. J. Comput. Phys. **161**(1), 114–139 (2000)

J.K. Dukowicz, J.D. Ramshaw, Tensor viscosity method for convection in numerical fluid dynamics. J. Comput. Phys. **32**(1), 71–79 (1979)

M.S. Engelman, R.L. Sani, Finite element simulation of incompressible fluid flows with a free/moving surface. Comput. Tech. Fluid. Flow **47**, 47–74 (1986)

C.A.J. Fletcher, *Fundamental and General Techniques*. Computational Techniques for Fluid Dynamics, vol. 1 (Springer, Berlin, 1997)
G. Fritsch, M. Giles, Second-order effects of unsteadiness on the performance of turbomachines, in *37th International Gas Turbine and Aeroengine Congress and Exposition* (1992). GT–92–389
L. Gamet, F. Ducros, F. Nicoud, T. Poinsot, Compact finite difference schemes on non-uniform meshes. application to direct numerical simulations of compressible flows. Int. J. Numer. Methods Fluids **29**(2), 159–191 (1999)
B.J. Geurts, D.D. Holm, Regularization modeling for large-eddy simulation. Phys. Fluids **15**(1), L13–L16 (2003)
S. Ghosal, An analysis of numerical errors in large-eddy simulations of turbulence. J. Comput. Phys. **125**(1), 187–206 (1996)
M.B. Giles, Nonreflecting boundary conditions for Euler equation calculations. AIAA J. **28**(12), 2050–2058 (1990)
M. Giles, The HYDRA user's guide. Version 0.06 (2004)
W. Glanfield, Personal communication (2000)
J. Glass, W. Rodi, A higher order numerical scheme for scalar transport. Comput. Methods Appl. Mech. Eng. **31**(3), 337–358 (1982)
P.M. Gresho, R.L. Lee, R.L. Sani, On the time-dependent solution of the incompressible Navier-Stokes equations in two and three dimensions. Recent Adv. Numer. Methods Fluids **1**, 27–79 (1980)
P.M. Gresho, S.T. Chan, R.L. Lee, C.D. Upson, A modified finite element method for solving the time-dependent, incompressible Navier-Stokes equations. Part 1: theory. Int. J. Numer. Methods Fluids **4**(6), 557–598 (1984)
A. Hadjadj, Large eddy simulation of shock/boundary layer interaction. AIAA J. **50**(12), 2919–2927 (2012)
T. Haga, H. Gao, Z.J. Wang, A high-order unifying discontinuous formulation for the Navier-Stokes equations on 3d mixed grids. Math. Model. Nat. Phenom. **6**(3), 28–56 (2011)
F.H. Harlow, J.E. Welch, Numerical calculation of time-dependent viscous incompressible flow of fluid with free surface. Phys. Fluids **8**, 2182–2189 (1965)
A. Harten, High resolution schemes for hyperbolic conservation laws. J. Comput. Phys. **49**(3), 357–393 (1983)
A. Harten, B. Engquist, S. Osher, S.R. Chakravarthy, Uniformly high order accurate essentially non-oscillatory schemes, III. J. Comput. Phys. **71**(2), 231–303 (1987)
L. He, D.X. Wang, Concurrent blade aerodynamic-aero-elastic design optimization using adjoint method. J. Turbomach. **133**, 011021 (2011)
L.S. Hedges, A.K. Travin, P.R. Spalart, Detached-eddy simulations over a simplified landing gear. J. Fluids Eng. **124**(2), 413–423 (2002)
R.A.W.M. Henkes, Natural-convection boundary layers. PhD thesis, Technical University Delft (1990)
B.P. Hignett, A.A. White, R.D. Carter, W.D.N. Jackson, R.M. Small, A comparison of laboratory measurements and numerical simulations of baroclinic wave flows in a rotating cylindrical annulus. Q. J. R. Meteorol. Soc. **111**(467), 131–154 (1985)
C.W. Hirt, A.A. Amsden, J.L. Cook, An arbitrary Lagrangian-Eulerian computing method for all flow speeds. J. Comput. Phys. **14**(3), 227–253 (1974)
R. Hixon, Prefactored small-stencil compact schemes. J. Comput. Phys. **165**(2), 522–541 (2000)
K. Horiuti, T. Itami, Truncation error analysis of the rotational form for the convective terms in the Navier–Stokes equation. J. Comput. Phys. **145**(2), 671–692 (1998)
F.Q. Hu, A stable, perfectly matched layer for linearized Euler equations in unsplit physical variables. J. Comput. Phys. **173**(2), 455–480 (2001)
F.Q. Hu, M.Y. Hussaini, J.L. Manthey, Low-dissipation and low-dispersion Runge–Kutta schemes for computational acoustics. J. Comput. Phys. **124**(1), 177–191 (1996)
A. Hujeirat, R. Rannacher, A method for computing compressible, highly stratified flows in astrophysics based on operator splitting. Int. J. Numer. Methods Fluids **28**(1), 1–22 (1998)

H.G. Im, Numerical studies of transient opposed-flow flames using adaptive time integration. KSME Int. J. **14**(1), 103–112 (2000)

A. Iserles, Generalized leapfrog methods. IMA J. Numer. Anal. **6**(4), 381–392 (1986)

R.I. Issa, Solution of the implicitly discretised fluid flow equations by operator-splitting. J. Comput. Phys. **62**(1), 40–65 (1986)

A. Jameson, Time dependent calculations using multigrid, with applications to unsteady flows past airfoils and wings. AIAA Pap. **1596**, 1991 (1991)

A. Jameson, The construction of discretely conservative finite volume schemes that also globally conserve energy or entropy. J. Sci. Comput. **34**(2), 152–187 (2008a)

A. Jameson, Formulation of kinetic energy preserving conservative schemes for gas dynamics and direct numerical simulation of one-dimensional viscous compressible flow in a shock tube using entropy and kinetic energy preserving schemes. J. Sci. Comput. **34**(2), 188–208 (2008b)

A. Jameson, W. Schmidt, E. Turkel, Numerical solutions of the Euler equations by finite volume methods using Runge-Kutta time-stepping schemes. AIAA Pap. **81**(125), 9 (1981)

R.J. Jefferson-Loveday, Numerical simulations of unsteady impinging jet flows. PhD thesis, Swansea University (2008)

W.P. Jones, A.J. Marquis, Calculation of axisymmetric re-circulating flows with a second order turbulence model, in *Proc. of the 5th Symp. on Turbulent Shear Flows*, Cornel University (1985), pp. 20.1–20.11

J. Joo, P. Durbin, Simulation of turbine blade trailing edge cooling. J. Fluids Eng. **131**, 021102 (2009)

S.A. Karabasov, V.M. Goloviznin, New efficient high-resolution method for nonlinear problems in aeroacoustics. AIAA J. **45**(12), 2861–2871 (2007)

S.A. Karabasov, V.M. Goloviznin, Compact accurately boundary-adjusting high-resolution technique for fluid dynamics. J. Comput. Phys. **228**(19), 7426–7451 (2009)

Z. Khatir, Discrete vortex modelling of near-wall flow structure in turbulent boundary layers. PhD thesis, Fluid Dynamics Research Centre, The University of Warwick (2000)

J.W. Kim, D.J. Lee, Optimized compact finite difference schemes with maximum resolution. AIAA J. **34**(5), 887–893 (1996)

J. Kim, P. Moin, Application of a fractional-step method to incompressible Navier-Stokes equations. J. Comput. Phys. **59**(2), 308–323 (1985)

M.H. Kobayashi, J.C.F. Pereira, Culation of incompressible laminar flows on a nonstaggered, nonorthogonal grid. Numer. Heat Transf., Part B, Fundam. **19**(2), 243–262 (1991)

J.A. Krakos, D.L. Darmofal, Effect of small-scale output unsteadiness on adjoint-based sensitivity. AIAA J. **48**(11), 2611–2623 (2010)

C. Lacor, *Industrial Computational Fluid Dynamics*. Lecture Series 1999-06 (von Karman Institute for Fluid Dynamics, Sint-Genesius-Rode, 1999)

S. Laizet, E. Lamballais, High-order compact schemes for incompressible flows: a simple and efficient method with quasi-spectral accuracy. J. Comput. Phys. **228**(16), 5989–6015 (2009)

K.R. Lee, J.H. Park, K.H. Kim, High-order interpolation method for overset grid based on finite volume method. AIAA J. **49**(7), 1387–1398 (2011)

C.E. Leith, Numerical simulation of the earth's atmosphere. University of California, Lawrence Radiation Laboratory (1964)

S.K. Lele, Compact finite difference schemes with spectral-like resolution. J. Comput. Phys. **103**(1), 16–42 (1992)

B.P. Leonard, A stable and accurate convective modelling procedure based on quadratic upstream interpolation. Comput. Methods Appl. Mech. Eng. **19**(1), 59–98 (1979)

M.S. Liou, C.J. Steffen, A new flux splitting scheme. J. Comput. Phys. **107**(1), 23–39 (1993)

X.D. Liu, S. Osher, T. Chan, Weighted essentially non-oscillatory schemes. J. Comput. Phys. **115**(1), 200–212 (1994)

Y. Liu, M. Vinokur, Z.J. Wang, Spectral (finite) volume method for conservation laws on unstructured grids v: extension to three-dimensional systems. J. Comput. Phys. **212**(2), 454–472 (2006)

D.P. Lockard, K.S. Brentner, H.L. Atkins, High-accuracy algorithms for computational aeroacoustics. AIAA J. **33**, 246–251 (1995)

S. Loiodice, P.G. Tucker, J. Watson, Coupled open rotor engine intake simulations, in *Proceedings of the 48th AIAA Aerospace Sciences Meeting and Exhibit*, Orlando, Florida, 4–7 January 2010. AIAA 2010–840

Y. Lu, W.N. Dawes, X. Yuan, Investigation of 3D internal flow using new flux-reconstruction higher order method, in *Proceedings of the ASME Turbo Expo*, Copenhagen, Denmark, 11–15 June 2012. GT2012–69270

K. Mani, D.J. Mavriplis, Discrete adjoint based time-step adaptation and error reduction in unsteady flow problems, in *18th AIAA Computational Fluid Dynamics Conference*, Miami, FL, 25–28 June 2007. AIAA 2007–3944

K. Mani, D.J. Mavriplis, Spatially non-uniform time-step adaptation for functional outputs in unsteady flow problems, in *48th AIAA Aerospace Sciences Meeting Including the New Horizons Forum and Aerospace Exposition*, Orlando, Florida, 4–7 January 2010. AIAA 2010–121

E. Manoha, B. Troff, P. Sagaut, Trailing-edge noise prediction using large-eddy simulation and acoustic analogy. AIAA J. **38**(4), 575–583 (2000)

J.C. Marongiu, F. Leboeuf, E. Parkinson, Numerical simulation of the flow in a Pelton turbine using the meshless method smoothed particle hydrodynamics: a new simple solid boundary treatment. Proc. Inst. Mech. Eng. A, J. Power Energy **221**(6), 849–856 (2007)

J.C. Marongiu, F. Leboeuf, J.Ë. Caro, E. Parkinson, Free surface flows simulations in Pelton turbines using an hybrid SPH-ALE method. J. Hydraul. Res. **48**(S1), 40–49 (2010)

I. Mary, P. Sagaut, Large eddy simulation of flow around an airfoil near stall. AIAA J. **40**(6), 1139–1145 (2002)

P.J. Morris, L.N. Long, A. Bangalore, Q. Wang, A parallel three-dimensional computational aeroacoustics method using nonlinear disturbance equations. J. Comput. Phys. **133**(1), 56–74 (1997)

C. Moulinec, S. Benhamadouche, D. Laurence, M. Peric, LES in a U-bend pipe meshed by polyhedral cells. Eng. Turbul. Model. Exp. **6**, 237–246 (2005)

B. Muhlbauer, B. Noll, M. Aigner, Numerical investigation of entropy noise and its acoustic sources in aero-engine, in *Proceedings of the ASME Turbo Expo*, Berlin, Germany, 9–13 June 2008. GT2008–50321

T. Muramatsu, H. Ninokata, Thermal striping temperature fluctuation analysis using the algebraic stress turbulence model in water and sodium. JSME Int. J., Ser. 2, Fluids Eng. Heat Transf. Power Combust. Thermophys. Prop. **35**(4), 486–496 (1992)

K. Nakahashi, F. Togashi, Unstructured overset grid method for flow simulation of complex multiple body problems, in *ICAS 2000 Congress*, (2000). ICAS 0263

A.G.F.A. Nasser, M.A. Leschziner, Computation of transient recirculating flow using spline approximations and time-space characteristics, in *Proc. 4th Int. Conf. on Numerical Methods in Laminar and Turbulent Flow*, Swansea (1985), pp. 480–491

P.D. Orkwis, M.G. Turner, J.W. Barter, Linear deterministic source terms for hot streak simulations. J. Propuls. Power **18**(2), 383–389 (2002)

S.A. Orszag, Accurate solution of the Orr-Sommerfeld stability equation. J. Fluid Mech. **50**(4), 689–703 (1971)

Y. Ozyoruk, L.N. Long, Multigrid acceleration of a high-resolution computational aeroacoustics scheme. AIAA J. **35**(3), 428–433 (1997)

U. Paliath, H. Shen, R. Avancha, C. Shieh, Large eddy simulation for jets from chevron and dual nozzles, in *Proceedings of 17th AIAA/CEAS Aeroacoustics Conference*, Portland, Oregon, 5–8 June 2011. AIAA 2011–2881

A.T. Patera, A spectral element method for fluid dynamics: laminar flow in a channel expansion. J. Comput. Phys. **54**(3), 468–488 (1984)

L.L. Pauley, P. Moin, W.C. Reynolds, The structure of two-dimensional separation. J. Fluid Mech. **220**(1), 397–411 (1990)

D.W. Peaceman, H.H. Rachford, The numerical solution of parabolic and elliptic differential equations. J. Soc. Ind. Appl. Math. **3**(1), 28–41 (1955)

A. Pinelli, I.Z. Naqavi, U. Piomelli, J. Favier, Immersed-boundary methods for general finite-difference and finite-volume Navier–Stokes solvers. J. Comput. Phys. **229**(24), 9073–9091 (2010)

H. Pitsch, Large-eddy simulation of turbulent combustion. Annu. Rev. Fluid Mech. **38**, 453–482 (2006)

W.H. Press, B.P. Flannery, S.A. Teukolsky, W.T. Vetterling, *Numerical Recipes: The Art of Scientific Computing (FORTRAN)* (Cambridge University Press, Cambridge, 1989)

D. Rayner, A numerical study into the heat transfer beneath the stator blade of an axial compressor. PhD thesis, School of Engineering, University of Sussex (1993)

D.T. Reindl, W.A. Beckman, J.W. Mitchell, C.J. Rutland, Benchmarking transient natural convection in an enclosure, in *National Heat Transfer Conference*, vol. 91 (1991), pp. 1–7

C.M. Rhie, W.L. Chow, Numerical study of the turbulent flow past an airfoil with trailing edge separation. AIAA J. **21**(11), 1525–1532 (1983)

D.P. Rizzetta, M.R. Visbal, P.E. Morgan, A high-order compact finite-difference scheme for large-eddy simulation of active flow control. Prog. Aerosp. Sci. **44**(6), 397–426 (2008)

P.J. Roache, A flux-based modified method of characteristics. Int. J. Numer. Methods Fluids **15**(11), 1259–1275 (1992)

P.L. Roe, Approximate Riemann solvers, parameter vectors, and difference schemes. J. Comput. Phys. **43**(2), 357–372 (1981)

P.L. Roe, Characteristic-based schemes for the Euler equations. Annu. Rev. Fluid Mech. **18**(1), 337–365 (1986)

S.E. Rogers, D. Kwak, Upwind differencing scheme for the time-accurate incompressible Navier-Stokes equations. AIAA J. **28**, 253–262 (1990)

S.E. Rogers, D. Kwak, J.L.C. Chang, Numerical solution of the incompressible Navier-Stokes equations in three-dimensional generalized curvilinear coordinates. NASA STI/Recon Tech. Rep. N **87**, 11964 (1986)

M.P. Rumpfkeil, D.W. Zingg, A general framework for the optimal control of unsteady flows with applications, in *Proceedings of the 45th AIAA Aerospace Meeting and Exhibit*. Reno, Nevada, 8–11 January 2007. AIAA 2010–1128

N.D. Sandham, Q. Li, H.C. Yee, Entropy splitting for high-order numerical simulation of compressible turbulence. J. Comput. Phys. **178**(2), 307–322 (2002)

L.J. Segerlind, *Applied Finite Element Analysis* (Wiley, New York, 1976)

V. Seidl, M. Peric, M. Schmidt, Space-and time-parallel Navier-Stokes solver for 3d block-adaptive Cartesian grids, in *Proceedings of the Parallel Computational Fluid Dynamics*, vol. 95 (1995), pp. 557–584

J.A. Sethian, Fast marching methods. SIAM Rev. **41**(2), 199–235 (1999)

M.L. Shur, P.R. Spalart, M.K. Strelets, A.K. Travin, Towards the prediction of noise from jet engines. Int. J. Heat Fluid Flow **24**(4), 551–561 (2003)

S. Skelboe, The control of order and steplength for backward differentiation methods. BIT Numer. Math. **17**(1), 91–107 (1977)

P. Spalart, L. Hedges, M. Shur, A. Travin, Simulation of active flow control on a stalled airfoil. Flow Turbul. Combust. **71**(1), 361–373 (2003)

E.T. Spyropoulos, G.A. Blaisdell, Large-eddy simulation of a spatially evolving supersonic turbulent boundary-layer flow. AIAA J. **36**(11), 1983–1990 (1998)

D. Stanescu, M.Y. Hussaini, F. Farassat, Aircraft engine noise scattering—a discontinuous spectral element approach. AIAA Pap. **800**, 2002 (2002)

A. Staniforth, J. Côté, Semi-Lagrangian integration schemes for atmospheric models—a review. Mon. Weather Rev. **119**(9), 2206–2223 (1991)

J.L. Steger, P. Kutler, Implicit finite-difference procedures for the computation of vortex wakes. AIAA J. **15**(4), 581–590 (1977)

J.L. Steger, R.F. Warming, Flux vector splitting of the inviscid gasdynamic equations with application to finite-difference methods. J. Comput. Phys. **40**(2), 263–293 (1981)

J. Steinhoff, D. Underhill, Modification of the Euler equations for "vorticity confinement": application to the computation of interacting vortex rings. Phys. Fluids **6**, 2738–2744 (1994)

M. Strelets, Detached eddy simulation of massively separated flows, in *Proceedings of 39th AIAA Aerospace Sciences Meeting and Exhibit*, Reno, NV, 8–11 January 2001. AIAA 2001–0879

R.C. Swanson, E. Turkel, On central-difference and upwind schemes. J. Comput. Phys. **101**(2), 292–306 (1992)
T. Talha, A numerical investigation of three-dimensional unsteady turbulent channel flow subjected to temporal acceleration. PhD thesis, School of Engineering, University of Warwick (2012)
C.K.W. Tam, Advances in numerical boundary conditions for computational aeroacoustics. J. Comput. Acoust. **6**(4), 377–402 (1998)
C.K.W. Tam, H. Shen, Direct computation of nonlinear acoustic pulses using high order finite difference schemes. AIAA Pap. **4325** (1993)
C.K.W. Tam, J.C. Webb, Dispersion-relation-preserving finite difference schemes for computational acoustics. J. Comput. Phys. **107**(2), 262–281 (1993)
C.K.W. Tam, C. Jay, D. Zhong, A study of the short wave components in computational acoustics. J. Comput. Acoust. **1**(01), 1–30 (1993)
L. Tang, J.D. Baeder, Uniformly accurate finite difference schemes for p-refinement. SIAM J. Sci. Comput. **20**(3), 1115–1131 (1998)
P.D. Thomas, C.K. Lombard, Geometric conservation law and its application to flow computations on moving grids. AIAA J. **17**, 1030–1037 (1979)
C. Tu, M. Deville, L. Dheur, L. Vanderschuren, Finite element simulation of pulsatile flow through arterial stenosis. J. Biomech. **25**(10), 1141–1152 (1992)
P.G. Tucker, Numerical precision and dissipation errors in rotating flows. Int. J. Numer. Methods Heat Fluid Flow **7**(7), 647–658 (1997)
P.G. Tucker, *Computation of Unsteady Internal Flows: Fundamental Methods with Case Studies* (Kluwer Academic, Dordrecht, 2001)
P.G. Tucker, Novel multigrid orientated solution adaptive time-step approaches. Int. J. Numer. Methods Fluids **40**(3–4), 507–519 (2002a)
P.G. Tucker, Temporal behavior of flow in rotating cavities. Numer. Heat Transf., Part A, Appl. **41**(6–7), 611–627 (2002b)
P.G. Tucker, Novel MILES computations for jet flows and noise. Int. J. Heat Fluid Flow **25**(4), 625–635 (2004)
P.G. Tucker, Turbulence modelling of problem aerospace flows. Int. J. Numer. Methods Fluids **51**(3), 261–283 (2006)
P.G. Tucker, The LES model's role in jet noise. Prog. Aerosp. Sci. **44**(6), 427–436 (2008)
A. Uzun, M.Y. Hussaini, Simulation of noise generation in near-nozzle region of a chevron nozzle jet. AIAA J. **47**(8), 1793–1810 (2009)
G.D. Van Albada, B. Van Leer, W.W. Roberts Jr., A comparative study of computational methods in cosmic gas dynamics. Astron. Astrophys. **108**, 76–84 (1982)
B. Van Leer, Towards the ultimate conservative difference scheme III. Upstream-centered finite-difference schemes for ideal compressible flow. J. Comput. Phys. **23**(3), 263–275 (1977)
B. Van Leer, Towards the ultimate conservative difference scheme. V. A second-order sequel to Godunov's method. J. Comput. Phys. **32**(1), 101–136 (1979)
B.C. Vermeire, J.S. Cagnone, S. Nadarajah, Iles using the correction procedure via reconstruction scheme, in *Proceedings of 51st AIAA Aerospace Sciences Meeting Including the New Horizons Forum and Aerospace Exposition*, Grapevine, Dallas/Ft. Worth Region, Texas, 7–10 January 2013. AIAA 2013-1001
M.R. Visbal, D.V. Gaitonde, On the use of higher-order finite-difference schemes on curvilinear and deforming meshes. J. Comput. Phys. **181**(1), 155–185 (2002)
S.G. Wallis, J.R. Manson, Accurate numerical simulation of advection using large time steps. Int. J. Numer. Methods Fluids **24**(2), 127–139 (1997)
Z.J. Wang, Y. Liu, G. May, A. Jameson, Spectral difference method for unstructured grids ii: extension to the Euler equations. J. Sci. Comput. **32**(1), 45–71 (2007)
J.M. Weiss, W.A. Smith, Preconditioning applied to variable and constant density flows. AIAA J. **33**(11), 2050–2057 (1995)
G. Yang, D.M. Causon, D.M. Ingram, R. Saunders, P. Batten, A Cartesian cut cell method for compressible flows. Part B: moving body problems. Aeronaut. J. **101**(1002), 57–65 (1997)

Chapter 3
Turbulence and Its Modelling

3.1 Introduction

Turbulence is an unsteady three-dimensional phenomenon. Its nature is such that large eddies are generated by the mean flow. Generally, the energy from these is transferred by successively smaller eddies (note, reverse energy cascades, i.e. transfer from smaller to larger scales, can also occur in complex turbulent flows). This is called the turbulence energy cascade. The larger eddies (integral length scales) are a strong function of the local geometry and flow system, but the smaller eddies (near the Kolmogorov scale, η) are more universal in nature. The time scales relating to the integral length scales are approximately equal to the ratio of the flow system characteristic size, L, to a characteristic velocity scale U_o (i.e. $t = L/U_o$). The Kolmogorov time scale can be evaluated as $\sqrt{\mu/\rho\varepsilon}$, where ε is the viscous dissipation. Most engineering turbulence models, based on solving the RANS (Reynolds Averaged Navier-Stokes) equations, try to model both the effects of the unsteady larger and smaller turbulence scales. Since the dynamics of the larger scales are not universal, their success is highly case dependent. This lack of universality of such models has resulted in the availability of a wide range of turbulence models and modelling strategies. No attempt is made here to discuss the wide range of potential RANS models available. A brief overview of the their nature and a range of palliative corrections, for RANS models, can be found in Appendix B.

Most modelling approaches use some form of averaging. This can either be temporal, spatial, or even a mix of these two. Hence, averaging procedures are discussed below. The averaging procedures, will be applied to the Navier-Stokes equations. To aid simplicity and compactness these are stated for an incompressible flow below in a tensor form

$$\frac{\partial u_i}{\partial t} + \frac{\partial u_i u_j}{\partial x_j} = -\frac{1}{\rho}\frac{\partial p}{\partial x_i} + \frac{\mu}{\rho}\frac{\partial}{\partial x_j}\left(\frac{\partial u_i}{\partial x_j} + \frac{\partial u_j}{\partial x_i}\right) \tag{3.1}$$

where, u is velocity, p, static pressure and x the spatial co-ordinates.

P.G. Tucker, *Unsteady Computational Fluid Dynamics in Aeronautics*,
Fluid Mechanics and Its Applications 104, DOI 10.1007/978-94-007-7049-2_3,

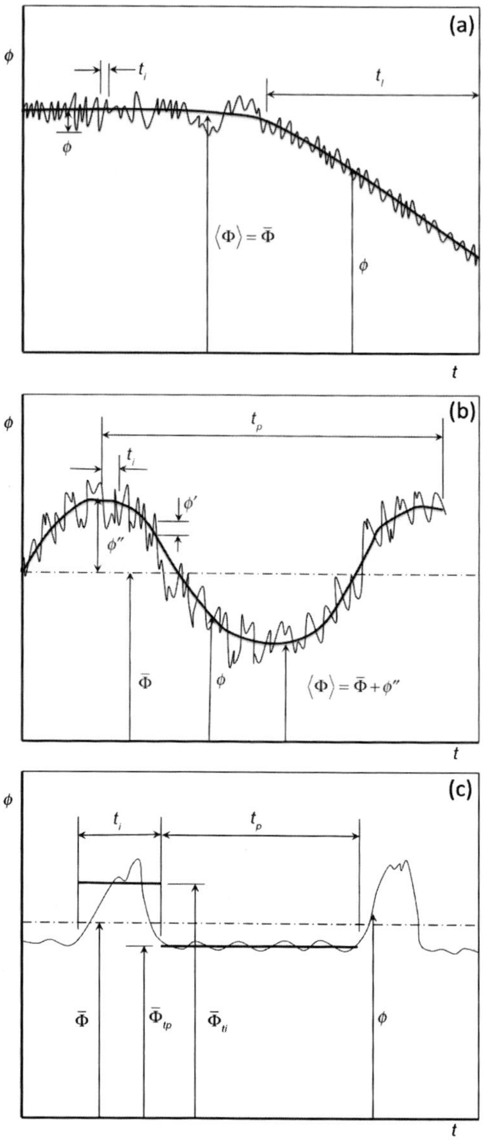

Fig. 3.1 Characteristic traces for the change in a general variable ϕ with time: (**a**) transient flow; (**b**) cyclic and (**c**) intermittent

3.2 Averaging Procedures

3.2.1 Time Based Averaging

For temporally focused averaging, it is convenient to classify flows as either transient, periodic or intermittent. Schematics of characteristic traces for the change in a general variable ϕ, with time, for these classifications are shown in Fig. 3.1. Frame (a) gives a transient flow trace, where t_i and t_t, represent turbulence and transient characteristic time scales, respectively. For this trace, the instantaneous velocity ϕ can be expressed as $\bar{\phi}+\phi'$ where $\bar{\phi}$ is an ensemble average $\bar{\phi}(t_n)=(\sum_1^N \phi(t_n))/N$ (where t_n is a fixed time for N tests and $N\rightarrow\infty$) and ϕ' the turbulence fluctuation. In forming the ensemble average it is necessary that $t_t \gg t_i$. More formally we can write

$$\frac{(\int_{-t_l/2}^{t_l/2}\phi dt-\int_{t_l/2}^{3t_l/2}\phi dt)}{\int_0^{t_1}\phi dt}\ll 1 \tag{3.2}$$

where t_l is a minimum general (not just for transient flows) period required to gain a satisfactory average.

Frame (b) shows the periodic flow trace, where t_p and ϕ'' represent the unsteadiness period and amplitude, respectively. The instantaneous velocity can now be expressed as

$$\phi=\bar{\Phi}+\phi''+\phi \tag{3.3}$$

where $\bar{\Phi}$ is the time mean. Alternatively, the following can be written

$$\phi=\langle\Phi\rangle+\phi' \tag{3.4}$$

where $\langle\Phi\rangle$is the phase averaged ($\langle\Phi\rangle(\alpha)=(\sum_1^N \phi(\alpha+2\pi(N-1)))/N$ where α is a fixed phase angle for N cycles and $N\rightarrow\infty$) component resolved by the numerical scheme. Therefore, $\langle\Phi\rangle$ can be expressed as

$$\langle\Phi\rangle=\overline{\Phi}+\phi'' \tag{3.5}$$

For transient flow $\phi''=0$ and so obviously

$$\langle\Phi\rangle=\overline{\Phi} \tag{3.6}$$

Similar to transient flow, a key URANS requirement is that $t_p \gg t_i$. Again, this is because it is mathematically impossible to form a valid time average for the turbulence if the base state changes as fast as the turbulence. This is again reflected through Eq. (3.2).

Frame (c) shows an intermittently unsteady flow trace. Such traces can be approximately observed around the outer edge of turbulent boundary layers and in turbulence transition (For a boundary layer, the peaks shown in Frame (c) would probably not be quite so distinct). At a particular point, as suggested in Frame (c), the flow may be quasi-laminar (assuming the trace applies to, for example, a turbulent boundary layer edge) for a period t_p and turbulent for a period t_i. Different averaged quantities (conditioned averages) can be derived depending on whether the whole trace is considered or just specific sections. When considering such traces the intermittency γ is a key parameter. It defines the ratio of the say turbulent flow duration to total trace duration i.e.

$$\gamma=\frac{\sum_1^N t_i}{(\sum_1^N t_i+\sum_1^N t_p)} \tag{3.7}$$

Using the intermittency, a conditioned average can be formed for say the Frame (c) quasi laminar period as follows

$$\overline{\Phi}_{tp} = \frac{1}{(1-\gamma)t_l} \int_{t-t_l/2}^{t+t_l/2} \phi\delta(t)dt \tag{3.8}$$

where

$$t_l = \sum_{1}^{N} t_i + \sum_{1}^{N} t_p \tag{3.9}$$

and $\delta(t)$ is a step function that assumes the value of unity for quasi-laminar periods and zero elsewhere. Conditioned averaging can also be hybridized with transient and periodic flows. When a condition is met, Eq. (3.8) is applied.[1]

3.2.2 Spatial Averaging/Filtering

With LES, generally a spatial filtering is used. However, temporal filtering is possible (Pruett 2000). Also space-time filtering is can be formulated. Dakhoul and Bedford (1986a,b) make compelling arguments for this mixed filtering. They note the implicit link between the space and time filtering elements. Dakhoul and Bedford observe that spatial filtering performs a temporal filtering role but that the temporal resolution in simulations can be much higher than the spatial creating an inconsistency in filtering. Here, just spatial filtering, which is the accepted LES standard approach is considered. The decomposition is simpler than with URANS, having just the two components given below

$$\phi = \overline{\phi} + \phi' \tag{3.10}$$

These involve a spatially filtered $\overline{\phi}$ component and a subgrid scale fluctuation ϕ'. Even though the filtering is spatial, the final equations are the same as those for URANS with $\langle\phi\rangle = \bar{\phi}$. The spatial filtering can be expressed as

$$\langle\phi\rangle = \bar{\phi} = \int_{-\infty}^{\infty} \phi(r)G(x-r)dr \tag{3.11}$$

The so called filter kernel, G, in the above, can take various forms. Table 3.1 gives potential filter functions expressed in physical and Fourier space, where w is wave number, w_c its cutoff and Δ, is the filter width.

---

[1]For hypersonic and highly compressible flows, where the turbulent fluctuations lead to substantial density fluctuations, it can be necessary to use Favre averaged RANS equations. This involves making density weighted averages. Hence, with this process, for example, $\bar{\phi} = \frac{\int_{-t_l/2}^{t_l/2} \rho(t)\phi(t)dt}{\int_{-t_l/2}^{t_l/2} \rho(t)dt} = \frac{\overline{\rho\phi}}{\bar{\rho}}$.

Table 3.1 Potential LES filter functions expressed in physical and Fourier space

Filter kernel, G	Physical space	Spectral space
Box filter	$\frac{1}{\Delta}$if $\lvert x-r\rvert<\frac{\Delta}{2}$ else 0	$\frac{\sin(\frac{1}{2}k\Delta)}{\frac{1}{2}k\Delta}$
Gaussian filter	$(\frac{6}{\pi\Delta^2})\exp(-\frac{6(x-r)^2}{\Delta^2})$	$\exp(-\frac{k^2\Delta^2}{24})$
Sharp spectral	$\frac{\sin(\pi(x-r)/\Delta)}{\pi(x-r)}$	1 if $\lvert w\rvert \le w_c$ else 0

A key constraint for the filter is that when filtering a constant ϕ field, $\overline{\phi}=\phi$. Therefore,

$$1=\int_{-\infty}^{\infty} G(x-r)dr \tag{3.12}$$

Also, because the filter width is assumed constant

$$\overline{\frac{\partial\phi}{\partial s}}=\frac{\partial\bar{\phi}}{\partial s} \quad s=x,t \tag{3.13}$$

Equation (3.11) can be written as the following convolution operation in a shorthand notation (Sagaut et al. 2007)

$$\langle\phi\rangle=\bar{\phi}=G*\phi \tag{3.14}$$

Using the above notation it can be seen that the fluctuation field can be recovered as

$$\phi'=(1-G)*\phi \tag{3.15}$$

Note, for highly compressible flows Favre filtering is needed. Also, for a non-uniform grid, the nice commutation property, expressed by Eq. (3.13), when the convolution filters we have discussed are used, is lost. This results in a compromising of the standard LES equations. Then, if it is desired to address the commutation error, specialized approaches are needed (van der Ven 1995). Alternatively, it must be acknowledged that additional closure terms arise (Vreman and Geurts 2002). Hence, rather than the usual situation where

$$\overline{\frac{\partial(\phi_i\phi_j)}{\partial x_j}}=\frac{\partial(\bar{\phi}_i\bar{\phi}_j)}{\partial x_j}+\frac{\partial(-\overline{\phi_i'\phi_j'})}{\partial x_j} \tag{3.16}$$

and $\phi=u$ (a velocity component), since $\overline{\partial\phi/\partial x}\neq\partial\bar{\phi}/\partial x$ we have

$$\overline{\frac{\partial(\phi_i\phi_j)}{\partial x_j}}=\frac{\partial(\bar{\phi}_i\bar{\phi}_j)}{\partial x_j}+\frac{\partial(-\overline{\phi_i'\phi_j'})}{\partial x_j}+\left[\overline{\frac{\partial(\phi_i\phi_j)}{\partial x_j}}-\frac{\partial\overline{(\phi_i\phi_j)}}{\partial x_j}\right] \tag{3.17}$$

The latter terms would ideally need to be modelled. Notably, the commutation error essentially vanishes if the equations are first transformed into computational space before filtering. This is because a uniform spacing can be defined in the computational plane (Jordan 1999). However, the metric terms then must be represented

Table 3.2 Discrete filter weighting functions

Wf_i	$Wf_{i\pm1}$	$Wf_{i\pm2}$	$Wf_{i\pm3}$	$Wf_{i\pm4}$	$Wf_{i\pm5}$	$\bar{\Delta}$
1/2	1/4	0	0	0	0	$\sqrt{6}\Delta x$
2/3	1/6	0	0	0	0	$2\Delta x$
1/2	9/32	0	−1/32	0	0	$2\Delta x$
1/2	75/256	0	−25/512	0	3/256	$2\Delta x$

as filtered quantities. Fortunately, this naturally takes place. However, a key drawback is that the physics of the subgrid scale modelling needs to be based on the contravariant velocity components. Then the physical basis for the model becomes unclear. The construction of commutative filters on unstructured meshes is outlined by Marsden et al. (2002). There are similar associated issues with URANS (Tucker 2001) arising through the temporal averaging.

3.2.3 Discrete Spatial Filters

Certain LES models, as will be shown later, require two levels of filtering and hence a discrete filter level. A one-dimensional, discrete, filter form is given below

$$\bar{\phi}_i = Wf_i\phi_i + \sum_{k=1,5} Wf_{i\pm k}\phi_{i\pm k} \tag{3.18}$$

Weighting function values and their corresponding filter widths are given in Table 3.2. The higher order (4th and 6th) filters are from Brandt (2008). Notably, as shown by Brandt, whether the filter is implemented as an implicit or explicit part of the solution process makes a substantial difference to solutions. Also, it is sometimes recommended to just apply explicit filters in homogeneous directions and hence the use of explicit filters seems to carry some uncertainties.

For multi-dimensional problems, the one-dimensional filter can be used in different directions. For complex grid topologies the differential form of the filter can be discretized. As noted by Jordan (1999), when using curvilinear coordinate system based approaches, explicit filtering in the uniform, computational domain, is more computationally expedient. Explicit filtering in a finite element framework is outlined by Hughes et al. (1998).

Note, as part of the DNS/LES post processing stage, temporal averages must still be formed. Again these can be simple time, ensemble or phase averages, depending on the physical situation being considered. In the work of Talha (2012), for a temporally accelerating boundary layer, ensemble averaging is used over a series of transient LES/DNS runs. What is more, popular DNS/LES test cases typically involve bi or even tri-periodic boundaries. Then spatial averaging in homogeneous flow directions can be used to supplement time averaging and thus gain better statistical averages over shorter time periods.

3.3 Governing Averaged Equations

3.3.1 (U)RANS Equations

Periodic flows occur frequently in many aerospace problems. These include gas turbine aero engines. With these, there are rows of rotating and stationary blade pairs. Also, the rotors in open rotor aircraft and helicopter rotors give periodic flow. Processes like shock buffet on wings and aeroelastic response can also be periodic. Therefore, following Johansson et al. (1993), Bosch and Rodi (1998) and Tatsumi et al. (1999) the equations are presented in a form that mainly reflects this potentially periodic flow nature. Subjecting Eq. (3.1) to phase averaging gives the URANS equations below

$$\frac{\partial \langle u_i \rangle}{\partial t} + \frac{\partial \langle u_i \rangle \langle u_j \rangle}{\partial x_j} = -\frac{1}{\rho}\frac{\partial \langle p \rangle}{\partial x_i} + \frac{\mu}{\rho}\frac{\partial}{\partial x_j}\left(\frac{\partial \langle u_i \rangle}{\partial x_j} + \frac{\partial \langle u_j \rangle}{\partial x_i}\right) + \frac{1}{\rho}\frac{\partial \langle \tau_{ij} \rangle}{\partial x_j} \tag{3.19}$$

The role of the turbulence model is to approximate the Reynolds stress terms,

$$\langle \tau_{ij} \rangle = \left\langle -\rho \overline{u_i' u_j'} \right\rangle \tag{3.20}$$

in Eq. (3.19). To approximate these, here, generally, the Boussinesq approximation is used

$$\left\langle -\overline{u_i' u_j'} \right\rangle = -n_0 \frac{2}{3} \langle k \rangle \delta_{ij} + \frac{2\mu_t}{\rho} S_{ij} \tag{3.21}$$

where δ_{ij} is the Kronecker delta ($\delta_{ij} = 1$ if $i = j$ and $\delta_{ij} = 0$ if $i \neq j$) and μ_t is the turbulent viscosity. Also, $\langle k \rangle$ is turbulent kinetic energy and $n_0 = 0$ or 1 depending on the model used or the chosen pressure formulation. The term S_{ij} (the mean strain rate) has the following form

$$S_{ij} = \frac{1}{2}\left(\frac{\partial \langle u_i \rangle}{\partial x_j} + \frac{\partial \langle u_j \rangle}{\partial x_i}\right) \tag{3.22}$$

Note, the equation's basic form remains the same for ensemble averaging.

3.3.2 LES Equations

The final form of the LES equations are essentially identical to Eq. (3.19). They are given below.

$$\frac{\partial \bar{u}_i}{\partial t} + \frac{\partial \bar{u}_i \bar{u}_j}{\partial x_j} = -\frac{1}{\rho}\frac{\partial \bar{p}}{\partial x_i} + \frac{\mu}{\rho}\frac{\partial}{\partial x_j}\left(\frac{\partial \bar{u}_i}{\partial x_j} + \frac{\partial \bar{u}_j}{\partial x_i}\right) + \frac{\partial \tau_{ij}}{\partial x_j} \tag{3.23}$$

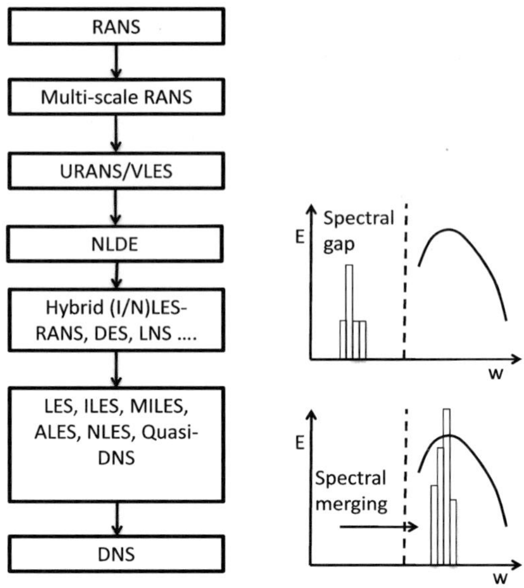

Fig. 3.2 Turbulence modelling hierarchy

The averages, used for URANS (and RANS), and identified here as $\langle\phi\rangle$, are replaced by spatially filtered variables, $\bar{\phi}$. The key major, initial, difference is that

$$\tau_{ij} = -\rho\overline{u_i' u_j'} + L_{ij} + C_{ij} \tag{3.24}$$

The above group of terms were derived by Leonard (1975). The first term is the usual Reynolds stress that also occurs with (U)RANS. The last two components are called the Leonard, L_{ij}, and Clark, C_{ij}, terms. They involve components indicative of interactions between the large scales, and large and small scales, respectively. The interactions between modelled and resolved scales complicates the modelling task. For RANS, more simply, $L_{ij} = C_{ij} = 0$. However, for RANS the modelling fidelity is a critical component.

3.4 VLES/URANS Modelling

Figure 3.2 presents a basic turbulence modelling hierarchy. Since they are not unsteady flow modelling techniques, RANS and multiscale RANS (where partitioned scales are used in the energy spectrum modelling) are not discussed here. Hence, URANS modelling will be discussed next. Then the various, broadly increasing in fidelity, strategies will be discussed.

The application of unsteady RANS models to flows with unsteady vortices arising from separation is sometimes called Very Large Eddy Simulation (VLES). This involves solving the RANS equations in their Unsteady form. Hence, it is perhaps more accurate to describe VLES as URANS (Spalart 2000).

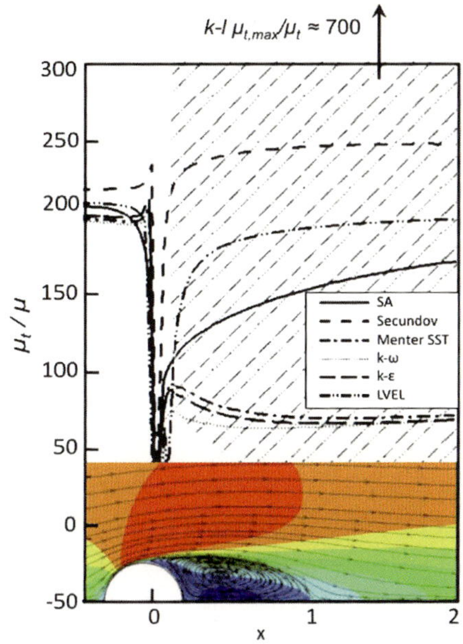

Fig. 3.3 Ratio of eddy viscosity to fluid viscosity for flow over a $Re = 25000$ cylinder (taken from Tucker and Liu 2007). Published with kind permission from Elsevier

A survey on the application of URANS to simple flows is given in Tucker (2001). From this, it is evident that, for flows with unsteady vortices, URANS predictions can give improved accuracy over RANS solutions (Orszag et al. 1997; Shur et al. 1996). However, there is strong temporal response sensitivity to turbulence model.

The $t_p \gg t_i$ requirement, noted in Sect. 3.2, implies there is a spectral gap—see upper right hand schematic in Fig. 3.2. However, such gaps seldom fortuitously appear. Alternatively, we might be able to compare the modelled length scales of the turbulence to the resolved. The former might be smaller than the resolved structures. However, again this would be purely fortuitous—constituting VLES. These fortuitous gaps indicate a clear demarcation between resolved and modelled turbulence. For a wake flow, strictly speaking, all structures in the wake are turbulence. Hence, the notion of having a RANS model to resolve them and unsteady modelling to pick up larger unsteady scales makes no sense. It is 'double accounting'. Nonetheless, the presence of a spectral gap in wakes is also tenuous (Lardeau and Leschziner 2005).

Indeed as shown by Tucker and Liu (2007), the eddy viscosity in a cylinder's wake can vary by over two orders of magnitude for different RANS models. This is shown in Fig. 3.3, which is taken from Tucker and Liu (2007). It plots the variation of eddy viscosity around a cylinder. The vertical extent of the hatched zone is the range of the eddy viscosities predicted by the 15 RANS models. If we assume that the modelled turbulence length scale for different models is approximately consistent, scaling with the cylinder diameter, this implies that the modelled time scales for different RANS models can vary by a factor of 200. Hence, any form of spectral gap is hard to guarantee. Assuming that the time scale is fixed will give a wide range of

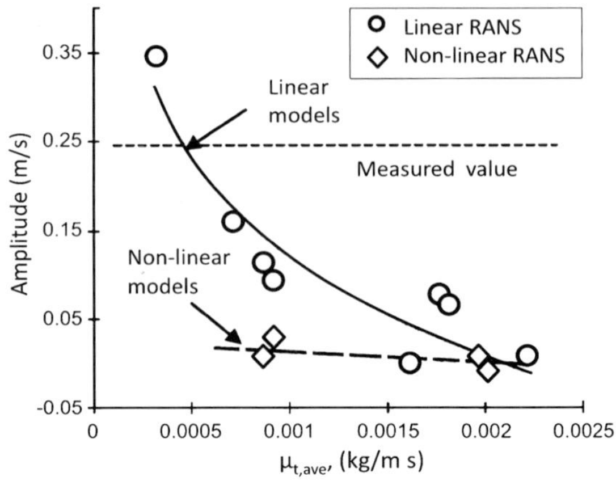

Fig. 3.4 Plot of average unsteadiness amplitude against average eddy viscosity for linear and non-linear (U)RANS models (from Liu et al. 2008). Published with kind permission from Elsevier

implied modelled length scales. The variation of the modelled time or length scale will in themselves be highly model dependent.

Figure 3.4 plots the predicted average unsteadiness amplitude against average domain μ_t for a complex geometry system (Liu et al. 2008) for a range of RANS models. The circles represent results for linear RANS models. The diamonds are for more advanced non-linear models making use of a non-linear type of Boussinesq approximation. Full lines, representing best fits to these numerical data sets are also shown. The horizontal dashed line is the measured unsteadiness amplitude. Clearly the model's nature strongly influences the predicted temporal content. With LES the situation is very different with regards to model influence. As can be seen from Fig. 3.3 (and implied earlier), the wide range of μ_t returned by RANS models means a wide range of unsteadiness amplitude can be predicted, depending on the RANS model choice.

Also, eddy viscosity models, unlike Reynolds stress, cannot account for some of the affects of stress-strain misalignment in unsteady flows. This can result in eddy viscosity models predicting too much turbulent viscosity in separated wake regions. To overcome this, Revell et al. (2011) present an approach that avoids the expense of solving a Reynolds stress model. It solves a transport equation for the dot product of S_{ij} and the turbulence anisotropy tensor given below

$$a_{ij} = \frac{\overline{u'_i u'_j}}{k} - \frac{2}{3}\delta_{ij} \tag{3.25}$$

This characterizes the lag between the stress and strain field responses. The method is combined with k–ϵ and Menter (1993) SST k–ω frameworks. It is applied to a range of flows, including a NACA0012 with an angle of attack that is sufficient for separated unsteady flow. Some encouraging results are shown. Notably, in the wake zone, of the NACA0012, substantial nonalignment of the stress and strain components is observed. This is consistent with the negative production of k, which simple eddy viscosity models cannot capture.

For attached unsteady boundary layer flows URANS may well perform well. Chapter 4 outlines examples of when URANS is expected to be helpful and when it is not. However, clearly in an industrial design context compromises have to be made.

Note, in the above, details of different RANS models are not given. Even, through the range is vast, in reality most practical simulations make use of a restricted range. These include the Spalart and Allmaras (1994), SST k–ω (Menter 1993) and k–ϵ models (Launder and Sharma 1974). Next LES, Implicit LES (ILES) and variants along with DNS will be discussed. This will be followed by hybrid RANS-LES approaches where the RANS and (I)LES approaches are combined.

3.5 (I)LES and DNS

An attractive idea, if not for cost, is to use temporal and spatial integration schemes of sufficient accuracy to resolve the larger more problematic turbulent scales. This leaves the turbulence (sub-grid scale) model to account for just the effects of the more universal and notionally isotropic smaller scales. This is the basis of Large Eddy Simulation (LES). LES models can be broadly classified as functional and structural. Functional models are discussed first. These are quite reminiscent of phenomenological RANS models. They try to capture the turbulence energy spectrum and hence the dissipation of turbulence at high wave numbers through dissipation induced by use of an eddy viscosity, μ_{SGS}.

3.5.1 Functional Models

Smagorinsky Model and Related Variants The Smagorinsky (1963) sub-grid scale (SGS) model is popular, but is crude and has several drawbacks. These are listed later. It is basically a mixing length model, the von Karman constant being replaced by Smagorinsky's constant, C_s (0.1–0.2). Hence, the model equation is as below

$$\mu_{SGS} = \rho(C_s\Delta)^2|S| \tag{3.26}$$

where

$$|S| = \sqrt{2\bar{S}_{ij}\bar{S}_{ij}} \tag{3.27}$$

Hence, effectively, Eq. (3.21) becomes

$$-\overline{u_i'u_j'} = -n_0\frac{2}{3}k_{SGS}\delta_{ij} + 2(C_s\Delta^2)|S|\bar{S}_{ij} \tag{3.28}$$

where $n_o = 0$. Note, the subgrid turbulent kinetic energy, k_{SGS}, if available (which it is not for this model) would generally be added to pressure thus redefining it. Akin to

RANS, even though there are a wide range of LES model variants, most simulations make use of either the Smagorinsky model in its standard or dynamic (see later) forms. The standard constant is derived by assuming a Kolmogorov spectrum—$E(w) = K_0 \varepsilon^{2/3} w^{-5/3}$. Here, w is wave number. Assuming a sharp cut-off filter with $K_0 \cong 1.4$ gives $C_s \cong 0.18$. However, for a shear flow a lower value is typically needed (Deardorff 1970). Indeed the required value will be heavily dependent on the filter choice (see later). Hence, there is a large published range for C_s. Another technique for shear flows is to use the splitting method of Schumann (1975). This involves two eddy viscosity type terms. The Smagorinsky model assumes a local turbulence production-destruction equilibrium (considered further later). To compensate for this defect Yoshizawa (1989) proposes the following relationship for the Smagorinsky 'constant'

$$C_S = C_{S0}\left[1 + C\bar{S}^{-2}\frac{D\bar{S}}{Dt}\right] \tag{3.29}$$

In the above C_{S0} and C are constants. For equilibrium $C_S = C_{S0}$. The Smagorinsky model does not have the correct near wall asymptotic behaviour and this has given rise to the WALE (wall-adapting local eddy viscosity model) model of Nicoud and Ducros (1999). Further Smagorinsky model defects are discussed later.

Germano Model The Germano model is sometimes know as the dynamic Smagorinsky model. Essentially it is the Smagorinsky model but with C_s computed based on a double filtering procedure. Note, however, the double filtering procedure is not restricted to the Smagorinsky model and can be combined with both functional, structural and mixed (see later) models. The filtering involves the usual LES filtering level resulting in a general flow variable field, $\bar{\phi}$. Also, a coarser test filter level $\overrightarrow{\phi}$ is used. Stress tensors are estimated around the two filter levels to give $\tau_{ij} = \overline{u_i u_j} - \bar{u}_i \bar{u}_j$ and $T_{ij} = \overrightarrow{\overline{u_i u_j}} - \vec{\bar{u}}_i \vec{\bar{u}}_j$. The difference of these stresses gives the exact what is called the Germano identity

$$L_{ij} (= \overrightarrow{\bar{u}_i \bar{u}_j} - \vec{\bar{u}}_i \vec{\bar{u}}_j) = T_{ij} - \vec{\tau}_{ij} \tag{3.30}$$

Figure 3.5 shows the area of the energy spectrum that these components represent. We can symbolically express the two stresses that we wish to model using some LES model (generally the Smagorinsky model) as

$$\tau_{ij} = C_s f_{ij}(\Delta, \bar{u}), T_{ij} = C_s f_{ij}(\vec{\Delta}, \vec{\bar{u}}) \tag{3.31}$$

Using the above and the Germano indentity a least squares minimization can be performed (Sagaut et al. 2007) to yield

$$C_s^2 = \frac{\langle L_{ij} M_{ij} \rangle}{\langle M_{ij} M_{ij} \rangle} \tag{3.32}$$

Fig. 3.5 Modelled stress components used in the Germano model to evaluate C_s

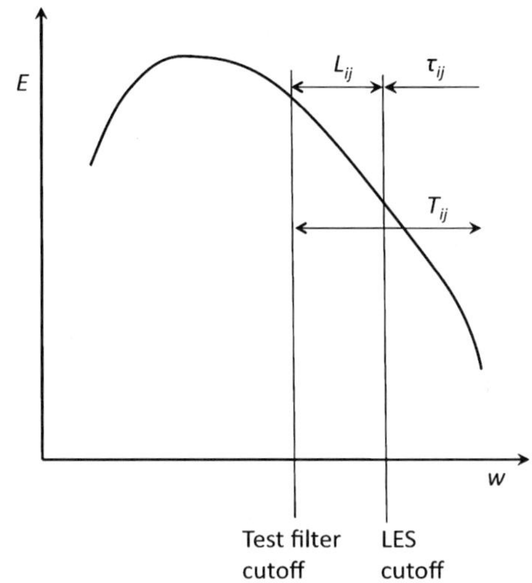

where

$$M_{ij} = f_{ij}\vec{\Delta}, \vec{\bar{u}} - f_{ij}(\vec{\Delta}, \bar{u}) \tag{3.33}$$

Note, the brackets in Eq. (3.32) indicate that an averaging procedure, discussed further later, is used. This assists with dealing with what are in fact an ill-posed set of equations. This aspect has resulted in several variants of the basic Germano model and also numerous ad hoc treatments to secure solutions.

Cited advantages of the Germano model are that no modifications are needed to the chosen base LES model as the wall is moved towards. Also, if the flow is sufficiently resolved, the scheme automatically reverts to DNS. The Germano model, evidently, has the potential to deal with transitional flows. Also it can be formulated to naturally compensate for numerical errors. However, the numerous averaging procedures (defined using $\langle\rangle$ in Eq. (3.32)) to deal with the potentially extreme values of computed C_s are doubtless likely to influence solutions for transitional flow. Procedures to deal with the extreme C_s values and their impact on solution stability, include averaging in homogeneous flow directions, local patch averaging, damping (Breuer and Rodi 1994), averaging along streamlines, averaging in time and clipping.

Multiscale Closures The Variational Multiscale (VMS) method was first introduced by Hughes et al. (2000) in a variational/finite element type framework. With this approach, three scales are considered. These are, for a general variable ϕ, the large, ϕ_L, small, ϕ_S (resolved) and modelled scales, ϕ'. Hence,

$$\phi = \phi_L + \phi_S + \phi' \tag{3.34}$$

where

$$\overline{\phi} = \phi_L + \phi_S \tag{3.35}$$

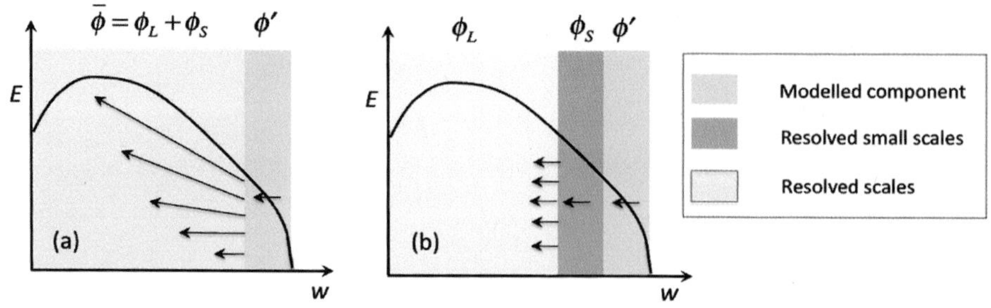

Fig. 3.6 Schematic of key difference between standard LES and VMS LES: (**a**) standard LES and (**b**) VMS (adapted from Gravemeier 2006)

In order to split the total resolved scales into large and small components again a test filter is needed. Vreman (2003) introduced an approach that made use of standard LES filtering. Hence, the method is relatively easy to implement. For example, it can be based around the Smagorinsky model with a top hat filter in a finite difference or finite volume solver. For wall bounded flows, the approach showed considerable improvements over the standard Smagorinsky model. As originally formulated the modelled scales only interacted with the smallest resolved scales.

Figure 3.6, taken from Gravemeier (2006), illustrates the difference between standard LES and VMS LES. With standard LES the modelled scales can act on all scales. Except, when the $-5/3$ zone is small, as occurs at lower Reynolds numbers, this is incorrect. With VMS, the modelled scales can interact with just the resolved small scale thus correcting this problem. A more complicated VMS variant, based on multiple grids, can be found in Terracol et al. (2001).

Yoshizawa Model As noted, the Smagorinsky model assumes locality of turbulence production and destruction. The locality aspect can be remedied using the Yoshizawa (1993) model. This solve a transport equation for subgrid scale turbulent kinetic energy as given below

$$\rho\frac{\partial k_{SGS}}{\partial t} + \rho\frac{\partial \bar{u}_i k_{SGS}}{\partial x_i} = \frac{\partial}{\partial x_j}\left[\left(\mu + \frac{\mu_{SGS}}{\sigma_k}\right)\left(\frac{\partial k_{SGS}}{\partial x_j}\right)\right] + P_k - \rho\varepsilon \tag{3.36}$$

In the above

$$P_k = 2\mu_{SGS}\bar{S}_{ij}\bar{S}_{ij} \tag{3.37}$$

and

$$\varepsilon = \frac{C_\varepsilon k_{SGS}^{\frac{3}{2}}}{\Delta} \tag{3.38}$$

are production and dissipation terms, respectively. The value of the constant $C_\mu = 0.07$ and $C_\epsilon = 1.05$. Since the model is k_{SGS} based, it follows that

$$\mu_t = \rho C_\mu \Delta k_{SGS}^{\frac{1}{2}} \tag{3.39}$$

As with the Smagorinsky model, the Yoshizawa can be implemented in a dynamic 'Germano model' form (Kim and Menon 1999). Such a model is used in later chapters. Structural models will now be discussed.

3.5.2 Structural Models

Where as the functional models have a more phenomenological outlook the structural have a mathematical basis. With them, the subgrid scale tensor $\tau_{ij} = \overline{u_i u_j} - \bar{u}_i \bar{u}_j$ is evaluated based on approximations (identified here by * superscripts) to the unfiltered solution u. Hence, these methods evaluate

$$\tau = \overline{u_i^* u_j^*} - \overline{u_i^*}\,\overline{u_j^*} \tag{3.40}$$

where, based in Eq. (3.14) above, the filtering procedure is inverted to give

$$u^* = G^{-1} * \bar{u} \tag{3.41}$$

The inverse filter, G^{-1}, can be defined as a series expansion G_n^{-1}. This can be truncated, depending on the level of mathematical exactness required. Hence, we have, again in a simple symbolic form

$$u^* = G_n^{-1} * \bar{u} \tag{3.42}$$

For $n = 0$ (Sagaut et al. 2007) the scale similarity model of Bardina et al. (1980) is recovered. An alternative procedure is to write the resolved velocity at a point x using Taylor's series as below

$$u(r) = u(x) + \sum_{n=1,\infty} \frac{(r-x)}{n!} \frac{d^n u}{dx^n} \tag{3.43}$$

The above is inserted in the usual filter convolution operator ($\overline{\phi^*} = \int_{-\infty}^{\infty} \phi^*(r) G(x - r) dr$). Then an inversion procedure is made to yield a model having the broad form

$$u^* = \left[\sum_{n=0,\dots} \frac{(-1)^n}{n!} \Delta^n \dots \frac{\partial^n}{\partial x^n} \right]^{-1} * \bar{u} \tag{3.44}$$

Hence, a model is created that involves evaluating derivatives of the resolved flow field. Clark's model (Clark et al. 1979), discussed further later, fits in with the above Taylor series, resolved velocity, based model (Sagaut et al. 2007). This type of model can involve the evaluation of a considerable number of gradient quantities. In practical coding terms the gradient tensors can be compactly evaluated through 'do loops'. Structural models tend to give good correlation (based on DNS evidence) with exact subgrid stresses. They also appear to model well turbulence anisotropy. This is reflected in later results.

3.5.3 LES Model Defects

As will be seen from the later chapters, for practical flows the basic Smagorinsky is the most popular model. Limitations of the Smagorinsky model are listed below:

(I) Tendency to be overly dissipative;
(II) Draining energy from all scales rather than just small;
(III) Non-vanishing eddy viscosity near walls and general inapplicability in this zone (Opiela et al. 2001);
(IV) Model constant being calibrated for a rather simple strain field (homogeneous decaying turbulence);
(V) Altering the flow's effective Reynolds number (Bogey and Bailly 2005);
(VI) Providing no back scatter;
(VII) Non-tensorial behaviour of eddy viscosity (Sagaut LES 2002) and hence inability to model turbulence anisotropy (Kosovic 1997);
(VIII) Locality in space and time (Sagaut LES 2002; Yoshizawa 1993);
(IX) Lack of transition prediction capability (the Smagorinsky constant must be set to zero in laminar flow zones);
(X) Insensitivity to buoyancy, rotation and streamline curvature (Holzäpfel 2004) (like many LES models);
(XI) Incomplete characterization of the pressure field (Yoshizawa and Horiuti 1985) and
(XII) Inapplicability at low Reynolds numbers where the inertial subrange is minimal (George and Tutkun 2009; Razafindralandy et al. 2007).

Hence, based on the above, there is the 'strong discouragement' of the use of the Smagorinsky model in the AIAA best LES practice guidelines (Georgiadis et al. 2010). With regards to Point (VIII), as noted by Sagaut LES (2002), for the locality assumption to be valid the characteristic length and time scales of the subgrid scales (l_{SGS} and t_{SGS}) must be much smaller than those for the resolved scales (LR and TR). Hence,

$$l_{SGS} \ll L_R \quad \text{and} \quad t_{SGS} \ll T_R \tag{3.45}$$

This can be alternatively viewed that, at the subgrid scale level, production is assumed equal to dissipation i.e. there is equilibrium (the same assumption as used with simpler RANS models). However, turbomachinery flows can be of relatively low Reynolds number. This is especially so for the rig scale data that is frequently used to study and validate numerical models for turbomachinery. Hence, a (spectral) gap is hard to guarantee. Transport equation based LES models seem a more sensible choice, such as the Yoshizawa (1993) model. This model can rationally account for the convection of the subgrid scale eddies by the grid scale eddies i.e. any lack of turbulence equilibrium.

The Smagorinsky and Germano models are based on the assumption of a long inertial subrange ($w^{-5/3}$ region), in the energy spectrum. This assumption leads to the universal Smagorinsky constant (Yoshizawa 1989). Using a model energy spectrum (Tyacke 2009), with two-equation RANS solutions, to obtain estimates of the

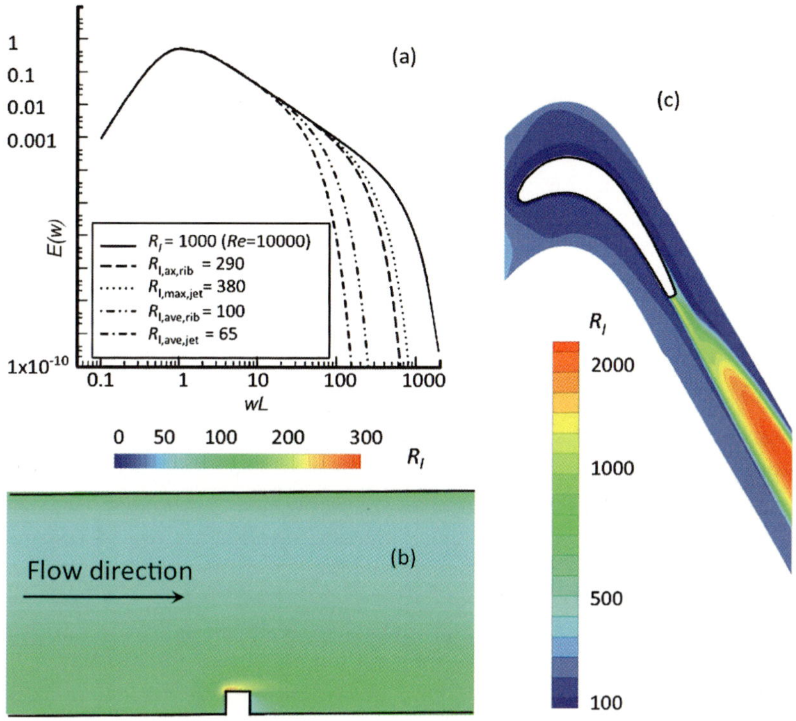

Fig. 3.7 Estimation of $w^{-5/3}$ region: (**a**) energy spectra based on R_l and (**b**, **c**) R_l field contours (Tucker et al. 2011). Published with kind permission from ASME

range of scales, it is found that there is little evidence of a long inertial subrange, particularly for low-pressure turbine (e.g. T106) profiles at Reynolds numbers typically considered. As suggested by George and Tutkun (2009), the ratio R_l of the largest to smallest eddies should be 1000 for a substantial $w^{-5/3}$ zone to exist. Estimates, from Tucker et al. (2011) for an impinging jet and ribbed channel are shown in Fig. 3.7a. These are expressed in terms of the maximum and average ratio of R_l. Figures 3.7b, c give R_l contours for a ribbed channel and high pressure turbine blade flow. These, along with the Frame (a) plot suggest the assumption of a long $w^{-5/3}$ region may again be invalid.

Further evidence is given by Tyacke et al. (2013) and plotted in Fig. 3.8. This shows estimates of the range of scales in the inertial subrange, relative to the inertial range of the fan for different regions of an engine. Again the estimates are RANS based and linked to a CFD simulation for a Trent 1000 engine. For high *Re* zones, such as the fan and high-pressure turbine, there is a significant inertial subrange region (around 20-25 % of the total scales). For the low-pressure turbine, as noted earlier, there is only a short $w^{-5/3}$ region.

This notion of a short $w^{-5/3}$ region is not only supported by numerical estimate, but also experimental evidence. Mydlarski and Warhaft (1996) proposed that a $w^{-5/3}$ region is found only in the limit of infinite Reynolds number, in reality

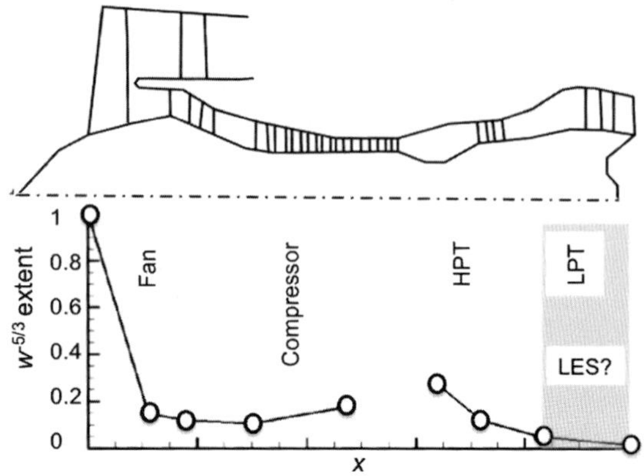

Fig. 3.8 Extent of $w^{-5/3}$ range (fraction of the fan $w^{-5/3}$ range of length scales)

being a function of *Re*. For all the Reynolds numbers of their experiments, the spectrum rolled-off more slowly than $w^{-5/3}$. As stated by Gamard and George (1999), there are few experiments that unequivocally show a $w^{-5/3}$ range. This lack of a $-5/3$ zone will change the turbulence physics, invalidating most LES models, hence, requiring the use of more specialized models. Such a model is proposed by Razafindralandy et al. (2007).

A point of concern is that despite the excessively dissipative nature of the Smagorinsky model, it still sees wide spread use with excessively dissipative (practical) flow solvers. The Germano model is not without a range of theoretical flaws (Ghosal et al. 1995). It intrinsically assumes an inertial subrange exists. As discussed above, in many instances, this is questionable. On the other hand, if it does exist, a universal constant can be derived rendering the dynamic model unnecessary. If there is no homogeneous direction to average the dynamic constant, the standard dynamic model is hard to utilise, as discussed earlier, needing ad hoc, smoothing related procedures.

A key problem with *structural* models is that they are well known to provide too little dissipation. They also assume locality of the energy transfer process between the resolved and small scales. On the other hand, the *functional* models do not display good correlation with the exact subgrid stresses. However, broadly they capture the dissipative processes. Also, they can provide modelling of the non-locality of the resolved-subgrid scale interactions. Hence, there is a strong attraction to combine functional and structural models to get the best of all worlds. Such mixed models are discussed next. Notably, in the ILES (Implicit LES) approach (see later), the numerical properties (such as dissipation) are used to act as a subgrid scale model. There is no reason why structural modelling cannot be combined with such approaches. However, Clark type terms can naturally occur for certain ILES.

3.5.4 Mixed-(Nonlinear) Models

As noted above, mixed LES models are attractive. They are similar to non-linear RANS models in that they have a usual linear Boussinesq based component (as

Table 3.3 Nonlinear stress expressions for different models using $\alpha = \Delta$

LES model	$\overline{NL}$ (Nonlinear part of τ_{ij})
Kosovic (1997)	$C_k\rho\alpha^2(0.5\frac{\partial\bar{u}_i}{\partial x_l}\frac{\partial\bar{u}_l}{\partial x_j}+1.5\frac{\partial\bar{u}_i}{\partial x_l}\frac{\partial\bar{u}_j}{\partial x_l}-0.5\frac{\partial\bar{u}_l}{\partial x_i}\frac{\partial\bar{u}_l}{\partial x_j})$
Leray (Geurts and Holm 2006)	$C_l\rho\alpha^2(\frac{\partial\bar{u}_i}{\partial x_l}\frac{\partial\bar{u}_l}{\partial x_j}+\frac{\partial\bar{u}_i}{\partial x_l}\frac{\partial\bar{u}_j}{\partial x_l})$
α (Domaradzki and Holm 2001)	$C_\alpha\rho\alpha^2(\frac{\partial\bar{u}_i}{\partial x_l}\frac{\partial\bar{u}_l}{\partial x_j}+\frac{\partial\bar{u}_i}{\partial x_l}\frac{\partial\bar{u}_j}{\partial x_l}+\frac{\partial\bar{u}_l}{\partial x_i}\frac{\partial\bar{u}_l}{\partial x_j})$
Clark et al. (1979)	$C_L\rho\alpha^2(\frac{\partial\bar{u}_i}{\partial x_l}\frac{\partial\bar{u}_l}{\partial x_j}+\frac{\partial\bar{u}_i}{\partial x_l}\frac{\partial\bar{u}_j}{\partial x_l})$

with the Smagorinsky model) and non-linear additional terms. Hence, they can be expressed as

$$\tau_{ij} = L + \overline{NL} \tag{3.46}$$

where L and $\overline{NL}$ represent the linear and non-linear terms, respectively. The linear stresses could be characterized using the Smagorinsky model. However, considering the Smagorinsky model issues noted above, any mixed model results reported here are based on the Yoshizawa (1993) model. Following, Tucker et al. (2006), $\overline{NL}$ for models used here can be expressed as shown in Table 3.3. There is implicit summation over l.

The Leray and LANS-α terms are equivalent to creating nonlinear terms out of the following estimate of the subgrid scale velocity fluctuations:

$$u' = -\alpha^2\nabla^2\bar{u} \tag{3.47}$$

where $\alpha = \Delta$. This is a practical and easy estimate of u' that avoids the double filtering needed in the dynamic Smagorinsky model. The Leray model, effectively involves using the following Helmholtz filter operator

$$G = \frac{1}{1-\alpha^2\nabla^2} \tag{3.48}$$

In the LANS-α model the *NL* terms arise from the convection of SGS vorticity. The derivation of the nonlinear Leray and LANS-α models in Table 3.3 can be found in Domaradzki and Holm (2001) and Geurts and Holm (2003).

3.5.5 MILES Approach and Numerical Influences in LES

It is important to guarantee that the numerical influences (e.g. numerical dispersion and dissipation) are substantially smaller than the contribution of the LES model. However, as demonstrated analytically by Ghosal (1996) and through a priori DNS (Direct Numerical Simulation—see later) by Chow and Moin (2003), without high order schemes (or large filters overlaying multiple grid points) this is difficult to guarantee. Approaches like MILES or ILES (Boris et al. 1992; Margolin et al. 2006) and NLES (Pope 2004) etc acknowledge that this can be the case and either select numerical schemes that favourably act as, and hence can replace the LES model or simply admit that the numerical impact is potentially substantial (NLES) and not necessarily favourable. With MILES (Monotone Integrated LES) schemes that maintain monotonicity i.e. prevent wiggles in solution variables are used. This is achieved through use of the following general class of schemes: TVD (total variation diminishing), shock limiters, shape preserving and essentially non-oscillatory schemes. Some of these schemes were discussed in Chap. 2. Although differing in detail, these schemes are aimed at convecting spatial variable distributions with the minimal of distortion from dissipation. For flows with shocks, the use of MILES becomes especially persuasive.

By way of example, for a flux-limited finite volume scheme the total numerical flux, J, could be expressed as

$$J = J^H - (1 - \varepsilon)\left[J^H - J^L\right] \tag{3.49}$$

In the above J^H is a high-order flux function and J^L a dissipative low-order. The parameter ε is a limiter. Modified equation analysis, based on the above, shows that for the correct limiter the Navier-Stokes equations are being solved but with a tensorial SGS eddy viscosity model. Limiters can yield elements akin to those found in scale similarity models (Sagaut et al. 2007). Hence, the MILES approach can yield, in a sense, mixed type models. An alternative approach that has similarities to MILES/ILES is the LES-RF (LES-Relaxation Filtering) approach. With this, filtering alone is used to replace the role of the LES model. The filter is designed to not influence the larger scales. Berland et al. (2011) explore the influence of filter shape in this type of procedure finding (for mixing layers) that it has minimal impact. Visbal (2011) applies a similar approach when performing high-order compact scheme simulations of deep dynamic stall in a plunging aerofoil.

3.5.6 LES of Compressible Flows

LES of compressible flows is partly challenging since there are a wide range of forms that the equations governing compressible flow can be expressed. The choice of solution variables will result in the appearance of different subgrid terms. This is because of the intrinsic non-linearity in the equation systems being used. Vreman (1995) (exploring compressible mixing layers with $M \le 1.2$), used the Favre averaged velocity. Vreman observed that usual Reynolds stress type tensor term, found in incompressible LES, is also the dominant term in compressible flow LES. However, a range of subgrid terms arose in the energy equation, but these did not appear to play a substantial role. Indeed some cancelled with each other. Hence, more accurate results could be gained by omitting these terms rather than just modelling isolated ones. With regards to the modification of subgrid scale models for compressible flow, this also does not seem to be a major issue. The subgrid scale

Mach numbers will generally be considerably lower than the turbulent Mach numbers associated with the larger scales. Hence, the major influences of compressibility take place in the larger scales. Therefore, for functional modelling, moving concepts of density weighting to standard subgrid closures seems adequate and straightforward. Similarly, the mathematical processes in structural models need to absorb density. In this work, such compressible flow terms/modifications are ignored, the Mach numbers being sufficiently low for this to be a satisfactory approximation. The greater LES challenges, for compressible flows, are associated with numerical modelling. Some of these challenges were touched on in Chap. 2. Note, the Navier-Stokes equations are only an appropriate description of the flow physics near shocks at low supersonic Mach numbers—$M < 1.3$—(Sagaut et al. 2007).

3.5.7 Direct Numerical Simulation

The next extreme is DNS. For this, high quality numerical schemes are used. Also, grids are sufficiently fine to resolve structures all the way down to the smallest (Kolmogorov) scales. Typically, for a DNS, grid spacing in wall units, in all directions will be around 10. As will be evident later, the computational cost of DNS is severe. Hence, its use has been limited to low Reynolds number components, such as the low-pressure turbine, and for small spanwise sections—around 10 % of chord. In such zones computing the highly complex transition mechanisms is key (Coull 2009) and the DNS benefits clear.

3.6 Near Wall (I)LES Modelling and Grid Requirements

A successful LES needs virtually DNS (see later) near wall grid resolution. This has given rise to the use of wall functions. These are much like those used in RANS modelling. With them, analytical relationships, which give $\tau_w = f(U, y)$ are used that assume turbulence equilibrium. Such functions are unlikely to perform well in adverse pressure gradient flows. The analytical relations are typically applied for $y^+ > 30$. At low Reynolds numbers this can be some considerable distance form solid surfaces. Initial pioneers of the application of wall functions in LES are Deardorff (1970) and Schumann (1975). As will be seen later, under the banner of hybrid RANS-LES, simplified RANS models can be used to predict shear stress at the first off-wall grid nodes in LES. Another, effective approach, for LES, is to use what is sometimes called the Lilly correction. With this, the usual modelled turbulence length scale, in the Smagorinsky model, $l = C_s\Delta$, is modified to

$$l = \min(\kappa d, C_s\Delta) \tag{3.50}$$

where, κ, is the von Karman constant and, *d,* the nearest wall distance. This, as will be seen later, has a close similarity to the length scale description used in the most popular hybrid RANS-LES approach termed DES (Spalart 2000).

Table 3.4 Grid requirements for different eddy resolving methods

Method	$\Delta x^+/\text{Min}(\Delta y^+)/\Delta z^+$	Number $y^+ < 10$ points
DNS	$(10–15)/1/5$	3–5
Wall resolved LES	$(50–130)/1/(15–30)$	3–5
LES with a wall model	$100–600/(30–150)/100–300$	–
Hybrid RANS-LES	$(100–600)/1/(100–300)$	2–5
RANS	$1000/1$	2–5

Table 3.5 Number of nodes used for outer part of boundary layer

Reference	δ/Δ
Chapman (1979)	1/13
Piomelli et al. (1989)	1/10
Schumann (1975)	1/5–1/10
Deardorff (1970)	1/7
Tucker and Davidson (2004)	1/11
Average	≈1/10

Near wall grid resolution requirements for wall-resolved and non-wall resolved LES, and DNS are given in Table 3.4. For wall resolved LES, the dynamics of the inner part of the boundary layer such as the events responsible for turbulence production are directly resolved. The coordinates x, y and z are the streamwise, wall normal and spanwise, respectively. Notably, for RANS, expressed for a two-dimensional flow in Table 3.4, high aspect ratio grids are tolerated. This is in strong contrast to LES. Note, the hybrid RANS-LES figures are taken from Davidson and Peng (2003). The wall resolved LES ranges are given by Choi and Moin (2012), who revisit Chapman (1979) landmark grid resolution estimates using more accurate correlations of boundary layer properties. For RANS and hybrid RANS-LES, the number of grid points in the $y^+ < 10$ zone will be strongly dependent on the RANS model choice. For example, a low Reynolds k–ϵ model formulation will need similar wall normal requirements to those of DNS.

For the outer region, the grid requirements are more weakly dependent on Re ($\delta/c \sim Re^{-0.2}$ where c is the chords). Table 3.5 summarizes the outer layer grid spacings. These are based on successful LES for a range of workers along with the recommendation of Chapman (1979) but might be regarded as modest. In the outer region, the grid spacings used tend to be more isotropic, hence a single typical grid scale/spacing Δ is given in Table 3.5. Also, the spacings are non-dimensionalized by the boundary layer thickness δ. Note, Chapman's recommendation is based on integration of an analytical energy spectrum. It ensures that 90 % of the turbulence is resolved. The values given for Schumann (1975) use just the channel flow results from this work.

For flows with shocks, grid requirements become severe. The shock width is less than η. Hence, it seems that ideally, at least DNS resolution is needed around

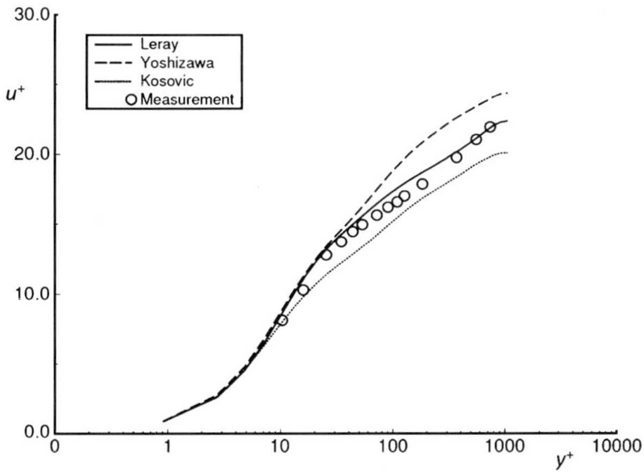

Fig. 3.9 'Law of the wall' for an excessively coarse cross-stream grid with different LES models

shocks. Also, the shock turbulence interaction causes wrinkling of the shock front. Therefore, the grid needs to be sufficiently fine to capture this wrinkling. Empirical evidence suggests that around shocks (Sagaut et al. 2007)

$$\frac{\Delta n}{\Delta y} \sim \frac{u'_n}{U_n} \tag{3.51}$$

In the above, n identifies quantities in the shock normal direction and Δy is the grid spacing parallel to the shock. Evidently, typical DNS have the above ratio as around 0.1.

As implied above, and indicated in Chap. 1, the inner layer cost is high. Hence, considerable savings can be made covering over fine scales populating the $y^+ <$ 100 zone (or even the whole boundary layer) with a RANS zone. This gives rise to hybrid RANS-LES approaches. These are discussed later.

To illustrate the importance of grid resolution in LES, Fig. 3.9 shows the "law of the wall" for $Re_\tau \cong 1000$—a Reynolds number based on friction velocity and boundary layer thickness. The grid is such that at first off-wall nodes $y^+ \approx 1$ with $\Delta x^+ \approx 200$ and $\Delta z^+ \approx 100$. The latter, deliberately constructed, spacing is inadequate for LES near wall streak resolution. The symbols represent accurate benchmark data. The long dashed line is for a linear (Yoshizawa) LES model. The full and short dashed lines are for two of the mixed LES models discussed above. As can be seen, and is to be expected, the linear LES "wall law" is poor. The near wall grid is too coarse to resolve the highly anisotropic near wall structures. Hence, there is insufficient turbulent stress and so the velocity is too high away from the wall. The terms in the mixed LES models improve agreement. In some sense, they are able to account for the near wall anisotropy. Note, for the mixed Leray model (Tucker et al. 2006), just selected terms are used in the *NL* component.

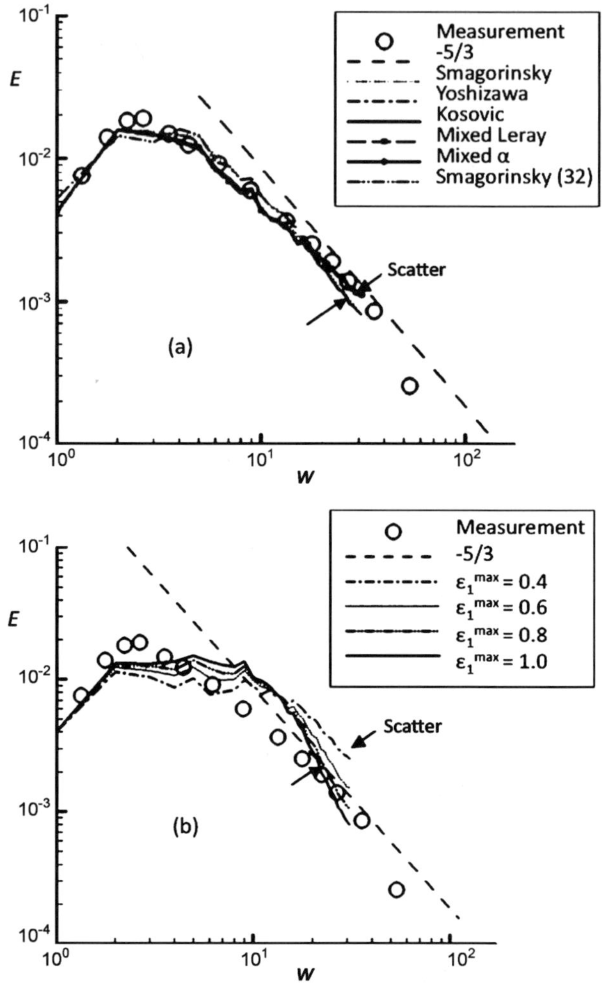

Fig. 3.10 Sensitivity to LES model and numerical scheme for homogeneous decaying turbulence: (**a**) solution for different LES models and (**b**) solutions for a different numerical parameter setting. Published with kind permission from ASME

3.7 LES Hierarchy

Next the application of the LES model away from walls is considered. Figure 3.10 shows the sensitivity of predictions to the LES model and numerical scheme for homogeneous decaying turbulence. The vertical axis, *E*, is the turbulence energy. The horizontal axis is wave number. This can be viewed as the inverse of eddy size. Frame (a) shows solutions for some of the different LES models, discussed above, using a pedigree LES solver. Full details of these simulations can be found in Liu et al. (2008). Frame (b) shows solutions where the LES model is omitted. Numerical dissipation drains energy instead of the explicit LES model. The spatial discretization uses a scheme specifically intended for LES (Mary and Sagaut 2002). This adapts numerical smoothing based on the flow state. The scheme is outlined in Chap. 2. The maximum user specified smoothing level is controlled by the pa-

Table 3.6 Average predicted primary shear stress error for five LES models and five numerical schemes

	Std. deviation	Ave. error
LES model	4.4 %	12.3 %
Numerical scheme	19.4 %	13.8 %

rameter ε_1^{max}. All Fig. 3.10 simulations use the same 65^3 grid, except for one set of Smagorinsky results. These use a 32^3 grid—identified as 'Smagorinsky (32).' The key point to observe is that numerical scheme sensitivity is greater than LES model sensitivity.

Next, the numerical model and LES model sensitivity is explored for shear layers. For these the expected maximum shear stress is around $0.01(\Delta U)^2$ where ΔU is the velocity difference across the layer. Table 3.6 plots the average 'LES' predicted errors for spatially developing shear layers. The errors are the average for essentially five different LES models (again using a pedigree LES solver) and also for five different 2nd order numerical schemes. The latter range from cell centered to cell vertex, and compressible and incompressible schemes. As will be seen from Chap. 2, the nature of compressible and incompressible schemes can be substantially different. Full details for these simulations can be found Liu et al. (2008) and Eastwood et al. (2009). The key thing to note is that the standard deviation for the five LES models is much lower than the five numerical schemes. This is even though the latter are more highly resolved. This reflects the well documented evidence, discussed above, of Ghosal (1996) and Chow and Moin (2003). This being that for 2nd order solvers (and standard filters) the numerical contribution is orders of magnitude higher than that of the LES model. The findings are consistent with the previous observations for homogeneous decaying turbulence. Next we will consider inflow conditions.

Figure 3.11 plots the shear stress distribution with respect to dimensionless axial distance for the free shear layer, NLES, just discussed above. However, now results are contrasted when the moving stream is seeded with synthetic turbulence, with a turbulence intensity (*Ti*) 10 % (full line) and 0 % (dashed line). The inflow turbulence is isotropic box turbulence—this is discussed later. A key point is that the inflow modelling influence appears greater than that of the LES model (cf. Table 3.6). Notably, for a real engineering system, the inflow complexity will be much more severe with typically flow from a range of upstream mechanical components.

Figure 3.12 gives a practical LES hierarchy. It is aimed at LES involving industrial type CFD programs. It is based partly on the above evidence and other experiences outlined in this text.

Notably, DeBonis (2006) classifies LES as rigorous and practical. With the former, high order codes are used for more canonical flows. Practical LES involves industrial RANS codes that are adapted for LES. Hence, they have low order quadrature, being designed with robustness and geometric flexibility in mind. Figure 3.12 relates to this type of LES. Notably the LES model is last. Note, the Fig. 3.12 hierarchy is not intended for transitional type flows where the LES model can have a key influence (if the simulation is not well resolved). Also, it is clear from above that

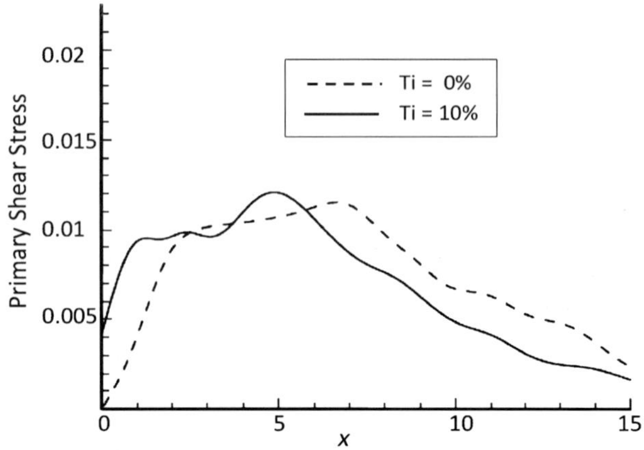

Fig. 3.11 Peak shear stress in a shear layer for different turbulence intensities (Tucker et al. 2011). Published with kind permission from ASME

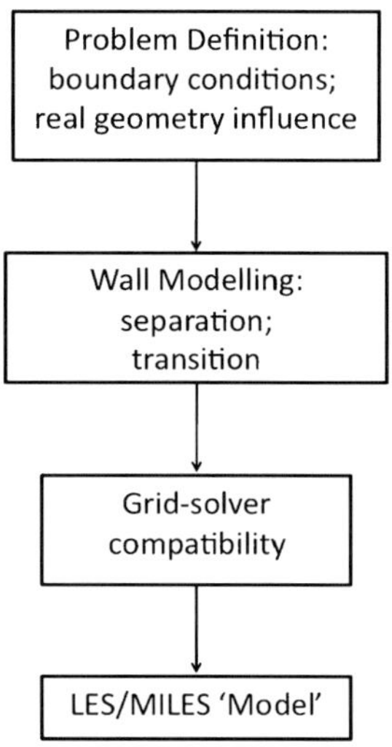

Fig. 3.12 Practical LES hierarchy

the LES model can have strong impact near walls. This might seem to contradict the hierarchy proposed above. However, the large variation is due to the excessively coarse grid i.e. DNS resolution grids are needed for wall resolved LES.

3.8 Hybrid RANS-LES and Related Methods

There are a range of methods that fit into the hybrid RANS-LES category to varying degrees. Some are closer to RANS and others to LES. Some restrict the RANS zone to just the near wall zone. Figure 3.13, shows the 'law of the wall' for the case discussed above (Fig. 3.9) with $\Delta z^+ = 100$ i.e. a grid that is too coarse for LES. This time a RANS zone of different extents is placed near the wall. As can be seen from the figure, if the RANS extent is sufficient to cover the fine near walls scales

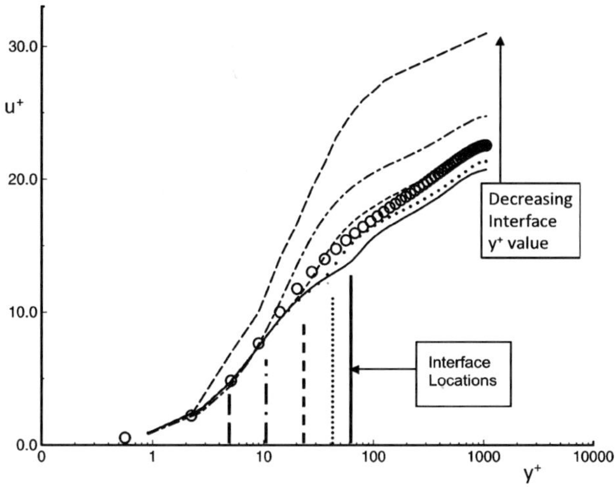

Fig. 3.13 'Law of the wall' for an excessively coarse grid with a near wall RANS layer of differing extents (Tucker and Davidson 2004). Published with kind permission from Elsevier

of the boundary layer (extending to about $y^+ \approx 100$) a satisfactory velocity profile (and hence agreement with the benchmark data given by symbols) can be gained on the coarse grid.

Other hybrid RANS-LES methods permit the RANS and LES zones to occur anywhere in the domain, adapting to grid suitability. Some methods are akin to LES in that the RANS solution is made on a separate grid. It is more designed to provide a shear stress boundary condition for the LES, this being applied as if a wall function LES were being performed. Of course, there is no reason why LES zones cannot be replaced with ILES, MILES or NLES. These differing genres of hybrid methods will be briefly discussed below. The discussion is in no way exhaustive.

3.8.1 DES Methods

The initial DES approach, termed here, DES97, is based around the Spalart-Allmaras (SA) RANS model. This solves a transport equation for the eddy viscosity function, $\tilde{\nu}$. The general form of the equation is given below

$$\frac{D\tilde{\nu}}{Dt} = \underbrace{c_1 \tilde{S}\tilde{\nu}}_{\text{Production term}} + \text{Diffusion terms} - \underbrace{c_2 f_w \left[\frac{\tilde{\nu}}{\tilde{d}}\right]^2}_{\text{Destruction term}} \tag{3.52}$$

where $\mu_t = \rho f_{\nu 1} \tilde{\nu}$. The damping function $f_{\nu 1}$ is designed to ensure that $\tilde{\nu} = \kappa u_\tau d$ (where u_τ is the friction velocity). The function f_w is designed to control behaviour in the outer part of the boundary layer. With DES97 the turbulence destruction uses a modified wall distance, $\tilde{d}$ (in the SA turbulence model $\tilde{d} = d$). This is given by the function below

$$\tilde{d} = \min(d, C_{DES}\Delta) \tag{3.53}$$

The constant $C_{DES} = 0.65$ and $\Delta = \max(\Delta x, \Delta y, \Delta z)$. Since it is Δ which controls the RANS-LES interface ($\tilde{d} = d$ and hence RANS is activated $d < C_{DES}\Delta$), the grid needs careful design. For turbulence in equilibrium (turbulence production equalling destruction) and away from walls $\nu_t \approx \Delta^2 S$ i.e. DES97 reduces to the Smagorinsky model. As recommended by Deck (2012), a further expedient modification for the LES zone is to set

$$f_{v1} = 1, \qquad f_{v2} = 0, \qquad f_w = 1 \tag{3.54}$$

Note, the f_{v2} function allows the vorticity magnitude S to be modified to $\tilde{S} = u_\tau/\kappa y$ keeping the log-law behaviour. The modifications (3.54), above, avoids the low eddy viscosity found in the LES zone being mistaken for wall proximity. It can also accelerate the growth of LES content.

If the grid structure is inappropriate, the LES zone can make an incursion into the boundary layer. To avoid this, Spalart et al. (2006) proposed delayed DES (DDES). This uses a boundary layer shielding function, preventing the LES mode being triggered in the boundary layer. For this purpose the quantity below is used

$$r_d = \frac{\nu_t + \nu}{\kappa^2 d^2 \sqrt{\bar{u}_{i,j}\bar{u}_{i,j}}} \tag{3.55}$$

where ν is kinematic viscosity. The tensor function in the above may be written more explicitly as

$$\left[\sum_{i,j}\left(\frac{\partial \bar{u}_i}{\partial x_j}\right)^2\right]^{1/2} \tag{3.56}$$

The function r_d takes a value of unity in the log layer and tends to zero towards the boundary layer edge. It is utilized in the function below

$$f_d = 1 - \tanh\left[(8r_d)^3\right] \tag{3.57}$$

Note, that in boundary layers $f_d = 0$ and outside them $f_d = 1$. Using this function the DDES length scale is re-expressed as

$$\tilde{d} = d - f_d \max(0, d - C_{DES}\Delta) \tag{3.58}$$

This function has the desired behaviour of ensuring that inside the boundary layer $\tilde{d} = d$ correctly supplying the RANS branch. Outside the boundary layer, $\tilde{d} = C_{DES}\Delta$—as it should be in the LES zone. Notably, yet more modelling constants emerge through use of this approach.

3.8.2 Menter SST Based DES

Most RANS models can be expressed in some form of DES framework. For, example when using a k–ω model, the destruction term in the k equation, given below, needs to be modified (Note, the phase averaged equation form is stayed with—in practice the near wall behaviour will require URANS, as discussed later).

$$\rho\frac{\partial\langle k\rangle}{\partial t}+\rho\frac{\partial\langle u_i\rangle\langle k\rangle}{\partial x_i}=\frac{\partial}{\partial x_j}\left[\left(\mu+\frac{\mu_t}{\sigma_k}\right)\left(\frac{\partial\langle k\rangle}{\partial x_j}\right)\right]+P_k-\frac{\rho\langle k^{3/2}\rangle}{l_{DES}} \tag{3.59}$$

The length scale modification is straight forwards (Strelets 2001) and as follows

$$l_{DES}=\min(l_{k-\omega},C_{DES}\Delta) \tag{3.60}$$

Hence, the length scale in the above is constructed so that it ranges between the RANS and LES limits. The former equates to $k^{1/2}(\beta^*\omega)$ where the modelling nomenclature is standard. For the k–ω based Menter SST model it is also possible to formulate a Delayed DES version where

$$l_{DDES}=\min\left[\frac{l_{DES}}{(1-F_{SST})},C_{DES}\Delta\right] \tag{3.61}$$

The function F_{SST} takes the form of F_2 (Menter 1993) in the SST model. It takes a value of zero outside the boundary layer and tends to unity inside it. For $F_{SST}=0$ the approach reduces to standard Menter SST DES.

3.8.3 DES Performance for Separated Flow

Separated flows present differing challenges for DES type methods. These are dependent on the nature of the separation. Figure 3.14 gives schematics of some cases involving separation that are of increasing challenge to RANS and hence hybrid RANS-LES techniques. It is the RANS layer that controls the separation point of the flow. Hence, this RANS limitation is inherited in any hybrid simulation.

For Case (I), shown in Fig. 3.14, the task for DES is relatively simple. The separation point is geometrically defined. Convection of the RANS layer, eddy viscosity, could delay shear layer development. However, there is no difficulty in capturing the separation point. For Case (II), with $\delta \ll h$, the separation will be strongly controlled by the pressure gradient. Both cases (I) and (II) will have strong shear layers that greatly dilute the influence of the upstream boundary layer. They both involve massively separated flows and so are ideal for DES. For Case (III), the dynamics of the incoming boundary layer are much more critical. The case presents a much greater DES challenge. There is no massive separation. Also, the separation point is much more challenging to capture. Once captured, the separated flow involves smaller resolved scales and hence needs a finer grid.

The Spalart-Allmaras, like many models will tend to under predict the separation onset. For mildly separated flows the limiter used in the Menter SST model is helpful (relative to the behaviour of the baseline model) in modelling separated flow behaviour. However, for more severe pressure gradients, there is evidence that, the limiter too strongly prevents eddy viscosity growth. Hence, for mildly separated flows a Menter SST based DES could be useful.

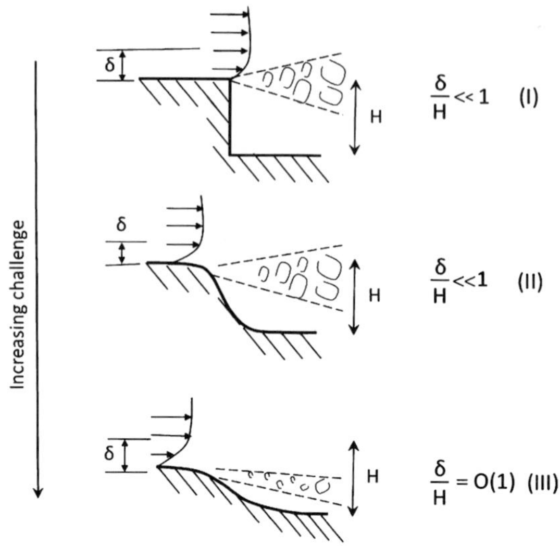

Fig. 3.14 Schematics of cases involving separation that are of increasing challenge (from frames (I–III)) to hybrid RANS-LES techniques (based on Deck 2012)

3.8.4 Explicitly Zonalised Methods

Laraufie et al. (2011) propose a multizone hybrid RANS-LES approach. With this, through marker zone variables RANS, DES and LES zones can be explicitly identified in a multi-block mesh environment. If:

$$N_{NB}^{I}=0 \quad \text{a RANS zone is defined and if}$$

$$N_{NB}^{I}=1 \quad \text{a DES zone is defined.}$$

In the above, *NB* is the block number. Another layer of logic follows the above. With this, for $y^+ < y_{int}^+$, where y_{int}^+ is say 100 then $N_{NB}^{II}=0$ forcing a RANS zone. For $y^+ \geq y_{int}^+$ and $N_{NB}^{II}=1$. The above flags are combined to overwrite the standard DES modelled length scale ($\tilde{d}=\min(d, C_{DES}\Delta)$) thus avoiding automatic loss of the RANS layer below y_{int}^+. Keeping a RANS-LES interface at a constant location means that the skin friction more sensibly converges with grid refinement (Deck et al. 2011). Using the above flags the standard DES length scale is re-expressed as

$$\begin{aligned}\tilde{d}_{NB} &= \left(1-N_{NB}^{I}\right)d \quad \text{RANS domain}\\ &+ N_{NB}^{I}N_{NB}^{II}\tilde{d} \quad \text{DES zone}\\ &+ N_{NB}^{I}\left(1-N_{NB}^{II}\right)d \quad \text{RANS within DES zone}\end{aligned} \tag{3.62}$$

Hence, explicit, extensive RANS zones can be defined and the near wall RANS zone shielded, based on wall unit distance estimates. At non-wall-normal RANS-LES interfaces, the modelled RANS turbulence levels can be converted into resolved scales using synthetic eddy generation techniques.

A further evolution of the above zonalization strategy can be found in Deck (2012). With this, again, multiple zones are defined. However, these are as follows:

(a) a pure RANS zone;
(b) a DES97 zone;
(c) a DDES zone and
(d) a wall resolved type LES zone.

Element (d) is suitable for dealing with Case (III) zones. On top of the above, different filter scales can be defined that are most appropriate for different flow zones. The strategy very much needs the user to engage with the flow physics and grid topologies found in different zones but seems sensible. It is consistent with the practical realities of using hybrid RANS-LES methods.

An interesting approach is that of Hamba (2001). With this the RANS-LES zonalization is reversed. Hence, the near wall RANS layer is replaced by LES and the LES zone by RANS.

The zonal RANS-ILES approach of Tucker (2004), Shur et al. (2006) and Morris et al. (2010) could be viewed as part of the explicitly zonalized LES family. With this approach, based on marked areas either a RANS layer or implicit LES zone is active. This method is used quite extensively here and so is outlined below.

3.8.5 Hybrid RANS-ILES Method

If the free stream turbulence levels are insufficient, the SA model can be prone to laminarization (Spalart and Rumsey 2007). If a NLES/MILES/ILES zone is used there will be no modelled turbulence outside the RANS zone. Hence, unless care is taken, there is the potential for laminarization of the RANS layer with SA. Therefore, *k–l* or even mixing length modelling is preferred with this approach for the RANS zone. The *k–l* based method is outlined below. With this, for the RANS part, the modelled turbulent kinetic energy, *k*, equation is, again expressed in phase averaged form, as follows:

$$\frac{\partial \rho\langle k_T\rangle}{\partial t} + \frac{\partial \rho\langle u_i\rangle\langle k_T\rangle}{\partial x_i} = \frac{\partial}{\partial x_j}\left[\left(\mu + \frac{\mu_T}{\sigma_k}\right)\left(\frac{\partial\langle k_T\rangle}{\partial x_j}\right)\right] + P_k - \frac{\rho\langle k_T\rangle^{3/2}}{l_\varepsilon} \tag{3.63}$$

For the (N/I/MI)LES zone $k_T = k_{SGS} = 0$. For the near-wall RANS region the *k–l* model of Wolfshtein (1969) is mostly used where

$$\mu_T = \rho C_\mu l_\mu \langle k_T\rangle^{1/2} \tag{3.64}$$

The length scale in Eq. (3.64) is

$$l_\mu = C_l \tilde{d}\left(1 - e^{-A_\mu y^*}\right) \tag{3.65}$$

For Eq. (3.63) the following is used

$$l_\varepsilon = C_l \tilde{d}\left(1 - e^{-A_\varepsilon y^*}\right) \tag{3.66}$$

and $y^* = \tilde{d}\rho k^{1/2} C_\mu^{1/4}/\mu$. Constants have the following standard values: $C_\mu = 0.09$, $C_l = 2.4$, $A_\mu = 0.016$, $A_\varepsilon = 0.263$ and $\sigma_k = 1$. For demarcation of zones, the distance function $\tilde{d}$ is used. This can be evaluated from a Hamilton-Jacobi (HJ) equation. This is given below

$$|\nabla \tilde{d}| = 1 + f(\tilde{d})\nabla^2 \tilde{d} + g(d) \tag{3.67}$$

Here

$$\tilde{f}(d) = \varepsilon_{HJ} \tilde{d} \tag{3.68}$$

and

$$g(d) = a\left(\sin(\pi R/2)\right)^{b/R} \tag{3.69}$$

or an alternative function. In the above $R = \tilde{d}/L_{RANS}$ where the length scale L_{RANS} is the distance from the wall to the (N/I/MI)LES region. Also, a and b are constants.

The left hand side of Eq. (3.67) is discretized using first order Godunov upwind type differences. For the Laplacian, second order central differences are used. With $\varepsilon_{HJ} = a = 0$ Eq. (3.67) reduces to the hyperbolic natured eikonal equation. Weak viscosity solution of this give exact nearest wall distances $d = \tilde{d}$. The Eikonal equation is solved by propagating fronts from solid surfaces (Sethian 1999). Here, for hybrid ILES-RANS solutions the equation is propagated towards the RANS-ILES interface. Then, at the interface, the condition $d = 0$ is set and Eq. (3.68) solved with $\varepsilon_{HJ}, a > 0$. The Laplacian enables a smooth transition between the modelled RANS length scale (that needs an accurate d) and the ILES zone (needing $d = 0$). The function $f(\tilde{d})$ forces the Laplacian to tend to zero near the walls. This ensures near wall distances are accurate. For hybrid ILES-RANS the function $g(d)$ controls the RANS length scale in the vicinity of the ILES zone. Figure 3.15 gives $\tilde{d}$ distributions for different values of b with $a = 1$. Despite the simple appearance of the hyperbolic natured eikonal/HJ equations they are not trivial to solve Tucker (2003). For example, when using transformed coordinate, curvilinear grids, transformed equation metric terms need to be discretized using offset metrics. Solution of the equation on unstructured grids is discussed in Tucker et al. (2011). For simulations presented here typically $\varepsilon_2 \simeq 0.2$ and $\varepsilon_3 \simeq 1.5$. Note, for some simulations, presented later, for convenience, just a curve fit to the HJ field is used. Also, where the SA model is used, to prevent laminarization the RANS layer can be frozen. Alternatively, in the ILES zone, $\tilde{d}$ is made a finite value that is sufficient to maintain the RANS layer but much smaller than an LES filter width.

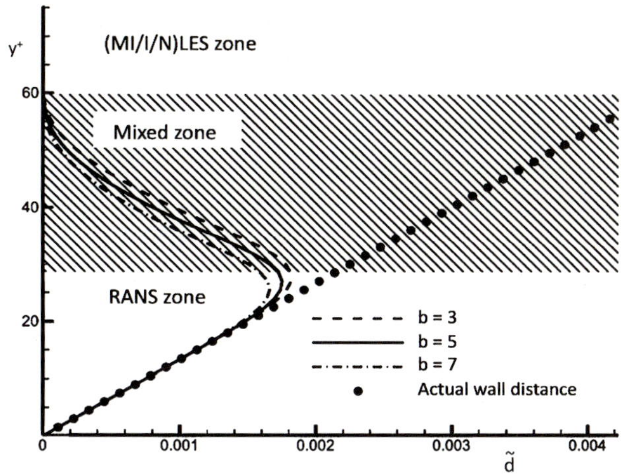

Fig. 3.15 Hybrid (N/I/MI)LES-RANS distance function

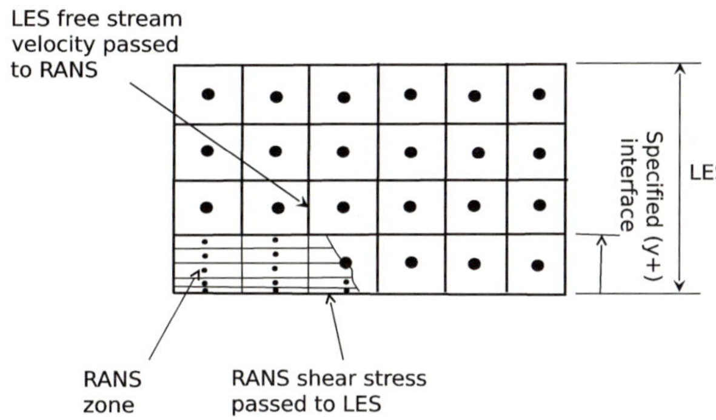

Fig. 3.16 Two distinct grids used in the two-layer model

3.8.6 Two-Layer Model

With the two-layer approach, described by Piomelli and Balaras (2002), as shown in Fig. 3.16, two-separate grids are used. The grids overlay each other. The LES grid is relatively isotropic. Beneath this is a RANS grid with high aspect ratio cells. The RANS grid solves simplified parabolized boundary layer equations. These take a uniform pressure field from the LES zone. This allows, a reduced set of one-dimensional near-wall governing equations to be solved. This would typically take place with a simple turbulence model like the mixing length. The near wall shear stress from the RANS is used as the boundary condition for the LES. The velocity field from the LES, located at the outer edge of the RANS layer, is used as a boundary condition for the RANS. A key assumption of this method is that the coupling between the inner and outer parts of the boundary layer is weak. A similar approach is outlined by Leschziner et al. (2009). However, in this work a strong correlation is found between wall shear stress and events some distance away from walls.

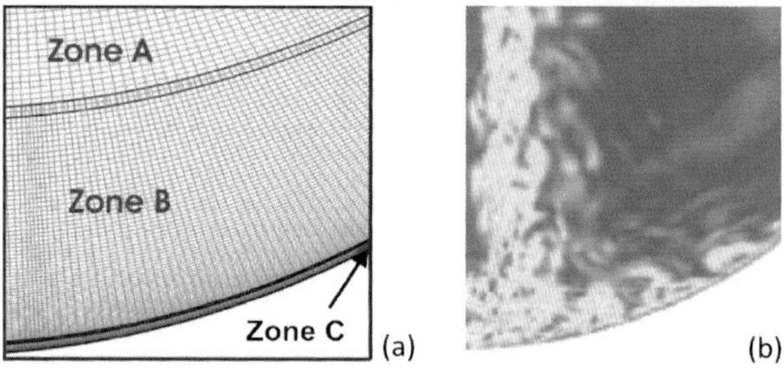

Fig. 3.17 Zonal grid embedding: (**a**) mesh structure and (**b**) instantaneous flow field (Jefferson-Loveday and Tucker 2011). Published with kind permission from Taylor and Francis

3.8.7 Grid Embedding

A further approach sometimes classed with hybrid approaches is the use of a fine embedded grid to resolve the near wall scales (Addad et al. 2008). Such a grid topology, taken from Jefferson-Loveday and Tucker (2011) is shown in Fig. 3.17. Frame (a) shows a three-zone grid. Zone C is the finest near wall grid. The grid is coarsened in Zone B and then A. Frame (b) gives an instantaneous flow field image. Grid embedding is certainly a sensible approach. However, it does not really alleviate the severe grid scaling requirements for LES. The LES scalings/estimates devised by Chapman (1979) already assume that the grids are nested in an idealized fashion. Addad et al. (2008) and Jefferson-Loveday and Tucker (2010, 2011) successfully apply this approach to flow over a controlled diffusion aerofoil and impinging jets, respectively. As noted by Addad et al. deciding on the optimal grid distribution needs considerable experience.

3.8.8 LNS and Related Methods

Essentially the Batten et al. (2002) LNS (Limited Numerical Scales) method is inspired by Speziale (1998). With LNS, the eddy viscosity takes the following form

$$\mu_t = \alpha \mu_{RANS} \tag{3.70}$$

where μ_{RANS} is the eddy viscosity that the base RANS model would provide (a cubic eddy viscosity model is used by Batten et al.) prior to multiplication by α. The variable, α, is called the latency parameter. This is defined as

$$\alpha = \frac{\min[(LV)_{LES}(LV)_{RANS}]}{(LV)_{RANS}} \tag{3.71}$$

In the above $(LV)_{LES} = C_s\Delta^2|\bar{S}|$ is the product of the LES velocity and length scales. Also, for RANS $(LV)_{RANS} = \delta + C_\mu k^2/\varepsilon$. Note, δ is a small number. As can be seen, this is formulated for a k–ϵmodel but this is not a rigid element. The model can either switch to a fully RANS or fully LES mode or some state between these two extremes. This depends on the level of turbulence resolution.

An eddy viscosity transport equation based model, of a broadly similar genre, is proposed by Menter et al. (2003). This model makes use of the von Karman turbulence length scale. It is called the Scale Adaptive Simulation (SAS) model. Joo and Durbin (2009) apply this approach when making hybrid RANS-LES for a cut-back film cooling flow representative of that found on a turbine blade.

For a simple boundary layer, the von Karman length scale takes the form below

$$L_{vK} = \left| \frac{\frac{\partial \bar{u}}{\partial y}}{\frac{\partial^2 \bar{u}}{\partial y^2}} \right| \tag{3.72}$$

For multi-dimensional calculations Menter gives the following formulation

$$L_{vK} = \sqrt{\frac{\frac{\partial \bar{u}_i}{\partial x_j}\frac{\partial \bar{u}_i}{\partial x_j}}{\frac{\partial^2 \bar{u}_j}{\partial x_l^2}\frac{\partial^2 \bar{u}_j}{\partial x_m^2}}} \tag{3.73}$$

The above length scale is inserted into a transport equation for eddy viscosity. This is very broadly akin to that used for the SA model or more closely aligned with the Nee and Kovasznay (1969) model. Hence, there is a $(\mu_t/L_{vK})^2$ destruction term (yielding, in the absence of convection and diffusion, the usual mixing length formulation). Care is needed to avoid a division through by zero.

3.8.9 NLDE

More recently, the use of so called nonlinear disturbance equations (NLDE) has been made. With this approach the mean flow, $\overline{\phi}$, is first solved for using RANS. Then a disturbance equation, for ϕ'', is solved. This retains non-linear terms. Notably these disturbances do not interact with the mean flow. Similar to URANS, there is potentially a three-level flow decomposition where

$$\phi = \overline{\phi} + \phi'' + \phi' \tag{3.74}$$

In the above ϕ' represents turbulent fluctuations at a subgrid scale. Labourasse and Sagaut (2002) incorporate ϕ' when modelling the flow over a low-pressure turbine blade section. The Smagorinsky LES model is used to estimate ϕ'. For jet noise, Morris et al. (1997) neglects the modelling of ϕ'. Notably, Batten et al. (2002, 2004) use a stochastic reconstruction of ϕ'—hence there is a dynamic forcing. The combination of the three layers can produce flows that appear rich in scales and akin to LES. Indeed such approaches are found in the LES literature and marketed as the most economical limiting extreme of LES. For a low pressure turbine, for example, as demonstrated by Labourass and Saguat, a two-dimensional RANS solution can be made. This can be interpolated onto a three-dimensional, coarsened, grid to solve the NLDE. The NLDE can be solved on a much coarser grid than needed

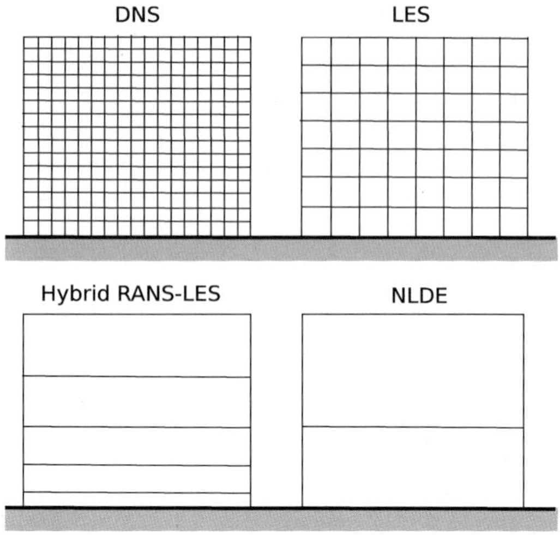

Fig. 3.18 Relative grid resolution requirements for eddy resolving methods

for normal LES. This is because the approach has a much higher modelled content. Belme et al. (2010) present a naturally zonalized NLDE approach, where areas of the flow are strictly RANS (when the grid is deemed too coarse) and others NLDE. The switch between zones is automatically defined. Encouraging results have been found, especially for acoustics problems.

Figure 3.18 summarizes the grid requirements for the eddy resolving methods noted above. As indicated, DNS has the highest grid resolution requirements and NLDE the lowest.

3.9 Hybrid RANS-LES Deficiencies

A disadvantage of the NLDE approach is that for a truly unsteady flow, where the approach is most attractive, a steady RANS solution is unlikely to be possible to iteratively converge. As discussed in Chap. 2, if a strong solver is used, to secure convergence, then physical realism can be lost. Also, the extrusion from two-dimensional to three-dimensional geometry reduces the class of problems for which the NLDE approach can be used.

LNS and approaches of this kind have the attractive element that they automatically pick the solution strategy appropriate to the grid density. However, a key drawback of them is that complex RANS-LES interfaces can be generated. Frequently these are such that LES zones are located downstream of RANS zones. Hence, unless very specialized treatments are used, the LES zone will initially have no resolved scales. For LNS, and approaches of this type, acoustics prediction is also complicated (Batten et al. 2007). For example, it is hard to convert the spatially and temporally varying differing levels of modelled and resolved scales into acoustics output. Figure 3.19 shows the RANS-LES zonalizations for the LNS approach of Batten et al. and DES. The flow considered is, three dimensional and recirculating.

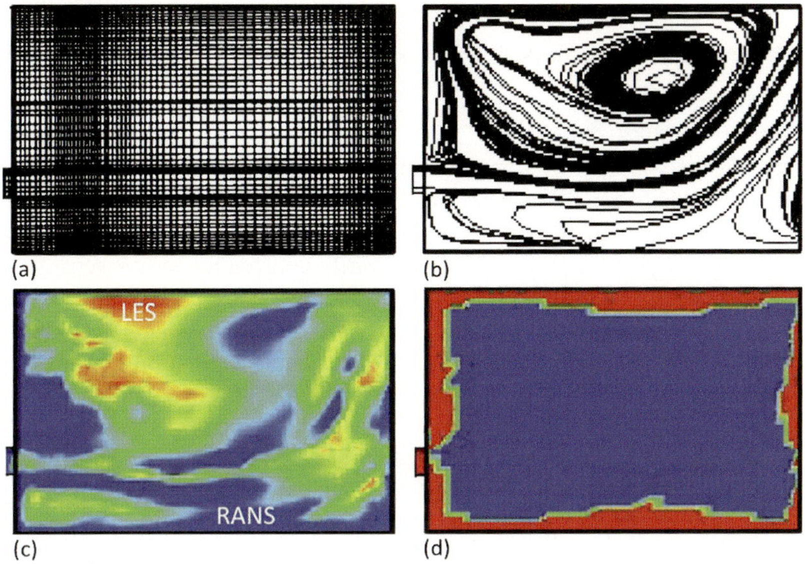

Fig. 3.19 Potential RANS-LES zonalisations for a recirculating flow: (**a**) grid; (**b**) flow streamlines; (**c**) LNS type interface and (**b**) DES (Tucker 2011). Published with kind permission from Elsevier

It arises from a jet of fluid entering the lower left hand side of the domain. However, the details of the flow are unimportant for the points to be illustrated. Frame (a) shows the grid. Frame (b) gives the mean flow field. Frame (c) shows the RANS and LES zones for the LNS approach. Finally, Frame (d) gives the RANS-LES interface for DES. It is clear that for the LNS approach (Frame (c)), the RANS-LES interface is highly non-uniform. As noted, it needs specialist treatments to avoid the underlying consequences of this. For this, and other reasons (Georgiadis et al. 2010) this type of approach and related methods, such as PANS—Partially Averaged Navier-Stokes—(Girimaji 2006), and partially integrated transport model—Chaouat and Schiestel (2005)—are not further considered here. They have found some use in, for example, combustion modelling. Nonetheless, as can be seen from Frame (d), the grid-controlled interface from DES can also have irregularities. This can result, as discussed earlier, in LES occurring inside the boundary layer where the grid resolution is inadequate.

The interface between the RANS and LES zones is know as the 'grey area'. When using this term it is mostly referring to the wall parallel zone between a near wall RANS layer and the LES region, as one moves in the wall normal direction. However, there is also a 'grey area' as flow emerges from a boundary layer into a free stream. This occurs for jets, where the flow inside the nozzle is treated in RANS mode and the flow emerges from these layers into the free shear layer. This zone will also occur at aerofoil trailing edges. Hence, we can discern two key types of 'grey area':

(a) RANS to LES in classical boundary layers and
(b) RANS to LES in free shear flow zones.

Mildly separated flows present a more complex form of the type (b) interface. The issue with the type (a) 'grey area' is that in the RANS layer there is a large eddy viscosity. As the wall is moved away from, the RANS layer gives way to the LES zone. In moving between these zones, there is a sudden loss of eddy viscosity. However, in moving from the RANS to LES zone, there is insufficient opportunity for resolved scales to develop (this can also be partly delayed by eddy viscosity 'leaking out' from the RANS zone). This results in a potential dramatic loss of wall shear stress, unless the mean velocity profile makes a strong adjustment. This gives rise to a kink in the velocity profile and a substantial under prediction of skin friction coefficient (Nikitin et al. 2000). Moving the inner layer inwards, and also reducing subgrid scale dissipation does not overcome the issue (Piomelli et al. 2003).

For the type (a) interface, to more rapidly energize resolved scales, Piomelli et al. (2003), Davidson and Billson (2006) and Davidson (2009) inject turbulence at the RANS-LES interface. Piomelli et al. (2003), and Davidson and Billson (2006) use synthetic turbulence. This is based on Gaussian random numbers and homogeneous decaying turbulence based on a modified von Karman spectrum, respectively. Davidson (2009) achieves energization through a scale-similarity subgrid model (Bardina et al. 1980). Backscatter from this is used to produce energetic scales.

An opposite issue is the excessive energization of the RANS zone. The resolved scales in the LES buffet the RANS layer. Consequently, although the expectation is that the RANS zone is steady it actually becomes a URANS zone. Hence, the use of the phase averaging in the above equations for the 'RANS' zone of hybrid RANS-LES is justified. The result of the buffeting is that when the resolved and modelled turbulence energies are added the RANS zone shows an excess of turbulence energy. Zhong and Tucker (2004) attempted to damp this unsteadiness through temporal filtering. In a different sense, an ad hoc spatial filtering approach is attempted by Hamba (2006). A key problem is that, although it is desirable to force a steady RANS layer thus avoiding double accounting; an unsteady boundary condition is needed at the RANS-LES interface for the LES. Temmerman and Leschziner (2002) take a further different tack, adjusting the RANS model calibration. The intention being to reduce RANS modelled turbulence levels, to ensure continuity of eddy viscosity between the RANS and LES zones. Gieseking et al. (2011), bases the RANS-LES switch over by looking at the outer layer and inner layer (a function of wall distance) length scales. The outer layer scale estimate is based on the resolved and modelled turbulence kinetic energies and frequencies. The inner and outer length scale ratio is used to construct a blending function. All of the above efforts have shown some promise but only in a preliminary and ad hoc sense.

For the type (b) 'grey area' the most appropriate solution appears to be that of Batten et al. (2004). With this, turbulence energy from an upstream RANS zone is converted into resolved turbulence through a synthetic turbulence reconstruction procedure. Downstream control planes can also be added. However, the coding complexity will greatly increase. As will be seen later there are a range of methods for

Table 3.7 Some potential filter options

Filter description	Filter length scale in model Δ
Quasi-isotropic Cartesian grids	$(\Delta x \Delta y \Delta z)^3$
Anisotropic grids	$\sqrt{(\Delta x^2 + \Delta y^2 + \Delta z^2)/3}$
Anisotropic grids	$\max(\Delta x, \Delta y, \Delta z)$
Anisotropic grids, Andersson et al. (2005)	$\min(\Delta x, \Delta y, \Delta z)$
Anisotropic grids, Zahrai et al. (1995)	$(\Delta x \Delta y \Delta z)^{1/9}(\Delta x_j)^{2/3}$
Anisotropic grids, Scotti et al. (1993)	$(\Delta x \Delta y \Delta z)^{1/3} \cosh(4[(\ln c_1)^2 + (\ln c_2)^2 - \ln c_1 \ln c_2]/27)$
Hybrid RANS-LES, Batten et al. (2011)	$2\max(\Delta x, \Delta y, \Delta z, \|\lambda\|\Delta t, L)$
Batten et al. (2007)	$2\max(\Delta x, \Delta y, \Delta z, \|u_i\|\Delta t, L_{vK})$
Hybrid RANS-LES, Mani (2004)	$2\max(\Delta x, \Delta y, \Delta z, \|u_i\|\Delta t, k\Delta t)$
Hybrid RANS-LES, Batten et al. (2007)	$2\max(\Delta x, \Delta y, \Delta z, L_{vK})$
Hybrid RANS-LES, Batten et al. (2007)	$2\max(\Delta x, \Delta y, \Delta z, L_{min})$
Vorticity aligned with a grid line, Chauvet et al. (2007)	$\sqrt{n_x^2 \Delta y \Delta z + n_y^2 \Delta x \Delta z + n_z^2 \Delta x \Delta y}$
General definition of Chauvet et al.'s—Deck (2012)	$\overline{A_\omega}$
Curvilinear finite difference	$(J \Delta\xi \Delta\eta \Delta\zeta)^{1/3}$
Finite volume	$(Vol)^{1/3}$
Unstructured control volume—Batten et al. (2007)	$4\max_{k=0,\dots,n}[\|x_c - x_k\|]$
Unstructured control volume, Spalart	Maximal circle or sphere that encompasses cell
Farge and Schneider (2001)	Wavelet based
NLES, ILES...	$\Delta \rightarrow 0$[a]

[a]As input to the modelled scales

generating resolved scales from RANS based information. The chosen approach can have a strong solution impact.

3.9.1 Filter Choices

For both hybrid RANS-LES and LES, grid based filters are generally defined. Typical filter choices are summarized in Table 3.7. For curvilinear, finite differences, as noted in Chap. 2, J, is the usual Jacobian. Also, $\Delta\xi$, $\Delta\eta$ and $\Delta\zeta$ the grid spacings in the computational plane. Furthermore, c_1 and c_2 are grid aspect ratios and *Vol* the cell volume. The filter of Zahrai et al. (1995) has a tensorial/matrix form. In the filter of Chauvet et al. (2007), $\mathbf{n}$ is the unit vector aligned with vorticity. In the Batten et al. (2011) filter, λ is wave speed. In Batten et al.'s unstructured mesh definition, n is the number of faces. Also, $\mathbf{x}_k$ is the midpoint of a face k and $\mathbf{x}_c$ is the cell centroid. The maximal circle or sphere concept of Spalart is unpublished but reported in Batten et al. (2007). Also, in Batten et al. (2007)—see table—the actual eddy resolvable based on the selected time step (and magnitude of local velocity, $|u_i|$) is taken into account. This is consistent with the notion of space-time filtering, discussed earlier. As noted in Batten et al. (2007), with the Mani (2004) definition, large, transient, spurious, k, values can contaminate solutions. In the Deck (2012) definition, $\overline{A_\omega}$ is the average cross-sectional area of a cell. This is normal to the vorticity vector, as utilized by Chauvet et al. Note, L_{min}, in Table 3.7, for hybrid RANS-LES (Batten et al. 2007) is a user prescribed, minimum cell length scale. This additional parameter can be used to shield a RANS layer zone that has dense grid clustering. This can occur around regions of high surface curvature. For high aspect ratio, aerodynamic, grids the cell aspect ratio can be 1000 to 1. The tabulated filter definitions hence would give a range of this order. Therefore, although there is frequent careful reporting of small variations in, for example, the Smagorinsky constant, it is the filter length scale that has an extremely powerful influence. This will potentially dominate the actual subgrid scale model choice.

3.10 Generation of Resolved Turbulent Inflow

The challenges for outflow conditions in eddy resolving simulations are outlined in Chap. 2. However, also, frequently the characterization of correct inflow is a critical issue for LES. This area, in itself, could be the subject of a book. Hence, the discussion below is brief. More extensive reviews of different approaches can be found in Jewkes (2008), Keating et al. (2004) and Tabor and Baba-Ahmadi (2010).

Fortunately, for jets, and many flow involving detached shear layers, there can be strong shearing. Hence, the highly unstable and inflectional velocity profiles will rapidly generate resolved velocity scales. Then, turbulence inflow becomes less critical, but as shown in Fig. 3.11, can have a tangible impact. There are numerous methods for generating inflow turbulence. A relatively convenient approach is to have an extended upstream domain, with some form of numerical trip. Mullenix et al. (2011) and Bisek et al. (2013) use a body force based trip. This produces a force in a direction that opposes the flow direction. The force generates a separation bubble. The arising shear layer generates turbulence. Blowing and suction has also been explored for this purpose by other workers. Obviously, the downside of such approaches is the increased computational cost.

3.10.1 Separate Inflow Simulations

Precursor simulations can be made and this data stored in a database. The data can undergo some scaling to make the turbulence more compatible with the desired inflow field—see later. Alternatively, a simulation can be run concurrently.

Wu and Hickey (2012), through a precursor simulation, create DNS inflow for a high-pressure turbine blade. This is achieved by, first, synthetically generating an initial flow field for homogeneous decaying turbulence. This is then decayed in time in a tri-periodic domain. Random numbers are used to produce 25 different initializations and hence realizations. The decaying of the turbulence is stopped when the turbulence reaches a specified intensity. The realizations are then linked together, as a database, to be marched though in time. When the end of the time series is reached a loop is made back to the start of the sequence.

A key method for generating turbulence inflow is through use of re-cycling procedures. This can involve simple pipe or channel flow LES/DNS with periodic flow boundary conditions in the streamwise direction, giving rise to fully developed boundary layers. Schluter et al. (2005) apply this approach to whole engine simulations. However, (for the compressor) RANS information is used to set the mean velocity and turbulence levels. The recycled pipe flow simulation, from a stored database, is used to convert this into unsteady inflow information. The expression below is used for this purpose

$$\bar{u}_{i,LES}(t) = \underbrace{\bar{u}_{i,RANS}}_{I} + \underbrace{\left[\bar{u}_{i,DB}(t) - \bar{\bar{u}}_{i,DB}\right]}_{II} \bullet \underbrace{\frac{\sqrt{\overline{\overline{u'^2_{i,RANS}}}}}{\sqrt{\overline{\overline{u'^2_{i,DB}}}}}}_{III} \tag{3.75}$$

In the above the subscript 'RANS' identifies RANS data and 'DB' data from the database. The double bars represent an ensemble average of the LES. Term (I) gives the mean RANS velocity profile data. Term (II) is the difference between the instantaneous and mean velocities at a point taken from the database. Term (III), effectively, scales the fluctuations from the database so that they are consistent with the RANS target.

The most popular, general procedure, that allows for developing boundary layers, and follows on from work of Spalart, is due to Lund et al. (1998). Figure 3.20 gives a schematic of Lund's recycling procedure. As shown in Fig. 3.20, there is an inflow boundary (I) (where a target momentum thickness is specified), a recycling station (II) and an outflow boundary (III). Data is taken from station (II) and recycled into the inflow plane. Since the boundary layer is developing, basic boundary layer scaling laws need to be applied. These are based on the expected inner and outer layer boundary layer dynamics. The resulting unsteady data can either be stored or the recycling and main flow simulations carried out concurrently (Stolz and Adams 2003). Lund's recycling approach does not have to be for classical smooth walls. It can, for example, include roughness (Nozawa and Tamura 2002). Clearly, there is the expectation of meeting some target state in terms of momentum thickness and turbulence levels. These can be based on measurements or even RANS based information. The adaptation of Lund's method to compressible flow is outlined by Stolz and Adams (2003). Although the basic approach sounds in essence simple, there are numerous potential implementation difficulties (Jewkes 2008) and hence variants. What is more, for many practical flows, there can be classical boundary layer

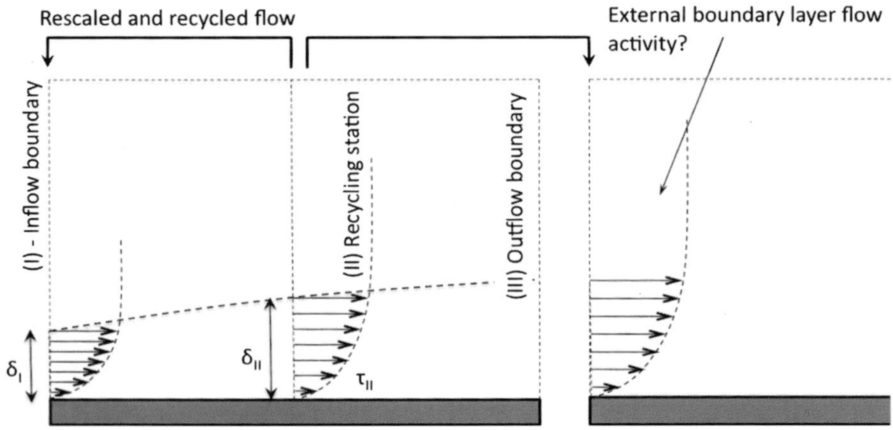

Fig. 3.20 Schematic of Lund's recycling procedure

content and also flow activity external to the boundary layer. This could include wakes from turbine blades along with inter-wake turbulence. When Lund's recycling is used, these two, coupled elements, would need to be blended and scaled in some plausible way. Wu (2010) terminates his wake modelling at some multiple of the momentum thickness away from the boundary layer. For high-pressure turbine simulations, Vadlamani (2013) combines Lund's recycling with wakes (from Wu) and also isotropic box turbulence (for the inter-wake zone external to the boundary layer). Figure 3.21 shows the merging of these components at the simulation inflow plane. In the figure, the blending of these components is less gradual. This helps identify the three different inflow components. For hybrid RANS-LES, if unmodified, Lund's recycling would force resolved scales into the RANS zone, which is undesirable.

3.10.2 Synthetic Turbulence

Synthetic turbulence can be generated based on RANS information or even measurements. One of the earliest approaches of this type was as developed by Kraichnan (1970) (for homogeneous isotropic turbulence). This is extended by Smirnov et al. (2001) to model anisotropic turbulence. The latter approach was subsequently applied to low-pressure turbine LES by O'Mahoney (2011). However, the turbulence showed too rapid a decay. This is a key problem with such approaches. Depending on the underlying RANS model, anisotropy can be specified when synthetically generating turbulence. With this, RANS based second moments and the rate of dissipation must be specified. Batten et al. (2004) (recently extended in Batten et al. 2011) presents a simplified version of Smirnov et al.'s approach. With this, the

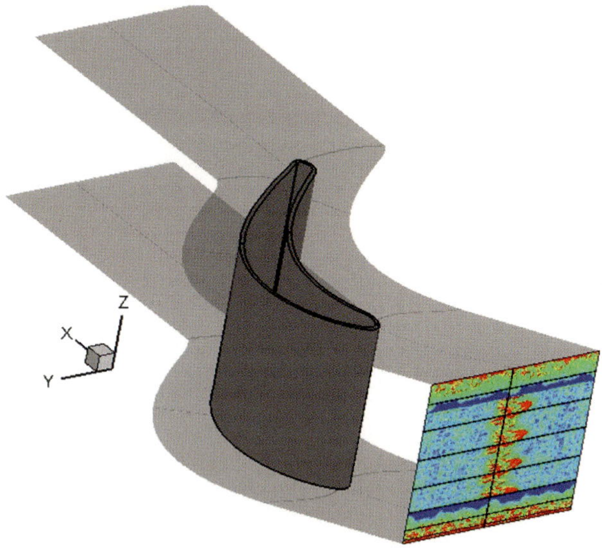

Fig. 3.21 Merging of wake turbulence, Lund's recycling and homogeneous turbulence from a high-pressure turbine simulation, at the simulation inflow plane (Vadlamani 2013)

velocity field is constructed using the following equation

$$u_i(x_j,t) = a_{ik}\sqrt{\frac{2}{N}}\sum_{n=1}^{N}\left[p_k^n \cos\left(\hat{d}_j^n \hat{x}_j + \omega^n \hat{t}\right) + q_k^n \sin\left(\hat{d}_j^n \hat{x}_j + \omega^n \hat{t}\right)\right] \qquad (3.76)$$

where

$$\hat{x}_j = \frac{2\pi x_j}{L}, \qquad \hat{t} = \frac{2\pi t}{\tau}, \qquad \hat{d}_j^n = \frac{d_j^n V}{c^n}, \qquad V = \frac{L}{\tau},$$

$$c^n = \sqrt{\frac{3}{2}\overline{u_l' u_m'}\frac{d_l^n d_m^n}{d_k^n d_k^n}}, \qquad p_i^n = \varepsilon_{ijk}\eta_j^n d_k^n, \qquad q_i^n = \varepsilon_{ijk}\xi_j^n d_k^n,$$

$$\eta_i^n, \xi_i^n = N(0,1), \qquad \omega^n = N(1,1), \qquad d_i^n = N(0,1/2)$$

and a_{ij} is given by

$$a_{ij} = \left\langle \begin{array}{ccc} \sqrt{\overline{u_1' u_1'}} & 0 & 0 \\ \overline{u_1' u'_2}/a_{11} & \sqrt{\overline{u_2' u_2'}\ a_{21}^2} & 0 \\ \overline{u_1' u'_3}/a_{11} & (\overline{u_2' u'_3}\ a_{21} a_{31})/a_{11} & \sqrt{\overline{u_3' u_3'}\ a_{31}^2\ a_{32}^2} \end{array} \right\rangle \qquad (3.77)$$

The elements in the above are real if the RANS generated Reynolds stresses are realizable. The time and distance components in the Fourier space model are scaled by a time scale, τ, and a velocity scale c^n. In the above $N(a,b)$ are normally distributed random numbers with a mean of a and standard deviation of b. Also,

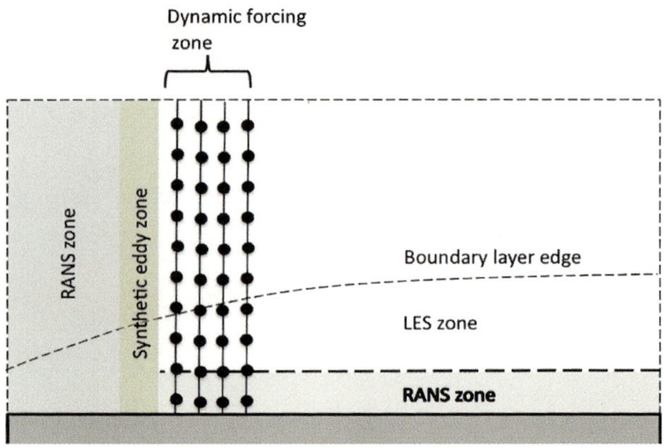

Fig. 3.22 Dynamic forcing method in conjunction with the synthetic eddy generation method and hybrid RANS-LES

ε_{ijk} is the alternating symbol. The problem with synthetically generated turbulence is that it can exhibit an initial decay. This could be partly because of the fidelity of the generated structures but also due to numerical errors and also LES model influences. To overcome this problem some workers have used downstream control zones where forcing is imposed (Keating et al. 2004). The approach of Batten et al. is very much framed around hybrid RANS-LES.

Again, with hybrid RANS-LES in mind Laraufie et al. (2011) explore the use of a dynamic forcing method in conjunction with the synthetic eddy generation method of Jarrin et al. (2009). An extensive upstream RANS zone is defined, as shown in Fig. 3.22. A synthetic eddy generation zone, that absorbs turbulence scale information from the RANS zone follows this. A zone containing dynamic forcing planes then follows. This is implemented based on the approach Spille-Kohoff and Kaltenbach (2001). This approach utilizes a source term in the momentum equation in the wall normal direction. A proportional and integral controller is used to pin the shear stress to a target distribution. However, in the approach of Laraufie et al. the self-adaptive process is found to perform better when the wall normal Reynolds stress is forced to a target value by the controller. This reduces the necessary development region by nearly 80 % compared to using just the synthetic eddy generation zone. The wall normal Reynolds stress is the most directly linked fluctuation component to shear stress production. Indeed for a simple boundary layer the main contributor to shear stress production is

$$P_{shear} = \overline{v'v'}\frac{dU}{dy} \tag{3.78}$$

Hence, directly controlling v' has a better link than controlling the shear stress $-\rho\overline{u'v'}$. Mathey et al. (2003) devise a simple synthetic vortex method. With this, two-dimensional vortices aligned with the inflow direction are generated. Fluctuations in the streamwise direction are expected to naturally develop.

The digital filtering approach of Klein et al. (2003) is also a popular method for generating turbulence inflow. With this, a random data series about a mean of zero is generated. This is then subjected to the digital filter. The filter coefficients can be generated based on a correlation function for the turbulence field that absorbs a modelled turbulence length scale. Proper Orthogonal Decomposition (PODs) can be used, for example, to take time resolved experimental measurements, such as PIV, and impose them at turbulence inflow boundaries (Perret et al. 2008; Druault et al. 2004) with minimal data storage. In the work of Perret et al. the PODs are just used to capture the low order modes. The smaller scales are imposed using Gaussian random numbers, adjusted to give a correct turbulence energy spectrum. Hence, the approach is hybrid. Druault et al. (2004) use proper orthogonal decomposition to generate inflow for shear layers. For such flows, POD can characterize the large scale coherence of the flow.

3.10.3 Advantages and Disadvantages

With respect to cost and user ease, synthetic turbulence generally seems the best option. As noted by Mathey et al. (2003), for the high Reynolds numbers found in engineering systems and typical multiple flow boundaries found in for example, gas turbines, recycling methods have limited practical applicability. However, for simple boundary layers, synthetic turbulence will be less authentic than recycled. Therefore, a longer development zone will generally be needed. Control plane forcing can be used to reduce the extent of the initial turbulence adjustment period. However, this adds another user complication. Synthetic turbulence construction procedures have the added advantage that they can also be used as a subgrid type model. Thus, they can specify unresolved turbulence. This could be used, for example, as input to acoustic models. This aspect is discussed in Chap. 6.

Figure 3.23 summarizes the potential inflow options noted above. Some key elements relating to the relative performance of different inflow techniques are:

(a) How long a development length is needed to establish turbulence;
(b) How long (for recycling methods) is needed to overcome initial transients and thus establish a developed turbulent state;
(c) How practically usable is the method;
(d) How Reynolds number limited is the method and
(e) How well does the output connect with real engineering systems.

With regards to element (b), Lund's recycling technique can establish a turbulent state relatively quickly. Sagaut et al. (2004) explore turbulence inflow techniques for a range of flows. For a wide cavity flow, it is noted that resolved turbulence in the inflow boundary layer is not important. This is evidently for the following reasons: (i) the separation point is geometrically defined; (ii) the cavity generates large pressure fluctuations that force unsteadiness much greater than that in the upstream turbulent boundary layer and (iii) the Kelvin-Helmholtz instability in the shear layer is sensitive to the mean velocity profile but not to the turbulence fluctuation upstream. It is pointed out that Kelvin-Helmholtz instability, magnifies two-dimensional instabilities and not the three-dimensional in the upstream boundary

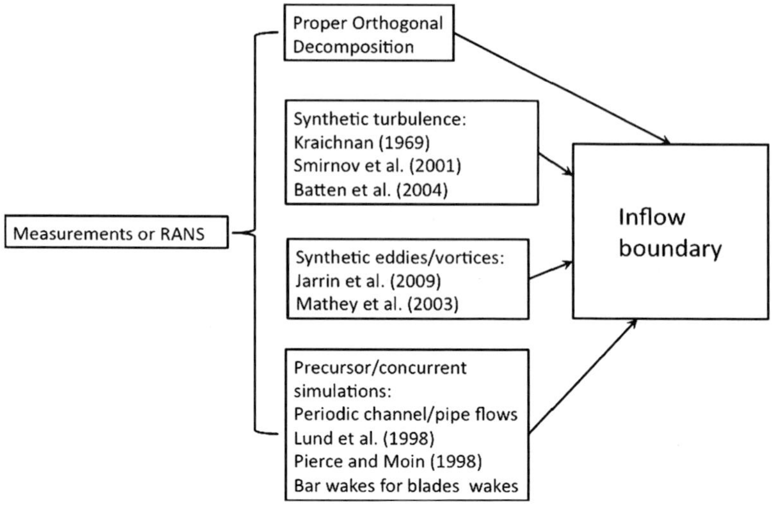

Fig. 3.23 Potential inflow options for idealized turbulence

layers. For the laminar separation, in a low pressure turbine blade, acoustic inflow information is found important in the upstream zone to ensure the correct flow development. Element (e), in the list above i.e. how well the output from idealized turbulence generation techniques connects with that found in real engineering systems, is discussed next.

3.10.4 Industrial Inflow

Many of the more idealized inflow approaches, noted above, in their raw form, have minimal relevance to complex engineering problems. For example, it might be tempting to consider applying Lund's recycling, to jet nozzle flows. For the replication laboratory experiments this has some value. However, with the nozzle used by Bridges and Wernet (2003), for example, the contraction is substantial. This geometry, which is not untypical, will give rise to partial flow relaminarization. Also, as outlined in Eastwood (2010), in a real engine, upstream of the nozzle, there is a compressor, combustor and turbine. There are also numerous other associated geometrical features. It has been supposed for some time that these may influence the jet noise itself (Moore 1977). At the rig scale, the influence of geometry in corrupting noise is clearly illustrated by Viswanathan (2003). However, there is little evidence exploring the influence of real geometrical features on the flow. Certainly, due to the endwall and hub casings, the flow into the nozzle will not be radially uniform. Guide vanes in the core passage, and structural A-frames with the gearbox shaft in the bypass passage will disrupt the flow. There can also be some residual swirl in the flow due to the later turbine stages. Through conservation of momentum, this will become magnified as the flow passes over the exhaust tail cone. In fact, the

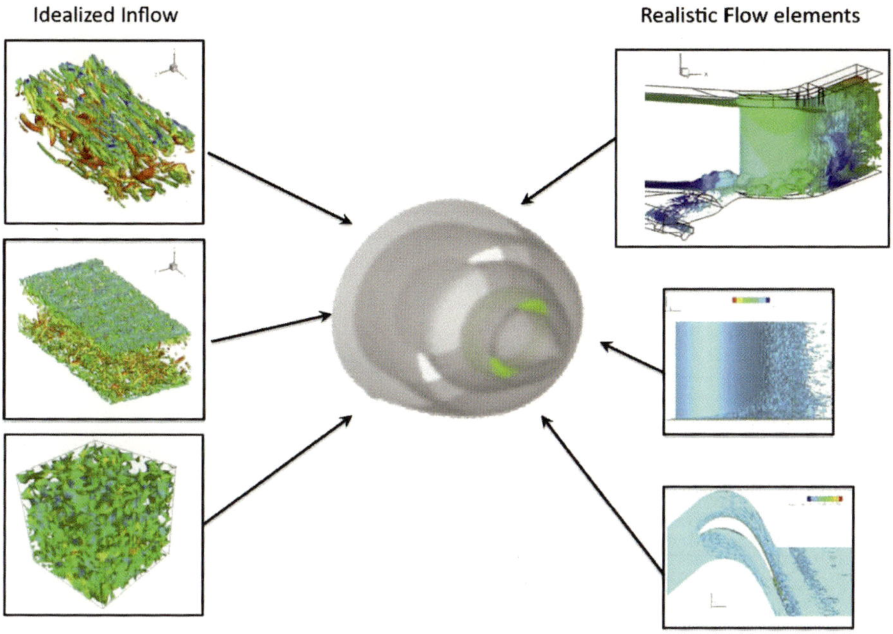

Fig. 3.24 Idealized and some potential realistic inputs to propulsive jet nozzle simulations

angular momentum magnification can be so large that it results in a substantial axial momentum drop. Wakes shed from blades and the listed components will propagate into the jet (Eastwood 2010). Various nozzle components are often badly fitting, causing flow disturbance. This is mainly due to the varying expansion rates of different materials with heating. To accommodate the rear-mounting strut, which also penetrates the flow, the nozzle exit, is sometimes non-circular. Probes used for temperature and pressure measurements also disrupt the flow. The flow will experience acceleration in the nozzle. The core flow is hot, having passed through the combustor. The combustion process can be unpredictable with hotspots and blocking where a combustor 'can' (there can be 20 around the circumference of the engine) does not burn properly. Aerodynamics surfaces can also be rough. All these flow obstructions and imperfections may influence the jet development. Indeed, as shown in the LES work of Eastwood (2010), the influence of upstream blade rows and the pylon can have a substantial influence on the jet development. Fortunately, their modelling with LES is feasible.

Figure 3.24 shows some idealized and potential realistic inputs to propulsive jet nozzle simulations at inflow boundaries. The left hand frames show recycling techniques for smooth and rough surfaces along with isotropic box turbulence. The right hand frames giving an indication of the complexity of the end wall flows generated by blades and sealant flows and also the extent of wake zones. The lower, right hand frames, are specifically for low-pressure turbine blades. These are closest to the jet nozzle.

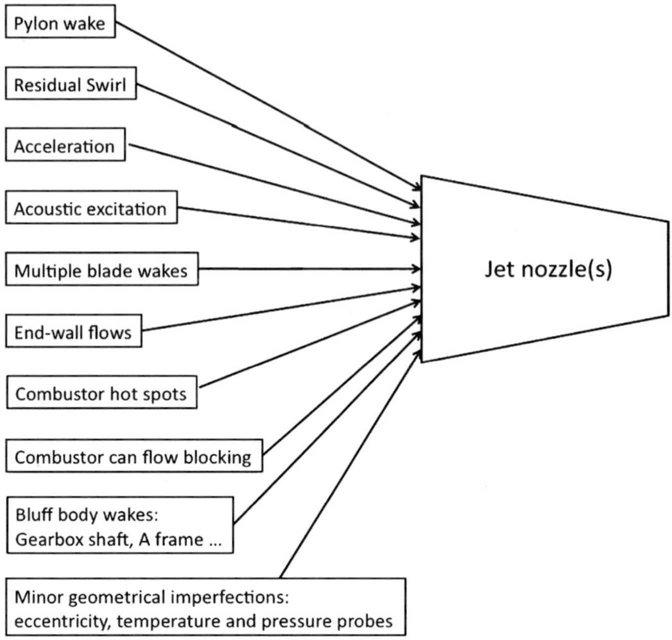

Fig. 3.25 Inputs to flow field in a real propulsive jet nozzle

Figure 3.25 summarises some of the inputs to flow field in a real propulsive jet nozzle that have been noted above. Clearly, imposing rather idealized turbulence fields has limitations in such a large-scale coupled, system as a gas turbine engine. The approach of Batten et al., perhaps, seems most realistic in a practical engineering context. Large-scale upstream (U)RANS simulations can be performed. These can be used to generate synthetic turbulence inflow.

Having discussed the propulsive jet, it also seems worth considering the intake, situated at the other end of an engine. For example, Loiodice et al. (2010) demonstrated that the intake duct flow, found in open rotor engines, is strongly influenced by both the upstream rotor(s) and also the downstream compressor. Through its demand for flow uniformity, the compressor redistributes the duct flow. The rotor wakes modify the turbulence field entering the duct and the rotor the mean flow field entering it.

In the work of Li (2012) the compressor outlet guide vane (OGV) and prediffuser zone of a gas turbine engine is considered. The latter connects to the combustion system. As shown by Li, for optimal diffuser design it is necessary to consider these two zones in combination. The performance of these two elements will, in turn have a strong impact on the combustor external aerodynamics. The wakes from the OGVs (and even the rotors upstream of these) will influence the diffuser performance, which in its optimal state will be sensitive to the form of incoming boundary layers. Li performed LES applying both recycling and a recycling with body force based procedure related to that of Pierce and Moin (1998). Pierce and

Moin used body forces to produce a swirl component. Along, similar lines, Eastwood (2010) used body forces to represent blades upstream of a propulsive jet in hybrid RANS-LES work. Through calibration, these provided realistic wake turbulence.

For the application of LES to realistic turbomachinery systems, the vision for the development of both inflow and outflow boundary conditions perhaps needs extending. It needs to reflect the coupled nature of components inside of gas turbine engines. It also needs to consider the complex, multi-component, nature of the turbulence fields produced. Having said this, many rig tests also suffer from the same lack of proper inflow and outflow characterization that LES does. This, in turn, hampers the assessment of LES. Certainly, most experimental engine intake duct studies do not include upstream rotors or a downstream compressor. The subject of coupled simulations is discussed in Chap. 7.

3.11 Conclusions

Averaging procedures used to form the URANS and LES equations have been outlined. Also, turbulence modelling hierarchies for unsteady flows have been presented. These ranged from URANS to DNS. The grid requirements for different approaches have also been discussed. The LES models that are used for results here were given, along with some details of others. Different types of LES filters/widths have been outlined. Their potential strong impact on results was noted. A hierarchy of key elements for industrial LES has been proposed. Notably, for flows without transition, the actual explicit LES model came low down. More the key element, is the numerical schemes. These were discussed in the numerical methods Chap. 3. The range of hybrid RANS-LES methods is extensive. Hence, just the key approaches have been given. The advantages and disadvantages of different approaches have been outlined. From the discussion, it is clear that considerable expertise is needed to safely use hybrid RANS-LES techniques. Hence, there seems the need for best practice guidelines. Also, a small sample of the extensive range methods for generating turbulence inflow was outlined. It was noted that many of these have limited applicability to complex engineering systems. For these, the Reynolds numbers are high and geometries complex. This can give rise to multiple inflow boundaries and system coupling (see Chap. 7) can occur. Some applications of the techniques outlined will follow in the remaining chapters.

References

Y. Addad, R. Prosser, D. Laurence, S. Moreau, F. Mendonca, On the use of embedded meshes in the les of external flows. Flow Turbul. Combust. **80**(3), 393–403 (2008)

N. Andersson, L.E. Eriksson, L. Davidson, LES prediction of flow and acoustic field of a coaxial jet, in *Proceedings of 11th AIAA/CEAS Aeroacoustics Conference*, Monterey, CA, 23–25 May 2005. AIAA Paper Number 2005-2884

J. Bardina, J.H. Ferziger, W.C. Reynolds, Improved subgrid-scale models for large-eddy simulation, in *Proceedings of the 13th AIAA Fluid and Plasma Dynamics Conference*, vol. 1, Snowmass, Colo., 14–16 July 1980. AIAA Paper Number 80-1357

P. Batten, U. Goldberg, S. Chakravarthy, LNS—An approach towards embedded LES. AIAA J. **427**, 2002 (2002)

P. Batten, U. Goldberg, S. Chakravarthy, Interfacing statistical turbulence closures with large-eddy simulation. AIAA J. **42**(3), 485–492 (2004)

P. Batten, P.R. Spalart, M. Terracol, Use of hybrid RANS-LES for acoustic source prediction, in *Large-Eddy Simulation for Acoustics*, ed. by T. Huttl, C. Wagner, P. Sagaut (2007)

P. Batten, U. Goldberg, E. Kang, S. Chakravarthy, Smart sub-grid-scale model for LES and hybrid RANS/LES, in *Proceedings of the 6th AIAA Theoretical Fluid Mechanics Conference*, Honolulu, HI, 27–30 June 2011, pp. 326–371. AIAA Paper Number 2011-3472

A. Belme, S. Wornom, A. Dervieux, B. Koobus, M.V. Salvetti, Application of hybrid and VMS-LES turbulent models to aerodynamic simulations, in *Proceedings of the 27th International Congress of the Aeronautical Sciences, ICAS* (2010), pp. 1–8

J. Berland, P. Lafon, F. Daude, F. Crouzet, C. Bogey, C. Bailly, Filter shape dependence and effective scale separation in large-eddy simulations based on relaxation filtering. Comput. Fluids **47**(1), 65–74 (2011)

N.J. Bisek, D.P. Rizzetta, J. Poggie, Plasma control of a turbulent shock boundary-layer, interaction. AIAA J. (2013). doi:10.2514/1.J052248

C. Bogey, C. Bailly, Decrease of the effective Reynolds number with eddy-viscosity subgrid-scale modeling. AIAA J. **43**(2), 437–439 (2005)

J.P. Boris, F.F. Grinstein, E.S. Oran, R.L. Kolbe, New insights into large eddy simulation. Fluid Dyn. Res. **10**(4–6), 199–228 (1992)

G. Bosch, W. Rodi, Simulation of vortex shedding past a square cylinder with different turbulence models. Int. J. Numer. Methods Fluids **28**(4), 601–616 (1998)

T.T. Brandt, Usability of explicit filtering in large eddy simulation with a low-order numerical scheme and different subgrid-scale models. Int. J. Numer. Methods Fluids **57**(7), 905–928 (2008)

M. Breuer, W. Rodi, Large-eddy simulation of turbulent flow through a straight square duct and a 180 bend, in *Direct and Large-Eddy Simulation I* (1994), pp. 273–285

J. Bridges, M.P. Wernet, Measurements of the aeroacoustic sound source in hot jets, in *Proceedings of the 9th AIAA/CEAS Aeroacoustics Conference*, South Carolina, 12–14 May 2003. AIAA Paper Number 2003-3130

B. Chaouat, R. Schiestel, A new partially integrated transport model for subgrid-scale stresses and dissipation rate for turbulent developing flows. Phys. Fluids **17**, 065106 (2005)

D.R. Chapman, Computational aerodynamics, development and outlook. AIAA J. **17**, 1293–1313 (1979)

N. Chauvet, S. Deck, L. Jacquin, Zonal detached eddy simulation of a controlled propulsive jet. AIAA J. **45**(10), 2458–2473 (2007)

H. Choi, P. Moin, Grid-point requirements for large eddy simulation: Chapman's estimates revisited. Phys. Fluids **24**(1), 011702 (2012)

F.K. Chow, P. Moin, A further study of numerical errors in large-eddy simulations. J. Comput. Phys. **184**(2), 366–380 (2003)

R.A. Clark, J.H. Ferziger, W.C. Reynolds, Evaluation of subgrid-scale models using an accurately simulated turbulent flow. J. Fluid Mech. **91**(1), 1–16 (1979)

J.D. Coull, Wake induced transition in low pressure turbines. PhD thesis, Department of Engineering, The University of Cambridge (2009)

Y.M. Dakhoul, K.W. Bedford, Improved averaging method for turbulent flow simulation. Part I: theoretical development and application to Burgers' transport equation. Int. J. Numer. Methods Fluids **6**(2), 49–64 (1986a)

Y.M. Dakhoul, K.W. Bedford, Improved averaging method for turbulent flow simulation. Part I: theoretical development and application to Burgers' transport equation. Int. J. Numer. Methods Fluids **6**(2), 65–82 (1986b)

L. Davidson, Hybrid LES-RANS: back scatter from a scale-similarity model used as forcing. Philos. Trans. R. Soc. A, Math. Phys. Eng. Sci. **367**(1899), 2905–2915 (2009)

L. Davidson, M. Billson, Hybrid LES-RANS using synthesized turbulent fluctuations for forcing in the interface region. Int. J. Heat Fluid Flow **27**(6), 1028–1042 (2006)

L. Davidson, S.H. Peng, Hybrid LES-RANS modelling: a one-equation SGS model combined with a k–ω model for predicting recirculating flows. Int. J. Numer. Methods Fluids **43**(9), 1003–1018 (2003)

J.W. Deardorff, A numerical study of three-dimensional turbulent channel flow at large Reynolds numbers. J. Fluid Mech. **41**(2), 453–480 (1970)

J.R. DeBonis, Progress towards large-eddy simulations for prediction of realistic nozzle systems, in *Proceedings of the 44th AIAA Aerospace Sciences Meeting and Exhibit*, Reno, NV, 9–12 January 2006. AIAA Paper Number 2006-487

S. Deck, Recent improvements in the zonal detached eddy simulation (ZDES) formulation. Theor. Comput. Fluid Dyn. **26**, 523–550 (2012)

S. Deck, P.É. Weiss, M. Pamiès, E. Garnier, Zonal detached eddy simulation of a spatially developing flat plate turbulent boundary layer. Comput. Fluids **48**(1), 1–15 (2011)

J.A. Domaradzki, D.D. Holm, Navier-Stokes-alpha model: LES equations with nonlinear dispersion, in *Modern Simulation Strategies for Turbulent Flow*, ed. by B.J. Geurts. ERCOFTAC Bulletin, vol. 107 (Edwards, Ann Arbor, 2001)

P. Druault, S. Lardeau, J.P. Bonnet, F. Coiffet, J. Delville, E. Lamballais, J.F. Largeau, L. Perret, M.N. Glauser, B. George et al., Generation of three-dimensional turbulent inlet conditions for large-eddy simulation. AIAA J. **42**(3), 447–456 (2004)

S. Eastwood, Hybrid LES-RANS of complex geometry jets. PhD thesis, University of Cambridge (2010)

S.J. Eastwood, P.G. Tucker, H. Xia, C. Klostermeier, Developing large eddy simulation for turbomachinery applications. Philos. Trans. R. Soc. A, Math. Phys. Eng. Sci. **367**(1899), 2999–3013 (2009)

M. Farge, K. Schneider, Coherent vortex simulation (CVS), a semi-deterministic turbulence model using wavelets. Flow Turbul. Combust. **66**(4), 393–426 (2001)

S. Gamard, W.K. George, Reynolds number dependence of energy spectra in the overlap region of isotropic turbulence. Flow Turbul. Combust. **63**, 443–477 (1999)

W.K. George, M. Tutkun, Mind the gap: a guideline for large eddy simulation. Philos. Trans. R. Soc. A, Math. Phys. Eng. Sci. **367**(1899), 2839–2847 (2009)

N.J. Georgiadis, D.P. Rizzetta, C. Fureby, Large-eddy simulation: current capabilities, recommended practices, and future research. AIAA J. **48**(8), 1772–1784 (2010)

B.J. Geurts, D.D. Holm, Regularization modeling for large-eddy simulation. Phys. Fluids **15**(1), L13–L16 (2003)

B.J. Geurts, D.D. Holm, Leray and LANS-α modelling of turbulent mixing. J. Turbul. **7**(10), 1–33 (2006)

S. Ghosal, An analysis of numerical errors in large-eddy simulations of turbulence. J. Comput. Phys. **125**(1), 187–206 (1996)

S. Ghosal, T.S. Lund, P. Moin, K. Akselvoll, A dynamic localization model for large-eddy simulation of turbulent flows. J. Fluid Mech. **286**(1), 229–255 (1995)

D.A. Gieseking, J.I. Choi, J.R. Edwards, H.A. Hassan, Compressible-flow simulations using a new large-eddy simulation/Reynolds-averaged Navier-Stokes model. AIAA J. **49**(10), 2194–2209 (2011)

S.S. Girimaji, Partially-averaged Navier-Stokes model for turbulence: a Reynolds-averaged Navier-Stokes to direct numerical simulation bridging method. J. Appl. Mech. **73**(3), 413–421 (2006)

V. Gravemeier, The variational multiscale method for laminar and turbulent flow. Arch. Comput. Methods Eng. **13**(2), 249–324 (2006)

F. Hamba, An attempt to combine large eddy simulation with the k–ε model in a channel-flow calculation. Theor. Comput. Fluid Dyn. **14**(5), 323–336 (2001)

F. Hamba, A hybrid RANS/LES simulation of high-Reynolds-number channel flow using additional filtering at the interface. Theor. Comput. Fluid Dyn. **20**(2), 89–101 (2006)
F. Holzäpfel, Adjustment of subgrid-scale parameterizations to strong streamline curvature. AIAA J. **42**(7), 1369–1377 (2004)
T.J.R. Hughes, G.R. Feijóo, L. Mazzei, J.B. Quincy, The variational multiscale method—a paradigm for computational mechanics. Comput. Methods Appl. Mech. Eng. **166**(1), 3–24 (1998)
T.J.R. Hughes, L. Mazzei, K.E. Jansen, Large eddy simulation and the variational multiscale method. Comput. Vis. Sci. **3**(1), 47–59 (2000)
N. Jarrin, R. Prösser, J.C. Uribe, S. Benhamadouche, D. Laurence, Reconstruction of turbulent fluctuations for hybrid RANS/LES simulations using a synthetic-eddy method. Int. J. Heat Fluid Flow **30**(3), 435–442 (2009)
R.J. Jefferson-Loveday, P.G. Tucker, LES of impingement heat transfer on a concave surface. Numer. Heat Transf., Part A, Appl. **58**(4), 247–271 (2010)
R.J. Jefferson-Loveday, P.G. Tucker, Wall-resolved LES and zonal LES of round jet impingement heat transfer on a flat plate. Numer. Heat Transf., Part B, Fundam. **59**(3), 190–208 (2011)
J. Jewkes, An improved turbulent boundary layer inflow condition, applied to the simulation of jets in cross-flow. PhD thesis, School of Engineering, University of Warwick (2008)
S.H. Johansson, L. Davidson, E. Olsson, Numerical simulation of vortex shedding past triangular cylinders at high Reynolds number using a k–ε turbulence model. Int. J. Numer. Methods Fluids **16**(10), 859–878 (1993)
J. Joo, P. Durbin, Simulation of turbine blade trailing edge cooling. J. Fluids Eng. **131**, 021102 (2009)
S.A. Jordan, A large-eddy simulation methodology in generalized curvilinear coordinates. J. Comput. Phys. **148**(2), 322–340 (1999)
A. Keating, U. Piomelli, E. Balaras, H.J. Kaltenbach, A priori and a posteriori tests of inflow conditions for large-eddy simulation. Phys. Fluids **16**, 4696–4712 (2004)
W.W. Kim, S. Menon, An unsteady incompressible Navier–Stokes solver for large eddy simulation of turbulent flows. Int. J. Numer. Methods Fluids **31**(6), 983–1017 (1999)
M. Klein, A. Sadiki, J. Janicka, A digital filter based generation of inflow data for spatially developing direct numerical or large eddy simulations. J. Comput. Phys. **186**(2), 652–665 (2003)
B. Kosovic, Subgrid-scale modelling for the large-eddy simulation of high-Reynolds-number boundary layers. J. Fluid Mech. **336**(1), 151–182 (1997)
R.H. Kraichnan, Diffusion by a random velocity field. Phys. Fluids **13**(1), 22–31 (1970)
E. Labourasse, P. Sagaut, Reconstruction of turbulent fluctuations using a hybrid RANS/LES approach. J. Comput. Phys. **182**(1), 301–336 (2002)
R. Laraufie, S. Deck, P. Sagaut, A dynamic forcing method for unsteady turbulent inflow conditions. J. Comput. Phys. **230**(23), 8647–8663 (2011)
S. Lardeau, M.A. Leschziner, Unsteady rans modelling of wake–blade interaction: computational requirements and limitations. Comput. Fluids **34**(1), 3–21 (2005)
B.E. Launder, B.I. Sharma, Application of the energy-dissipation model of turbulence to the calculation of flow near a spinning disc. Lett. Heat Mass Transf. **1**, 131–137 (1974)
A. Leonard, Energy cascade in large-eddy simulations of turbulent fluid flows. Adv. Geophys. **18**, 237–248 (1975)
M. Leschziner, N. Li, F. Tessicini, M. Leschziner, N. Li, F. Tessicini, Simulating flow separation from continuous surfaces: routes to overcoming the Reynolds number barrier. Philos. Trans. R. Soc. A, Math. Phys. Eng. Sci. **367**(1777), 2885–2903 (2009)
J. Li, Hybrid RANS/LES modelling of OGV/prediffuser flow. PhD thesis, Department of Aeronautical and Automotive Engineering, Loughborough University (2012)
Y. Liu, P.G. Tucker, R.M. Kerr, Linear and nonlinear model large-eddy simulations of a plane jet. Comput. Fluids **37**(4), 439–449 (2008)
S. Loiodice, P.G. Tucker, J. Watson, Coupled open rotor engine intake simulations, in *Proceedings of the 48th AIAA Aerospace Sciences Meeting and Exhibit*, Orlando, FL, 4–7 January 2010. AIAA Paper Number 2010–840

T.S. Lund, X. Wu, K.D. Squires, Generation of turbulent inflow data for spatially-developing boundary layer simulations. J. Comput. Phys. **140**(2), 233–258 (1998)
M. Mani, Hybrid turbulence models for unsteady simulation. J. Aircr. **41**(1), 110–118 (2004)
L.G. Margolin, W.J. Rider, F.F. Grinstein, Modeling turbulent flow with implicit LES. J. Turbul. **7**(15), 1–27 (2006)
A.L. Marsden, O.V. Vasilyev, P. Moin, Construction of commutative filters for LES on unstructured meshes. J. Comput. Phys. **175**(2), 584–603 (2002)
I. Mary, P. Sagaut, Large eddy simulation of flow around an airfoil near stall. AIAA J. **40**, 1139–1145 (2002)
F. Mathey, D. Cokljat, J.P. Bertoglio, E. Sergent, Specification of LES inlet boundary condition using vortex method, in *Proceedings of the 4th International Symposium on Turbulence, Heat and Mass Transfer*, Antalya, Turkey, 12–17 October 2003
F.R. Menter, Zonal two equation k–ω turbulence models for aerodynamic flows, in *Proceedings of the AIAA Fluid Dynamics Conference, 24th*, Orlando, FL, USA, 6–9 July 1993. AIAA Paper Number 93-2906
F.R. Menter, M. Kuntz, R. Bender, A scale-adaptive simulation model for turbulent flow predictions, in *Proceedings of the 41th AIAA Aerospace Sciences Meeting and Exhibit*, vol. 767, Reno, NV, USA, 6–9 January 2003
C.J. Moore, The role of shear-layer instability waves in jet exhaust noise. J. Fluid Mech. **80**(02), 321–367 (1977)
P.J. Morris, L.N. Long, A. Bangalore, Q. Wang, A parallel three-dimensional computational aeroacoustics method using nonlinear disturbance equations. J. Comput. Phys. **133**(1), 56–74 (1997)
P.J. Morris, Y. Du, K. Kara, Jet noise simulations for realistic jet nozzle geometries, in *Proceedings of the IUTAM Symposium on Computational Aero-Acoustics for Aircraft Noise Prediction, Procedia Engineering*, vol. 6, 29–31 March (Elsevier, Amsterdam, 2010), pp. 28–37
N.J. Mullenix, D.V. Gaitonde, M.R. Visbal, A plasma-actuator-based method to generate a supersonic turbulent boundary layer inflow condition for numerical simulation, in *Proceedings of the 20th AIAA Computational Fluid Dynamics Conference*, Honolulu, HI, USA, 27–30 June 2011. AIAA Paper Number 2011-3556
L. Mydlarski, Z. Warhaft, On the onset of high-Reynolds-number grid-generated wind tunnel turbulence. J. Fluid Mech. **320**, 331–368 (1996)
V. Nee, L. Kovasznay, Simple phenomenological theory of turbulent shear flows. Phys. Fluids **12**, 473 (1969)
F. Nicoud, F. Ducros, Subgrid-scale stress modelling based on the square of the velocity gradient tensor. Flow Turbul. Combust. **62**(3), 183–200 (1999)
N.V. Nikitin, F. Nicoud, B. Wasistho, K.D. Squires, P.R. Spalart, An approach to wall modeling in large-eddy simulations. Phys. Fluids **12**, 1629 (2000)
K. Nozawa, T. Tamura, Large eddy simulation of the flow around a low-rise building immersed in a rough-wall turbulent boundary layer. J. Wind Eng. Ind. Aerodyn. **90**(10), 1151–1162 (2002)
T.S.D. O'Mahoney, Large-eddy simulation of turbine rim seals. PhD thesis, School of Engineering, Surrey University (2011)
M. Opiela, R. Keinke, W. Schroder, LES of wake–blade interaction, in *Proceedings of the 4th European Conference on Turbomachinery Fluid Dynamics and Thermodynamics*, Firenze, Italy, 20–23 March 2001
S.A. Orszag, V. Borue, W.S. Flannery, A.G. Tomboulides, Recent successes, current problems, and future prospects of CFD, in *Proceedings of the 35th AIAA Aerospace Sciences Meeting and Exhibit*, Reno, NV, USA, 6–9 January 1997. AIAA Paper Number 97-0431
L. Perret, J. Delville, R. Manceau, J.P. Bonnet, Turbulent inflow conditions for large-eddy simulation based on low-order empirical model. Phys. Fluids **20**, 075107 (2008)
C.D. Pierce, P. Moin, Method for generating equilibrium swirling inflow conditions. AIAA J. **36**(7), 1325–1327 (1998)
U. Piomelli, E. Balaras, Wall-layer models for large-eddy simulations. Annu. Rev. Fluid Mech. **34**(1), 349–374 (2002)

U. Piomelli, J. Ferziger, P. Moin, J. Kim, New approximate boundary conditions for large eddy simulations of wall-bounded flows. Phys. Fluids **1**, 1061–1068 (1989)

U. Piomelli, E. Balaras, H. Pasinato, K.D. Squires, P.R. Spalart, The inner–outer layer interface in large-eddy simulations with wall-layer models. Int. J. Heat Fluid Flow **24**(4), 538–550 (2003)

S.B. Pope, Ten questions concerning the large-eddy simulation of turbulent flows. New J. Phys. **6**(1), 35 (2004)

C.D. Pruett, Eulerian time-domain filtering for spatial large-eddy simulation. AIAA J. **38**(9), 1634–1642 (2000)

D. Razafindralandy, A. Hamdouni, C. Béghein, A class of subgrid-scale models preserving the symmetry group of Navier–Stokes equations. Commun. Nonlinear Sci. Numer. Simul. **12**(3), 243–253 (2007)

A.J. Revell, T.J. Craft, D.R. Laurence, Turbulence modelling of unsteady turbulent flows using the stress strain lag model. Flow Turbul. Combust. **86**(1), 129–151 (2011)

P. Sagaut, LES and DNS simulation of turbomachinery flows, in *Recent Developments in Numerical Methods for Turbomachinery Flows*, ed. by T. Arts. VKI Lecture Series (2002)

P. Sagaut, E. Garnier, E. Tromeur, L. Larcheveque, E. Labourasse, Turbulent inflow conditions for large-eddy simulation of compressible wall-bounded flows. AIAA J. **42**(3), 469–477 (2004)

P. Sagaut, T. Hüttl, C. Wagner, *Large-Eddy Simulation for Acoustics* (Cambridge University Press, Cambridge, 2007)

J.U. Schluter, X. Wu, S. Kim, S. Shankaran, J.J. Alonso, H. Pitsch, A framework for coupling Reynolds-averaged with large-eddy simulations for gas turbine applications. Trans. Am. Soc. Mech. Eng. J. Fluids Eng. **127**(4), 806 (2005)

U. Schumann, Subgrid scale model for finite difference simulations of turbulent flows in plane channels and annuli. J. Comput. Phys. **18**(4), 376–404 (1975)

A. Scotti, C. Meneveau, D.K. Lilly, Generalized smagorinsky model for anisotropic grids. Phys. Fluids A, Fluid Dyn. **5**, 2306 (1993)

J.A. Sethian, Fast marching methods. SIAM Rev. **41**(2), 199–235 (1999)

M. Shur, P. Spalart, M. Strelets, A. Travin, Navier-Stokes simulation of shedding turbulent flow past a circular cylinder and a cylinder with backward splitter plate, in *ECCOMAS Computational Fluid Dynamics Conference*, Paris, 9–13 September 1996, pp. 676–682

M.L. Shur, P.R. Spalart, M.K. Strelets, A.V. Garbaruk, Further steps in LES-based noise prediction for complex jets, in *Proceedings of the 44th AIAA Aerospace Sciences Meeting and Exhibit*, Reno, NV, 9–12 January 2006. AIAA Paper Number 2006-485

J. Smagorinsky, General circulation experiments with the primitive equations. Mon. Weather Rev. **91**(3), 99–164 (1963)

A. Smirnov, S. Shi, I. Celik, Random flow generation technique for large eddy simulations and particle-dynamics modeling. Trans. ASME I J. Fluids Eng. **123**(2), 359–371 (2001)

P.R. Spalart, S.R. Allmaras, A one-equation turbulence model for aerodynamic flows. Rech. Aérosp. **1**, 5–21 (1994)

P.R. Spalart, Strategies for turbulence modelling and simulations. Int. J. Heat Fluid Flow **21**(3), 252–263 (2000)

P.R. Spalart, C.L. Rumsey, Effective inflow conditions for turbulence models in aerodynamic calculations. AIAA J. **45**(10), 2544–2553 (2007)

P.R. Spalart, S. Deck, M.L. Shur, K.D. Squires, M.K. Strelets, A. Travin, A new version of detached-eddy simulation, resistant to ambiguous grid densities. Theor. Comput. Fluid Dyn. **20**(3), 181–195 (2006)

C.G. Speziale, Turbulence modeling for time-dependent RANS and VLES: a review. AIAA J. **36**(2), 173–184 (1998)

A. Spille-Kohoff, H.J. Kaltenbach, Generation of turbulent inflow data with a prescribed shear-stress profile, in *3rd AFOSR International Conference on DNS/LES Arlington, TX*, ed. by L. Sakell, T. Beutner, C. Liu. DNS/LES Progress and Challenges, Greyden, Columbus, OH, 5–9 August 2001, pp. 326–371

S. Stolz, N.A. Adams, Large-eddy simulation of high-Reynolds-number supersonic boundary layers using the approximate deconvolution model and a rescaling and recycling technique. Phys. Fluids **15**(8), 2398–2412 (2003)

M. Strelets, Detached eddy simulation of massively separated flows, in *Proceedings of the 39th AIAA Aerospace Sinces Meeting and Exhibit*, Reno, NV, USA, 8–11 January 2001. AIAA Paper Number 2001-0879

G.R. Tabor, M.H. Baba-Ahmadi, Inlet conditions for large eddy simulation: a review. Comput. Fluids **39**(4), 553–567 (2010)

T. Talha, A numerical investigation of three-dimensional unsteady turbulent channel flow subjected to temporal acceleration. PhD thesis, University of Warwick (2012)

K. Tatsumi, H. Iwai, E.C. Neo, K. Inaoka, K. Suzuki, Prediction of time-mean characteristics and periodical fluctuation of velocity and thermal fields of a backward-facing step, in *Proceedings of the 1st Internal Symposium on Turbulence and Shear Flow*, ed. by S. Banerjee, J.K. Eaton, Santa Barbara, USA, 12–15 September 1999, pp. 139–144

L. Temmerman, M. Leschziner, A priori studies of near wall RANS model within a hybrid LES/RANS scheme, in *Engineering Turbulence Modelling and Experiments, 5*, ed. by W. Rodi, N. Fueyo (2002), pp. 326–371

M. Terracol, P. Sagaut, C. Basdevant, A multilevel algorithm for large-eddy simulation of turbulent compressible flows. J. Comput. Phys. **167**(2), 439–474 (2001)

P.G. Tucker, *Computation of Unsteady Internal Flows: Fundamental Methods with Case Studies* (Kluwer Academic, Dordrecht, 2001)

P.G. Tucker, Differential equation-based wall distance computation for DES and RANS. J. Comput. Phys. **190**(1), 229–248 (2003)

P.G. Tucker, Novel MILES computations for jet flows and noise. Int. J. Heat Fluid Flow **25**(4), 625–635 (2004)

P.G. Tucker, Computation of unsteady turbomachinery flows: Part 2—LES and hybrids. Prog. Aerosp. Sci. **47**, 546–569 (2011)

P.G. Tucker, L. Davidson, Zonal *k–l* based large eddy simulations. Comput. Fluids **33**(2), 267–287 (2004)

P.G. Tucker, Y. Liu, Turbulence modelling for flows around convex features giving rapid eddy distortion. Int. J. Heat Fluid Flow **28**(5), 1073–1091 (2007)

P.G. Tucker, Y. Liu, R.M. Kerr, Plane jet simulations using LES with linear and nonlinear models, in *Proceedings of the 44th AIAA Aerospace Sciences Meeting and Exhibit*, Reno, NV, USA, 10–13 January 2006. AIAA Paper Number 2006-0702

P.G. Tucker, S. Eastwood, C. Klostermeier, R. Jefferson-Loveday, J. Tyacke, Y. Liu, Hybrid LES approach for practical turbomachinery flows—part I: hierarchy and example simulations. J. Turbomach. Trans. ASME **134**(2), 021023 (2011)

J.C. Tyacke, Low Reynolds number heat transfer prediction employing large eddy simulation for electronics geometries. PhD thesis, Swansea University (2009)

J. Tyacke, P.G. Tucker, R.J. Jefferson-Loveday, R. Vadlamani, R. Watson, I. Naqavi, X. Yang, Les for turbines: methodologies, cost and future outlooks, in *Proceedings of ASME Turbo Expo 2013 GT2013*, San Antonio, TX, USA, 3–7 June 2013. ASME Paper Number GT2013-94416

R. Vadlamani, High fidelity large eddy simulation of turbines: current status and future outlook. PhD thesis, School of Engineering, The University of Cambridge (2013)

H. van der Ven, A family of large eddy simulation (LES) filters with nonuniform filter widths. Phys. Fluids **7**, 1171 (1995)

M.R. Visbal, Numerical investigation of deep dynamic stall of a plunging airfoil. AIAA J. **49**(10), 2152–2170 (2011)

K. Viswanathan, Jet aeroacoustic testing: issues and implications. AIAA J. **41**(9), 1674–1689 (2003)

B. Vreman, Direct and large-eddy simulation of the compressible turbulent mixing layer. PhD thesis, Twente University (1995)

A.W. Vreman, The filtering analog of the variational multiscale method in large-eddy simulation. Phys. Fluids **15**(8), L61–L64 (2003)

A.W. Vreman, B.J. Geurts, A new treatment of commutation-errors in large-eddy simulation, in *Advances in Turbulence IX*, ed. by I.P. Castro, P.E. Hancock. Proceedings of the 9th European Turbulence Conference Southampton, UK, 2–5 July 2002

M. Wolfshtein, The velocity and temperature distribution in one-dimensional flow with turbulence augmentation and pressure gradient. Int. J. Heat Mass Transf. **12**(3), 301–318 (1969)

X. Wu, Establishing the generality of three phenomena using a boundary layer with free-stream passing wakes. J. Fluid Mech. **664**, 193–219 (2010)

X. Wu, J.P. Hickey, Visualization of continuous stream of grid turbulence past the Langston turbine cascade. AIAA J. **50**(1), 215–225 (2012)

A. Yoshizawa, Subgrid-scale modeling with a variable length scale. Phys. Fluids A, Fluid Dyn. **1**(7), 1293–1295 (1989)

A. Yoshizawa, Bridging between eddy-viscosity-type and second-order models using a two-scale dia, in *9th International Symposium on Turbulent Shear Flows*, vol. 18, Kyoto, Japan, 16–18 August 1993

A. Yoshizawa, K. Horiuti, A statistically-derived subgrid-scale kinetic energy model for the large-eddy simulation of turbulent flows. J. Phys. Soc. Jpn. **54**, 2834–2839 (1985)

S. Zahrai, F.H. Bark, R.I. Karlsson, On anisotropic subgrid modeling. Eur. J. Mech. B, Fluids **14**(4), 459–486 (1995)

B. Zhong, P.G. Tucker, *k–l* based hybrid LES/RANS approach and its application to heat transfer simulation. Int. J. Numer. Methods Fluids **46**, 983–1005 (2004)

Chapter 4
Computational Aerodynamics Methods

4.1 Introduction

In this chapter, specific unsteady modelling methods for external and internal aerodynamics will be discussed in the areas of turbomachinery and airframes. The former is dealt with first and greater emphasis is given to it. This is because, the outlook of this text is more centred around the power plant and then how this interacts with the airframe. However, airframe, component, performance predictive capabilities are also explored. For turbomachinery, since, even at design, the processes in the main gas path, are intrinsically unsteady there are a wide range of methods. For example, Fig. 4.1 is a schematic of the various levels of fidelity used to model unsteady turbomachinery and, to an extent, other flows. This is adapted from Tucker (2011). The left hand, dashed, box considers approaches where the unsteady flow is reduced to a steady problem through body forces or, in the case of turbomachinery blade rows, circumferential averaging based procedures (Tucker 2011). Here, the main focus is on the application of the unsteady methods in the 2nd column. We will start with harmonic methods.

4.2 Perturbation Equation Based Methods

4.2.1 Linear Harmonic Methods

The 1st level of fidelity, where unsteady fluctuations are actually partly resolved is the linear harmonic method. With this approach, essentially we have the following three-component decomposition

$$\phi = \bar{\Phi} + \phi'' + \phi' \qquad (4.1)$$

In the above, $\bar{\Phi}$ is the time mean of the flow. If possible (i.e. a steady solution can be secured) this can be computed using RANS. If RANS is used to solve for $\bar{\Phi}$, the

P.G. Tucker, *Unsteady Computational Fluid Dynamics in Aeronautics*, Fluid Mechanics and Its Applications 104, DOI 10.1007/978-94-007-7049-2_4,

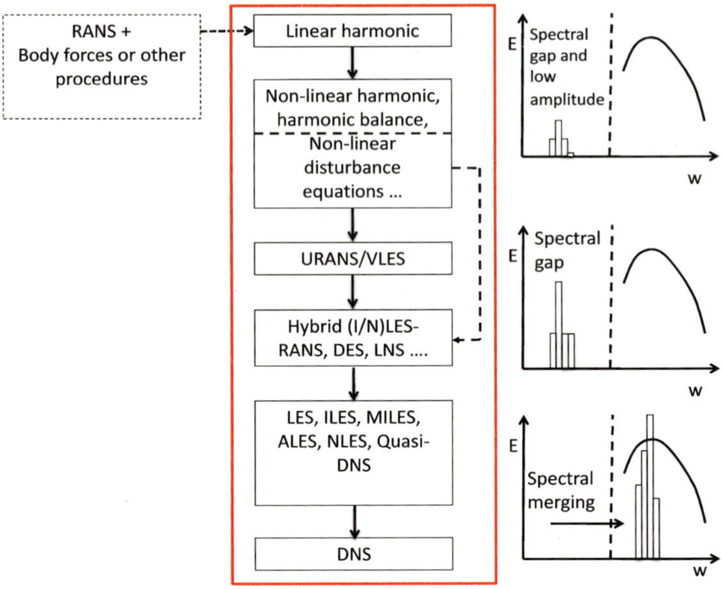

Fig. 4.1 Schematic of unsteady modelling hierarchy (based on Tucker 2011). Published with kind permission from Elsevier

final ϕ' term will naturally, via, for example, eddy viscosity, be modelled. Then an equation for *linear* perturbations, ϕ'', about the mean is solved. This can take place either in the time or frequency domains. The frequency(s), ω, of the perturbations are specified. The superposition of these solutions gives the complete solution.

4.2.2 Non-linear Harmonic Methods

To understand non-linear harmonic methods it is helpful to assume that the flow variables, ϕ, vary as

$$\phi(x,t) = \bar{\Phi}(x) + B(x)\cos(\omega t) + C(x)\sin(\omega t) \tag{4.2}$$

Hence, for a specified ω, the unsteady flow time history is described if $\bar{\Phi}(x)$, $B(x)$ and $C(x)$ are defined. If the system is assumed to be *linear* then $\bar{\Phi}(x)$ will be identical to that for the steady flow problem. Then, $B(x)$ and $C(x)$ just need to be evaluated. If the unsteady discretized equation takes the form

$$\frac{\partial \phi(x,t)}{\partial t} = A(\phi) \tag{4.3}$$

combining Eqs. (4.2) and (4.3) gives

$$\omega\big(-B(x)\sin(\omega t)+C(x)\cos(\omega t)\big)=A\big(\bar{\Phi}(x)+B(x)\cos(\omega t)+C(x)\sin(\omega t)\big) \tag{4.4}$$

Allowing *non-linearity* gives *three* unknowns—B, C and $\bar{\Phi}(x)$. However, if we specify three ωt values i.e. corresponding (for a fixed ω) to three different points in time, *t*, of a cycle we will end up with three equations. These will have the form of (4.3). They have unsteady source terms of $-\omega B$, ωC and ωB. This allows *t* in Eq. (4.3) to be replaced by a pseudo time τ. This can then be solved to a steady state yielding the three unknowns. A more general, non-linear, procedure for multiple harmonics (ω values) is given by Hall et al. (2002). In this procedure Eq. (4.4) is extended to

$$\sum_{n=1}^{Nf}\big[n\omega\big(-B_n(x)\sin(n\omega t)+C_n(x)\cos(n\omega t)\big)\big]$$
$$=A\left(\bar{\Phi}(x)+\sum_{n=1}^{Nf}B_n(x)\cos(n\omega t)+C_n(x)\sin(n\omega t)\right) \tag{4.5}$$

where *Nf* is the number of Fourier modes considered. If we wish to solve for *Nf* harmonics $2Nf+1$ steady system equations need to be solved.

The forerunner of the above approach is that of He and Ning (1998). This solved a coupled set of equations. One is for the time mean and the other for disturbances about this. Hence, the time dependent fluctuation equation provided resolved stresses to be fed into the mean flow equations. The level of non-linearity provided by this approach is restricted. Notably, in a similar fashion to He, Giles (1992) proposed a three-stage method. A linear perturbation equation about the mean flow is solved. As with He's method these perturbations are then imposed on the mean flow. The approach of Hall et al. solves a fully coupled set of equations for harmonics and the time mean rather than the loosely coupled approaches of He and Ning.

4.2.3 Advantages and Applications of Fourier Based Methods

As implied above, turbomachinery blade row calculations, like many other flows, naturally have temporal periodicity. Therefore, one can take the full unsteady flow equations and discretize them spectrally in time using Fourier series. Working in the frequency domain has the following advantages:

(I) The flow governing equations become effectively steady and so efficient steady state solution methods (see Chap. 2) can be used;

(II) To further save computational expense a few selected key modes can be considered in the Fourier series;

(III) The periodic steady state can be moved to more quickly (avoiding the resolution of potentially long initial transients that would be resolved by, for example, URANS);
(IV) A spectral gap can potentially be enforced between the resolved and modelled scales (As noted in Chap. 3, this is necessary for any RANS modelling to be theoretically valid);
(V) Adjoint based design optimization methods can be readily applied and
(VI) Numerical stiffness from conjugate problems can be removed.

With regards to linear perturbation equation solutions, for flows where the perturbation magnitude is small the solution of a linear perturbation equation can be more accurate than solving the full non-linear equations. This is especially so for acoustics—see Chap. 6. Advantages (I–III), above, can produce an order of magnitude saving in computational effort. Advantages (V) and (VI) are discussed next. The reduction of unsteady problems to steady makes application of the adjoint approach (see Chap. 2) to unsteady flows much more practical. For any design variable, α, Eq. (4.2) can be re-expressed as

$$\phi(x,t,\alpha) = \bar{\Phi}(x,\alpha) + B(x,\alpha)\cos(\omega t) + C(x,\alpha)\sin(\omega t) \tag{4.6}$$

and the optimization problem reduced to one involving steady equations. The actual adjoint equations(s) can be dealt with in a similar way to the flow equations. Hence, they can be solved in pseudo time to a steady state. Using the above, harmonic, approach He and Wang (2011) carry out concurrent aerodynamic and aeroelastic adjoint based design optimization of flutter stabilization for a transonic compressor fan. In relation to advantage (VI), He (2011) uses a harmonic method for the removal of numerical stiffness when making conjugate heat transfer problems. As would be expected, the thermal time scales in solid components can be orders of magnitude larger than the fluid. To overcome this, He (2010) uses a Fourier based temporal discretization for the solid zone of a turbine blade. A time-domain discretization is used for the fluid. The use of the former allows a small set of steady conduction related equations to be solved for in the solid domain.

Hall et al. (2002) apply a non-linear harmonic method to the front stage rotor of a high-pressure compressor exploring transonic flutter. Good accuracy (agreement with full nonlinear solutions) is found even when considering just a few harmonics. A more numerically efficient version of the method of Hall et al. is presented by McMullen et al. (2006). McMullen et al. apply their approach to transonic pitching aerofoils. Encouraging agreement is found with data using less than 3 modes. Campobasso and Baba-Ahmadi apply the non-linear harmonic (balance) method to wind turbines. They find a factor of ten speed up relative to using URANS.

Weiss et al. (2011) apply an unstructured grid harmonic balance method to a two-dimensional compressor stage. Relative to solving the full unsteady equations, they find substantial computational time savings. Also, accuracy is acceptable when just resolving 2–3 harmonics.

4.2.4 Some Performance Issues

With many turbomachinery flows the key frequencies of interest can be established from the blade counts, *Nb*, and angular velocities. For example, following, Custer et al., for three blade rows, in a stator-rotor-stator configuration, with blade numbers of Nb_1, Nb_2 and Nb_3, respectively, the following equation for circumferential wave numbers (nodal diameters) can be written

$$N = n_1 Nb_1 + n_2 Nb_2 + n_3 Nb_3 \tag{4.7}$$

In the above n are integers of any value. In the rotor frame of reference there are the following frequencies

$$\omega = -(n_1 Nb_1 + n_3 Nb_3)\Omega \tag{4.8}$$

where Ω is the rotor angular velocity. In the stationary frame

$$\omega = n_2 Nb_2 \Omega \tag{4.9}$$

It can be inferred that with more stages a wide range of harmonics can come into play. This is problematic for the non-linear harmonic methods. Then such approaches are be best used with a low number of harmonics as an initial guess to say URANS (Gopinath et al. 2007). However, in the commercial STAR-CCM+ code there is a zonalization approach. This limits the range of influence of harmonic interactions between stages. Such an approach allows convergence whilst retaining a tractable number of solution harmonics.

Olausson (2011) and Corral and Crespo (2011) note that due to instability growth, with the harmonic balance method convergence is not always guaranteed and can be slow. Olausson damps instabilities using a more advanced simultaneous equation solver. The work also suggests that adding extra frequencies beyond those of interest may be needed to avoid divergence arising from aliasing errors. He (2008) also addresses convergence issues, avoiding the cost of advanced simultaneous equation solvers, but this time for the linear harmonic method.

With flows involving wakes, the number of energetic wake local modes will increase with Reynolds number. Hence more modes/computational overhead is needed. Mosahebi and Nadarajah (2011) partially addressed this problem by efficiently dealing with more modes through an adaptive approach. This automatically adds modes where high temporal resolution is required. A notable aspect of practical harmonic balance methods is that reduced computing time comes at the cost of increased storage, solutions being stored at several time instances in the periodic cycle.

4.2.5 NLDE

More recently the use of so called non-linear disturbance equations has been made. With this approach the mean flow, $\bar{\Phi}$, is first solved for using RANS. Then, as with the work of He and Ning, a disturbance equation is solved. However, with the NLDE approach, there is no specified frequency. The disturbance equation retains non-linear terms. More notably these disturbances do not interact with the mean flow. As, with the above, there is still a three-level flow decomposition. The fluctuations ϕ'' about $\bar{\Phi}$ are provided by NLDEs—and they no longer need be periodic functions. Distinctly, this time ϕ' are turbulent fluctuations but at a subgrid scale. Labourasse and Sagaut (2002) apply this approach to a low-pressure turbine blade section. More discussion of this approach can be found in Chaps. 3 and 6.

4.3 URANS/VLES

The next, level of fidelity considered is URANS. Even, though the fidelity level is higher, in practical programming terms URANS is simpler than the perturbation approaches discussed above. URANS simply involves making unsteady RANS simulations. The expectation is that the increased computational cost should yield accuracy benefits that are consistent with these costs. As noted in Chap. 3, even though the URANS equations are identical to the RANS different forms of averaging are implicit in them. Care must be taken to ensure the validity of these averages.

4.3.1 Spectral Gaps

As noted in Chap. 3, a key requirement for this is that the time scales of any resolved unsteadiness, t_p is much greater than the turbulence time scale, t_i (or that there is a corresponding length scale separation) i.e. there is a spectral gap. This is implied in the top two spectral schematics in the far right of Fig. 4.1. Next spectral gaps will be explored. Mostly this is for turbomachinery applications. First blade passages will be considered.

To characterize the ratio of the approximate period for a fluid particle to pass through a blade passage to the wake passing period, the dimensionless so-called reduced frequency can be defined as below.

$$\tilde{f}_p = \frac{1}{t_p}\frac{c}{V} = \frac{t_{fp}}{t_p} \approx 1 \tag{4.10}$$

In the above, c, is the chord and V, a relative velocity scale for the blade passage. Also, t_p is the wake passing time period. The ratio $\tilde{f}_p$ is indicative of the number of wakes in a blade passage at one particular time instant. It can be viewed as representing the ratio of the time, t_{fp}, that it would take a fluid particle to pass through the blade passage to the time that it would take the wake. Table 4.1 summarises reduced frequencies in different engine zones. The zones are the compressor, low-pressure turbine (LPT), high-pressure turbine (HPT) and at the HPT rotor

Table 4.1 Reduced frequencies in different turbine engine zones

Engine zone	$\tilde{f}_p$
Compressors	0.8–1.2
LPT	0.3–1.0
HPT	0.34
HPTR/IPTS	3.7

(HPTR)—intermediate pressure turbine (IPT) stator (IPTS) interface. The exact location of these zones is not critical. The key point is that $\tilde{f}_p \sim 1$.

Another key time scale is that for the wake from upstream blades. This will have an approximate dimensionless shedding period (Strouhal number) given below.

$$\tilde{f}_b = \frac{1}{t}\frac{L}{V} \approx 0.3 \tag{4.11}$$

The variable L is a representative blade trailing edge length scale ($L \sim 0.1c$) and t the shedding period. Note, Sieverding et al. (2004) find $\tilde{f}_b = 0.22$ for a turbine blade wake. Furthermore, there is the Kelvin-Helmholtz time scale for any detached shear layers to be considered. Ho and Huerre (1984) present the following empirical expression for this

$$f_{KH} = \tilde{f}_{KH}\frac{V}{\theta} = 0.032\frac{V}{\theta} \tag{4.12}$$

In the above, θ is the momentum thickness and V the mean velocity of the two shear layer streams. Also, $\tilde{f}_{KH}$ is the Strouhal number based on θ and V. Adapting this expression to a separation bubble's, separated shear layer (Tucker 2011) the following can be gained.

$$\tilde{f}_{KH} \approx f_{KH}\frac{c}{V} \approx 0.025 Re_c^{0.5} \tag{4.13}$$

In the above, simple mid-chord boundary layer estimates are used. For $Re_c = 1 \times 10^4$, LPT, LES, Mittal et al. (2001) get $\tilde{f}_{KH} \approx 1$. Equation (4.13) gives $\tilde{f}_{KH} = 2.5$. Hence, there is reasonable accord. Also, LPT measurements for $Re_c = 4.5 \times 10^4$ of Coull (2009) give $\tilde{f}_{KH} = 6$ and Eq. (4.13) gives $\tilde{f}_{KH} = 5.3$. Hence, again there is reasonable agreement with Eq. (4.13) estimate. Based on this and the Table 4.1 data, at low Re_c the spectral gap is tenuous with respect to $\tilde{f}_p$ and $\tilde{f}_{KH}$. Equation (4.11) suggests $\tilde{f}_b \sim \tilde{f}_p$ creating another potential clash of time scales. Looking at attached blade boundary layer zones, simple boundary layer thickness and eddy scale estimates give for the mid-chord zone (Tucker 2011).

$$\tilde{f}_{BL} \approx f_{BL}\frac{c}{V} \approx 30 Re_c^{1/5} \tag{4.14}$$

In the above, $\tilde{f}_{BL}$, is the typical outer part of the boundary turbulence frequency. Hence, with a typical $Re_c \sim 1 \times 10^5$ for a gas turbine, $\tilde{f}_{BL} \gg \tilde{f}_p$. Therefore, a spectral gap between the attached boundary layers and wake passing seems clear.

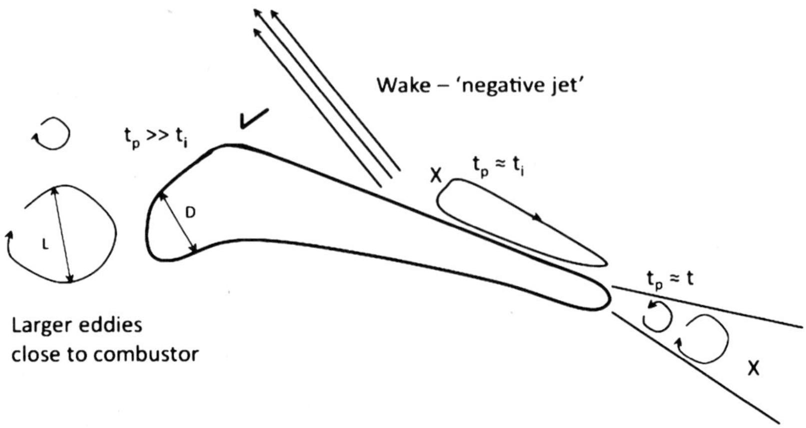

Fig. 4.2 Expectations with regards to spectral gap on a blade

However, as shown by, for example, Lardeau and Leschziner (2005) in wake zones $t_i \approx t$ (see Eq. (4.11)) and so t_p. Hence, there is no spectral gap (see Chap. 3). Another potentially key time scale is an acoustic one. It relates to the time it takes sound waves generated in the wake region to reach any separated flow on the suction surface. It is postulated by Raverdy et al. (2003) that this feedback excites separated shear layers. The frequency, f_a, of the 1st harmonic of this feedback loop can evidently be estimated from the expression given by Raverdy (based on the work of Tam 1978) given below

$$f_a = \frac{1}{(1/|V-c| + 1V)L} \tag{4.15}$$

Note, f_a is dimensional. Also, L is the distance between the separation point and the trailing edge and c is now the speed of sound. Figure 4.2 summarises expectations with regards to spectral gaps on a blade. The '×' marks indicates where the gap is tenuous.

4.3.2 Stall and Intake Distortion

As noted in Chap. 1, there seem to be two key acknowledged routes to stall. These are modal and spike . Modal inception involves relatively long wavelengths in the tangential direction of say L_θ. These are much larger than the blade pitch defined here as, L. In the absolute frame of reference, the modes travel at about 40 % of the rotor speed. Hence, they will rotate at around $\alpha \Omega R$ where R is a typical radial location being considered, Ω the rotational speed of the blade and $\alpha = 0.4$. Hence,

Table 4.2 Fraction of the blade speed that the stall cell rotates at

Stall type	α
Part span	0.5–0.8
Full span	0.3–0.5

the modal time scale, t_p, is $O(L_\theta/\Omega R)$ or more precisely

$$t_p \sim \frac{L_\theta}{\alpha \Omega R} \tag{4.16}$$

For a typical compressor $V_x/\Omega R \cong 0.5$ where V_x is axial velocity. Hence, the boundary layer turbulence time scale, t_i is $O(\delta/\Omega R)$. Since $\delta \ll L_\theta$, $t_p \gg t_i$ and there is a spectral gap. Note, however, modal stall inception does not occur in multistage compressors. For spike inception, L_θ, is much smaller. The other time scales stay roughly the same and so the potential for a spectral gap lessens.

Stall cells (see Chap. 1), that follow on from the above, are a precessing zones of axially quasi-stagnant flow. This stagnant zone will typically occupy multiple, say N, blade passages. Therefore, $L_\theta \sim NL$. Once a stall cell has established (see Chap. 1) the time scale that the *blade* will see is L_θ divided by the velocity at which the stall cell passes over the blade. If α is the fraction of the blade speed that the stall cell rotates at Eq. (4.16) again applies.

Table 4.2 gives typical α values for stall. Since, again $\delta \ll L_\theta$ and t_i is $O(\delta/\Omega R)$ there should be a spectral gap. This is particularly so for part span stall but in this regime N (the number of passages the occupy a single stall cell) can be smaller. The former estimate neglects that the boundary layer within the stall cell zone is nolonger a classical layer and that in this zone $V_x \sim 0$ (in a classical boundary layer the time scale will be $O(\delta/V_x)$). Nonetheless, broadly speaking the stall cell precession time scale should be relatively distinct from much of the turbulence time scales in the passages that the cell moves into. However, the time scales of the modelled turbulence and any resolved turbulence in the separated flow zones arising from the cells approach is less likely to have a spectral gap.

For low observability, military intakes, the engine can be inside a convoluted duct. Such a configuration is shown in Fig. 4.3a. This would give rise to eddies of the scale of the radius of curvature of the bend, R. For more standard, less convoluted intakes, as shown in Fig. 4.3c, eddies of the scale of roughly $L \sim R/2$, where R is this time the duct radius, can interact with the compressor/fan. Eddies of scale R would be large but are possible if a separation is triggered by a flight manoeuvre. Also, as noted by Defoe and Spakovszky, for boundary layer ingesting aircraft engines, the boundary layer can be 30 % of the intake diameter. Assuming eddies of this scale, the ratio of the turbulence time scales of the *intake structures* to blade passing will roughly as given below

$$\frac{t_i}{t_p} \sim \frac{Nb}{2\pi} \tag{4.17}$$

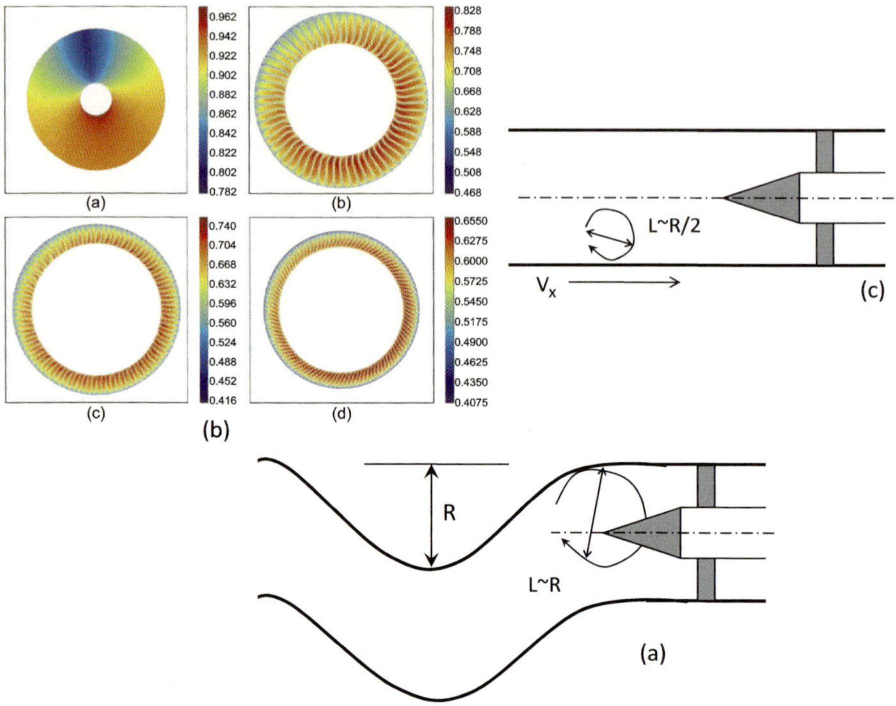

Fig. 4.3 Spectral gap for inlet distortion modelling in compressors: (**a**) schematic of a potential un-manned, combat, aircraft intake; (**b**) entropy contours for a distortion transfer calculation and (**c**) schematic indicating vortex entering a straight intake

Hence, from the viewpoint of the intake turbulent structures the spectral gap is tenuous. Indeed, there appears to be a gap reversion i.e. the resolved time scales are much faster than those of the dominant turbulent scales. This aspect is discussed next.

4.3.3 Rapid Eddy Distortion

For intake distortion, blended wing bodies and HPTs the Lagrangian eddies of scale, L, approaching the blades are of significant size relative to the blade leading edge effective diameter, D, scale. For a HPT, see Fig. 4.2, $L/D \sim 2$ and a compressor $L/D > 10$ is plausible. As shown by Wang et al. (1999) and Maslov et al. (1999), this leading edge region, where there is stagnation, presents some important turbulence modelling challenges. The interaction of the upstream ambient turbulence on the stagnation region just upstream of cylinder 'leading edges' is theoretically analysed by Hunt (1973) using the 'rapid distortion' theory simplification. With this, the eddy life time $L/\sqrt{2k/3}$ (k is magnitude of the *upstream* turbulence kinetic energy) is considered large compared with the mean flow time scale, centred on the leading edge. The latter scales as D/V_{rel}, where V_{rel} is the fluid relative velocity. This implies the following ratio

$$\psi = \frac{\sqrt{2k/3}}{V_{rel}} \frac{D}{L} \ll 1 \tag{4.18}$$

Hence the turbulence is assumed to be subjected to such a rapid distortion that it is modified by the mean flow but not (non-linear) self interactions between different turbulent scales. Hunt (1973) explored the limiting states of $D/L \to \infty$ and $D/L \to 0$, the latter being the rapid distortion limit. The theoretical analysis of Hunt suggests for $D/L \to 0$ the normal Reynolds stress near the leading edge is damped by the blocking effect of the edge. In the stagnation region, the normal Reynolds stress greatly controls heat transfer. Hence, Wang et al. (1999) find for large L (and a high free stream turbulence intensity) apparently suppressed heat transfer. This is consistent with the heat transfer correlation of Dullenkopf and Mayle (1995) which (importantly) includes the influence of L. The stagnation region modelling can have a key influence on URANS performance (Lodefier and Dick 2006). Also, as discussed, it appears to be a zone where classical URANS averaging ideas are challenged.

4.3.4 Shock Buffet

Shock buffet is the large-scale sustained oscillation of a shock over an aerofoil. The shock induces separation. The vortices from the separated flow interact with the aerofoil trailing edge, giving rise to acoustic waves from the trailing edge. This results in a strong feedback mechanism. The time scale of the shock buffet links to the time scale of downstream information propagation from shocks. It also connects to the upstream propagation of information from the intense acoustic event arising from the interaction of vortices (from the separated flow region) with the trailing edge. Fortunately, when modelling shock buffet, there is a distinct spectral gap (Garnier and Deck 2008), the turbulent eddy time scales being much shorter than the shock oscillation period. With shock buffet, the frequency of the feedback loop is approximately given by the expression below of Lee (1990)

$$f = \frac{(V + c) + |V - c|}{L} \tag{4.19}$$

This time, L, is the mean distance between the trailing edge and the shock. Equation (4.19) relates to the two time scales noted above.

4.3.5 URANS Applications and Outlook

A range of URANS failures, for separated flows, is reported in Tucker (2001). As noted by Deck et al. (2005), for URANS, sometimes the angle of attack needs to be artificially increased to induce shock buffet. This being dependent on the chosen RANS model. Colin et al. (2007) explore the modelling of intake distortion generated by cross winds. They note that to capture hysteresis in the separated zone on nacelle lips, URANS simulations are needed. The RANS model choice is found to be key for acceptable predictive accuracy. Loiodice et al. (2010) explore the use of Spalart-Allmaras based URANS when modelling open rotor intakes. The intake(s) are downstream of a rotor. The application of URANS is found essential to get a realistic connection to measurements (not reported for commercial confidence reasons). Loiodice et al. note that the spectral gap is tenuous for these intake simulations.

Michelassi et al. (2003a) explore unsteady heat transfer in a transonic turbine with rotating and stationary blade interactions (see next section). They concluded that k–ω URANS is essential to get sensible heat transfer. Michelassi et al. (2003b) make k–ω based URANS simulations for an LPT representative profile. Both DNS and LES are used to refine the URANS modelling. The turbulence in the leading edge stagnation region is found to be nearly an order of magnitude higher than that suggested by the LES and DNS. Also, careful application of transition modelling is needed to gain credible results. Nürnberger and Greza (2002) study a range of LPT related flows. Even with substantial modelling efforts, the level of transition in the separated shear layer is under predicted. For an LPT profile, with a transport equation for intermittency and k–ω based URANS modelling, Suzen and Huang (2005) capture general trends. However, when studying the same profile, Lodefier and Dick (2006) need strong, double action to deal with the leading edge stagnation point's excessive turbulence generation. This can have a substantial downstream influence. Yang et al. (2007) perform URANS for an LPT cascade. Moving bars replicate upstream wakes. With appropriate transition modelling, broad accord with expected flow physics is found.

Yao et al. (2008) explore the use of k–ϵ based multistage URANS for computing compressor distortion transfer. Encouraging agreement is found with measurements. Montomoli et al. (2011) carry out Spalart-Allmaras model based URANS computations for a four-stage compressor. Unsteadiness is found to play a key role in the development of the end wall flows and so gives clear accuracy benefits. However, this substantial improvement, in predictive accuracy, came at over two orders of magnitude higher cost. Gourdain et al. (2006) study rotating stall in a subsonic compressor using quasi three-dimensional URANS. The broad features of rotating stall are captured. However, it is concluded that the turbulence model is inadequate. Gourdain et al. (2010) make three-dimensional URANS. Rotating stall cells are captured but they differ in number to those observed in experiments.

Especially for separated flows or strongly sheared detached flows, URANS can frequently improve predictive accuracy relative to RANS. This is especially so if

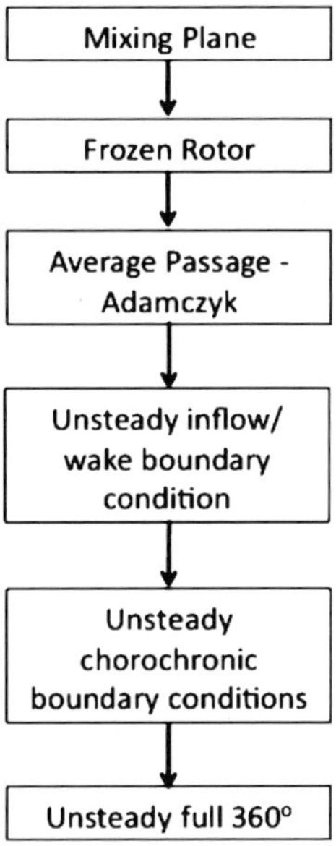

Fig. 4.4 Blade row modelling fidelities (Tucker 2011). Published with kind permission from Elsevier

performed in three-dimensional mode. Also, for flows that involve substantial large-scale unsteadiness, steady state CFD solution iterative convergence will be hard to secure without specialized solution procedures or body forces. With a steady solution process, unless the solver is overly dissipative, or strong, or the turbulence model has at least sufficient dissipation to model all the unsteady scales, a steady solution cannot be gained. Notably, as discussed in Chap. 2, strong solvers can yield less plausible solutions than the semi-converged from weaker solvers. Hence, in this numerical convergence context, URANS again becomes attractive. Clearly, if there is boundary movement URANS becomes unavoidable. However, the unsteadiness content in results will be highly dependent on the chosen base RANS model. It is of note that only one of the studies above explores the theoretical solution validity in terms of the level of any spectral gap. When URANS is used for multistage turbomachinery calculations, sliding planes will generally be used. This aspect is discussed next.

4.4 Making Blade Row Calculations

As noted in Chap. 1, turbomachines consist of multiple rows of blades having relative rotation. Figure 4.4 gives the broad hierarchy for blade row modelling fidelities. This excludes the modelling of so called end wall features (see later) and their connection with the internal air systems. Again, the Fig. 4.4 hierarchy starts with steady practices, followed by those that attempt to model and then explicitly resolve unsteadiness. Starting with the 1st Fig. 4.4 entry—standard turbomachinery blade

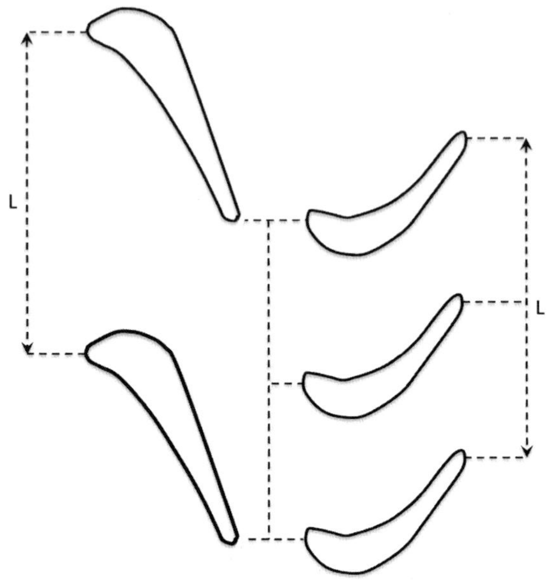

Fig. 4.5 Blade configuration that naturally allows use of periodic boundary conditions. Published with kind permission from Elsevier

row design calculations are generally steady. So called mixing planes, pioneered by Denton (1992), link the stationary and rotating blade row domains. These are located about half way between the upstream blades trailing edge and downstream blades leading edge. The planes connect the two domains through transferring circumferentially averaged solution variables.

The steady flow modelling fidelity can be potentially increased using what is called the frozen rotor assumption. With this, wakes are passed to downstream blade rows (domains). However, there is no temporal (relative) wake movement. For distortion transfer, acoustics and general high fidelity calculations sliding planes need to be used. With sliding planes, the rotor mesh(s) slide/rotate relative to the stator(s). They are incremented in the tangential direction every time step. Clearly an unsteady 360° annulus calculation (i.e. including all blades) is expensive. However, if the number of rotors and stators are the same, their domain will have the same circumferential extent (pitch). Then sliding planes with single meshed blades can be used. This assumes something like distortion transfer is not being explored, since then this would not be possible. For the single passage calculations, circumferentially periodic boundary conditions are applied at the domain faces coincident with the neglected adjacent blades. Alternatively un-equal pitches can sometimes be dealt with through a modest increase in modelled blade count. This can also be combined with (see later) adjustments in pitch on a particular blade row, thus avoiding resolving a 360° sector. However, for acoustics, preserving the blade topology is more crucial for meaningful simulations. Figure 4.5 gives a schematic of an appropriately pitch matched domain. To increase accuracy, but reduce computational cost (see Montomoli et al. 2011), for multistage machines, a mixture of sliding and mixing planes can be used.

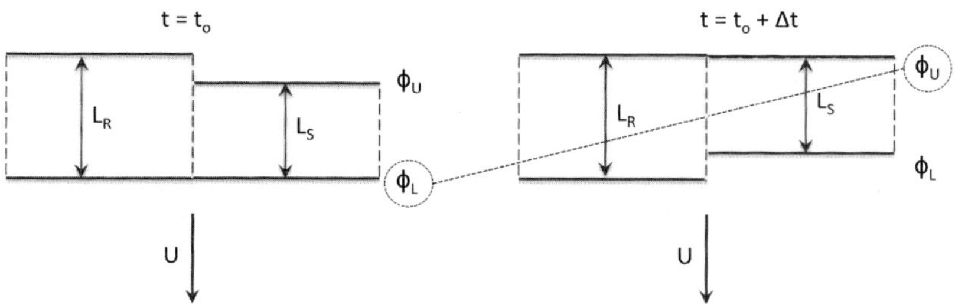

Fig. 4.6 Enforcing periodicity through a phase shift (adapted from Biesinger et al. 2010)

4.4.1 Phase Lagged Boundary Conditions

Unfortunately, in practical systems the number of rotors and stators are unlikely to allow modelling low blade counts (acoustics constraints are a key reason). Then, to enable reduced circumferential domain calculations chorochronic (in Greek choros relates space and chronos to time) periodicity can be used. Figure 4.6 gives a schematic to help illustrate the general concept. In this figure the blade pitches for the rotor, L_R, and stator, L_S, and the tangential velocity, U are identified. With this approach use is made of the observation that for the Fig. 4.6 system, for the stator (downstream of the rotor) at $t = t_o + \Delta t$ (note, Δt, is not the numerical time step as used in Chap. 2) the property ϕ^U is equal to ϕ^L at $t = t_o$ i.e

$$\phi^L(t_o) = \phi^U(t_o + \Delta t) \tag{4.20}$$

Using the above relationship, with some form of data storage, and knowing that

$$\Delta t = \frac{L_R - L_S}{U} \tag{4.21}$$

allows periodicity to be enforced using Eq. (4.20). The time history at the periodic boundaries is either directly stored (Erdos et al. 1976) or stored as a Fourier series (Gerolymos et al. 2002) and used in a time-lagged fashion. This is sometimes called Fourier-transformation. As noted, the lag is proportional to the mismatch between the rotor and stator domains. However, as noted further later, when there are more than two blade rows, with different L, see Fig. 4.7 (where an, Inlet Guide Vane—IGV—has been added), the relationship/coupling becomes more complex (the IGV footprint as processed by the rotor will now be seen at the stator inlet).

An alternative approach is 'time inclining' (Giles 1988, 1991) or time-transformation. With this, the computational 'domain' is inclined in time in the circumferential direction. Therefore, use is made of a transformed time coordinate—see Chap. 2. This is a linear function of circumferential location, blade pitch and rotor angular velocity. The resulting time transformation complicates post processing. The contoured solution, when looking in the radial direction, will be at different points in time - unless an inverse transformation is applied. With techniques like the harmonic balance, time shifted information is readily accessible and no special treatment is needed. Only the appropriate phase shift need be applied. With the

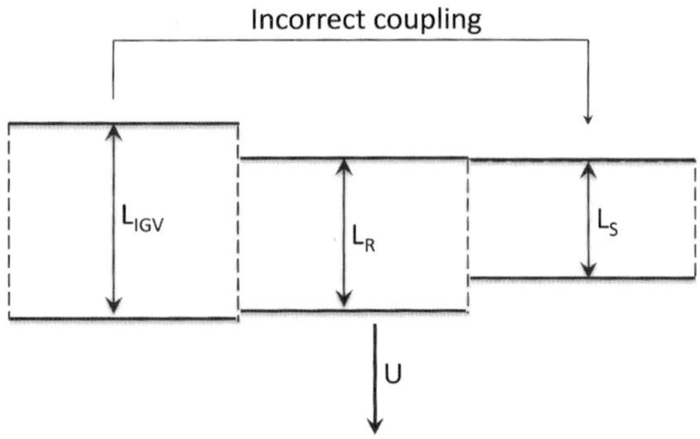

Fig. 4.7 Phase lagged boundary conditions for more than two blade rows

time-inclining, following Giles, the following transformation is made

$$\begin{aligned} \xi &= x \\ \eta &= r\theta \\ \tau &= t - \lambda r\theta \end{aligned} \tag{4.22}$$

In the above, with reference to Fig. 4.6,

$$\lambda = \frac{\Delta t}{L_S} \tag{4.23}$$

Using the above transformed variables in the convective terms gives

$$\frac{\partial(\phi - \lambda G)}{\partial \tau} + \frac{\partial F}{\partial \xi} + \frac{\partial G}{\partial \eta} = 0 \tag{4.24}$$

In the above, ϕ, is the usual solution vector. The transformation changes this to $\phi - \lambda G$. In the new coordinate system, each computational plane corresponds to $\tau =$ constant. As noted by Giles, another way to view the transformation, is to look at a solution control volume in the (non-transformed) $r\theta$–t plane (see Fig. 4.8). In this, the new solution vector, can be shown to be decomposed as follows

$$(\phi - \lambda G) r \Delta\theta = \phi(r\Delta\theta) - G\Delta t \tag{4.25}$$

i.e. in terms of a flux through a space-time face.

Hence, the time inclining approach could be viewed as utilizing space-time control volumes as discussed in Chap. 2. For unsteady flows, phase lagged approaches can compromise accuracy. For, example, as noted by Giles (1988), enforcing spatial periodicity means that the sign of the vorticity shed from adjacent blades rows must be identical. This can yield a mathematically valid solution but one that is physically implausible. Notably, for LES, the enforcement of spatial or temporal synchronicity will have a contaminating influence. Also, due to computational nuances, time-savings do not scale linearly with the reduction in the number of blade passages allowed through use of phase lagged conditions. Indeed, the approach can carry a high overhead and even prevent iterative convergence (Olausson 2011). Also, it can be unstable, without damping, having stability restrictions Connell et al. (2011). For

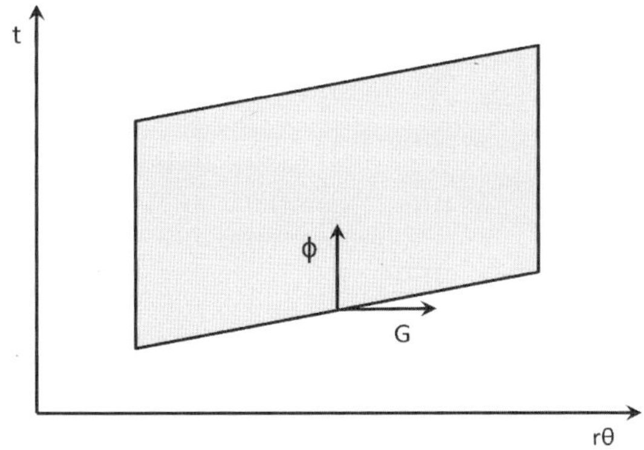

Fig. 4.8 Space-time type element in $r\theta$–t plane

URANS and hybrid RANS-LES simulations Olausson (2011) use temporal damping, for stability. This is achieved via Fourier coefficients, over a buffer zone at periodic boundaries. On overall accuracy grounds, temporal smoothing seems preferable to spatial. As shown by Van Zante et al. (2008) the use of phase lagged conditions can give rise to an order of magnitude increase in the necessary computational time steps (relative to resolving half the circumferential extent of an annulus). Also, for multistage calculations inaccurate results are shown. This is due to the phase lag boundary condition being unable to account for changes in the mean flow field. This in turn is due the physical constraints imposed by the method. Furthermore, List et al. (2008) also note that aliasing errors can occur for higher harmonics, arising from blade row interactions, when using phase lagged boundary conditions.

Blumenthal et al. (2011) contrast an extended form of Giles' time transformation with Fourier transformation and also the automatic rescaling of flow profiles (Galpin et al. 1995)—profile transformation. The latter simply overcomes the difficulty of uneven pitches by rescaling the flow profile communicated between the adjacent blade rows. Differences in the results for the methods are found. However, the differences, relative to measurements, are too small to comment on the relative merits of the methods. Nonetheless, such calculations were found to be an order of magnitude faster than making calculations for a 180° sector. They were also an order of magnitude slower than steady calculations. Notably, transition modelling had as large an impact on results as moving to unsteady modelling.

Connell et al. (2011) contrast time transformation, profile transformation and Fourier transformation for high and low pressure turbine stages. Key observations are that the time-transformation approach has the same convergence rate as using standard periodic boundary conditions. However, Fourier transformation, which treats the periodic update in an explicit lagged fashion, shows slower convergence in an implicit solver. This, slowdown, would not be observed if an explicit solver is used. With the Fourier transform the dominant frequency (blade passing) and its harmonics are selected. This results in, as shown by Connell et al., filtering of frequencies that are lower than the dominant. This is not an issue for time transformation. Connell et al. see the profile transformation as a potentially sensible compromise between computational cost and accuracy. It is also noted by these authors that,

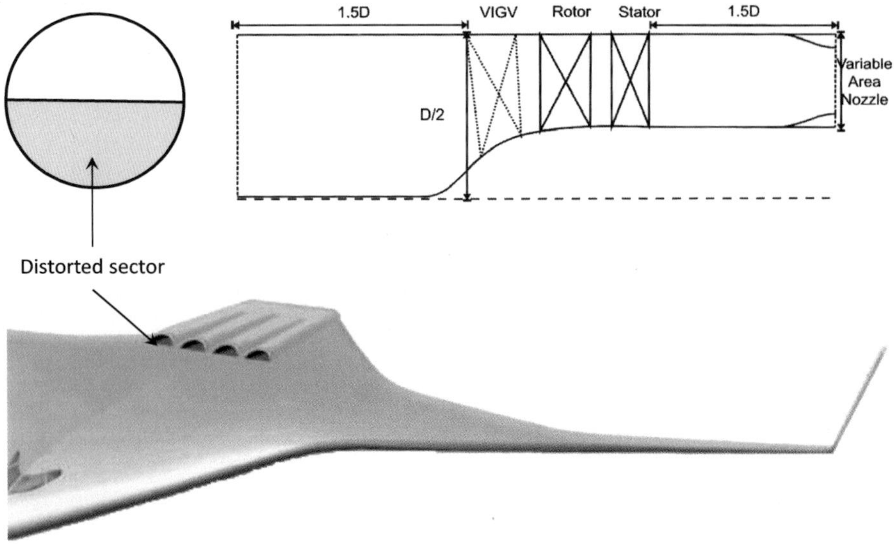

Fig. 4.9 Simple schematic of a transonic fan rig with imposed distortion along with a representation of a configuration that could generate strong intake distortion (adapted from Shaw et al. 2013)

bearing in mind the aforementioned potential deficiencies, increasing blade counts, to allow the use of standard periodic boundary conditions, can be a sensible path. The application of time-inclining to LPT LES type computations is advocated by Connell et al. As noted, this approach does not corrupt accuracy.

4.4.2 Example of High Fidelity Blade Row Calculations

As noted, compressors can have intakes downstream of complex intake ducts. This can give rise to a complex coupled problem. Also, due to a potentially large (30 % of fan diameter) fuselage boundary layer, for blended wing body aircraft the compressor needs to absorb distortion. To deal with the distortion inlet guide vanes can be used. Hence, next some simulations exploring the coupling between these and a transonic fan are presented. Complete 360° calculations are necessarily made with sliding planes. To reduce the coupling challenge, to represent the distortion, in both the CFD and experiments a sector of total pressure loss is imposed upstream of the fan. Note, here the distortion is steady but in practical system this can be time varying. For a fan downstream of a duct it certainly will. For a manoeuvring aircraft this will also be the case. The sector of low total pressure is shown in the left hand frame of Fig. 4.9. The objective is to study the local coupling between the IGVs (variable angle) and the fan. Full details of the set-up for these SA model based URANS simulations can be found in Shaw et al. (2013). For the H–O–H hexahedral grid $N = 80\text{–}90 \times 10^6$, the finer grid being used with simulations having the IGVs.

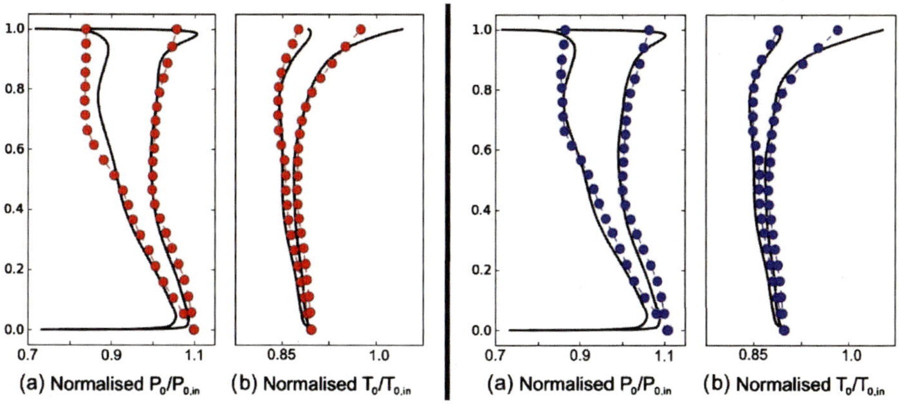

Fig. 4.10 Radial variation of normalized total pressure and temperature. *Left hand frames* with IGVs and *right hand* no IGVs—measurements (*circles*) and time averaged URANS (*solid lines*)—from Shaw et al. (2013)

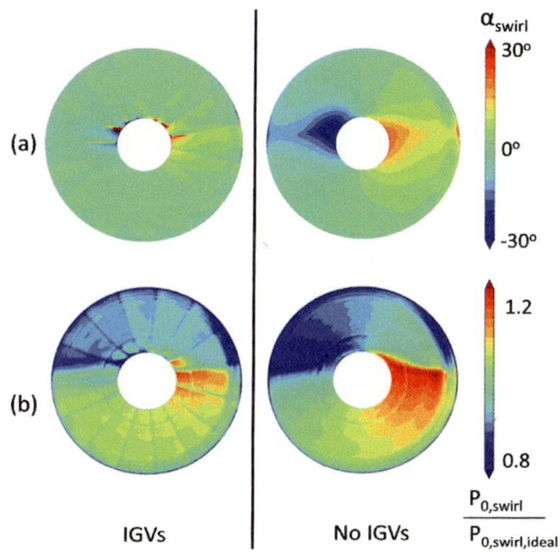

Fig. 4.11 Contours ahead of rotor of time averaged: (**a**) swirl angle and (**b**) relative stagnation pressure (with IGVs, *left*, without VIGVs, *right*) from Shaw et al. (2013)

Figure 4.10 shows radial profiles of total pressures and temperatures ahead of the stator with a clean inlet flow. The left hand frames are for the geometry with IGVs and the right hand without. Profiles are shown near stall and near choke at similar flow rates. The profiles show encouraging agreement with the high speed rig measurements. Shaw et al. show similar comparisons for when there is distortion. Radial profiles are considered through clean and distorted sectors. Broadly, a similar level of agreement is found for the undistorted profiles shown, with just a slight deterioration between the URANS and measurements. Hence, even though this flow is transonic with shocks and the URANS based theoretical questions noted above, meaningful solutions appear to be gained.

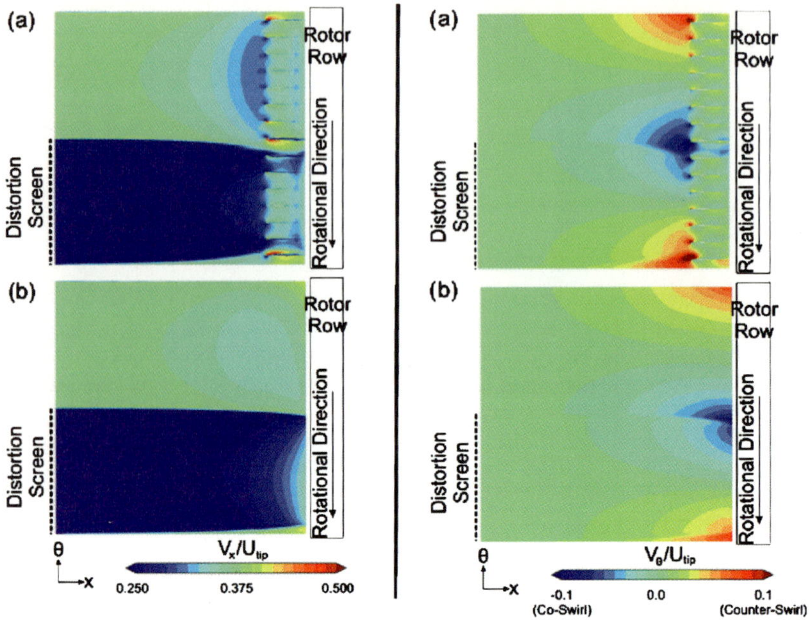

Fig. 4.12 Unwrapped time averaged contours of the flow at the blade midspan: (**a**) with IGVs and (**b**) without IGVs. *Left hand frames*—axial velocity and the *right frames*—tangential velocity (from Shaw et al. 2013)

Figure 4.11 gives contours ahead of the rotor for time averaged properties. Frame (a) gives the absolute swirl angle (see by the rotor) and (b) relative stagnation pressure. The left hand plots are for when there are IGVs and the right when there are none. For no IGVs there are clear zones of co and counter swirl. These will give rise to extremes of flow incidence onto the rotor which will be challenging for the rotor to deal with. The distorted flow sector is clear from the contours of total pressure. Also, it is clear that the IGVs give a cleaner flow thus allowing safer fan performance.

Figure 4.12 gives an unwrapped view of the time averaged flow at the blade midspan. The left hand frames give normalized axial velocity contours and the right tangential. The upper frames are when there are IGVs and the lower without them. The key thing to, again, note is that the IGVs give a cleaner flow.

4.5 Body Force and Mixed Fidelity Modelling

The above simulations, even though Graphical Processor Unit based, are relatively expensive. Hence, for the pressing industrial time scales reduced order models, such as body force, are also needed. Figure 4.13 gives a hierarchy for unsteady body force models. The schematics of spectral content on the right indicate that with

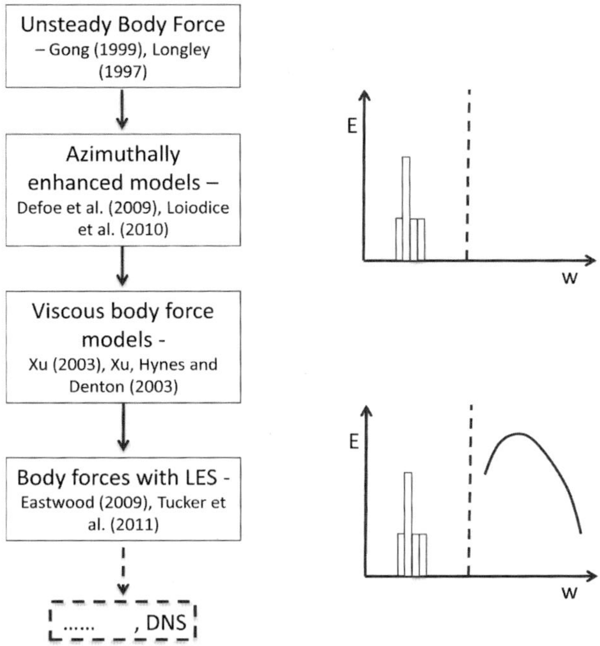

Fig. 4.13 Hierarchy for unsteady body force models (from Tucker 2011). Published with kind permission from Elsevier

increasing model refinement both turbulence and large-scale system unsteadiness are both accounted for. Longley (1997) explores the use of body force models for compressor stability. Gong (1999) applies them to unsteady distortion transfer in compressors and stall inception.

With the body force approach, typically there are no resolved blades within the annulus (the viscous approach of Xu (2003) and Xu et al. (2003) being the exception). Instead, the Euler equations are solved in a quasi-axisymmetric form. Then source/force terms are applied within the axial extent of the blade locations. Generally the source term field is axisymmetric. They are designed to introduce appropriate levels of flow turning, essentially forcing the flow to follow local blade metal angles. Calibration is typically used to match system characteristics. In the model of Gong (1999) the body forces are typically divided into components parallel and normal to the blade passage. In cylindrical coordinates these components have the following form

$$F_{n,x} = \frac{K_n}{L}\frac{V_\theta}{V}(V_x \cos\alpha + V_\theta \sin\alpha)(V_\theta \cos\alpha - V_x \sin\alpha) + n_p \frac{1}{\rho}\frac{\partial p}{\partial x}\sin^2\alpha$$

$$F_{n,\theta} = \frac{K_n}{L}\frac{V_x}{V}(V_x \cos\alpha + V_\theta \sin\alpha)(V_\theta \cos\alpha - V_x \sin\alpha) + n_p \frac{1}{\rho}\frac{\partial p}{\partial x}\sin\alpha\cos\alpha \tag{4.26}$$

$$F_{n,x} = f_{sw}\frac{K_{vd}}{L}V_x V, \qquad F_{p,\theta} = -f_{sw}\frac{K_{vd}}{L}V_\theta V, \qquad F_{p,r} = -f_{sw}\frac{K_{vd}}{L}V_r V \tag{4.27}$$

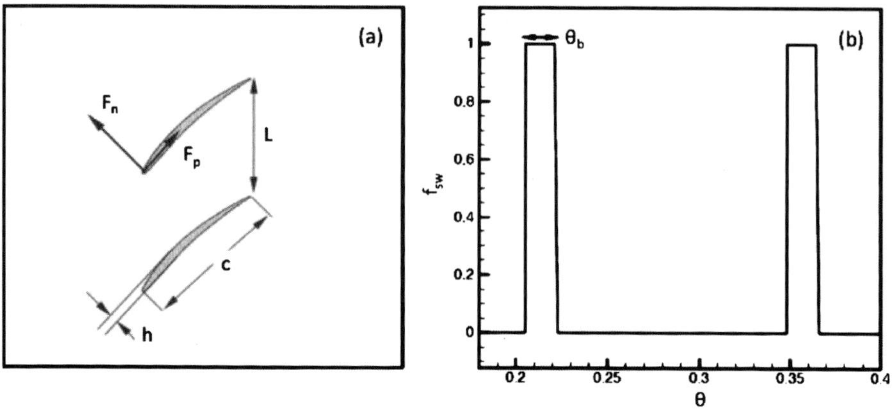

Fig. 4.14 Nomenclature and duty cycle for body force model: (**a**) nomenclature relative to blades and (**b**) viscous force duty cycle

In the above V_x, V_r and V_θ are the velocity components in the axial, radial and tangential directions, respectively. Also, V is the fluid velocity magnitude relative to the blades. In Gong's model $n_p = f_w = 1$. As indicated in Fig. 4.14(a), α now represents the blade metal angle and s is the blade pitch. Also, F_n and F_p are forces normal (Eqs. (4.26)) and parallel (Eqs. (4.27)) to the blade passage. They could be considered akin to lift and drag forces. They are uniformly distributed in the volume defined by the blade row chord and span. The equations assume an infinite number of blades and locally axisymmetric flow. The above equations have two calibration constants. The constant K_{vd} multiplies the viscous drag terms and K_n the inertial force related terms.

4.5.1 Modelling Wakes

To extend the capability of body force models it can be useful to generate wake zones and the corresponding momentum deficit and increased turbulence levels associated with these areas. It is possible to create such zones by locally increasing source terms. The use of localised source terms, as opposed to a uniform distribution, can be found in computational approaches known as immersed boundary methods (Pinelli et al. 2010). The main feature of this method is that typically the entire simulation is carried out on a Cartesian grid. This does not conform to the geometry. The presence of the geometry (or boundary), immersed in the flow, is simulated using local source terms in the Navier-Stokes equations. The magnitude of such terms is found by imposing the no-slip and impermeability conditions along the immersed boundary surface. Loiodice et al. (2010) use a simpler approach. However, it broadly follows the aforementioned ideas of enhanced sources to generate blade wake zones. This is approximately done by defining a square wave or one

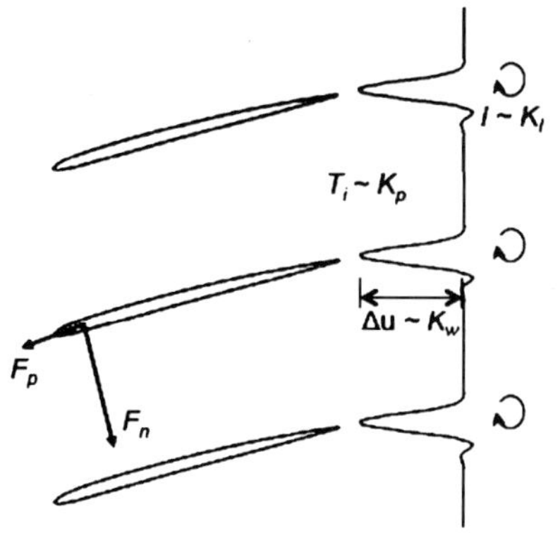

Fig. 4.15 Schematic of wake control parameters and areas of influence

with Gaussian content etc. This has a duty cycle equal $Nbh/2\pi R$ proportional to the blade thickness, h, multiplied by the number of blades Nb and normalized by radius R. The square wave function is shown in Fig. 4.14b, in terms of angular position, θ, in radians.

The function f_{sw}, shown in Fig. 4.14(b), is used to multiply all the forces parallel to the flow (Eqs. (4.27)). Hence, these sources will only be active when $f_{sw} > 0$. Therefore, in a generic form, F_p is revised to reflect the wake localized viscous force and also the inter-passage force.

$$F_p = \left[K_W f_{sw} + K_{vd}(1 - f_{sw})\right]\frac{\rho V^2}{L} \tag{4.28}$$

After a careful calibration of K_W, using the above function, it is possible to closely simulate localized wakes. One further change within the proposed model is required to ensure rational interblade and wake turbulence length scales, l. In the case of the k–ϵ turbulence model, this is relatively simple. It is achieved by modifying l such that:

$$l = 0.08 K_l h f_{sw} + 0.09\frac{L}{2}(1 - f_{sw}) \tag{4.29}$$

In the above K_l is a parameter used to modify the wake length scales. The l values can be directly used to compute the turbulence dissipation $\varepsilon \propto k^{3/2}/l$. In the case of the Spalart-Allmaras turbulence model there is no way to directly express and turbulence length scale. Then a solution is to locally reduce the model to that of Nee and Kovasznay (1969). With this, the direct prescription of a turbulence length scale is possible. This process is performed by Loiodice et al. The Nee-Kovasnay model being used just in the rotating mesh zone used for the rotor area. The stationary mesh zone uses the Spalart-Allmaras model. The modified Nee-Kovasnay equation is given below

$$\frac{D\nu_t}{Dt} = C_b\left[K_p f_{sw} + (1 - f_{sw})\right]S\nu_t + \frac{1}{\sigma}\{\nabla\cdot[(\nu + \nu_t)\nabla\nu_t]\} - C_w\left(\frac{\nu_t}{l}\right)^2 \tag{4.30}$$

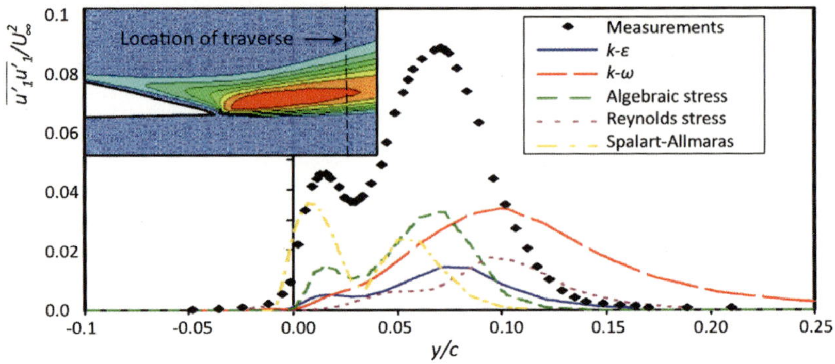

Fig. 4.16 Computed Reynolds stresses in with wake of a NACA4412 aerofoil for different turbulence models

The nomenclature is broadly as given in Chap. 3 and details of the constants are not given here. The Fig. 4.15 schematic summarises the wake control parameters. In summary K_l, K_w and K_p control the turbulence length scale, wake momentum deficit and turbulence intensity.

Model Validity

To set this process into context, Fig. 4.16 gives the computed Reynolds stresses in with wake of a NACA4412 aerofoil for different turbulence models. As can be seen, there is a substantial scatter between different turbulence models. What is more, none of them come close the measured (Coles and Wadcock 1979) Reynolds stress given by the symbols. However, there is a reasonable body of literature on expected wake turbulence properties. Hence, the use of this to calibrate the body force model is likely to yield as meaningful results as high fidelity RANS.

Notably, Gong solves the axisymmetric Euler equations. Results made with the current implementation, to be discussed, are based on the full Navier-Stokes equations. Also, there is no explicit forcing of axis-symmetry. As noted by Defoe et al. (2009) this axis-symmetry assumption is valid if the reduced frequency

$$\tilde{f}_p = \frac{2\pi \Omega R c}{\lambda V_x} \sim 1 \tag{4.31}$$

is of the order unity. In the above, λ, is now redefined as the wavelength of an azimuthal disturbances. Also, R, is the blade radius and Ω blade angular velocity. Basically, $\tilde{f}_p$ is the ratio for the time that it takes an azimuthal disturbance wave/profile to tangentially move through λ over the time that it would take a disturbance/wave front to pass through. Equation (4.31) requirement ensures consistency between the quasi-axisymmetric and full three-dimensional equation sets.

Fig. 4.17 Characteristic for the NASA 35 rotor (Inset stage efficiency against mass flow), from Tucker (2011). Published with kind permission from Elsevier

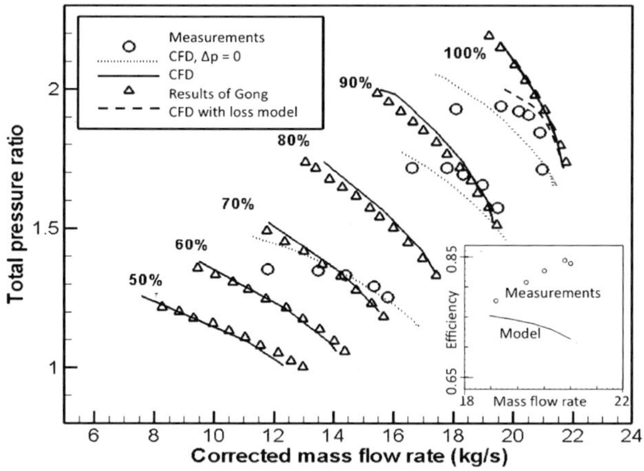

4.5.2 Gong Body Force Model Examples

Here, the performance of the Gong body force model without wakes ($f_{sw} = 1$) is first considered. Figure 4.17 shows a plot of compressor pressure ratio against (corrected) mass flow rate. Results are shown for various rotational speeds. These simulations are for the so called NASA 35 rotor. The circle symbols are measurements. The triangles are Gong's predictions. The lines are results with this model implemented in an unstructured cell vertex based program (HYDRA—Lapworth 2004). The full black line is for the full Gong model. The dotted lines are computations when none of the blade pressure force based terms used by Gong are implemented i.e $n_p = 0$. The rationale for these terms is debatable for subsonic flow. For the full model there is generally encouraging agreement with the Gong's predictions. This suggests the method has little sensitivity to code implementation. Despite the non-rigorous enforcement of zero tangential momentum fluxes (relating to Eq. (4.28)), as can be seen from Fig. 4.17, the model gives satisfactory performance. However, the compressor characteristic curve does not 'roll over' (curve gradient tend to zero or of a negative sign) at lower mass flow rates. Indeed, one could argue that the results for the Gong model with $n_p = 0$ are better in this respect. Some correction for deviation (the flow not following the blade metal angle) makes the characteristic curve 'roll over'—results not shown here. However, tests show that the Gong model predicts an efficiency increase as it is progressed towards stall. This is wrong—a decrease must occur. The efficiency increase is shown in the inset to Fig. 4.17. The predicted efficiency looks most disappointing. The Fig. 4.18 line is a polynomial fit to loss against incidence measurements from Howell (2095). The latter are given by the symbols. This polynomial fit is used to modify the viscous loss term in Gong's model. For stability, the incidence used in the deviation curve is radially and axially integrated. The axial integration zone is roughly around 5 % of c. As can be seen from the dashed, Fig. 4.17, line this corrects the excessive climb in pressure as the compressor moves towards stall. As can be seen from all of the above, Gong's

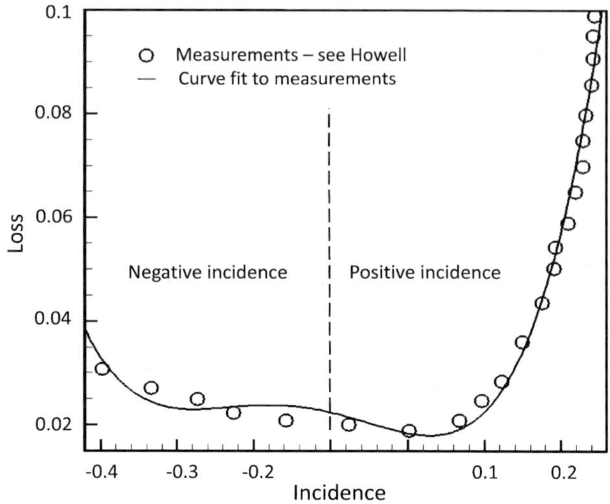

Fig. 4.18 Plot of incidence against loss (adapted from Tucker 2011). Published with kind permission from Elsevier

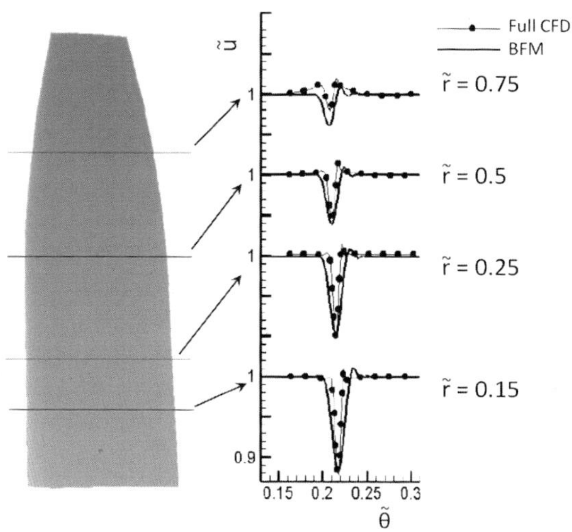

Fig. 4.19 Comparison of azimuthally enhanced body force model with high fidelity CFD in local wake zone (from Tucker 2011)

models appear to offer a reasonably effective, conceptually simple, computational framework.

Its use with the additional wake loss term, discussed above, is considered next. They are used to explore the interaction of the wakes from open rotors with intake ducts upstream of a compressor. The full simulations are discussed further in Chap. 7, which focuses on coupled simulations. The intake ducts in open rotor engines are positioned around the hub region of the rotor. Here 'solidity' (the degree of flow confinement by solid surfaces) is high. This is most ideal for Gong's model. The accuracy of the model at capturing rotor wakes is illustrated in Fig. 4.19. Simulations involve the Spalart-Allmaras model reduced to the Nee-Kovasnay in the rotor zone. The rotor dimensionless axial velocity is plotted at various radial loca-

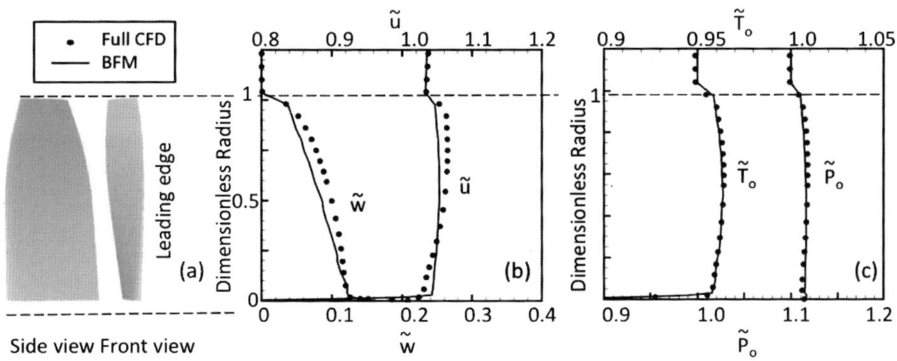

Fig. 4.20 Comparison of downstream radial profiles from body force model with high fidelity CFD: (**a**) rotor geometry (**b**) dimensionless velocity profiles and (**c**) dimensionless total pressure and temperature

tions. The symbol with lines are the high fidelity validated CFD of Zachariadis and Hall (2011). As can be seen, there is encouraging agreement. Finally, Fig. 4.20 compares circumferentially averaged radial profiles one chord downstream of the rotor trailing edge. Frame (a) gives views of the rotor geometry. It is worth noting that this is quite primitive. Frame (b) plots the radial variation of dimensionless axial and radial velocity. Frame (c) shows the radial variation of total pressure and temperature. Again, the agreement is encouraging.

4.6 Deterministic Stresses

As noted earlier, deterministic stresses can be used to render steady RANS solutions, hence avoiding the cost of URANS. Some processes for evaluating deterministic stresses are outlined in this section. Figure 4.21 summarises potential ways of estimating deterministic stress. These will be further outlined later.

4.6.1 Mean Source Terms (MST)

Following Ning and He (2001), looking symbolically at the Navier-Stokes equations and defining the unsteady Navier-Stokes operator, *NS*, we can write the following as true

$$NS(\phi) = 0 \tag{4.32}$$

The above can be integrated in an *unsteady* fashion using the operator, *NS* for a time varying solution, ϕ. This can be time averaged over some period t_l as defined

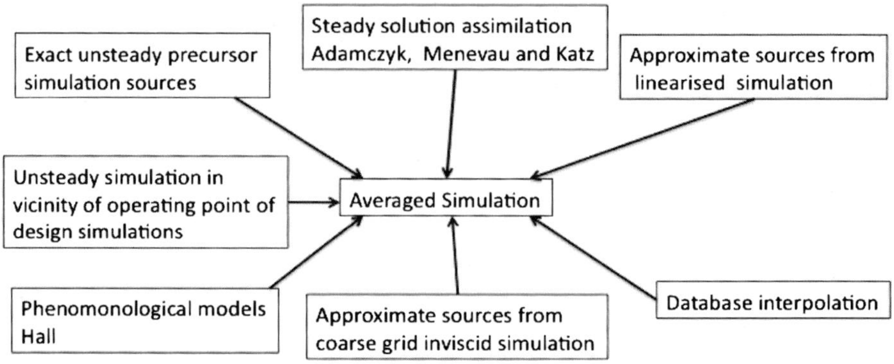

Fig. 4.21 Source inputs for deterministic stress simulations (from Tucker 2011). Published with kind permission from Elsevier

in Chap. 3.

$$\overline{\Phi} = \frac{1}{t_l}\int_{t-t_l/2}^{t+t_l/2} \phi dt \tag{4.33}$$

The former assumes the usual Eq. (4.1) decomposition $\phi = \bar{\Phi} + \phi'' + \phi'$. We could produce a (residual) source term, S, by applying the steady Navier-Stokes operator, NS^s, to the computed mean

$$S = NS^s(\overline{\Phi}) \tag{4.34}$$

The source, S, accounts for the stresses arising from the fluctuating quantities $\phi'' + \phi'$. If we know S, (not an easy task in practice) then we can solve directly for the mean by solving the following equation

$$NS^s(\overline{\Phi}) - S = 0 \tag{4.35}$$

We could also do this in a URANS to steady RANS context using the following symbolic operators, *URANS* and *RANS*, respectively. Hence, we solve

$$URANS(\langle\Phi\rangle) = 0 \tag{4.36}$$

and this time average $\langle\Phi\rangle$. We then have $S = RANS(\overline{\Phi})$ where *RANS* is a symbolic steady RANS operator. For RANS/URANS, ϕ', is accounted for by the RANS model. The source accounts for ϕ''. $RANS(\overline{\Phi}) - S = 0$ is then the final steady equation system.

4.6.2 Deterministic Stress Modelling (DSM)

Alternatively, more practically with URANS, we could produce steady state solutions using deterministic stress modelling. With this, we have the usual decomposi-

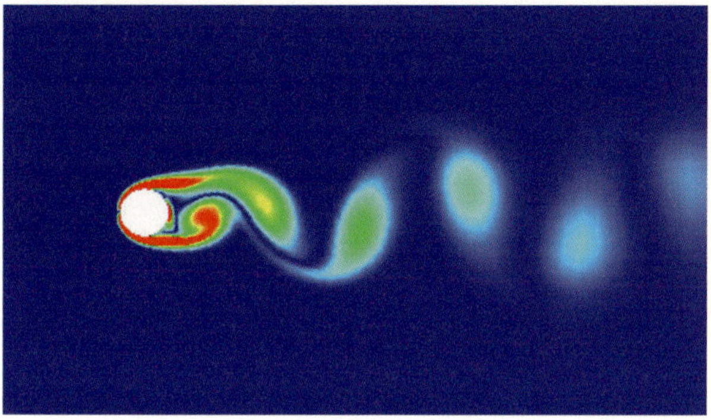

Fig. 4.22 Instantaneous flow over a cylinder for $Re = 200$. (The deterministic stress results and images were kindly provided by Dr. Richard Jefferson-Loveday)

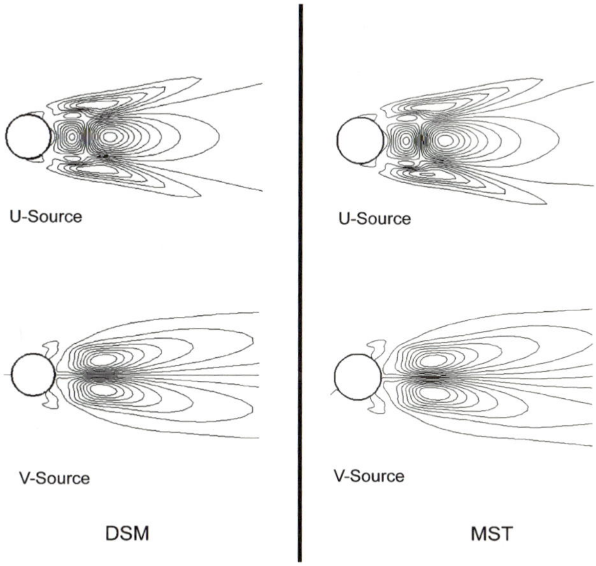

Fig. 4.23 Ultimate source terms arising from the DSM and MST approaches for flow over a cylinder at a Reynolds number of 200

tion $\phi = \overline{\Phi} + \phi'' + \phi'$. With URANS we solve for $\langle \Phi \rangle = \overline{\Phi} + \phi''$

$$\frac{\partial \langle \Phi \rangle}{\partial t} + RANS\big(\langle \Phi \rangle\big) = 0 \tag{4.37}$$

Full closure of the above system will have terms of the following form $\overline{\phi_i' \phi_j'}$, $\overline{\phi_i'' \phi_j''}$ and other non-linear terms. With deterministic stress modelling the key task is to devise a model for the deterministic stress terms $\overline{\phi_i'' \phi_j''}$. These are analogous the Reynolds stresses. Like them, they enter the time-averaged equations in a differentiated form—$\partial \overline{\phi_i'' \phi_j''} / \partial x_j$. As noted above, $\partial \overline{\phi_i' \phi_j'} / \partial x_j$ is accounted for by the RANS model. Following Ning and He (2001), initially, we will explore these approaches for laminar flow ($\overline{\phi' \phi'} = 0$) over a cylinder. The Reynolds number is 200, based on the cylinder diameter. The instantaneous flow is shown in Fig. 4.22. Figure 4.23

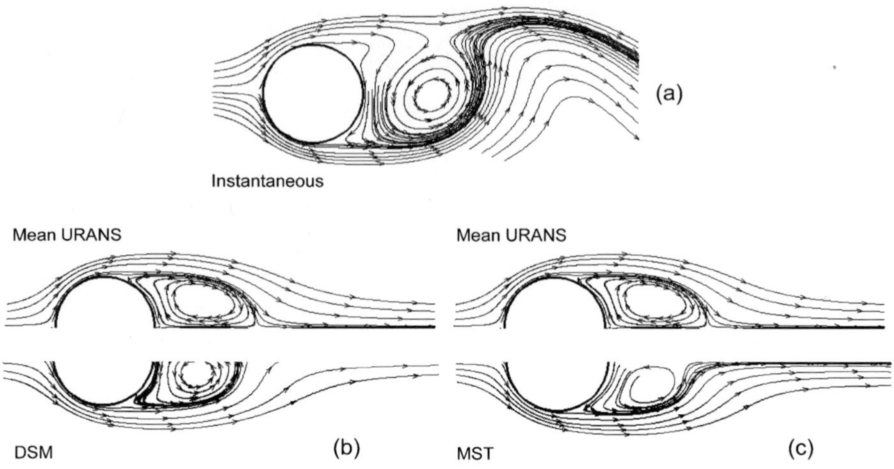

Fig. 4.24 Contrasting flow from time averaged URANS solution with MST and DSM approaches: (**a**) lines seeded in instantaneous snapshot of the flow field; (**b**) comparison of time-averaged URANS solution with DSM solution and (**c**) comparison of time-averaged URANS solution with MST approach

shows the ultimate source terms arising from the *MST* and *DSM* approaches discussed above. The upper frames correspond to the x-momentum sources and the lower the y. As would be expected the ultimate sources look similar.

Next, we move to URANS. Figure 4.24, Frame (a) shows lines seeded in an instantaneous snapshot of the flow field and integrated through time. Frames (b) and (c) compare time-averaged URANS results with the DSM and MST results. There are subtle differences between the time averaged URANS, MST and DSM approaches. A key reason is presumably because, with the DSM approach, the Reynolds stress analogous source components need differentiation. This added stage introduces numerical error.

Figure 4.25 contrasts the iterative convergence of the differing approaches and also centreline velocities in the cylinder's wake. Frame (a) gives the convergence histories. As would be expected, the steady RANS solution will not iteratively converge. However, correctly, the DSM and MST approaches do. Frame (b) contrasts the velocity distributions. Notably, the time averaged URANS solution and MST results are identical, as would be expected. The DSM result is again slightly different.

Clearly both approaches are effective in yielding steady solutions. However, as already noted, alone, they are of no practical use. The ultimate source terms cannot be determined a priori.

Jefferson-Loveday et al. (2012) use a differential equation to 'algebraically' specify wake turbulence length scales in the base RANS model used. These scales are specified through Dirichlet boundary conditions in a Hamilton-Jacobi differential equation. This equation is discussed further in Chap. 7. However, the basic form of

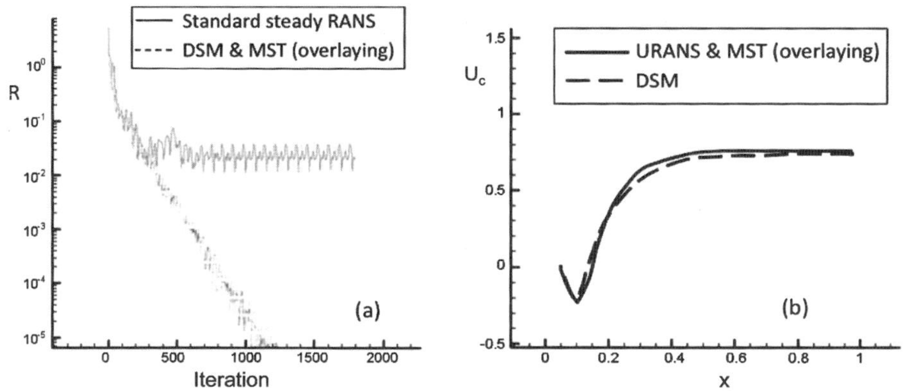

Fig. 4.25 Comparison of the iterative convergence of the differing approaches and also centreline velocities in the wake of the cylinder: (**a**) convergence histories and (**b**) velocity distributions

the equation, in the current context, is given below:

$$|\nabla \tilde{L}| = 1 + \Gamma \nabla^2 \tilde{L} \tag{4.38}$$

In the above $\tilde{L}$ is a turbulence length scale. Normally this would be $\tilde{L} = l/\kappa = d$. However, where there is a clear expected integral scale, $\tilde{L}$ can be set to this. It is set via a wall based Dirichlet condition. Since a plausible integral scale is enforced (in the case of the cylinder this scales as cylinder diameter), steady solutions arise. As with the body force strategy above there might be reservations about accuracy. However, as can be see from Fig. 3.3, in Chap. 3, as with aerofoil wakes, for cylinders, as with many wake flows, RANS model performances can be poor. Hence, as shown in Jefferson-Loveday et al., application of the approach to cylinder wake and basic cavity flows, with the Nee-Kovasnay model, can show considerably improved accuracy relative to a wide range of RANS models.

The motivation for the above work is dealing with turbine rim seal flows. This zone is located between rotating and stationary blade rows. Essentially, the geometry is a complex cavity flow. However, the upstream and downstream faces have rotation. These cavity flows (O'Mahoney 2011) are characterized by large scale unsteady eddies. Figure 4.26 shows a solution of the Hamilton-Jacobi equation for relatively complex turbine geometry. The key thing to note is that $\tilde{L}$ is high in the groove regions. This gives a high integral scale giving a steady solution.

There is little high quality detailed validation date for turbine blade flows with rim seals. Hence, a hybrid NLES has been performed. In this, the rim seal zone is treated in pure NLES mode. The main blade passage is treated using hybrid NLES-RANS. This is acceptable, especially bearing in mind that the main passage area is not the key zone of interest. Figure 4.27, Frame (a) shows an instantaneous NLES zone predicted vorticity field for a rim-seal cavity. A view of the mesh used is shown in Frame (b). For validation purposes the hybrid NLES is validated against blade surface pressure measurements, flow visualization and also the radial variation of

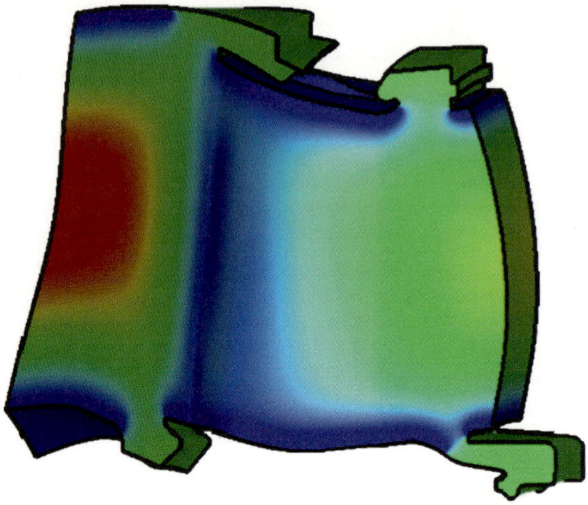

Fig. 4.26 Hamilton-Jacobi equation solution ($\tilde{L}$) for a turbine geometry

the mass weighted total pressure loss coefficient downstream of blade(s). This is carried out for a case with no cavity where there are measurements. The radial variation of total pressure loss is shown in Fig. 4.28a. Encouraging agreement is found with all the measurements. These simulations are discussed further in Chap. 5. Figure 4.28c shows the general flow complexity through instantaneous isosurfaces of vorticity magnitude. Figure 4.28b contrasts the hybrid NLES results (symbols) with standard model results (dashed line) and the Hamilton-Jacobi equation based simulations (full lines). As can be seen the additional stress has given a more accurate and importantly a steady solution. Note, the standard model results are semi-converged *steady* simulations.

4.6.3 Blade Row Specific Approaches

The relative rotation between blade rows in turbomachinery means that the wake relative tangential movements gives rise to deterministic stresses. The Adamczyk (1985) type modelling, later revisited by Hall (1998), falls into this area of deterministic stress modelling. With this approach, stresses (spatial correlations) arising from fluctuations of the scale ϕ'' caused by wake passing etc., are estimated using a series of averaging operators. These average out turbulence, periodic fluctuations, finally giving a circumferential, passage average. With the approach of Adamczyk overlapping grids are needed. A body force gridded zone axially flanks each blade row. This zone does not explicitly resolve blade geometry but overlays a separate grid that does so. The difference between the axisymmetric body force zone solution and the geometry resolved solutions gives an effective circumferential velocity perturbation field. Using a reversed Taylor type hypothesis (having blade relative velocity as the velocity scale) this spatial distribution can be converted into a temporal trace. From this, a root-mean-square perturbation, ϕ'', can be formed and hence a

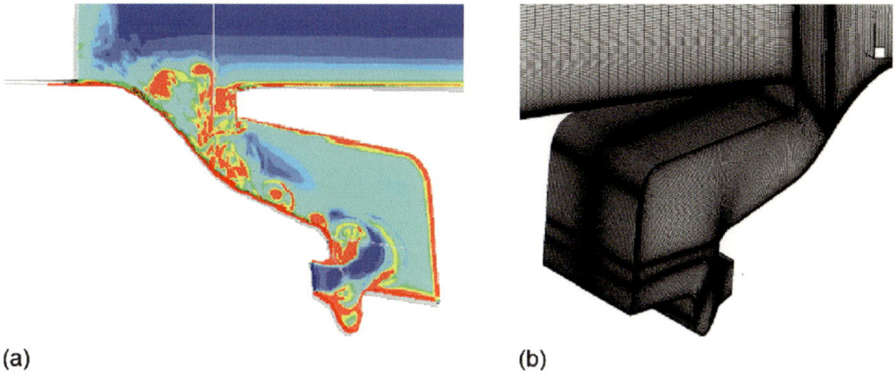

Fig. 4.27 Hybrid NLES-RANS of rim seal: (**a**) rim seal vorticity contours, (**b**) rim seal NLES grid

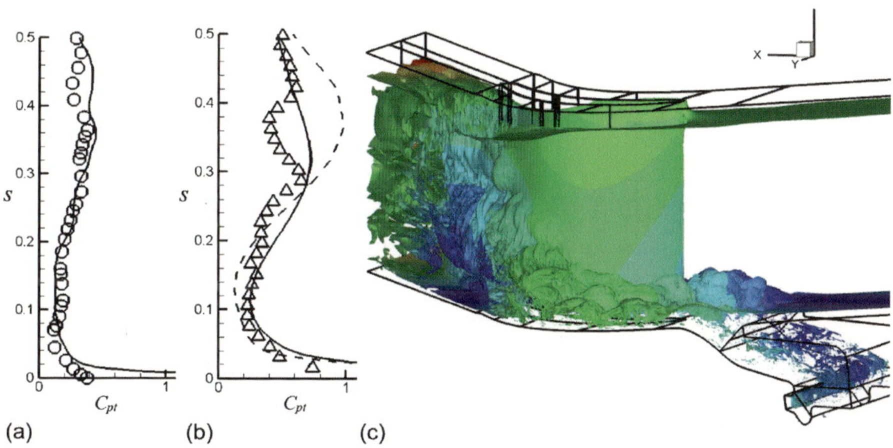

Fig. 4.28 Validation and assessment for Hamilton-Jacobi equation applied to a turbine rim-seal flows: (**a**) Pitchwise averaged total pressure loss coefficient for baseline, no-cavity, validation case: circular symbols measurements, *solid line* RANS-NLES; (**b**) modified integral length scale & standard SA RANS *vs.* high order RANS-NLES for cavity case *triangular symbols*: RANS-NLES, *solid line*: integral length scale modified-RANS, *dashed line*: RANS, (**c**) vorticity magnitude isosurfaces for cavity geometry

deterministic stress evaluated. Notably, the deterministic stress does not vary azimuthally. The influence of downstream components on upstream can be accounted for. However, as noted by Charbonnier and Leboeuf (2003), since the deterministic stresses can be strongly localized to the trailing edge-leading edge zone the grid overlap can be restricted. Rhie et al. (1998) neglect upstream influences in their application of Adamczyk's model. Figure 4.29 shows the components of an Adamczyk analysis but in the manner implemented by Rhie et al. i.e. the one-way propagation of deterministic stresses. The schematic shows the four separate grids needed for a two-stage analysis. Rhie et al. also, include extra deterministic terms in their k–ϵ

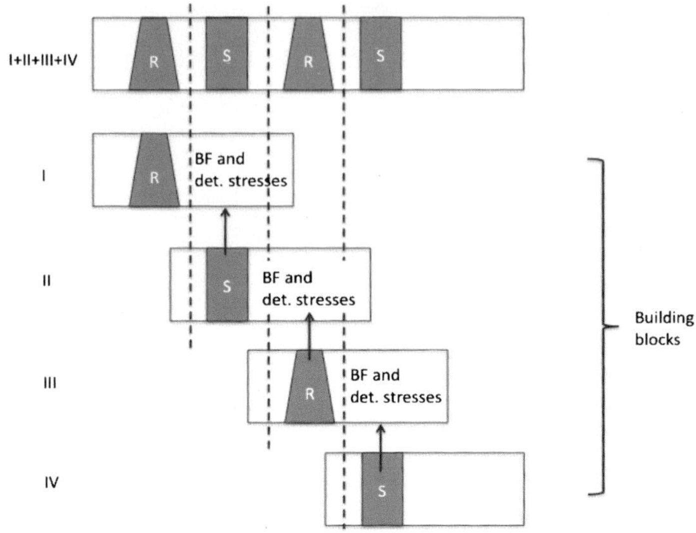

Fig. 4.29 Component organization in Adamczyk modelling (from Tucker 2011). Published with kind permission from Elsevier

turbulence model. This allows the enhanced production arising from wakes passing through downstream blades rows (from the upstream) to the captured in the downstream rows. The approach showed enhancements when considering predictions of total pressure ratio and circumferentially averaged radial distributions of variables.

However, the Adamczyk approach appears slightly cumbersome to implement and not readily understandable by a design engineer. The method of Hall (1998) is conceptually simpler. It also appears easier to implement not requiring the extended grid zones. Figure 4.30 gives a schematic relating to Hall's approach. Hall takes standard wake correlations. The first is for wake half width

$$\frac{w}{L} = \frac{1.63732(L/c)C_D^{1/8} - 0.01944}{5.5755(L/c)^{1/8} + 1} \tag{4.39}$$

In the above, L, is the blade pitch and C_D the aerofoil section drag coefficient. The wake, centreline, velocity deficit is given by

$$\left(\frac{\Delta V}{V}\right)\left(\frac{1}{C_D^{1/4}}\right) = \frac{1.17543(L/c) - 1.28626}{10.79857(L/c) + 1} \tag{4.40}$$

where V is the relative free stream velocity. Based on measurements, the above is modified by Hall to account for the increased wake dissipation/spreading due to, for example, chopping by downstream blades

$$\frac{w}{s} = \frac{1.17543(L/c) - 1.28626}{21.0(L/c) + 1} \tag{4.41}$$

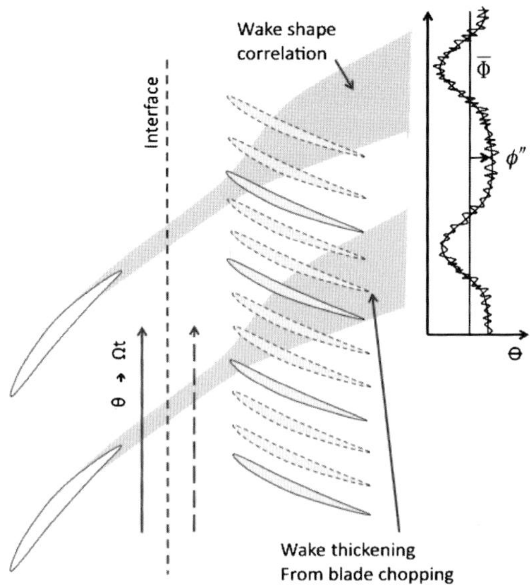

Fig. 4.30 Schematic of deterministic model of Hall (from Tucker 2011). Published with kind permission from Elsevier

Note, *s* is the streamwise distance from the rotor trailing edge. As discussed above, in relation to the body force modelling, for wakes the use of correlations has the potential to rival wake modelled CFD. The CFD can give a wide scatter in solutions. None of which might be that close to the measurements. The expressions used by Hall are relatively simple to apply. The grey shaded zone in Fig. 4.30 shows the increase in wake width. The modified wake correlation, describing a momentum deficit is swept around at an appropriate relative speed. This, when referenced to a mean state, gives a velocity fluctuation estimate that can be converted into a deterministic stress. Hence, the approach of Hall is based around phenomenological modelling for the deterministic stresses.

Both the Adamczyk and Hall models assume that the deterministic stress distributions are axisymmetric. However, they can have extreme azimuthal variations. To account for these Meneveau and Katz (2002) make multiple RANS simulations. The deterministic stresses are formulated by averaging these simulations with appropriate weightings to reflect time durations which the steady state is most representative of.

Bardoux and Leboeuf (2001) use URANS to evaluate the exact source terms. The source terms are shown to be decomposed into spatial, temporal-spatial and purely temporal correlations. The Adamczyk model neglects the latter two. Bardoux and Lebouf's simulations show that just having the spatial correlations can give worse results than when steady mixing plane solutions are made. Bardoux et al. (1999) note that wake segregation effects (see Chap. 1) cannot be captured with just spatial correlations. Charbonnier and Leboeuf (2003, 2005) consider a transport equation for deterministic stresses. They propose a model for the velocity-pressure gradient term in this equation. Hence, like Hall, phenomenological modelling is used albeit at a deeper level, intended to account for temporal deterministic stresses.

Busby et al. (2000) explore the use of deterministic stress modelling for hot streak segregation in turbines. Coarser grid inviscid simulations are used to evaluate the deterministic stress terms. These stresses are used to secure steady *viscous* simulations. The time averaged temperature in these simulations is found to be substantially improved (5 %) with an over 20 % reduction in computational cost. Similarly, Sondak et al. (1996) look at turbine rotor-stator interactions using this approach. For transonic flow, when the rotor and stator are in close proximity the unsteady interaction terms are found important. Orkwis et al. (2002) extend the work of Busby et al. by evaluating the deterministic stress terms based on linearised unsteady inviscid solutions. The migration of the hot fluid to the rotor pressure surface was successfully captured.

Stollenwerk and Kugeler (2011) explore the use of deterministic stress modelling for a multi-stage transonic compressor with strong blade row interactions. Deterministic stresses were extracted at one operating point from URANS. These stresses are fixed and used in mixing plane calculations for a range of operating points. Improved accuracy (greater accord with URANS) is observed for global performance and more detailed parameters.

4.6.4 Generic Approaches

Kulisa et al. (1992) explore the modelling of film cooling holes with deterministic type stresses. A three-dimensional, integral, jet model is used to compute source terms. These are fed into two-dimensional near wall viscous equations. There is an iterative exchange of information. Encouraging agreement is found with flat plate measurements.

Leading on from the Hamilton-Jacobi equation work discussed above; inspired by the need to model endwall cavity flows in turbomachinery, Gangwar et al. (2001) and Lukovic (2002) explore deterministic stress modelling of idealized cavity flows. Lukovic uses URANS simulations as training data for a neural network source model. Hence, the neural network is used as a form of interpolator in the design space. Global performance (cavity drag) is found to be predicted within around 10 % of the URANS. Also, this is at two orders of magnitude lower cost. Further work on the application of neural networks to cavity flows can be found in Gangwar et al. (2001), Lukovic et al. (2001, 2002b). The latter explores geometric variations and the ability of neural network modelling to predict this variation. Lukovic et al. (2002a) includes a wall tangential velocity component. This brings greater connection to aero engine rim seal flows. The application of neural networks to a cylinder flow, as explored earlier, is discussed by Benning et al. (2001).

As can be seen from Fig. 4.21, there is wide range of potential input sources for deterministic stresses. The user must appreciate which is the most appropriate. Hence, it needs to born in mind if, for example, viscous processes are important or not. If not, then inviscid solutions on coarser meshes can be used. The work of Bardoux and Lebouf is indicative of the challenging task that an engineer faces in grappling with complex non-linear flow physics and the most appropriate model for it. For the rim seal flow, hybrid LES type simulations were shown to be useful to validate a deterministic stress model. There is no reason why such high fidelity simulations cannot be used to calibrate such models. They could also be used, for example, to calibrate the wake body force model. However, such a calibration would perhaps be best carried out using blade section LES calculations. Otherwise the cost could become prohibitive.

4.7 Conclusions

The various unsteady modelling hierarchies in aerospace have been reviewed. These range from linear harmonic to DNS. Unsteady reduced order modelling encompassing deterministic stresses and body forces have been discussed. Hierarchies were presented for different modelling lineages and fidelity levels. For URANS, the occurrence of a spectral gap in certain turbomachinery zones is found tenuous. Deterministic stresses can be used to account for unsteadiness. However, the a priori determination of such stresses presents an extreme modelling challenge. Harmonic based methods offer the potential for high temporal accuracy. Also, truncating the number of modelled harmonics can offer substantial computational savings with minimal accuracy penalty, albeit with substantially increased storage. However, the excessive truncation of the number of harmonics can prevent convergence. Then special convergence measures are needed. For specialist purposes non-linear and linear disturbance equations can also be solved. However, it is necessary to compute a steady base flow. Forcing the steady state with a strong solver could result in a physically inaccurate base state.

With turbomachinery specifically in mind, for un-equal blade pitches scaling decisions need to be made or phase shift boundary conditions used. Various forms of these exist with different advantages, defects and stability limits. Also, such boundary conditions can impact on the physical realism of results.

For unsteady flows the user needs to select, for example, appropriate:

(I) Phenomenological models, be they RANS with a transition model and potentially a deterministic stress level to characterize coherent unsteady components or even LES, MILES, ILES, NLES, hybrid LES-RANS—see Chap. 3;
(II) Accurate and efficient numerical schemes—see Chap. 2;
(III) Appropriate field specific methods e.g. levels of harmonic filtering or linearisation, phase shift boundary condition implementations etc. and
(IV) Problem definition, boundary conditions e.g. passage rescaling, stage interface connection strategy etc.

Clearly many of the above elements have specific applicability and also stability ranges. In terms of user assimilation, the potential number of modelling and numerical variants poses a serious challenge. This is especially so when reported experiences for the same scheme can differ dramatically. For calibration of reduced order models such as body force and deterministic stress, LES related computations could be used.

References

J.J. Adamczyk, Model equation for simulating flows in multistage turbomachinery, in *ASME, 30th International Gas Turbine Conference and Exhibit*, Houston, Texas (1985). ASME Paper No. 85-GT-226

F. Bardoux, F. Leboeuf, Impact of deterministic correlations on the steady flow field. Proc. Inst. Mech. Eng. A, J. Power Energy **215**(6), 687–698 (2001)

F. Bardoux, F. Leboeuf, C. Dano, C. Toussaint, Characterization of deterministic correlations for turbine stage. Part 1: time average flow analysis, in *ASME* (1999). ASME Paper No. 99–GT–100

R.M. Benning, T.M. Becker, A. Delgado, Initial studies of predicting flow fields with an ANN hybrid. Adv. Eng. Softw. **32**(12), pp. 895–901 (2001)

T. Biesinger, C. Cornelius, C. Rube, A. Braune, R. Campregher, P.G. Godin, G. Schmid, L. Zori, Unsteady CFD methods in a commercial solver for turbomachinery applications, in *ASME Conference Proceedings* (2010), 2441–2452. ASME Paper No. GT2010-22762

R. Blumenthal, B. Hutchinson, L. Zori, Investigation of transient CFD methods applied to a transonic compressor stage, in *Proceedings of the ASME Turbo Expo. 2011: Power for Land, Sea and Air, GT2011*, Vancouver, BC, Canada, 6–11 June 2011. ASME Paper No. GT2011–46635

J. Busby, D. Sondak, B. Staubach, R. Davis, Deterministic stress modeling of hot gas segregation in a turbine. J. Turbomach. **122**(1), 62–67 (2000)

M.S. Campobasso, M.H. Baba-Ahmadi, Analysis of unsteady flows past horizontal axis wind turbine airfoils based on harmonic balance compressible Navier-Stokes equations with low-speed preconditioning (2011). ASME Paper No. GT2011-45303

D. Charbonnier, F. Leboeuf, Development of a model for the deterministic stresses in a transonic turbine stage, in *5th European Conference on Turbomachinery*, Prague (2003)

D. Charbonnier, F. Leboeuf, Steady flow simulation of rotor-stator interactions with an unsteady deterministic model, in *Proceedings of the 6th European Conference on Turbomachinery*, Lille (2005). Paper No. 033–05/151

D. Coles, A.J. Wadcock, Flying-hot-wire study of flow past an NACA 4412 airfoil at maximum lift. AIAA J. **17**, 321–329 (1979)

Y. Colin, B. Aupoix, J. Boussuge, P. Chanez, Numerical simulation of the distortion generated by crosswind inlet flows, in *Proceedings of the 18th ISABE Conference*, Beijing, China (2007). Paper No. ISABE-2007-1210

S. Connell, M. Braaten, L. Zori, R. Steed, B. Hutchinson, G. Cox, A comparison of advanced numerical techniques to model transient flow in turbomachinery blade rows, in *Proceedings of the ASME Turbo Expo 2011: Power for Land, Sea and Air, GT2011*, Vancouver, BC, Canada, 6–11 June 2011. ASME Paper No. GT2011-45820

R. Corral, J. Crespo, Development of an edge-based harmonic balance method for turbomachinery flows, in *Proceedings of the ASME Turbo Expo 2011: Power for Land, Sea and Air, GT2011*, Vancouver, BC, Canada, 6–11 June 2011. ASME Paper No. GT2011–41570

J.D. Coull, Wake induced transition in low pressure turbines. PhD thesis, Department of Engineering, The University of Cambridge (2009)

C.H. Custer, J.M. Weiss, V. Subramanian, W.S. Clark, K.C. Hall, Unsteady simulation of a 1.5 stage turbine using an implicitly coupled nonlinear harmonic balance method, in *Proceedings of ASME Turbo Expo 2012 GT2012*, Copenhagen, Denmark, 11–15 June 2012. ASME Paper No. GT2012–69690

S. Deck, P. Spalart, P. Sagaut, J. Forsythe, K.D. Squires, Numerical simulation of transonic buffet over a supercritical airfoil. AIAA J. **43**(7), 1556–1566 (2005)

J.J. Defoe, Z.S. Spakovszky, Effects of boundary layer ingestion on the aero-acoustics of transonic fan rotors, in *Proceedings of ASME Turbo Expo 2012 GT2012*, Copenhagen, Denmark, 11–15 June 2012. ASME Paper No. GT2012–68503

J. Defoe, A. Narkaj, Z.S. Spakovszky, A novel MPT noise prediction methodology for highly-integrated propulsion systems with inlet flow distortion, in *15th AIAA/CEAS Aeroacoustics*

Conference (30th AIAA Aeroacoustics Conference), Miami, Florida, 11–13 May 2009. Paper No. AIAA 2009–3366
J.D. Denton, The calculation of three-dimensional viscous flow through multistage turbomachines. J. Turbomach. **114**(1), 18–26 (1992)
K. Dullenkopf, R.E. Mayle, An account of free-stream-turbulence length scale on laminar heat transfer. J. Turbomach. **117**(3), 401–406 (1995)
J.I. Erdos, E. Alzner, W. Mcnally, Numerical solution of periodic transonic flow through a fan stage, in *American Institute of Aeronautics and Astronautics, 9th Fluid and Plasma Dynamics Conference*, San Diego, California (1976). AIAA Paper No. AIAA 76-369
G. Fritsch, M. Giles, Second-order effects of unsteadiness on the performance of turbomachines, in *37th International Gas Turbine and Aeroengine Congress and Exposition* (1992). ASME Paper No. 92–GT–389
P.F. Galpin, R.B. Broberg, B.R. Hutchinson, Three-dimensional Navier-Stokes predictions of steady state rotor/stator interaction with pitch change, in *Proceedings of 3rd Annual Conference of the CFD Society of Canada* (1995)
A. Gangwar, B. Lukovic, P. Orkwis, B. Sekar, Modeling unsteadiness in steady cavity simulations, Part I: parametric solutions, in *39th AIAA Aerospace Sciences Meeting and Exhibit*, Reno, Nevada (2001). Paper No. AIAA–2001–0153
E. Garnier, S. Deck, Large-eddy simulation of transonic buffet over a supercritical airfoil, in *Direct and Large Eddy Simulation VII ERCOFTAC Series 12*, ed. by V. Aemenio et al. (2008). doi:10.1007/978-90-481-3652-0-81
G.A. Gerolymos, G.J. Michon, J. Neubauer, Analysis and application of chorochronic periodicity in turbomachinery rotor/stator interaction computations. J. Propuls. Power **18**(6), 1139–1152 (2002)
M.B. Giles, Calculation of unsteady wake/rotor interaction. J. Propuls. Power **4**(4), 356–362 (1988)
M. Giles, UNSFLO: A numerical method for the calculation of unsteady flow in turbomachinery. Gas Turbine Laboratory, Massachusetts Institute of Technology (1991)
M.B. Giles, An approach for multi-stage calculations incorporating unsteadiness. ASME Paper No. 92-GT-282 (1992)
Y. Gong, A computational model for rotating stall and inlet distortions in multistage compressors. PhD thesis, Massachusetts Institute of Technology (1999)
A. Gopinath, E. Van Der Weide, J.J. Alonso, A. Jameson, K. Ekici, K.C. Hall, Three-dimensional unsteady multi-stage turbomachinery simulations using the harmonic balance technique. AIAA Pap. **892**, 2007 (2007)
N. Gourdain, S. Burguburu, F. Leboeuf, H. Miton, Numerical simulation of rotating stall in a subsonic compressor. Aerosp. Sci. Technol. **10**(1), 9–18 (2006)
N. Gourdain, S. Burguburu, F. Leboeuf, G.J. Michon, Simulation of rotating stall in a whole stage of an axial compressor. Comput. Fluids **39**(9), 1644–1655 (2010)
E.J. Hall, Aerodynamic modelling of multistage compressor flow fields. Part 2: modelling deterministic stresses. J. Aerosp. Eng. **212**(2), 91–107 (1998)
K.C. Hall, J.P. Thomas, W.S. Clark, Computation of unsteady nonlinear flows in cascades using a harmonic balance technique. AIAA J. **40**(5), 879–886 (2002)
L. He, Harmonic solution of unsteady flow around blades with separation. AIAA J. **46**(6), 1299–1307 (2008)
L. He, Fourier methods for turbomachinery applications. Prog. Aerosp. Sci. **46**(8), 329–341 (2010)
L. He, Efficient computational model for nonaxisymmetric flow and heat transfer in rotating cavity. J Turbomach. **133**(2) (2011)
L. He, W. Ning, Efficient approach for analysis of unsteady viscous flows in turbomachines. AIAA J. **36**(11), 2005–2012 (1998)
L. He, D.X. Wang, Concurrent blade aerodynamic-aero-elastic design optimization using adjoint method. J. Turbomach. **133**(1) (2011)
C.M. Ho, P. Huerre, Perturbed free shear layers. Annu. Rev. Fluid Mech. **16**(1), 365–422 (1984)

A.R. Howell, The present basis of axial flow compressor design: Part I—cascade theory and performance. Aeronautical Research Council R and M 2095 (1942)
J.C.R. Hunt, A theory of turbulent flow round two-dimensional bluff bodies. J. Fluid Mech. **61**(4), 625–706 (1973)
R.J. Jefferson-Loveday, P.G. Tucker, V. Nagabhushana Rao, J.D. Northall, Differential equation specification of integral turbulence length scales, in *Proceedings of ASME Turbo Expo 2012 GT2012*, Copenhagen, Denmark, 11–15 June 2012. ASME Paper No. GT2012–68091
P. Kulisa, F. Leboeuf, G. Perrin, Computation of a wall boundary layer with discrete jet injections. J. Turbomach. **114**, 756–764 (1992)
E. Labourasse, P. Sagaut, Reconstruction of turbulent fluctuations using a hybrid RANS/LES approach. J. Comput. Phys. **182**(1), 301–336 (2002)
L. Lapworth, HYDRA CFD: a framework for collaborative CFD development, in *International Conference on Scientific and Engineering Computation (IC-SEC)*, Singapore (2004)
S. Lardeau, M.A. Leschziner, Unsteady RANS modelling of wake–blade interaction: computational requirements and limitations. Comput. Fluids **34**(1), 3–21 (2005)
B.H.K. Lee, Transonic buffet on a supercritical airfoil. Aeronaut. J. **94**(935), 143–152 (1990)
M.G. List, S.E. Gorrell, M.G. Turner, Investigation of loss generation in an embedded transonic fan at several gaps using high-fidelity, time accurate CFD, in *Proceedings of the ASME Turbo Expo Power for Land, Sea and Air* (2008). ASME Paper No. GT-2008–51220
K. Lodefier, E. Dick, Modelling of unsteady transition in low-pressure turbine blade flows with two dynamic intermittency equations. Flow Turbul. Combust. **76**(2), 103–132 (2006)
S. Loiodice, P.G. Tucker, J. Watson, Coupled open rotor engine intake simulations, in *Proceedings of the 48th AIAA Aerospace Sciences Meeting and Exhibit*, Orlando, Florida (2010). Paper No. AIAA-2010-840
J.P. Longley, Calculating the flowfield behaviour of high-speed multi-stage compressors, in *International Gas Turbine and Aeroengine Congress and Exhibition*, (1997). ASME Paper No. 97–GT–468
B. Lukovic, Modeling unsteadiness in steady simulations with neural network generated lumped deterministic source terms. PhD thesis, Department of Aerospace Engineering, University of Cincinnati (2002)
B. Lukovic, A. Gangwar, P. Orkwis, B. Sekar, Modeling unsteadiness in steady cavity simulations, Part II: neural network modeling, in *39th AIAA Aerospace Sciences Meeting and Exhibit*, Reno, Nevada (2001). Paper No. AIAA–2001–0154
B. Lukovic, P. Orkwis, M. Turner, Modeling unsteady cavity flows with translating walls, in *32nd AIAA Fluid Dynamics Conference and Exhibit*, Missouri (2002a). Paper No. AIAA 2002–3288
B. Lukovic, P. Orkwis, M. Turner, B. Sekar, Effect of cavity L/D variations on neural network-based deterministic unsteadiness source terms, in *AIAA Aerospace Sciences Meeting & Exhibit, 40th*, Reno, NV (2002b). AIAA Paper No. AIAA 2002–0857
V.P. Maslov, B.I. Mineev, K.N. Pichkov, A.N. Secundov, A.N. Vorobiev, M.K. Strelets, A.K. Travin, Turbulence intensity, length scale, and heat transfer around stagnation line of cylinder and model blade, in *ASME, International Gas Turbine and Aeroengine Congress and Exhibition*, Indianapolis (1999). ASME Paper No. 99-GT-423
M. McMullen, A. Jameson, J. Alonso, Demonstration of nonlinear frequency domain methods. AIAA J. **44**(7), 1428–1435 (2006)
C. Meneveau, J. Katz, A deterministic stress model for rotor-stator interactions in simulations of average-passage flow. J. Fluids Eng. Trans. ASME **124**(2), 550–554 (2002)
V. Michelassi, J.G. Wissink, J. Fröhlich, W. Rodi, Large-eddy simulation of flow around low-pressure turbine blade with incoming wakes. AIAA J. **41**(11), 2143–2156 (2003a)
V. Michelassi, J.G. Wissink, W. Rodi, Direct numerical simulation, large eddy simulation and unsteady Reynolds-averaged Navier-Stokes simulations of periodic unsteady flow in a low-pressure turbine cascade: a comparison. Proc. Inst. Mech. Eng. A, J. Power Energy **217**(4), 403–411 (2003b)
R. Mittal, S. Venkatasubramanian, F.M. Najjar, Large-eddy simulation of flow through a low-pressure turbine cascade. AIAA Pap. 2001–2560 (2001)

F. Montomoli, H.P. Hodson, L. Lapworth, RANS–URANS in axial compressor, a design methodology. Proc. Inst. Mech. Eng. A, J. Power Energy **225**(3), 363–374 (2011)

A. Mosahebi, S.K. Nadarajah, An implicit adaptive non-linear frequency domain method (pNLFD) for viscous periodic steady state flows on deformable grids, in *49th Aerospace Sciences Meeting*, Orlando, Florida, January 2011. Paper No. AIAA 2011-0775

V.W. Nee, L.S.G. Kovasznay, Simple phenomenological theory of turbulent shear flows. Phys. Fluids **12**, 473 (1969)

W.E.I. Ning, L. He, Some modeling issues on trailing-edge vortex shedding. AIAA J. **39**(5), 787–793 (2001)

D. Nürnberger, H. Greza, Numerical investigation of unsteady transitional flows in turbomachinery components based on a RANS approach. Flow Turbul. Combust. **69**(3), 331–353 (2002)

M. Olausson, Turbomachinery Aeroacoustic Calculations using Nonlinear Methods. PhD thesis, Chalmers University of Technology (2011)

T.S.D. O'Mahoney, Large-eddy simulation of turbine rim seals. PhD thesis, University of Surrey (2011)

P.D. Orkwis, M.G. Turner, J.W. Barter, Linear deterministic source terms for hot streak simulations. J. Propuls. Power **18**(2), 383–389 (2002)

A. Pinelli, I.Z. Naqavi, U. Piomelli, J. Favier, Immersed-boundary methods for general finite-difference and finite-volume Navier-Stokes solvers. J. Comput. Phys. **229**(24), 9073–9091 (2010)

B. Raverdy, I. Mary, P. Sagaut, N. Liamis, High-resolution large-eddy simulation of flow around low-pressure turbine blade. AIAA J. **41**(3), 390–397 (2003)

C.M. Rhie, A.J. Gleixner, D.A. Spear, C.J. Fischberg, R.M. Zacharias, Development and application of a multistage Navier-Stokes solver: Part I-multistage modeling using bodyforces and deterministic stresses. J. Turbomach. **120**(2), 205–214 (1998)

M.J. Shaw, P. Hield, P.G. Tucker, The effect of inlet guide vanes on inlet flow distortion transfer and transonic fan stability, in *Proceedings of the ASME Turbo Expo 2013* (2013). ASME Paper No. GT2013–94998

C.H. Sieverding, D. Ottolia, C. Bagnera, A. Comadoro, J.F. Brouckaert, J.M. Desse, Unsteady turbine blade wake characteristics. J. Turbomach. **126**(4), 551–559 (2004)

D.L. Sondak, D.J. Dorney, R.L. Davis, Modeling turbomachinery unsteadiness with lumped deterministic stresses, in *32nd AIAA, ASME, SAE, and ASEE, Joint Propulsion Conference and Exhibit* Lake Buena Vista, FL (1996)

S. Stollenwerk, E. Kugeler, Unsteady flow modelling using deterministic flux terms, in *ETC 2011, 9th European Turbomachinery Conference*, vol. 161, Istanbul, Turkey, 21–25 March 2011

Y.B. Suzen, P.G. Huang, Numerical simulation of unsteady wake/blade interactions in low-pressure turbine flows using an intermittency transport equation. J. Turbomach. **127**(3), 431–444 (2005)

C.K.W. Tam, Excitation of instability waves in a two-dimensional shear layer by sound. J. Fluid Mech. **89**(2), 357–371 (1978)

P.G. Tucker, *Computation of Unsteady Internal Flows: Fundamental Methods with Case Studies* (Kluwer Academic, Dordrecht, 2001)

P.G. Tucker, Computation of unsteady turbomachinery flows: Part 1—progress and challenges. Prog. Aerosp. Sci. **47**(7), 522–545 (2011)

D. Van Zante, J. Chen, M. Hathaway, R. Chriss, The influence of compressor blade row interaction modeling on performance estimates from time-accurate, multistage, Navier-Stokes simulations. J. Turbomach. **130**(1), 11009 (2008)

H.P. Wang, R.J. Goldstein, S.J. Olson, Effect of high free-stream turbulence with large length scale on blade heat/mass transfer. J. Turbomach. **121**(2), 217–224 (1999)

J.M. Weiss, V. Subramanian, K.C. Hall, Simulation of unsteady turbomachinery flows using an implicitly coupled nonlinear harmonic balance method, in *Proceedings of the ASME Turbo Expo 2011: Power for Land, Sea and Air, GT2011*, Vancouver, BC, Canada, 1–11 June 2011. ASME Paper No. GT2011-46367

L. Xu, Assessing viscous body forces for unsteady calculations. J. Turbomach. **125**(3), 425–432 (2003)

L. Xu, T.P. Hynes, J.D. Denton, Towards long length scale unsteady modelling in turbomachines. Proc. Inst. Mech. Eng. A, J. Power Energy **217**(1), 75–82 (2003)

H. Yang, T. Roeber, D. Kozulovic, Hybrid-grid simulation of unsteady wake-boundary layer interaction on a high lift low pressure turbine airfoil, in *Proceedings of GT2007 ASME Turbo Expo Power for Land, Sea and Air*, Montreal, Canada, 14–17 May 2007. ASME Paper No. GT2007–28111

J. Yao, S.E. Gorrell, A.R. Wadia, High-fidelity numerical analysis of per-rev-type inlet distortion transfer in multistage fans: Part I—simulations with selected blade rows, in *ASME Turbo Expo 2008: Power for Land, Sea, and Air (GT2008)*, Berlin, Germany, 9–13 June 2008. ASME No. GT2008-50812

A. Zachariadis, C.A. Hall, Application of a Navier-Stokes solver to the study of open rotor aerodynamics. J. Turbomach. **133**(3) (2011)

Chapter 5
Applications of Eddy Resolving Methods

5.1 Introduction

In Chap. 4 an overview of unsteady methods specific to computational aerodynamics is given. The emphasis is on turbomachinery. Chapter 3 outlined the various LES and hybrids RANS-LES models. In this chapter an overview of the application of hybrid RANS-LES, LES and DNS in aerospace is given. Hence, the turbulence fidelity level is higher than that considered in Chap. 4. Typical grid counts, numerical methods used, validation levels and solution accuracies are outlined. The frequency of use of various solution approaches in different aerospace zones is detailed. Data available for the validation of eddy resolving simulations is discussed along with its level of appropriateness. The work reviewed in this chapter covers most areas of gas turbine propulsion along with airframes. Examples of eddy resolving approaches for acoustics and coupled simulations are not presented here. These are discussed in Chaps. 6 and 7. Gas turbine propulsion work is discussed first and then airframe eddy resolving simulations.

5.2 Propulsive Systems

Table 5.1 gives a summary of eddy resolving turbomachinery simulations. It includes grid densities, Reynolds numbers and validation levels for the work to be discussed below.

Table 5.2, extends that given by Sagaut and Deck (2009) defining levels of validation for eddy resolving simulations. This grading system is used to explore the level of validation currently used for various areas of propulsion unit and airframe design.

P.G. Tucker, *Unsteady Computational Fluid Dynamics in Aeronautics*,
Fluid Mechanics and Its Applications 104, DOI 10.1007/978-94-007-7049-2_5,

Table 5.1 Summary of eddy resolving simulations with typical grid densities, Reynolds numbers and validation levels in turbomachinery (extended from Tucker 2011a)

Engine zone	Author(s)	Method	$Re_c/10^5$	$N/10^6$	Validation level
Compressor/ Pump	Kato et al. (2003b,a)	LES	5.7	2.5	2
Compressor—tip flow	You et al. (2003, 2004, 2006b)	LES	4	20–25	1, 2, 3
Radial compressor impeller	Byskov (2000)	LES	1.4	0.385	2
Radial compressor	Mendonca et al. (2012)	DES	–	9	0
Compressor—end wall flow	Klostermeier (2008)	LES	2.3	5.2	2
Compressor—end wall flow	Wang and Yuan (2013)	Hybrid RANS-LES (SAS)	3.82	4.5	2
Compressor rotor	Hah (2009), Hahn and Drikakis (2009)	LES	10	14	1, 2
Compressor	Lardeau et al. (2012)	LES	1.4	5–10	DNS comparison to level 2, 3
Compressor	"	DNS	"	80	—
Compressor	McMullan and Page (2011)	LES	7	8–40	2
Compressor OGV–diffuser	Li (2012)	LES and hybrid RANS-LES	1.6	7.1	2
Turbine (LPT)	Mittal et al. (2001)	LES	0.1–0.25	1.2–3.5	0
Turbine (LPT)	Michelassi et al. (2003a)	LES	0.5–2	2–10	2
Turbine (LPT)	"	DNS	"	17	2
Turbine (LPT)	Wissink and Rodi (2004)	DNS	0.6	32–56	
Turbine (LPT)	Wu et al. (1999), Wu and Durbin (2001)	DNS	1.48	25–57	2

Table 5.1 (Continued)

Engine zone	Author(s)	Method	$Re_c/10^5$	$N/10^6$	Validation level
Turbine (LPT)—embedded	Raverdy et al. (2003)	MILES	1.5	3.5	2, 3
Turbine (LPT)	Opiela et al. (2001)	ILES	1.6	1.86	2
Turbine (LPT)	Medic and Sharma (2012)	LES	0.3–1.5	35–70	2
Turbine (LPT)	Connell et al. (2011)	Hybrid RANS-LES(SAS model)	$\sim$1	$\sim$10	0
Turbine (HPT)	Mei and Guha (2005)	ILES	15	0.016	2
Turbine	Tyagi and Acharya (2005)	LES	0.05	0.6	0
Turbine (LPT)	Wissink (2003)	DNS	0.72	93	2
Turbine (HPT) nozzle guide vane	O'Mahoney (2011)	LES	6.23	3.5	–
Turbine (LPT)—high order	Matsuura and Kato (2006)	LES	5	6	2
Turbine (LPT)—high order	De Wiart et al. (2012)	LES/DNS	0.85	–	0
Turbine (LPT)—high order	Balzer and Fasel (2013a)	DNS	0.25	0.31 × 160 collocation points with 97 spectral modes	2
Turbine (LPT)—high order with flow control	Balzer and Fasel (2013b)	DNS	0.25	"	2
Turbine (HPT), 3 nozzle guides, vanes, 6 rotors and rim seal cavity	O'Mahoney (2011)	LES	13.0	16.5	1

Table 5.1 (Continued)

Engine zone	Author(s)	Method	$Re_c/10^5$	$N/10^6$	Validation level
Turbine (HPT)	Wu et al. (2009)		0.1	85	2
Turbine (HPT)	Wu and Hickey (2012)	DNS	0.5	85	2
Turbine (LPT)	O'Mahoney (2011)	LES	5.5	7	0
Turbine (LPT)	Sarkar (2007, 2008, 2009), Sarkar and Voke (2003, 2006)	LES	1.6	2.3	2, 3
Turbine (LPT)	Rizzetta et al. (2008)	ILES/NLES	0.25	12	2
Turbine (LPT)	Lardeau et al. (2012)	LES	0.6	7	DNS validation to level 3
Turbine	Klostermeier (2008)	LES	5.9	5.4	2
Turbine (HPT)	Jimbo et al. (2008)	LES	3.7	17	2
Turbine (HPT)	Gourdain et al. (2012)	LES	10–20	29.7	2
Turbine—cut back trailing edge (highly idealized)	Ahlman et al. (2007)	DNS	0.02	10	0
Turbine—cut back trailing edge	Schneider et al. (2010)	LES	0.0625	19.2	2, 3
Turbine—cut back trailing edge	Martini et al. (2006a)	DES	–	2	2
Turbine—cut back trailing edge	Joo and Durbin (2009)	Hybrid RANS-LES	–	8–20	2
Shock-cooling-film interaction	Konopka et al. (2012)	MILES	0.046	34	2
Turbine—leading edge cooling	Rozati and Tafti (2007, 2008)		1.0 (based on leading edge diameter)	3	2

Table 5.1 (Continued)

Engine zone	Author(s)	Method	$Re_c/10^5$	$N/10^6$	Validation level
Idealized film cooling hole in cross flow	Fujimoto (2012)	LES	–	0.84–6.2	2
Turbine (HPT)	Dufour et al. (2009)	LES	2.8	6.3	2
Turbine (HPT)	Murari et al. (2012)	LES, DES, SAS	–	18.2	1, 2
Turbine (HPT)—conjugate	Dufour et al. (2009)	LES	3.8	6.5	2
Supersonic intake	Trapier et al. (2008)	DDES	29 (based on inlet height)	20	2
Square sectioned, curved intake duct	Laraufie and Deck (2012)	Zonal LES	4 (based on 'step' height in curved duct)	50	–
Idealized, round sectioned, curved intake duct	Petit and Mallet (2009)	DES, DDES	–	–	1, 2

5.2.1 Turbine Blade Simulations

Low Pressure Turbine

With the lower densities, at high altitudes, the LPT (Low Pressure Turbine) blade Reynolds numbers are low. Notably, the Reynolds number can change by around a factor of two between takeoff and cruise and capturing these altitude influences can be challenging. At altitude, LPTs have much laminar and transitional flow zones. In fact, most of the blade surface flow is laminar. The complex (bypass—because of the presence of wakes) transition modelling need, combined with the complex physics arising from wake interactions makes LPT blades poorly suited to URANS modelling. As noted in Chap. 4, in the suction surface separation zone, present in LPT blades, and blade trailing edge zones, the spectral gap needed for URANS is tenuous. However, the low Reynolds number found in LPTs is helpful for LES grid requirements (see Chap. 1), if not perhaps for LES models. Also, the lower Mach numbers, M, and hence absence of shocks is numerically helpful. The use of shock capturing with LES is a relatively immature field. LES and DNS studies of LPTs can

Table 5.2 Level of simulation validation (extended version of that due to Sagaut and Deck 2009)

Grade	Validation level
1	Integrated forces on structure—lift, drag and moments etc. Characteristic curve points or flow coefficient, sealing effectiveness etc.
2	Time averaged velocity, pressure, temperature, surface adiabatic film cooling effectiveness and Nusselt number profiles. Also, could include exit flow angle, total pressure loss profiles and surface shear stress
3	Second order statistics (r.m.s quantities) typically for turbulence
4	One-point spectral analysis
5	Two-point space/time-correlations
7	High-order space/time correlations

be broadly grouped into those that use flat plates (Wissink and Rodi 2004; Lardeau et al. 2012) with an external pressure gradient and those that involve explicit blade sections. For the former, typically the pressure gradient is imposed using a contoured wall as, shown in Fig. 5.1. Typical LPT experiments use moving bars to represent wakes from upstream blade rows. To capture these in LES, the wakes from cylinders etc. can be computed and stored as part of a separate simulation. This unsteady database can then be swept across the downstream blade row.

LPT Flat Plate Simulations

Wissink and Rodi (2004) perform DNS of a flat plate to represent the dynamics of a highly loaded LPT with $N = 32–56 \times 10^6$. The inviscid upper boundary of the domain is specifically contoured to impose an adverse streamwise pressure gradient.

Wissink and Rodi apply a slip wall condition in the region upstream of the leading edge of the lower boundary of the flat plate ($-0.5 < x < 0.0$). They performed two simulations, without and with incoming free-stream turbulence. The incoming turbulence intensity is set to 4 % in the latter. Modified box type turbulence is used. This is rescaled to incorporate anisotropy.

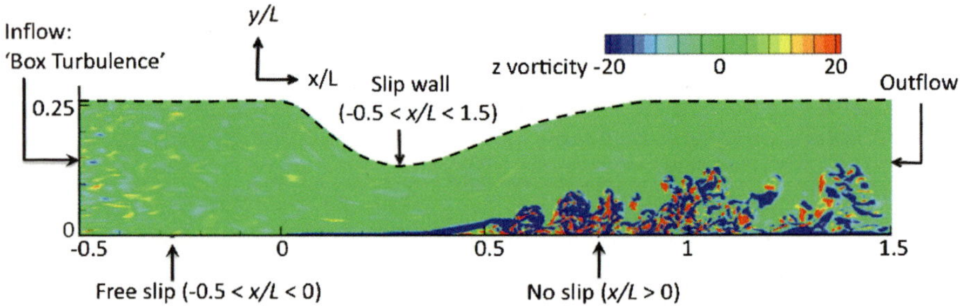

Fig. 5.1 Computational domain and the boundary conditions. Contours show the *z*-vorticity with incoming box turbulence

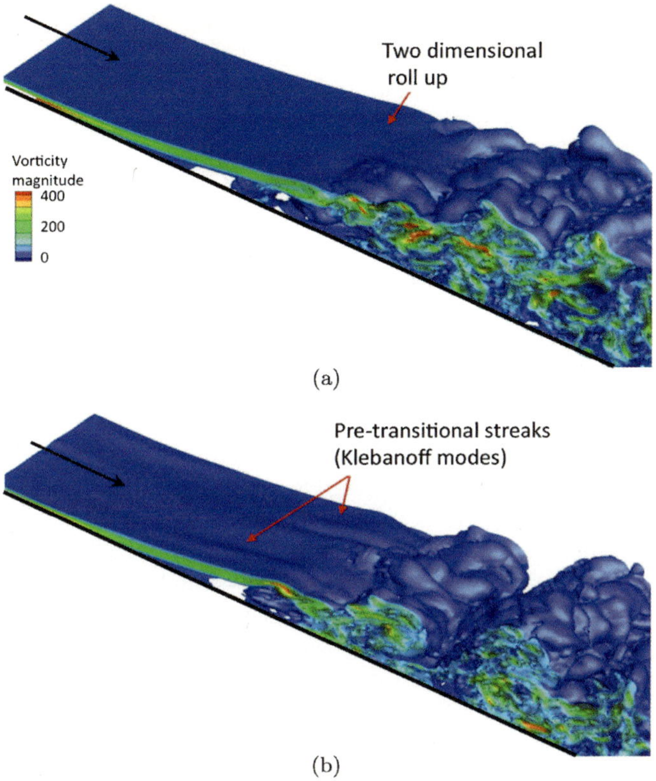

Fig. 5.2 Iso-surfaces of vorticity magnitude for simulations: (**a**) laminar inflow and (**b**) turbulent inflow

Figures 5.2a and 5.2b show results that repeat the simulations of Wissink and Rodi but for NLES with a 6th order flow solver and $N = 4.2 \times 10^6$. These simulations are more completely reported in Tyacke et al. (2012b). The figures show iso-surfaces of vorticity magnitude. While the separated shear layer prior to transition is fairly two-dimensional for laminar inflow (see Frame (a)), spanwise undulations clearly appear when there is turbulent inflow (Frame (b)). These undulations are the footprints of what are called Klebanoff modes. These modes are a result of the incoming free stream turbulence. This promotes early transition and stimulates early reattachment. Figure 5.3 compares NLES streamwise Reynolds stress profiles with the DNS of Wissink and Rodi (2004). Comparisons are made at seven different streamwise locations. Even though a much coarser grid is used, the NLES (solid line) is broadly in agreement with the DNS (symbols). The mean velocity profiles (not presented here) also show a similar level of agreement.

Lardeau et al. (2012) also perform LES for the Wissink and Rodi case with $N = 4.2 \times 10^6$. The Germano LES model is used with an incompressible, curvilinear, collocated, fractional step solver for pressure. Convergence of the Poisson equation for pressure is accelerated using a multigrid method. A 3rd order back-

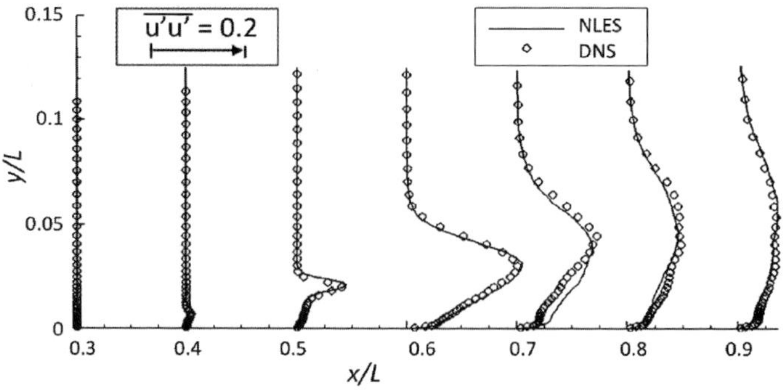

Fig. 5.3 Comparison of Reynolds stress profiles with DNS of Wissink and Rodi (2004)

wards difference (BD) temporal scheme is used with 2nd order central differences for space. Pressure field smoothing is used to prevent pressure velocity decoupling (Rhie and Chow 1983). The work uses the LES to assess a RANS transition modelling strategy. This seems a good LES application. Although the flat plate configuration is both experimentally and computationally simpler, the blade leading edge dynamics, for example, are absent (unless the geometry of Coull 2009 is used).

Figure 5.4 shows complex incoming wake induced transition occurring for an LES for the geometry reported in Coull (2009). This is again intended to be representative of a highly loaded LPT pressure distribution. This simulation captures the propagation and spreading of separate wakes before they influence the laminar boundary layer on both sides of the 'blade'. Significant differences can be seen in the developing flow structures on each side. Note, the wakes are imposed from a separate DNS simulation and box turbulence is introduced between the wakes. This is discussed in Chap. 3. For results shown in frame (b) there is close agreement with mean velocity profiles. Also, the time varying trailing edge momentum thickness variations along with a range of other parameters show encouraging agreement.

Non-flat Plate LPT Simulations

Mittal et al. (2001) explore the use of LES with the Germano model, fractional step method for pressure and staggered grids. The spanwise domain uses a Fourier/spectral discretisation with finite differences in the other coordinate directions. Sarkar (2007, 2008, 2009), Sarkar and Voke (2003, 2006) explore LES modelling for the T106 LPT representative profile using a modified Smagorinsky model and a broadly similar numerical framework. The upstream wakes had a critical impact on the pressure surface flow physics. Michelassi et al. (2003a,b) also explore the T106 profile using DNS and LES. The grid size for the LES reached 1×10^6 with that for the DNS increasing by a factor of 5. An incompressible, 2nd order, pressure correction based solver is used with Rhie and Chow (1983) interpolation and the Germano LES model. Solutions are advanced in time with a 3 stage

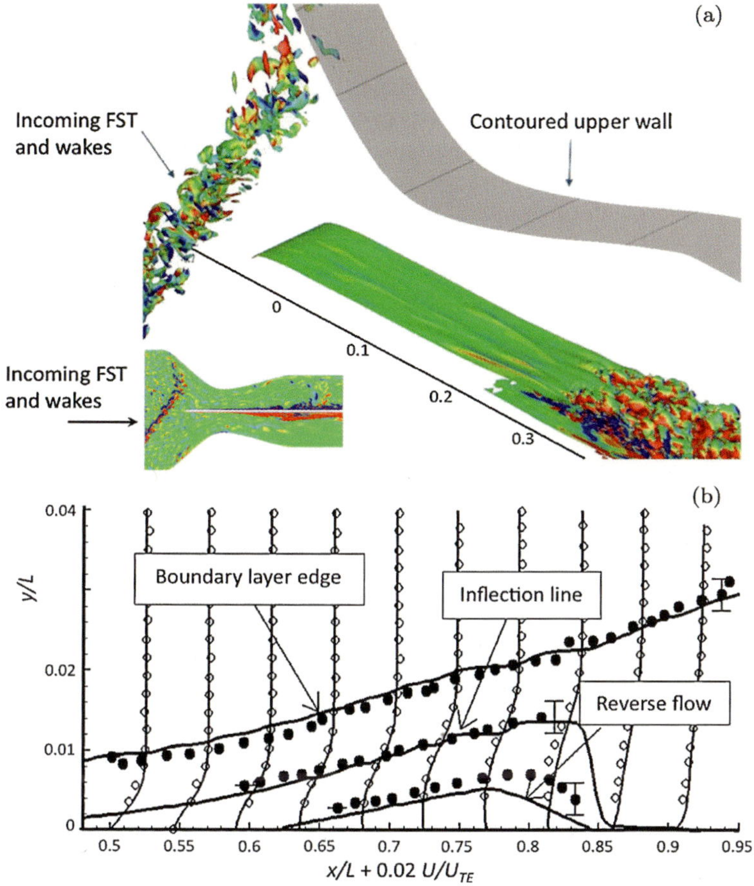

Fig. 5.4 Wake induced transition on a flat plate with imposed pressure gradient (Vadlamani 2013): (**a**) flow structures and (**b**) mean flow profiles

explicit Runge-Kutta (RK) scheme. The above work shows useful flow physics insights. However, evidence suggests that the Germano model can be defective when predicting the transition process. As noted by Sagaut (2001), relative to DNS, LES transition can be faster. Hence, it seems sensible to use grids tending to DNS resolution for LPT type computations. Medic and Sharma (2012) found this practice (see later) gave encouraging predictions. Michelassi et al., like Lardeau et al., use their LES to refine RANS modelling. This lower order model refinement is a key potential application for LES in industrial design.

Raverdy et al. (2003) used MILES to study a T106 profile for laminar inflow, in a compressible code with a 3rd order BD temporal disctetization. An embedded grid, with $N = 3.5 \times 10^6$ is used. This allows refinement in just the trailing edge zone. Results suggest acoustic waves, from trailing edge turbulence, energize the separated shear layer. Opiela et al. (2001) also make ILES type simulations for a

T106 LPT profile. A modified, low dissipation, 2nd order AUSM convective scheme (see Chap. 2) is used with an explicit 5 stage RK scheme. Turbulent upstream wakes are generated using bars and a sliding plane. Time averaged velocity profiles shown encouraging agreement with measurements.

Tyagi and Acharya (2005) use an immersed boundary method to model a rotor-stator system. Computations use a mixed dynamic LES model with a 2nd order explicit Adams-Bashforth temporal scheme. No comparison is made with measurements.

Matsuura and Kato (2006) make 6th order compact scheme studies with 10th order filtering for the T106 profile. Both, no incoming turbulence and also an, isotropic box, turbulence level of around 5 % are considered. Like Raverdy et al. they observe, for no free stream turbulence, an intense acoustic feedback. There are a substantial vast number of LES (and DNS) studies for the T106 profile and the above is by no means exhaustive.

Rizzetta et al. (2008) explore flow control strategies for the so called Pratt and Whitney, PakB, blades. The flow control measures include the use of plasma. Simulations use a 4th order compact scheme with a 6th order spatial filter. The latter is used to replace the LES model. Hence, these simulations could be viewed as being of the NLES/ILES genre. A similar study is also performed by Balzer and Fasel (2013a,b) for a profile in the PakB blade series. To eliminate pressure as a solution variable, these DNS involve a vorticity-velocity formulation. The spanwise direction is discretized using a spectral approach. The remaining directions use a 4th order compact scheme. Pulsed vortex generator jets, activated at the frequency of the shear layer's natural instability frequency, are show to reduce the extent of suction surface separation.

Dufour et al. (2009) perform conjugate LES for heat transfer from a turbine blade with film cooling holes (T120 cascade). The Re_c is 3.8×10^5 and $N = 6.5$ million (tetrahedral elements). These simulations are discussed further in Chap. 7.

De Wiart et al. (2012) apply the discontinuous Galerkin method to an LPT profile in a quasi-DNS mode. The unstructured hybrid grid has 133500 near wall hexahedral cells and 788500 prisms. The simulation is run at 4th order accuracy with 2nd order BD temporal differencing. The computation is found to scale well up to around 16000 cores on a BlueGene/P computer. No comparison is made with measurements. The general computational performance seems competitive with a more standard finite volume approach on similar grid resolutions. However, obviously there is the benefit of having a higher order scheme.

Medic and Sharma (2012) set out to assess LES performance in capturing the degradation in performance of LTPs with altitude. The work a focused on three-aerofoils geometries (Pack109, 110 and 111) for a range of inflow turbulence levels. The Reynolds number range explored (axial chord and exit velocity based) varied between 0.3×10^5 and 1.5×10^5. The WALE LES model is used with a 2nd order scheme that is implicit in time. A compact scheme based filtering is used to suppress high frequency numerical instabilities. A key element of the tests is that they are effectively blind comparisons—the person running the LES had not sight of the measurements until after the LES is performed. Notably, LES caught key trends in blade surface and total pressure loss variations. In many instances it is largely within the experimental error. Also, the job turnaround time is 14 days on modest computational resource. Hence, the simulations demonstrated that LES could now be used in house, for this type of flow.

Connell et al. (2012) apply the SAS hybrid RANS-LES approach. The key aspect of this work is that it proposes time-inclining (see Chap. 4) to avoid problems of modelling multiple passages and the excessive LES costs that this presents. The filtering with harmonic based phase shift methods and other defects that prevent their use for LES are noted. Importantly, for all of the above LPT simulations the spanwise domain extent is around $0.1c$.

High Pressure Turbine Simulations

Dufour et al. (2009) perform conjugate LES for a cascade HPT stator. The Re_c is 2.8×10^6 and the WALE LES model used. These simulations are discussed further in Chap. 7. Wu et al. (2009), Wu and Hickey (2012) perform DNS for Langston's HPT blade profile with H type grids and $N = 85 \times 10^6$. A structured, curvilinear, incompressible, fractional step, flow solver is used. This is of 2nd order accuracy with central differences and staggered grids. Although the study of Langston focused on end wall influences these studies explored the dynamics at the mid span having spanwise periodic boundary conditions. Away from the suction surface separation zone, reasonable agreement is found with midspan surface pressure coefficient measurements. Homogeneous decaying turbulence is used to describe the turbulence inflow. The inflow was constructed so that the turbulence could either directly impact on the blade leading edge or pass between blades. On the pressure side, (when turbulence interacted with the blade leading edge) elongated vortices are observed. They have similarities to Klebanoff type longitudinal vortices but have a different production mechanism. In the past it had been proposed that these were related to a Goertler type instability. However, the work suggests that the generation process is related to straining of the incoming turbulence.

Gourdain et al. (2012) perform LES of a HPT blade using a 4th order centered spatial scheme in a compressible, multiblock solver. The WALE LES model is used. Inflow turbulence was generated using a simplified synthetic eddy generation method—see Chap. 3. This is used to control the eddy size approaching the blade and also the turbulence intensity. A range of Mach and Reynolds numbers are considered. The focus is on heat transfer. For cases where boundary layers are fully turbulent, it proved hard to find conclusive LES accuracy gains relative to RANS. However, for conditions that involved transition and substantial blade surface areas with laminar zones the LES showed substantially improved agreement in predictive accuracy. The work suggests the need for better measurements that characterize turbulence length scales approaching blades. Again, for spanwise periodic simulations domain extents are around $0.1c$.

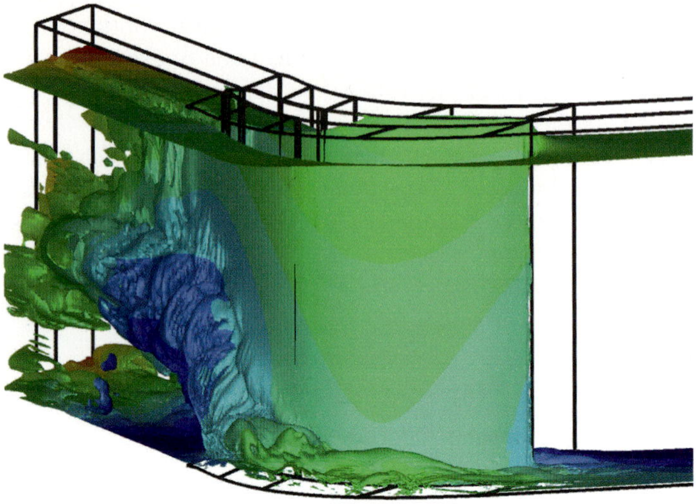

Fig. 5.5 Flow in HPT end wall region using high order RANS-NLES (Tyacke et al. 2012b)

LPT/HPT Simulations with Real Annulus Features

Jimbo et al. (2008) use LES to explore the development of vortical passage structures in a steam turbine, exploring the endwall flow. Klostermeier (2008) studies the LES modelling of turbine end wall flows. A compressible flow solver with the WALE LES model is used. Comparison is made with measurements of Langston et al. (1977). Endwall flows have the added demand of resolving both the blade and endwall boundary layers. However, for the actual blade profiles, to reduce computational cost Klostermeier uses wall functions. The lack of both resolved and modelled end wall turbulence resulted in some qualitative flow discrepancies.

Murari et al. (2012) make Smagorinsky LES, DES and SAS computations for a HPT with cooling holes and real annulus features meshed. The number of rotor blades is marginally increased. This allows two stator and three rotor blades to naturally provide periodicity. A single stage of a two-stage rig is modelled. Steady inflow data is provided to the 2nd stage based on a RANS simulation. Comparisons are made with radial profiles of total temperature and pressure for a full scale but low temperature test rig. Considering the extremely modest grids used the LES provides surprisingly encouraging agreement.

Tyacke et al. (2012b) repeat the case of Langston et al. with a 6th order flow solver using hybrid RANS-NLES. The artificial compressibility method, discussed in Chap. 2 is used to predict the pressure field. The spanwise pressure coefficient variation is considered. The variation in profiles across the span is due to the three-dimensional unsteady vortices at the hub endwall. The flow also separates on the suction side giving rise to a wake region. The instantaneous flow is shown in Fig. 5.5. Zones of separated, large scale, vortical structures contaminating the basic blade flow can be seen.

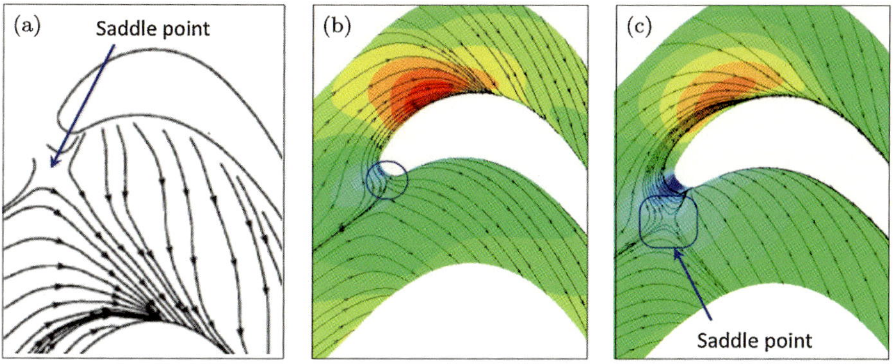

Fig. 5.6 Velocity contours and streamlines at the endwall: (**a**) Measured saddle point location (Langston et al. 1977), (**b**) NLES with incorrect saddle point, (**c**) RANS-NLES capturing the correct saddle point (Tyacke et al. 2012b)

In Fig. 5.6a, showing the flow visualization of Langston et al. (1977) a saddle point near the leading edge marks the origin of the endwall flow separation. For pure NLES, when the boundary layer region is under-resolved, the saddle point at the initial separation is not captured correctly. This is shown in Fig. 5.6b. Adding a RANS layer recovers correct wall behaviour and the correct saddle point, as shown in Fig. 5.6c.

A pressure coefficient plot for the RANS-NLES computation is provided in Fig. 5.7. Profiles are taken at 2.3, 12.5, 25 and 50 % span. The RANS-NLES results are again compared with the measurements of Langston et al. (1977). Encouraging agreement is observed for this complex flow at all span locations. The use of the RANS layer makes modelling of this high Reynolds flow economical. The NLES region captures more complex dynamics outside of the inner boundary layer region.

Figure 5.8 shows the instantaneous flow field for an LES of the endwall flow for an LPT (T106A) blade. Again, the simulation is 6th order. However, this time the VMS LES model is used. The simulation shows the complexity of the end wall flow and blockage that it creates. Such simulations offer the potential for greater understanding of endwall flow physics.

5.2.2 Compressor and Fan Simulations

For radial compressors centrifugal, Coriolis forces and streamline curvature will strongly influence turbulence. Also, secondary flow losses can be substantial. Hence, radial, like axial, compressors present substantial challenges for URANS models, especially as off design conditions are approached. LES should have potential for such flows. However, at off-design conditions, where the greatest benefits are expected, periodicity is lost and more passages require modelling. Byskov (2000) perform LES of a centrifugal pump impeller at design and off-design. The Germano

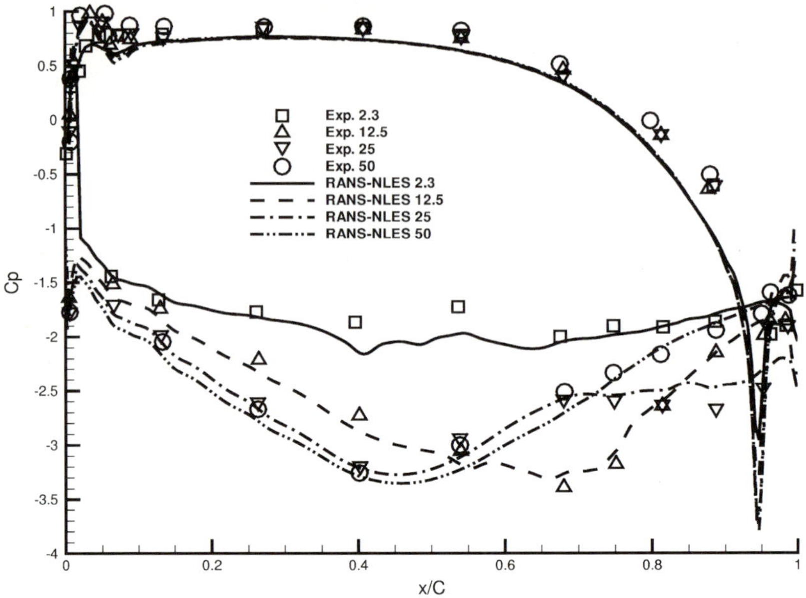

Fig. 5.7 Pressure coefficient profiles at different spanwise locations using high-order hybrid RANS-NLES (Tyacke et al. 2012b)

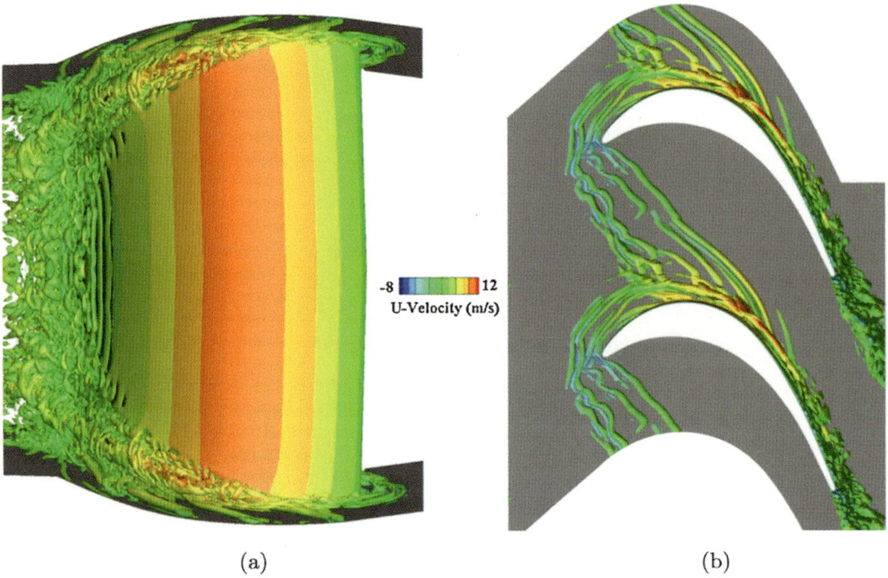

Fig. 5.8 High order, LES of end wall flow for LPT blade (Vadlamani 2013)

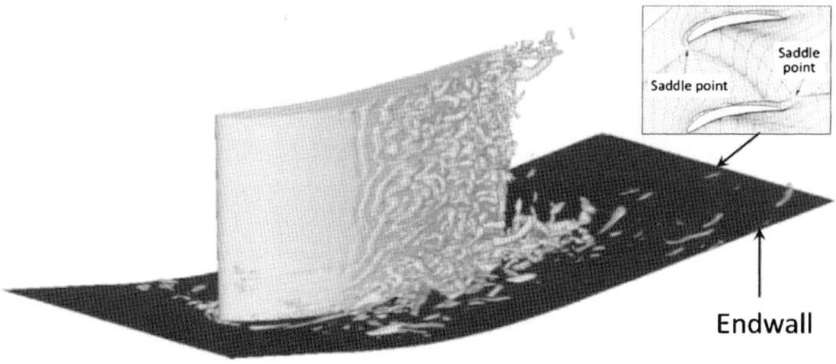

Fig. 5.9 λ_2 isosurfaces for the compressor blade (Tucker 2011a). Published with kind permission from Elsevier

LES model is used with a 2nd order Centred Difference (CD) finite volume method that makes use of a Jameson type stabilization. Solutions are advanced in time using a 4 stage RK scheme. The blade heights are low and secondary flow losses are likely to be substantial. Even though the grids used are modest some encouraging results are found. Mendonca et al. (2012) make centrifugal fan, Menter SST based DES type simulations exploring acoustics. The unstructured, commercial, pressure based STAR_CCM+ code is used. These, acoustics motivated simulations are discussed further in Chap. 6.

Kato et al. (2003a,b) explore LES modelling of a pump. A finite element discretization is used with streamline upwinding and the Smagorinsky LES model. Despite the modest grids ($N = 5 \times 10^6$ divided over two blade rows) there is encouraging agreement with mean flow measurements.

You et al. (2003, 2004, 2006b) look at LES modelling of tip clearance vortices. A combined curvilinear grid and immersed boundary method is used. The latter assists resolving the tip zone. Comparisons are made with linear cascade measurements having moving end walls. The Smagorinsky LES model is used in a Lagrangian dynamic form. Also, an incompressible flow solver is used with a Poisson equation for pressure, the Crank-Nicolson (CN) temporal scheme and second order central spatial differences. Broadly encouraging agreement is found with measurements including wake profiles.

Klostermeier (2008) makes LES and hybrid RANS-NLES of a compressor end wall flow. Figure 5.9 shows λ_2 isosurfaces for a hybrid simulation. Broadly, encouraging agreement is found with measured profiles for total pressure loss and exit flow angle. Hah (2009) and Hah et al. (2009) make dynamic Smagorinsky LES of a low aspect ratio transonic compressor rotor (NASA rotor 37). A pressure based, multiblock, curvilinear flow solver with a 3rd order spatial discretization and 2nd order, implicit, temporal discretization, is used. The LES is within around 5 % of the measured compressor characteristic. The benefits of the LES, in capturing the complex intermittent dynamics of the shock-tip vortex-wake interaction, relative to RANS is clear from these simulations. Results suggest that the stall inception can be strongly related to blade tip events.

Examples of the application of DNS to compressors can be found in Zaki et al. (2006, 2009). For these simulation the $Re_c = 138{,}500$ and $N = 84 \times 10^6$. A staggered grid incompressible flow solver is used and the physics of transition explored. Lardeau et al. (2012) LES model the transitional flow over a compressor blade comparing with DNS. Two LES models are tested. These being the Germano and a mixed time scale one (Inagaki et al. 2002). For this transitional flow, inconsistently with the hierarchy noted in Chap. 3, results are found sensitive to the choice of LES model. Hence, again, as with LPT LES, grids tending to DNS resolution seem to be necessary.

Tauveron (2010) explore the modelling of a stalled compressor blade using a two layer model (see Chap. 3). The flow equations are solved using a mixed finite-volume/finite-element approach. There is no validation against a detailed measurement set.

McMullan and Page (2011) apply Smagorinsky model based LES to a compressor blade ($N = 8$–40 million). A small spanwise domain (to address the high Reynolds number) and boundary conditions caused contamination of the predicted loss. However, the surface pressure distributions are in accord with measurements.

Li (2012) perform pure LES and LES with an upstream RANS zone for a coupled compressor OGV (Outlet Guide Vane)–diffuser system. The LES makes use of the Smagorinsky model in a structured flow solver with $N = 5 \times 10^6$ for each blade (two blades are modelled) and $N = 2.1 \times 10^6$ for the pre-diffuser. A MUSCL type convective scheme is used. For the inflow of the LES, recycling procedures are used. The accuracy of the LES is found to be comparable to that of high Reynolds number k–ϵ model RANS simulations. Failure of the LES, motivated hybrid RANS-LES. A RANS zone is used for the complete OGV zone and LES for the diffuser. The RANS zone, mean velocity and turbulence levels, are used to generate inflow turbulence for the LES. This is achieved via the recycling procedure with body forces. Again, total pressure loss and static pressure rise (for the diffuser) coefficients and comparisons of mean velocities are inconclusive with regards to hybrid RANS-LES accuracy benefits. Interestingly, the hybrid RANS-LES are used as part of a design optimization process for the OGV–diffuser system. Note, various aero engine fan LES related simulations are given in Chap. 6.

Wang and Yuan (2013) explore the end wall flow physics for a compressor blade at different incidences. The $Re_c = 0.382 \times 10^6$ with $N = 4.5 \times 10^6$ hexahedral cells. The commercial CFX code with the SAS hybrid RANS-LES model is used. The Karman length scale is shown to operate in a near LES mode in the corner separation zone. Some sensitivity of results to grid is observed. However, broadly encouraging agreement is found with blade surface pressure measurements within and away from the end wall flow. Also, downstream total pressure loss contours show reasonable agreement with measurements. As would be expected and quantitatively demonstrated there is substantial misalignment of the Reynolds stress with the shear strain. This occurs upstream of the corner separation and is convected into the separation. Also, substantial anisotropy is observed. The substantial lack of turbulence non-equilibrium is quantitatively defined. There is found to be the potential for a strong resonant type interaction between the trailing edge vortices and the corner separation.

5.2.3 Rotating Cylindrical Cavity Related Systems

As noted in Chap. 1, the internal air systems of gas turbine engines have a range of rotating disc configurations. These can involve co-rotating disks, rotating disks adjacent to stationary disks and cavities where all surfaces rotate. Radial inflow or outflow and axial through flow can also be imposed on these systems. The flow physics in these systems presents substantial challenges to RANS models (see Chap. 1).

Wu and Squires (2000) carry out LES for an isolated rotating disk. This has a three-dimensional boundary layer. A 2nd order, CD, staggered grid, incompressible (Poisson equation for pressure) flow solver is used with mixed implicit/explicit time advancement. The simulations are bi-periodic (tangential and radial directions). The Reynolds number is 6.5×10^5 with $N \cong 0.6 \times 10^6$. The Germano, NLES and two other subgrid scale modelling strategies are compared. As would be expected, for sufficiently fine grids, results are insensitive to the LES model. Following on from earlier DNS studies (Lygren and Andersson 2001, 2002) Andersson and Lygren (2006) perform LES for a rotor stator system. Bi-periodicity is again initialised. The discretization is similar to that of Wu and Squires, except that Adam's-Bashforth explicit time stepping is used. Again, three dimensionality of the flow is observed. For the finest grid, $N \cong 3 \times 10^6$ and 1.6×10^6 is the maximum Reynolds number considered. Again, a dynamic mixed LES model is tested along with the Germano. Three-dimensionality in the core flow is found to influence near wall turbulence production. Again for a rotor-stator system, Séverac and Serre (2007) make ILES related spectral predictions for a maximum Re of 7×10^5. O'Mahoney (2011) carries out LES under the same conditions as Andersson and Lygren. Both compressible and incompressible (commercial FLUENT program) solvers are used along with the Smagorinsky and Germano LES models. Broadly, the above LES based rotor stator studies find encouraging agreement for velocity profiles. For Reynolds stresses the agreement is much less encouraging, even for more advanced LES models. Compressible solver results are especially poor. Presumably the disparity in Mach number scales being problematic for such solvers. Such issues are discussed in Chap. 3. Notably, most internal air system CFD uses incompressible solvers.

The rotating cavity with an axial through flow is representative of the flow in the high-pressure compressor drum of a modern aero engine. The top image in Fig. 5.10 shows the real and an idealized cavity geometry. The flow, during an engine acceleration, is characterized by cyclones and anticyclones and there can be vacillation between different regimes. Tucker (2002) explores finite volume modelling of this case and He (2011) modelling with a spectral representation in the tangential direction. No subgrid scale modelling is used in these works. Results are in accord with qualitative expectations. For example, Fig. 5.10 contrasts NLES with experimental smoke flow visualization. Two different time instances are shown after an initial release of smoke in the coolant stream. The smoke enters the cavity through a single radial arm that is flanked by cyclones and anticyclones. Notably, for the condition considered, the flow can vacillate between a state of having one zone of radial out flow and then two. Hence, such a comparison is challenging.

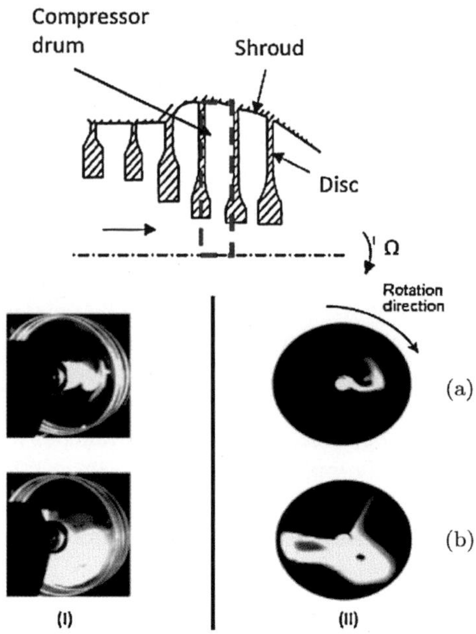

Fig. 5.10 Geometry and flow in a high-pressure compressor drum. Published with kind permission from the ASME

O'Mahoney et al. (2010), O'Mahoney (2011) explore the ingestion of hot external gasses into a rim seal (see Chap. 4, Fig. 4.27 for a typical geometry) using LES. A compressible flow solver with the Smagorinsky model is used with $N = 5.5 \times 10^5$. The rotational and main annulus Reynolds number are around 1.5×10^6. The simulations include blade geometry. The LES is found to more accurately predict hot gas ingestion (replicated through use of a tracer gas in the experiments) relative to URANS.

5.2.4 Turbine Blade Internal Cooling

Ribbed Passages

Ribbed passages are widely used for turbine blade internal cooling. However, as shown by Tucker et al. (2012), a variation of around 250 % can be found between different RANS models when predicting ribbed channel flow heat transfer. This can be seen from Fig. 5.11. Comparison is made with the heat transfer measurements of Acharya et al. (1993) for a biperiodic channel flow case. The scatter in the Frame (a) RANS results is substantial. This creates a great problem for design engineers. They, hence, also rely on lower order models instead of RANS. The former can have greater reliability. Notably, for the case just considered, the more advanced the RANS model the worse the predictive accuracy. Frame (b) shows results for a wide range of LES and hybrid LES approaches with different numerical schemes. As can be seen the scatter is much less. This flow is forgiving for LES and related methods.

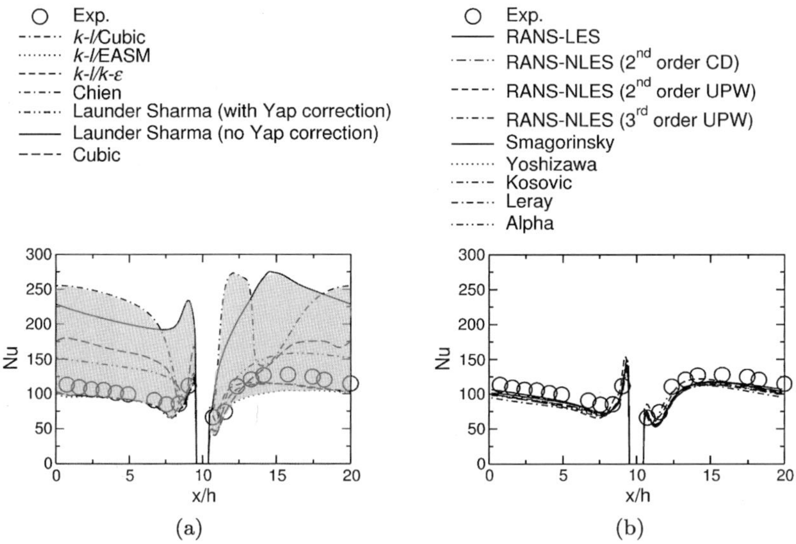

Fig. 5.11 Nusselt number distribution along a periodic ribbed channel floor: (**a**) RANS, (**b**) LES and hybrid RANS-NLES

Liu et al. (2006) make LES, SA based DES and hybrid RANS-LES for a ribbed channel flow. A staggered grid, second order CD and incompressible flow solver is used with the CN scheme. The simulations focused on a single rib, fully developed flow. This is extensive in the cross-stream direction. Hence, bi-periodicity is used. Grids ranged from around $N = 0.44 \times 10^6$ to 0.48×10^6 (for LES). The DES and hybrid RANS-LES methods showed encouraging agreement with heat transfer, mean velocity and Reynolds stresses measurements.

Tafti (2005) make NLES and dynamic Smagorinsky LES for a rib roughened passage that is bi-periodic. A similar structured solver to Liu et al. is used with grids of $N = 0.8 \times 10^6$ and 2.1×10^6. All computations capture the major flow elements accurately. However, the dynamic Smagorinsky model, gave greater consistency on the two grids and also improved the prediction of heat transfer.

Watanabe and Takahashi (2002) make LES of a non-isothermal, bi-periodic, ribbed channel flow with $N \cong 6 \times 10^5$. Again similar numerical approaches to Liu et al. are used. Both the standard and dynamic Smagorinsky models are tested. Encouraging agreement is found with mean velocity and heat transfer measurements.

Murata and Mochizuki (2000) make bi-periodic dynamic Smagorinsky LES of a rotating ribbed passage with $N = 91{,}125$–$274{,}625$. The bulk Reynolds number ranges between 4,000 and 9,000. Again, similar numerical schemes to Liu et al. are used but with Adams-Bashforth for the convective term time advancement. For the case of no rotation, close agreement is found with heat transfer measurements.

Saha and Acharya (2005) look at periodic ribbed channel heat transfer with rotation. The dynamic Smagorinsky model is utilized and $N \cong 0.9 \times 10^6$. A staggered-grid incompressible flow solver is utilized with 3rd order upwinding for the con-

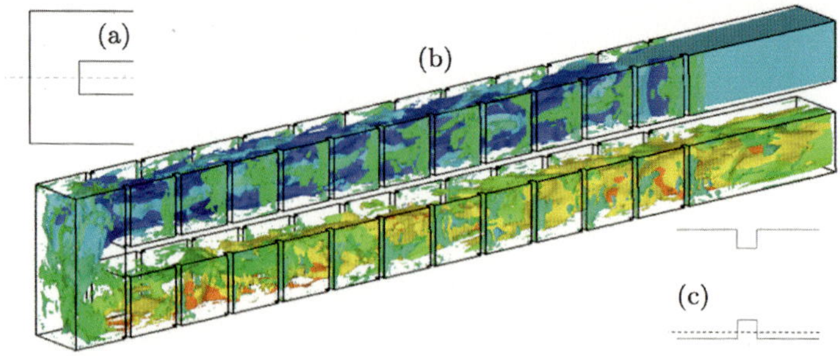

Fig. 5.12 Internal cooling duct: (**a**) velocity profile location in centre of duct, (**b**) geometry with vorticity isosurfaces, (**c**) Nusselt number profile location on lower wall

Fig. 5.13 Locations in terms of y values for the velocity profiles shown in Fig. 5.12

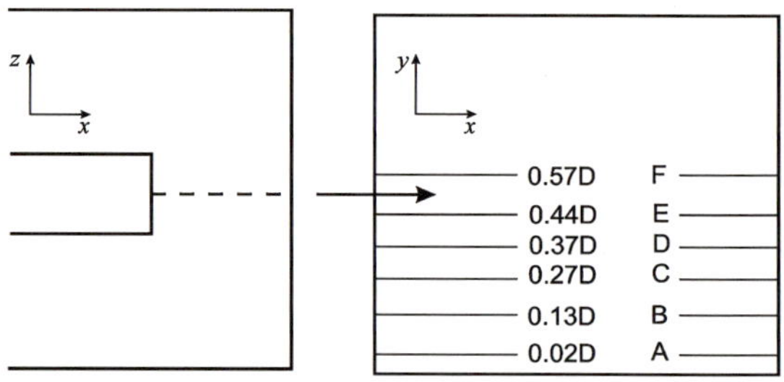

vective terms. Adams-Bashforth temporal differencing is used. The predicted heat transfer is within 20 % of measurements. Viswanathan and Tafti (2006) make k–ω based DES type simulations for a 180° bent duct populated with ribs, comparing with LES and measurements. An incompressible, 2nd order fractional step solver is used. For both the DES and LES $N \approx 7.5 \times 10^6$. The DES results compare well with velocity and heat transfer measurements. However, relative to LES, the development in the initial region to the duct is slower.

Some results, reported in Tyacke et al. (2012a) for the case considered by Viswanathan and Tafti (2006) are discussed below. Figure 5.12b gives instantaneous vorticity contours. The inset schematics in Fig. 5.12 give profile locations in the middle of the 180° bend and the bottom of the ribbed passage. Figure 5.13 gives the exact velocity profile locations. For the ribbed duct, a fairly coarse grid ($N = 7 \times 10^6$) is used. This is because the geometry contains large-scale bluff geometrical features. These define the largest scales that are of the order of the rib height.

Figures 5.14a–5.14f display the z-velocity profiles in the 180° bend for the locations shown in Fig. 5.13. Note, the Yoshizawa and LANS-α LES models are labelled as Yosh. and Alp., respectively. The hybrid RANS-NLES and LES results are in encouraging agreement with the measurements. This is in stark contrast to the Spalart-Allmaras RANS simulation given by the dotted line. Note, the RANS layer has minimal impact for this flow. This is because it has little classical boundary layer content. Average percentage errors for the velocity and Nusselt number profiles are

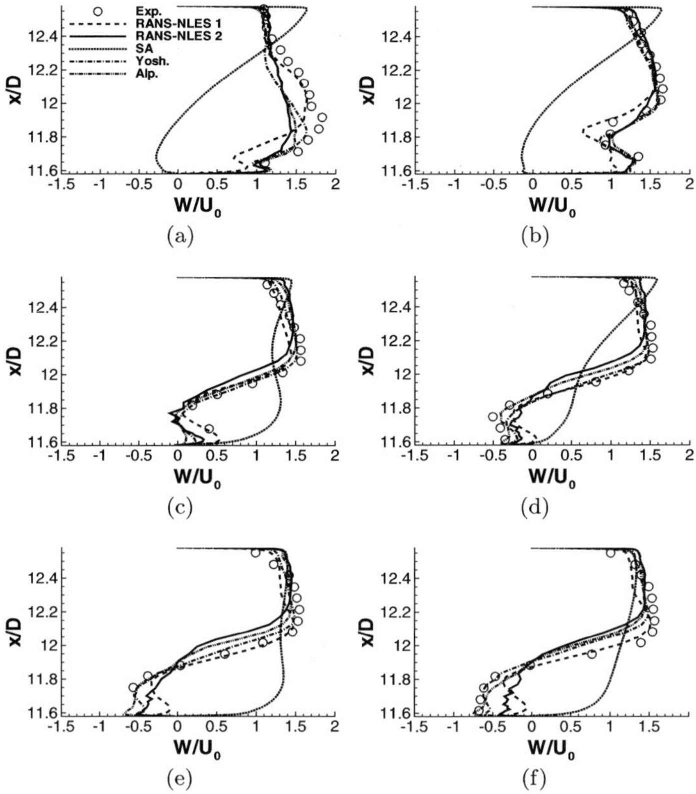

Fig. 5.14 Velocity profiles at increasing *y*-locations in the 180° bend

Table 5.3 Velocity and *Nu* errors (%)

Model	RANS-NLES	Yosh.	α
Velocity	10	6	7
Nu	16	8	12

given in Table 5.3. These are based on Eqs. (5.1) and (5.2)

$$Error(W) = \frac{\sum_{\exp} |W_{\exp} - W_{num}|}{\sum_{\exp} U_0} \times 100 \tag{5.1}$$

$$Error(Nu) = \frac{\sum_{\exp} |Nu_{\exp} - Nu_{num}|}{\sum_{\exp} Nu_{\exp}} \times 100 \tag{5.2}$$

for ribbed channel with 180° bend. Errors for the different modelling approaches are relatively similar.

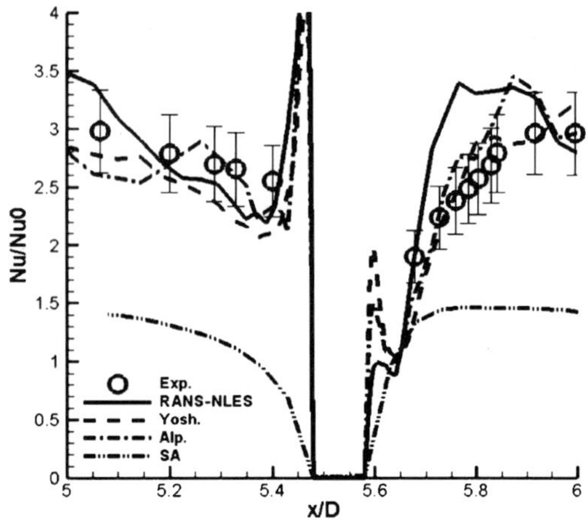

Fig. 5.15 Lower wall *Nu* distribution for ribbed passage with 180° bend

Figure 5.15 gives predicted *Nu* distributions in the center of the ribbed channel. As can be seen, the heat transfer is greatly improved using LES. Again, the Spalart-Allmaras RANS model fails to accurately predict heat transfer. Notably, the LES related methods are also in reasonable agreement with each other.

As noted above, the key structures in ribbed channel flows are large scales with a size of proportional to the rib height. These strongly influence the heat transfer. Hence, such flows are not that challenging for LES methods. Notably, they are also relatively Reynolds number independent. Hence, generally, the Reynolds numbers are not given in the discussion above. However, the bulk velocity based Reynolds numbers range from 10–20 $\times$ 10^3 in the work reviewed. Ribbed channel flows naturally avoid the severe Reynolds number scaling of LES. Hence, in terms accuracy gains (relative to RANS) normalized by cost they are ideal flows for the application of LES methods.

Cutback Trailing Edges

The cutback trailing edges used on turbine blades involve a shear layer between the hot and cold streams. This geometry is shown in Fig. 1.16d. The cold stream is used to shield the trailing edge. With sufficient velocity difference, the shear layer tends to produce large unsteady Kelvin-Helmholtz vortices. As will be shown later, the heat transfer arising from these is difficult to predict with RANS models. As discussed in Chap. 3, a spectral gap between modelled and resolved scales is hard to guarantee when URANS modelling. However, Martini et al. (2006b) demonstrated good predictive accuracy using DES. Figure 5.16a shows a grid and Fig. 5.16b shows the instantaneous flow field for a geometry with internal pedastalls, considered by Martini et al. The Frame (b) simulation uses the hybrid RANS-NLES type approach outlined in Chap. 3. The complexity of this flow is clear from the Frame (b) instantaneous flow field. The internal pedestals generate wakes. There will also be some

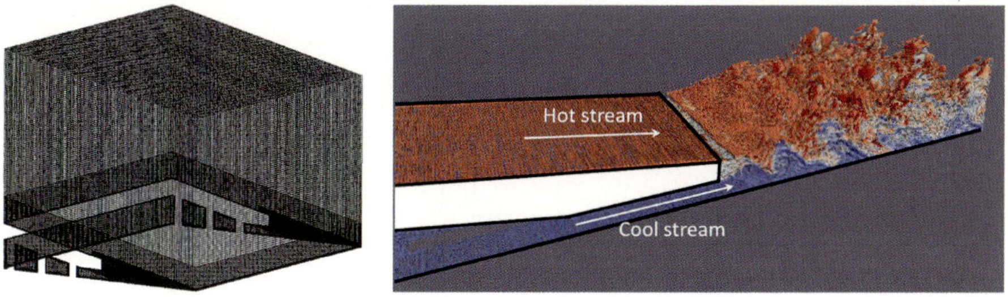

(a) (b)

Fig. 5.16 Simulations for the cutback trailing edge geometry of Martini et al. (2006a): (**a**) Octree, hexahedral grid with internal pedestals and (**b**) instantaneous flow structure (Tyacke et al. 2012a)

upstream boundary layer content. Both of these interact with the Kelvin-Helmholtz shear layer instability. Hence, there is the potential for considerable flow complexity. It is also worth noting that flows of this type can also involve complex low frequency modulations in global flow structure (Tucker et al. 2012). Figures 5.17a and 5.17b show the variation of surface adiabatic film cooling effectiveness, η, with axial location. The measurements are taken from Martini et al. (2006a). They are represented by symbols. The total range in RANS results for this flow is shown by the shaded area in Fig. 5.17a. This is for a blowing ratio of 0.5 i.e. the cooling stream mass flux is half of that of the flow over the pressure side of the blade. Neither of the RANS models tested (Menter SST and Spalart-Allmaras) give sensible results. In contrast, the solid black lines in Fig. 5.17, show that hybrid RANS-NLES is in encouraging agreement with the measurements. Note, the frame (b) simulation is for a blowing ratio of 0.35.

Ahlman et al. (2007) perform DNS of a highly idealized cut back trailing edge flow. The Reynolds number is 2000 and the external stream velocity is 10 % of the internal. The simulations use 4th order spatial CDs with $N \approx 10 \times 10^6$. The temporal integration uses a 4 stage, 3rd order RK scheme. At inflow, use is made of the digital filtering procedure of Klein et al described in Chap. 4.

Schneider et al. (2010) perform Smagorinsky LES of Martini et al.'s cutback trailing edge geometry with $N = 19.1 \times 10^6$ and *Re* (based on slot height of) 0.0625×10^5. They extend the DES work of Martini et al. (2006a) that used $N \approx 2 \times 10^6$. Agreement with measurements is gained for η and velocity profiles. To prescribe more realistic turbulence inflow statistics, Schneider et al. use recycling. This is applied at both the coolant slot and also for the mainstream. It is based around measured profiles. A curvilinear pressure correction code is used with a 2nd order spatial CDs and a low storage RK time integration. Momentum interpolation is used to avoid pressure field oscillations. A high, 1.1, and a low blowing ratio of 0.5 are considered. Reasonable agreement is found with measurements for η, velocity profiles and Reynolds stresses. Non-linear, complex, cooling effectiveness behaviour is observed. This depends on the blowing ratio and if the coolant flow is assumed laminar or turbulent.

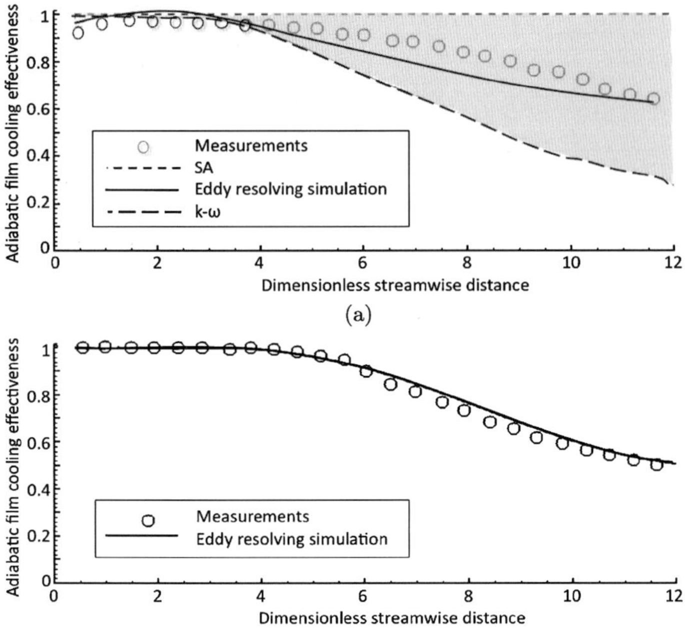

(b)

Fig. 5.17 Variation of surface adiabatic film cooling effectiveness for: (**a**) blowing ratio of 0.5, (**b**) blowing ratio of 0.35 (Tyacke et al. 2012a)

Joo and Durbin (2009) use a compressible multiblock solver to compute η for a realistic cut-back trailing edge geometry. A 2nd order, BD, temporal discretization is used. Upwind biasing is used for smoothing/stability along with a compact scheme based filtering. The skew-symmetric form of the convective term is also used to gain stability. For the simulations $N = 8 \times 10^6$–20×10^6. For turbulence inflow a type of Lund recycling is used to produce fluctuations pinned around a RANS predicted mean flow field. The scale-adaptive hybrid RANS-LES type technique discussed in Chap. 3 is used. Considerably improved agreement over RANS, for blowing ratios of both 1 and 1.5, is found for η.

Tyacke et al. (2012a) compare with the idealized cutback trailing edge data of Kacker and Whitelaw (1971). Lund's recycling is used to generate incoming boundary layer profiles. Both dynamic Smagorinsky based LES and also hybrid RANS-NLES are considered. Relative, to results for a range of RANS models, that include non-linear quadratic and cubic formulations, encouraging agreement is found with η for the RANS-NLES. The second order CD based LES ($N \cong 5$–23×10^6) generally shows encouraging agreement (for both grids) with mean velocity profile, Reynolds stress components and boundary layer shape factor development measurements. The resulting mean streamwise velocity and also Reynolds stress profiles at ten slot heights from the outlet are compared with measurements in Fig. 5.18. The measurements are given by the symbols. The lines show the LES. Blowing ratios of

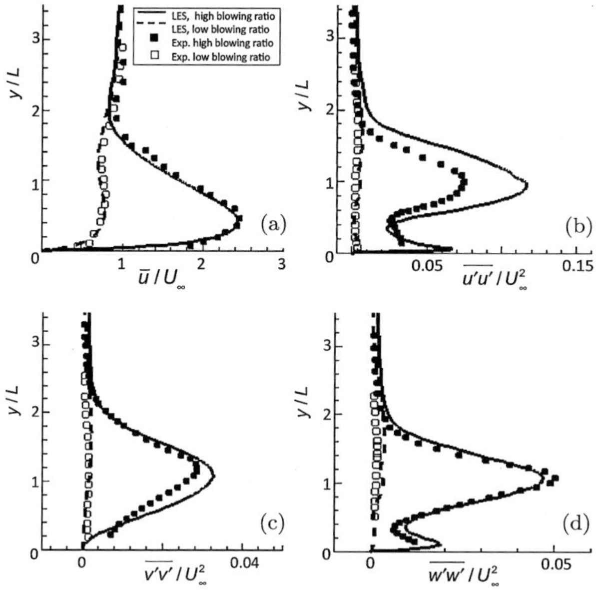

Fig. 5.18 Comparison of LES with measured mean velocity and turbulent stress profiles for an idealized cut-back trailing edge configuration (measurements represented by *symbols*, *full line* blowing ratio of 2.3 and *dashed* 0.75)

0.75 and 2.3 are considered. The dashed line is for the lower blowing ratio. As can be seen, the LES captures the trends convincingly. There is some discrepancy for the streamwise Reynolds stress for the high blowing ratio. However, this is largely attributed to rectification of the hot-wire signal in the large Kelvin-Helmholtz 'wake' zone.

Konopka et al. (2012) make supersonic flow simulations with shocks for a similar configuration to that reported above. However, this high speed flow work is motivated by cooling in scramjet engines. MILES is used in an AUSM framework with $N = 34 \times 10^6$ and $Re = 4600$—based on inner slot height. Compressible Lund's recycling is used to generate inflow data. Where comparison is made with measurements, the adiabatic film cooling effectiveness decrease is overpredicted but velocity profiles are in close accord. Strong (0.5–2) turbulent Prandtl number variations within the flow are observed.

Labyrinth Seals

The computation of labyrinth seal flows is relevant to the modelling of shrouded rotors and also general seals inside engines. For the labyrinth seal geometry shown in Fig. 5.19, Tyacke et al. (2012a) explore the performance of hybrid RANS-NLES and NLES. The accuracy of the following LES models, discussed in Chap. 3, is also studied: Yoshizawa (1993); mixed LANS-α (Geurts and Holm 2006; Leray 1934; Kosovic 1997). Due to small clearances between the seals and the shaft, although the mean flow field in Fig. 5.19 appears relatively simple, the flow physics is extreme.

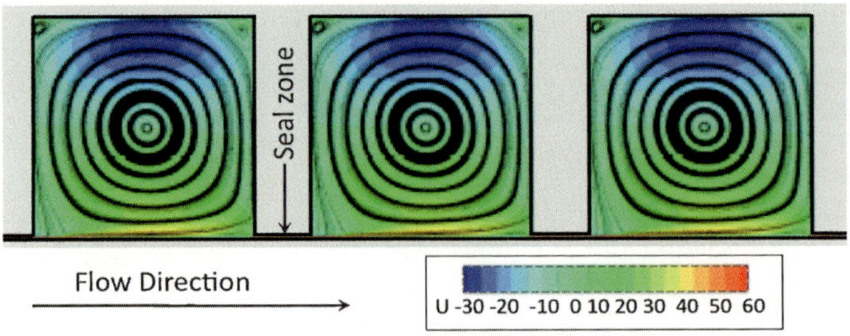

Fig. 5.19 Time averaged axial velocity field (constructed from periodic LES data Tyacke et al. 2012a)

Table 5.4 Mass flow coefficient errors (%) for LES and RANS simulations

Smag.	Yosh.	NLES	RANS-NLES	Ler.	Alp.	Kos.	RANS
−9.8	−0.98	−0.15	−9.3	0.94	1.42	−6.19	−9.14–28.56

For example, high values of the acceleration parameter K are found near the clearances (up to 1×10^{-4}). Values of $K = 3 \times 10^{-6}$ and above correspond to regions where the flow will tend towards re-laminarisation.

Table 5.4 shows the percentage mass flow coefficient errors. These are based on the measurements of Gamal and Vance (2008). The LES abbreviations are Smag. (Smagorinsky), Yosh. (Yoshizawa), Ler. (Leray), and Kos. (Kosović). The (N)LES results are accurate, whereas the RANS (for a range of established RANS models) errors are more than ten times larger.

Tyacke et al. (2012a) also considered the more complex seal geometry shown in Fig. 5.20. It is representative of those found at blade tips/shrouds and throughout the internal air system. The flow physics in this seal is again complex. It includes regions of high acceleration, separation, recirculation of different scales and a high velocity rotating lower wall. The latter gives rise to intense near wall shearing and some classical boundary layer content.

An LES grid ($N = 9 \times 10^6$) topology used is shown in Fig. 5.21. However other topologies were also tested. In the interest of saving space these are not shown here. The axial Reynolds number is 10,000 ($Re = \dot{m}/\mu\pi R_{ave}$, where R_{ave} is an average radius). The rotational (Couette) Reynolds number ($Re_C = \rho\Omega R_S(R_S - R_R)/\mu$, where Ω, R_S and R_R are angular velocity, stator and rotor radius, respectively) is 118,000. The divergent flow direction (flow right to left) resembles seals found at various locations in turbines.

Figure 5.22 gives vorticity contours. There is high vorticity near the rotating wall, showing where there is a boundary layer type flow. Vortices are shed from the tooth tips. Hence, this flow contains both wake and boundary layer type turbulence.

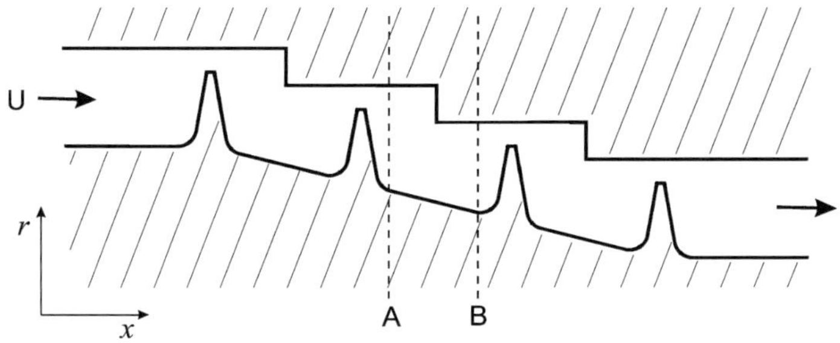

Fig. 5.20 Seal with *convergent* flow direction indicated along with velocity profile locations for measurements (Tyacke et al. 2012a)

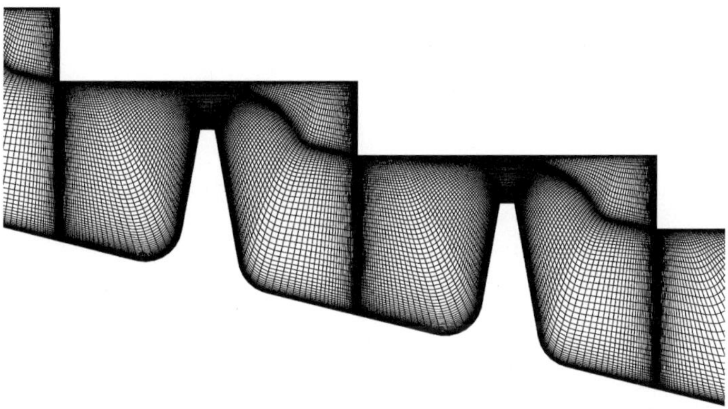

Fig. 5.21 LES grid (every other grid line shown)—(Tyacke et al. 2012a)

Figure 5.23 shows the swirl velocity profiles captured with RANS and also the hybrid RANS-NLES and LES approaches. These are for the convergent flow direction. The LES models used are the dynamic Smagorinsly model (labelled Germ.) and dynamic Yoshizawa related model (LDKM) (Kim and Menon 1999). Profiles are compared to the measurements of Denecke et al. (2005) at locations A and B (see Fig. 5.20). The RANS, in Frame (a), is for a range of models and flow solvers (identified by the labels in brackets and outlined in Tyacke et al. (2012a). As can be seen, there is potentially a large variation in RANS model results. However, the RANS-NLES and LES results, shown in Fig. 5.23b, display encouraging agreement with measurements. Axial velocities are also captured accurately and are presented in Tyacke et al. (2012a). Frames (c) and (d) show a similar disparity between RANS and LES results at profile location B. As can be seen in Tyacke et al. (2012a) similar traits are observed for the divergent flow direction.

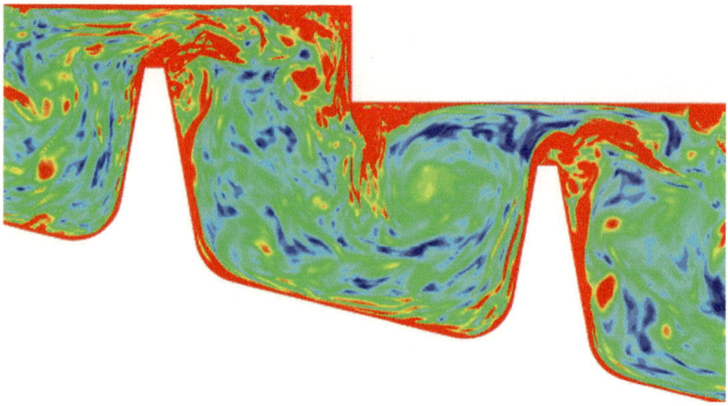

Fig. 5.22 Vorticity contours for RANS-NLES (Tyacke et al. 2012a)

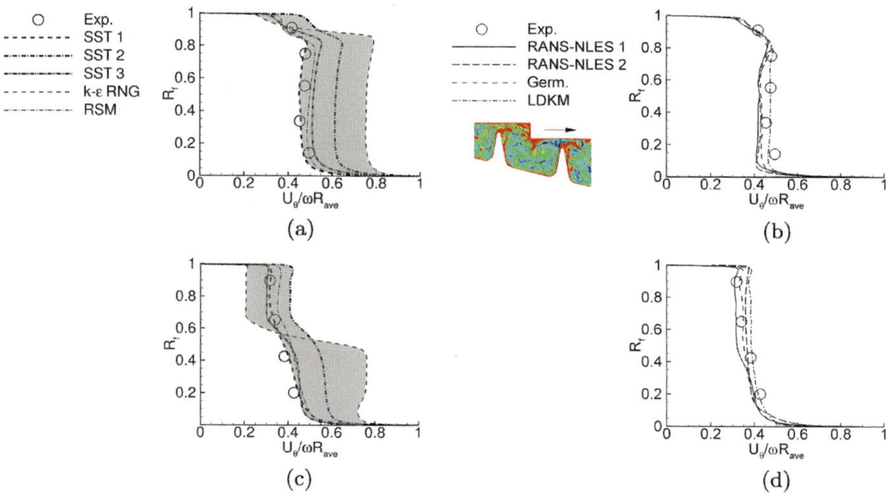

Fig. 5.23 Convergent flow velocity profiles: (**a**), (**b**) RANS and LES tangential velocity profiles at location A, (**c**), (**d**) RANS and LES tangential velocity profiles at location B. Note, the differing codes used are identified trough number in the line legends (Tyacke et al. 2012a)

Table 5.5 gives the errors in velocity profiles for all LES and RANS-NLES results. Errors are calculated using Eq. (5.3) below.

$$Error(U) = \frac{\sum_{\exp} |U_{j,\exp} - U_{j,num}|}{\sum_{\exp} \Omega R_{ave}} \times 100 \tag{5.3}$$

Table 5.5 show the wide range of RANS-NLES and LES results are consistent with each other and accurately match the measurements. Hybrid RANS-NLES errors for two very different flow solvers are tabulated. In strong contrast, the range of RANS velocity profile errors are up to 23 %.

Table 5.5 LES average velocity errors for both flow directions

SGS model	RANS-NLES	Germ.	LDKM
Convergent			
Axial	2.4–3.8	1.2	1.3
Swirl	3.3–3.4	2.9	2.2
Divergent			
Axial	4.1–5.1	4.9	5.0
Swirl	2.2–5.4	5.1	5.3

5.2.5 LES of Jet Flows

Film Cooling Holes

Downstream of a combustor the flow will be transonic or high subsonic. However the Mach number in film cooling holes will be low and hence most effectively modelled using an incompressible flow solver. Therefore, Iourokina and Lele (2005) couple two different flow solvers to study film cooling. This work is discussed in Chap. 7.

Tyagi and Acharya (2005) use an immersed boundary method to explore an inclined film-cooling hole in a flat plate. A mixed dynamic model with $N \cong 1 \times 10^6$ is used. Even though laminar inflow is used, encouraging agreement is found with velocity and heat transfer data. Jouhaud et al. (2007) again compute a jet in cross flow. The Smagorinsky LES model is used in a relatively standard compressible flow solver with unstructured cell topologies. Encouraging agreement is found with velocity and turbulence kinetic energy measurements for relatively modest grids ($N = 0.5$–2.7×10^6). The jet Reynolds number is 94,000, but this should be a relatively insensitive quantity in terms of LES resolution. Rozati and Tafti (2007, 2008) look at leading edge film cooling using LES with a structured, curvilinear, 2nd order, CD, incompressible solver for $N \cong 3 \times 10^6$. The dynamic Smagorinsky LES model is used. A precursor simulation captures turbulence inflow. The *Re*, based on the leading edge diameter is 100,000. The agreement with heat transfer measurements is encouraging.

Roy et al. (2003) explore the use of SA DES for film cooling. A single row of 35° holes in a flat plate is considered. A compressible flow solver with $N \cong 900{,}000$ is used. Notably, a symmetry plane is also employed. This is probably the key reason for the DES having worse heat transfer accuracy than RANS. A notable issue with DES is that, in its normal mode of operation, the cooling holes will be fully populated with RANS. Hence, the high level of eddy viscosity from this zone will convect into the jet shear layers and substantially suppress the resolved turbulence energy in the vicinity of the jets-flat plate interaction zone.

Fujimoto (2012) explore film cooling for an inclined hole in a flat plate. The OpenFOAM open source code is used. A 2nd order BD temporal scheme is used with 2nd order CD for the spatial disctretization. The Yoshizawa LES model is used. Simulations contrast the use of hexahedral octree meshes and traditional curvilinear grids. Similar results are found for both grids. Simulations are made for $N = 0.84 \times 10^6$ to 6.2×10^6. For high ratios of coolant jet velocity to cross stream, encouraging agreement is found with measurements for η. However, at lower values, Lund recycling is needed at the cross-flow, inflow boundary to gain agreement. This is presumably because at this blowing ratio the jet shear layer instability and growth is weakened and the cross flow has a dominant heat transfer influence.

General Jets

Chauvet et al. (2007), make zonal DES of a complex underexpanded jet with flow control via four radial injectors. To capture the shear layer growth the standard LES filter scale is modified to ignore azimuthal grid spacings. Such a strategy is likely to have helped in the film cooling DES of Roy et al. (2003) discussed above. Comparison is made with mean velocity and pressure profiles. Encouraging agreement is found. Hence, the approach appears reliable for this controlled jet configuration. Notably, the geometrical complexity is consistent with that found in industrial systems.

Schlüter (2000) make idealized LES of fuel injectors in a gas turbine burner. Essentially the system involves fuel-laden jets in a cross-flow. The study explores the level of fuel mixing. The Smagorinsky model is used with wall functions in a 2nd order unstructured finite volume solver for $330{,}000 \leq N \leq 600{,}000$. This and the work of Yuan et al. (1999) show the need to model the upstream geometry supplying the jet. For example, recirculation can occur in the feeding pipe. Also, the truncated domain, arising from not modelling the this, can result in reflections and a lack of resolved inflow turbulence.

There is a vast range of LES studies on basic jets, see for example Debonis and Scot (2002), Bodony and Leley (2004), Bodony and Lele (2006), DeBonis (2006). Much of this is related to acoustics and is addressed further in Chap. 6.

Jet impingement cooling is used inside gas turbine blades—see Chap. 1. Jefferson-Loveday and Tucker (2010, 2011) explore (I)LES ($N = 6 \times 10^6$) and hybrid RANS-ILES ($N = 3 \times 10^6$) modelling of jets impinging on curved surfaces with $Re = 23000$ (based on the jet diameter). These simulations make use of 6th order adaptive dissipation level (see Chap. 2) CD finite differences. An artificial compressibility flow solver is used with a Galerkin (see Chap. 2) time scheme. For the pure (I)LES, a highly refined near wall grid is used. Encouraging agreement found with heat transfer measurements.

5.2.6 Combustor LES

Reviews on the application of LES to gas turbine combustors along with those in ramjet and scramjet engines are given by Fureby (2009), Pitsch (2006), McGuirk and Taylor (2002), Swaminathan and Bray (2011). The application of LES to combustors has shown promise (Menzies 2009; Sengissen et al. 2007; McGuirk 2008; Boudier et al. 2007). Combustors have the LES advantage that their walls do not have classical boundary layers, cooling air being passed through them for protective shielding purposes via small holes. Hence, they do not suffer from the severe $Re^{2.5}$ scaling issue. Also, a key element is large scale mixing. RANS offers poor potential when dealing with this. Typically in the above work modest grids with $N = 1\text{–}3 \times 10^6$ are used and meaningful results gained at rig scale.

The geometries of combustors are tremendously complex. Boudier et al. (2007) shows need to account for full geometry such as cooling holes. Hence, there is the need for unstructured meshes, even though these have a substantial computational overhead and generally are of lower numerical order. Notably, hexahedral rich meshes are most ideal for LES accuracy. They have good energy conservation properties. Although certain types of hexahedral mesh types will not resolve the wall gradients well, this is of secondary importance for combustors. Tyagi and Acharya (2005) use an immersed boundary method for combustors.

A key aspect with combustion is that, as noted earlier, the small-scale mixing, where the chemical reactions that constitute combustion take place must be accurately modelled. Hence, a critical combustion challenge is dealing with the wide range of length scales. This makes subgrid scale modelling more crucial for these applications. It also strengthens the case for using high order numerical schemes (Pitsch 2006).

5.2.7 Engine Intakes

There seems to be little LES or hybrids on engines intakes. The Reynolds numbers in this zone will be high and hence this is clearly a limiting factor. Trapier et al. (2008) make DDES (to prevent LES incursions into the boundary layer) of flow in a supersonic inlet. Shock oscillations, linked to acoustic excitation mechanisms are found. With explicitly zonalized hybrid RANS-LES, LES zone incursions into the boundary layer are prevented. However, the extreme shock movements and changes in boundary layer flow dynamics makes the RANS zone hard to specify in this fashion. A multiblock, structured flow solver is used with a Roe scheme and the MUSCL approach to extend to 3rd order. An implicit 2nd order BD temporal scheme is used. Impressive agreement is found with surface pressure measurements and Schlieren visualizations.

Laraufie and Deck (2012) make zonal LES for a square sectioned curved duct. Upstream of the duct there is a full RANS zone. The is followed by a RANS-LES zone. Here the RANS extent is restricted to a small section of the boundary layer. The rational behind this is that the flow in the curved duct zone has a shallow separation. The more challenging flow physics in this zone needs greater LES content. Downstream of this is an area where DDES is used. The RANS zone is converted into resolved scales using dynamic forcing (see Chap. 3, Sect. 3.10.2).

Table 5.6 summarizes validation data sources for gas turbines with eddy resolving simulations in mind. The final column of the table gives validation data grading levels. These will be contrasted later with validation levels used for airframe related simulations.

5.3 Review of LES and Hybrids for Airframes

Next the use of eddy resolving simulations in the context of airframes is explored. These areas considered include stationary and oscillating/plunging aerofoils, trailing edges, wings, flaps, slats, full aircraft configurations, helicopters, landing gear, after bodies, base flows and cavities. The cases to be discussed are summarized in Table 5.7. This gives typical grid densities, Reynolds numbers and validation levels.

5.3.1 Aerofoil Flows

The DES of Spalart and coworkers (Shur et al. 1999) inspired great activity in applying eddy resolving approaches in aerospace engineering. This initial work involved a NACA0012 at an angle of attack of 60° creating a massively separated flow. These NACA0012 simulations are for a $Re_c = 1 \times 10^5$ and are made using 5th order upwinding. This allows the use of a modest O grid with $N = 0.225 \times 10^6$ and a spanwise domain of c. Strelets (2001) extended the NACA0012 work of Shur et al. This time the Menter SST based DES framework and also more sophisticated, flow based, control of numerical smoothing is used. A typical instantaneous vorticity field arising from this type of simulation is show in Fig. 5.24, taken from Tucker (2006). This simulation is a hybrid RANS-NLES.

Table 5.8 from Tucker (2006) gives lift and drag coefficient errors for different solution strategies. The parameter, N_{UP} is the order of the upwind component used in solutions. Also, N_{CD}, is the order of central differencing. The parameter ϵ_1 (see Chap. 2) gives the level of weighting of these two components. For $\epsilon_1 = 1$ a fully upwind solution is gained. A notable aspect, evident from Table 5.7, is that the Menter SST based URANS solution is poor. Standard SA based DES, with 5th order upwinding, is considerably more accurate. However, the control of numerical dissipation in the wake, by using a scheme heavily biased to CD in this zone, shows the best result. Use of the Menter SST model, based DES, yields further improvements relative to standard DES. However, notably, NLES, with no near wall RANS modelling, does surprisingly well. This is discussed further later.

Dahlström and Davidson (2001) perform LES for the Aerospatiale A–aerofoil with $Re_c = 2.1 \times 10^6$, $N = 1.54 \times 10^6$ and an angle of attack of 13.3°. The flow is challenging. It involves mildly separated boundary layer zones with transition. Specifically, on the suction surface, in the streamwise direction, there is first laminar flow. This results in separation and turbulence transition to a fully turbulent

Table 5.6 LES validation data sources for propulsion systems and validation grades

Engine zone	Nature of data	Data sources	General comments	Validation data level
Intake	Surface pressures (some unsteady), total pressures at fan face, Schlieren Note, data of Gardarin et al. has both PIV, hot-wire data, spectral information, Reynolds stresses, surface flow visualization and pressures	RAE M2129 Intake—May et al. (1993). NASA S—duct—Wellborn et al. (1993) Gardarin et al. (2008)—rectangular section duct, investigated in the S19Ch wind tunnel of ONERA F1 ONERA Fauga wind tunnel—Colin et al. (2007) Small scale Whittle Laboratory rig (cross wind conditions)—Hall and Hynes (2006) ONERA supersonic wind-tunnel S3MA (Modane, France), supersonic intake—Trapier et al. (2006, 2007) Boundary layer ingesting inlet—Owens et al. (2008) (mean velocity profile and pressure data including at engine inlet plane)	More data needed—Forthcoming Whittle Laboratory data should provide turbulence statistics—validation levels 1–3	1, 2, 3, 4 Approximate average 2
Fan	Characteristic curve data, total pressure radial profiles, wake profiles, some blade tip data, spectral information for noise and turbulence, mean velocities, Reynolds stresses. Both hot-wire and LDA data	Boeing 18″ fan rig—Ganz et al. (1998b,a), DLR low speed fan—(2001), NASA rotor 37—Suder and Celestina (1994), Suder et al. (1995)	Fan section type data would be useful but would compromise flow physics in terms of shock dynamics	1, 2, 3, 4

Table 5.6 (Continued)

Engine zone	Nature of data	Data sources	General comments	Validation data level
Rod-aerofoil	PIV and hot-wire measurements. Far field pressure spectra, velocity profiles, Reynolds stresses	Jacob et al. (2005)	This case is a very generic representation of interactions of a blade with an upstream wake and has been used, for example, in the context of fan noise. Note—some evidence of asymmetry in measurements	1, 2, 3, 4
IPC/HPC	Flow visualization, total pressure loss, exit flow angle, hot-wire and PIV data of wake boundary layer interaction (from peak suction to trailing edge), momentum thickness data, surface Kulite data, surface shear stress data, mean velocity and turbulence intensity profiles for boundary layer, wake data	Whittle Laboratory data (Deverson rig)—Gdadebo (2003), Goodhand and Miller (2010), Goodhand (2010), Wheeler (2007) GE data—multi-stage rig—Halstead et al. (1997) DLR cascade—Köller et al. (2000), Küsters et al. (2000) Naval Postgraduate School, Monterey—cascade data, Hobson et al. (2001), Elazar and Shreeve (1990)	Low Reynolds number of much of the measurements makes transition modelling a dominant issue. SMURF rig and Whittle laboratory should provide tip leakage flow interaction data in the future. Little differentiation in data between HPC and IPC but most tests intended to represent HPC. Monterey data has clearly defined geometry	1, 2, 3 Approximate average around 2
Compressor endwall flows	Blade surface pressures, total pressure loss downstream of the blade, surface flow visualization	Ma et al. (2011), Gdadebo (2003)	–	1, 2
Compressor representa-tive section NACA 6512–63	Surface pressure measurements, far field sound spectra, mean velocities and Reynolds stresses in wake along with turbulence spectral information	Winkler and Moreau (2008)	Most representative of a highly loaded, high-speed compressor blade. Has acoustic data	1, 2, 3, 4

Table 5.6 (Continued)

Engine zone	Nature of data	Data sources	General comments	Validation data level
IGV/ Diffuser	Mean velocities at the OGV and pre-diffuser exits. Total pressure loss and pressure recovery coefficients	EU, LOPOCOTEP project data—Reported in Li (2012)	For OGV inlet 2D map of velocity but 1D radial profiles of turbulence kinetic energy and length scales. No rotor, unsteady wake data	2
Combustor	Mean flow and turbulence statistics for a range of isothermal building block geometries of increasing geometrical complexity, species mass fractions, temperatures for non-premixed combustion	Barlow and Frank (1998), Dally et al. (1998), McGuirk (2008)	Data base for premixed combustion needed (2006)	1, 2, 3, 4
HPT/IPT	Flow visualization, total pressure loss, exit flow angle, velocity profiles turbulent stresses, heat transfer data	Aerodynamic data—Holley et al. (2006), Langston et al. (1977), Langston (1980) Heat transfer data—reviewed in Helms (2011)	The real HPT has film cooling and ideally this would need to be included with the rather complex inflow turbulence field from the combustor	1, 2, 3 Average around 2
LPT	System characteristic curves and radial traverses of total temperature and pressure and exit flow angle	Halila et al. (1982)	Two stage, full scale, low temperature rig with real geometry features	1, 2
LPT	Surface pressures, detailed LDA, PIV, hot-wire data, hot film, mean velocity profiles, Reynolds stresses, surface shear stress measurements	Coull (2009), Wissink and Rodi (2004), Wissink et al. (2006), Wu et al. (1999), Stieger (2002), Stieger and Hodson (2003), Halstead et al. (1997), Howell (1999), Zhang (2005), Schulte (1995), Morales (2004), Opoka (2007), Himmel (2010), Schuler et al. (2011) (explores rim seal influence)	Extensive data set including DNS with deep level of flow physics analysis	1, 2, 3

Table 5.6 (Continued)

Engine zone	Nature of data	Data sources	General comments	Validation data level
Shrouded blade tip	Pressure drop, LDA based velocity profiles, heat transfer related data, static pressure distributions	Gamal and Vance (2008), Denecke et al. (2005), Pfau et al. (2001, 2007), Rosic (2006), Rehder and Dannhauer (2007), Lehmann (2009)	Lack of turbulence intensity data. Data of (2008) and (2005) is for labyrinth seals	1, 2
Unshrouded blade tip	Detailed turbulence statistics and spectral information	Muthanna (1998), Wang and Devenport (2004)	Compressor cascade with moving wall	2, 3, 4
Propulsive jet	Spectral acoustic information, 4th order space-time turbulence velocity correlations, limited velocity-temperature correlations	Harper-Bourne (2007), Tanna (1977), Bridges and Wernet (2004), Bridges and Brown (2004, 2005), Bridges and Wernet (2007), Panda et al. (2004), Panda and Seasholtz (2004), Panda (2006), Viswanathan (2004), Eastwood et al. (2010)	Extensive range of studies for a wide range of geometries. Lack of characterization of inflow conditions. Need more data to explore interaction with wings and flap, flight stream, Mach number and heating influences	1, 2, 3, 4, 5, 6, 7
Rim seals	Surface pressure, concentration measurements, LDA	Popovic (2010), Gentilhomme (2004), Gentilhomme et al. (2010), Bohn et al. (2003, 2006)	LDA does not show mass conservation	2
Internal air system cavities	Generally high quality heat transfer, flow visualization, hot-wire anemometry and LDA data for idealized geometries	Farthing et al. (1992a,b), Long and Tucker (1994), Daily and Nece (1960), Itoh et al. (1992), Itoh (1995), Bohn et al. (2003, 2006)	Some LDA shows cross plotting error i.e. data at same point but different plots have different values	2, 3
Engine control unit	Heat transfer, mean velocity and turbulence intensity data from general field of electronics cooling	Data sources can be found in (2004) and (2001)	Conjugate heat transfer data available	2, 3

Table 5.6 (Continued)

Engine zone	Nature of data	Data sources	General comments	Validation data level
Ribbed cooling passages and impingement cooling	Heat transfer, mean velocities, Reynolds stresses, some DNS data	Ribbed passages—Acharya et al. (1993), Han et al. (1988), Rau et al. (1988), Sewall and Tafti (2004), Nagano et al. (2004) Impingement cooling—Lee et al. (1999), Geers et al. (2004), Kim (2005), Yan (1993), Lytle and Webb (1994), Cooper et al. (1993), Baughn et al. (1992), Fenot (2004)	For ribbed channels there is a good range of geometric complexity from simple bi-periodic geometries to developing flow in bent rotating passages. For impingement cooling there is good canonical geometry data but little real geometry data	2, 3
Cut back trailing edges	Adiabatic wall film cooling effectiveness, mean velocity profiles, turbulence intensities	Imperial College—Kacker and Whitelaw (1968a,b, 1969, 1971), Nicoll and Whitelaw (1967) University of Arizone—Zhou and Wygnanski (1993) University of Karlsruhe—Martini et al. (2003), Martini and Schulz (2004), Martini et al. (2006b) Clemson University—Holloway et al. (2002a,b)	The Mach numbers for the Imperial College and University of Karlsruhe data are low and this is problematic for LES with compressible flow solvers	1, 2, 3
Turbine (HPT) wakes	Smoke flow visualizations, holographic interferometry, white light differential interferometry, total pressure wake traverses probe, Kulite pressure measurements, thermocouple probe and a cold wire resistance probe for temperature measurements	Sieverding et al. (2004)	–	2

Table 5.6 (Continued)

Engine zone	Nature of data	Data sources	General comments	Validation data level
Film cooling	Cold wire anemometry and PIV	Lalizel et al. (2012)	PIV is near injection hole. Pulsation of film cooling is explored	2, 3
Film cooling	Adiabatic surface film cooling effectiveness, Reynolds stresses	Pietrzyk et al. (1990), Sinha et al. (1991)	–	2, 3
Interducts	Hot-wire, LDA. Flow visualization, surface static pressure and total pressure loss profiles. Spectral, turbulence intensity and turbulence length scale information	Turbine interduct—Axelsson and George (2008), Göttlich et al. (2007), Johansson and Axelsson (2009) Compressor induct—Karakasis et al. (2010), Britchford et al. (1994)	–	1, 2, 3, 4 Approximate average grade 2.5

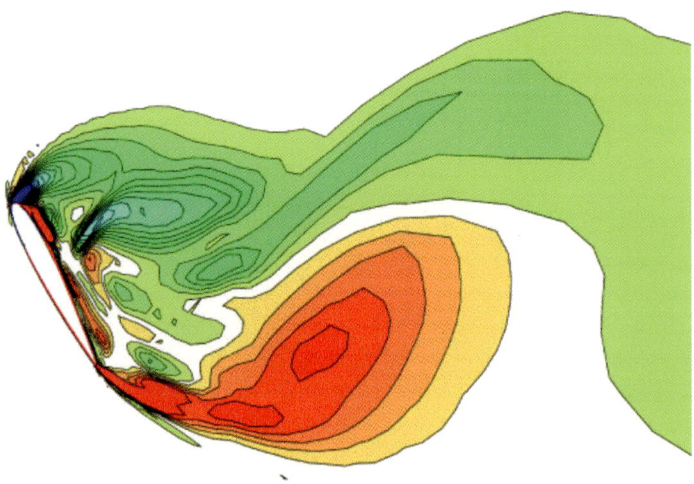

Fig. 5.24 Hybrid RANS-NLES giving instantaneous vorticity contours for a NACA0012 at a 60° angle of attack (Tucker 2006). Published with kind permission from Wiley and Sons

boundary layer state. The boundary layer then later separates. Dahlstrom and Davidson use an incompressible CD (blended with upwinding in unstable flow zones) based flow solver with collocated grids and the CN time scheme. For the modest

Table 5.7 Summary of aerospace eddy resolving simulations with typical grid densities, Reynolds numbers and validation levels

Component	Author(s)	Method	$Re_c/10^6$	$N/10^6$	Validation level
NACA0012	Shur et al. (1999)	Standard DES	0.1	0.225	1
Asymmetrically beveled trailing edge	Wang and Moin (2000)	LES	2.15	7.1	2, 3
Asymmetrically beveled trailing edge	Leschziner et al. (2009)	Two layer model	2.15	1.5	2, 3
Asymmetrically beveled trailing edge	Leschziner and Tessicini (2009)	SA DES, Menter SST DES, hybrid RANS-LES	2.15	1.5	2, 3
NACA0012	Strelets (2001)	Menter SST based DES	0.1	0.225	1
Aerospatiale A–Aerofoil	Mary and Sagaut (2002)	LES/ILES	2.1	7.6	2, 3
Aerospatiale A–Aerofoil	Dahlström and Davidson (2001)	LES with wall functions	2.1	1.5	2, 3
Aerospatiale A–Aerofoil	Reported in Mellen et al. (2003)—Lessons from LESFOIL—CERFACS	LES (WALE model)	2	0.29	2, 3
Aerospatiale A–Aerofoil	Reported in Mellen et al. (2003)—Lessons from LESFOIL—Dassault	LES (Smagorinsky, wall functions)	2	0.21	2, 3
Aerospatiale A–Aerofoil	Reported in Mellen et al. (2003)—Lessons from LESFOIL—FLUENT	DES	2	0.37	2, 3
Aerospatiale A–Aerofoil	Reported in Mellen et al. (2003)—Lessons from LESFOIL—IFH	LES—dynamic Smagorinsky	2	18.5	2, 3

Table 5.7 (Continued)

Component	Author(s)	Method	$Re_c/10^6$	$N/10^6$	Validation level
Aerospatiale A–Aerofoil	Reported in Mellen et al. (2003)—Lessons from LESFOIL—ONERA	LES—mixed scale model	2	17.9	2, 3
Aerospatiale A–Aerofoil	Reported in Mellen et al. (2003)—Lessons from LESFOIL—UMIST/QMW	LES—Smagorinsky,	2	6.39	2, 3
Aerospatiale A–Aerofoil	Reported in Mellen et al. (2003)—Lessons from LESFOIL—Schmidt and Thiele	DES	2	0.44	2, 3
NACA0012	Tucker (2006)	Hybrid RANS-NLES	0.1	0.225	1
OAT15A aerofoil	Deck (2005b)	DES, Zonal DES	3	2	1, 2, 3, 4
OAT15A aerofoil	Garnier and Deck (2008)	LES	3	20.8	1, 2, 3, 4
NACA0021	Garbaruk et al. (2009a)	DES (range of base RANS models, SST	0.27	0.5–5.2	1, 4 (C_L spectra)
Plunging SD7003 aerofoil	Visbal (2011)	ILES	0.06	2.4–18.6	1, 2, 3
Oscillating NACA0012	Knopp et al. (2009)	Standard DES, Menter SST DES and SAS, EARSM DES...	0.98	0.56–2.78	1
Wing-flap	Tucker (2006)	Hybrid RANS-NLES	23	3	1, 2
Spoiler	Gand (2012)	DES, Zonal DES	5	31	2, 4
Idealized tilt rotor wing in downwash	Spalart et al. (2003)	DES	0.3	0.6	1, 2

Table 5.7 (Continued)

Component	Author(s)	Method	$Re_c/10^6$	$N/10^6$	Validation level
Idealized tilt rotor wing in downwash	Kjellgren et al. (2000)	LES	0.13	0.256	1, 2
Double delta wing	Tucker (2006)	Hybrid RANS-NLES	2.2	1.2	1, 2
Double delta wing	Ceresola (2009)	DES	–	5.2 (structured)	1, 2, 3
Double delta wing	Ceresola (2009)	DES	–	13 (unstructured)	1, 2, 3
Heavily swept wing	Li and Leschziner (2007), Leschziner et al. (2009)	Two-layer model	0.21	23.6	1, 2, 3
"	Hahn and Drikakis (2009)	MILES	–	12.5	1, 2, 3
F–15—fighter aircraft	Forsythe et al. (2004)	DES	13.6	2.85–10	1
F/A–18—fighter aircraft	Morton et al. (2007)	DES	13	3.9	1, 2, 3
FA–5—fighter	Leicher (2009)	SST–SAS, SA DES, EASM DES	2.8	11.4 (structured) 13 (unstructured)	2, 3, 4
Helicopter fuselage	Le Chuiton et al. (2009)	Menter SST, SAS and DES ...	2.48	2.2–14.3	1, 2
Nose landing gear cavity	Langtry and Spalart (2009)	DES	–	6	0
Landing gear	Hedges et al. (2002)	DES	0.6	2.5	1, 2
Landing gear	Lockard et al. (2004)	DES	–	13.3	0
Landing gear	Soemarwoto et al. (2009)	Hybrid RANS-LES, DES	0.6	2–3	1, 2
After body	Deck and Thorigny (2007)	Zonal DES	1.1	8.3	4
Prolate spheroid	Fureby et al. (2004)	LES	4.2	0.8–1.2	1, 2

Table 5.7 (Continued)

Component	Author(s)	Method	$Re_c/10^6$	$N/10^6$	Validation level
Civil aircraft	Brunet and Deck (2008)	Hybrid RANS-LES	2.8	10	1, 2, 3
Wing-body junction (NACA0020 based)	Fu et al. (2007)	DES, DDES (k–ω based)	0.115	0.7	2, 3
Wing-body junction (TN D–712)	Fu et al. (2007)	DES, DDES (k–ω based)	7.5	2.2	2
Three-element aerofoil	Deck (2005b, 2009b)	Zonalized DES	1.7	5.5	1, 2, 3
Three-element aerofoil	Cummings et al. (2004)	DES	3.7	2.4–5.6 (unstructured)	1
Rocket nozzle	Deck (2009b)	DES	21	11	4
Supersonic base flow	Garbaruk et al. (2009b)	DES and zonal DES	3.2 (based on based diameter)	0.9–14	1, 2, 3
Supersonic base flow	Fureby et al. (1999)	LES, MILES	2.2	1.35–2.1	2, 3
Supersonic base flow	Forsythe et al. (2002)	DES	2.86 (based on base diameter)	~0.33–2.7	1, 2, 3
Supersonic base flow	Baurle et al. (2003)	MILES	2.86 (based on base diameter)	2.2	1, 2, 3
Supersonic base flow	Sandberg and Fasel (2005, 2006b,a)	DNS	0.1	5.84–43	–
Supersonic base flow	Kawai and Fujii (2005)	Hybrid RANS-LES, LES, MILES	2.86 (based on base diameter)	1.26–5.84	1, 2
Supersonic base flow	Simon et al. (2006)	DES, Zonal DES, LES, hybrid RANS-MILES	"	5–13.5	1, 2, 3
Supersonic base flow	Simon et al. (2007b)	Zonal DES	"	20.3	1, 2, 3
Supersonic base flow	Sandberg (2012)	DNS	0.1 (based on base diameter)	78–177	1, 2, 3
Subsonic and transonic missile flow	Simon et al. (2007a)	Zonal DES	–	7	2

Table 5.7 (Continued)

Component	Author(s)	Method	$Re_c/10^6$	$N/10^6$	Validation level
Deep supersonic cavity	Thomber and Dfikakis (2008)	ILES	0.86 (based on cavity length)	0.8–3	2, 3, 4
Deep supersonic cavity	Larchevêque et al. (2003)	LES (wall modelled)	0.86 (based on cavity length)	2.3	2, 3, 4
Subsonic cavity	Jefferson-Loveday et al. (2012)	LES	0.05(based on cavity height)	7	2, 3
Supersonic Cavity	Hamed et al. (2003)	DES (Menter SST based)	4	0.77	3
Three-dimensional cavity	Shieh and Morris (2001)	DES	0.2	1.7	2
M219 cavity flow with and without doors	Allen et al. (2005)	DES—SA, k–ϵ and k–ω based	13.5	1.1	4
M219 cavity flow	Peng (2009)	LES, DES SST SAS, hybrid RANS-LES	13.5	~1–6	2, 3, 4
M219 cavity flow with doors	Lawson et al. (2009)	DES, LES	13.5	4.5–8.5	2, 3, 4
Supersonic cavity with doors	Rizzetta and Visbal (2003) Rizzetta et al. (2002)	ILES	1.86	21	4
Supersonic compression ramp	Bisek et al. (2013)	MILES	0.015	1.3–95	–
Idealized wing tip vortex	Sreedhar and Ragab (1994)	LES	0.1 (based on initial vortex radius and maximum initial velocity)	0.19–0.43	2
Idealized wing tip vortex	Duraisamy and Lele (2008)	DNS	0.008	19	–
Wing tip vortex	Uzun et al. (2006)	ILES	0.5	26.2	2, 3
Vortex generators	You et al. (2006a)	LES	0.13 (based on vortex generator height)	1.2–3.27	2, 3

Table 5.8 Summary of NACA0012 C_L and C_D errors

Model	Discretization	Error $\%C_L$	Error $\%C_D$
Menter SST (URANS)	$\epsilon_1 = 1, N_{UP} = 5$	72	64
SA DES	$\epsilon_1 = 1, N_{UP} = 5$	19	13
SA DES	$\epsilon_1 = 0.1, N_{UP} = 5, N_{CD} = 4$	9	3
Menter SST DES	$\epsilon_1 = 1, N_{UP} = 5$	12	5
ILES	$\epsilon_1 = 1, N_{UP} = 3$	12	4

grids, wall functions are used, but found inadequate for this complex and transitional flow. Mary and Sagaut (2002) perform both LES and ILES for this case, but with $N = 7.6 \times 10^6$ and $M = 0.15$. A so-called selective, mixed scale, LES model is tested. The grid is hybrid with a two-dimensional zone for the pressure surface and a three-dimensional grid on the suction surface. The flow considered has a laminar separation bubble on the latter. The detached shear layers transition. There is reattachment followed by another separation further downstream. Hence, the flow is relatively complex. The spatial scheme is AUSM based with a 3rd order BD in time in a multi-block flow solver. Broadly, there is encouraging agreement with velocity, Reynolds stress and surface skin friction distributions.

The above Aerospatiale A–aerofoil work is part of the European Union funded LESFOIL project. The intention of the project was to assess the performance of LES and to an extent DES for aerofoil flows. The project involved seven partners. The approaches and grid densities used are partly summarized in Table 5.7. The work is fully reviewed by Mellen et al. (2003). The spanwise domain extents used ranged between 1.2 and 12 % of chord. The key aspect is that there is a wide scatter in results. Hence, the potential for LES looks disappointing. However, as pointed out by Mellen et al., for this transitional flow with shallow separations, adequate results can be found with LES but for much denser grids than typically used in the project simulations. To achieve this, some compromising of the ideal spanwise domain extent can be helpful. Mellen et al. reported that the LES model has a subordinate role in success relative to adequate grid resolution. DES is also carried out as part of LESFOIL. When considering the flow physics, this case does not seem well suited to DES.

Garbaruk et al. (2009a) report a range of eddy resolving predictions for the flow over at NACA0012 aerofoil at 60° incidence, $Re_c = 2.7 \times 10^5$ and $M = 0.1$. Both unstructured and structured solvers are used with $N = 0.5 \times 10^6$ to 5.2×10^6. A range of DES approaches are tested with several base RANS models (including EASM, k–ω) along with the SAS approach of Menter et al. Spanwise domain extents of c–$4c$ are considered. For this flow, which has massive separation, results are found insensitive to the RANS model used. Also, since large vortical structures are generated, modest grids ($N = 2 \times 10^6$) are found sufficient for gaining grid independent mean properties. As would be expected, considerable improvements in results, relative to URANS, are found. However, it is noted that to get converged statistics, simulations needed to be run for a long time. Typically 300–400 dimensionless time units, where time is made dimensionless using the c and free stream velocity. The SST model showed encouraging performance. Note, for results reported by Garbaruk et al., data comparisons are based on surface pressure distributions and lift coefficient spectra.

Deck (2005a) perform SA based zonal LES of shock boundary layer interaction for the OAT15A aerofoil at $Re_c = 3 \times 10^6$. A 2nd order structured solver is used with the AUSM convective scheme. A BD temporal scheme is used in a structured multiblock solver. To avoid modelled stress depletion and hence grid induced separation, zonal LES is preferred to DES in this work. The RANS layer is forced to cover the whole boundary layer. A spanwise domain of $0.26c$ is used with $N = 2 \times 10^6$. The shock motion time scale is found to agree closely with the analytical estimate of Lee (1990). Predicted pressure spectra agreed well with measurements. Notably, standard DES failed to capture the shock oscillation characteristic of buffet.

Garnier and Deck (2008) apply LES to the same case and at an identical Reynolds number. The same solver is also used but with the Roe scheme and adaptive numerical dissipation. A mixed scale LES model is used. Notably, to reduce the grid density, the pressure side of the aerofoil is treated entirely in RANS mode. Even so $N = 20.8 \times 10^6$ is needed for adequate LES grid resolution with a spanwise domain of just $0.0365c$. Broadly, encouraging agreement for mean velocities, surface pressures and the spectra of these, is found with measurements. Also, turbulent fluctuations agree well with measurements. However, the shock location is not that well captured. There is evidence of the small spanwise domain having an impact on the results. The shock oscillation frequency is within around 25 % of Lee's (1990) estimate. However, this estimate is idealized.

Visbal (2011) performs 6th order, compact scheme, ILES (no LES model but filtering for small scales) for deep dynamic stall of a plunging (SD7003) aerofoil with $Re_c = 6 \times 10^4$. Grids range from $2.4 \times 10^6 < N < 18.6 \times 10^6$ and the results found to be relatively insensitive to grid choice. The span-wise domain is $0.4c$. Encouraging agreement is found with measurements for the initial stages of stall. However, at later stages, the level of agreement between the two sets of available measurements and also the LES, breakdown. This appears to be due to inconsistencies in end wall boundary conditions.

5.3.2 *Trailing Edge Flows*

Wang and Moin (2000) make LES of the trailing edge flow from an asymmetrically beveled strut. The $Re_c = 2.15 \times 10^6$ and $N = 7.1 \times 10^6$. The dynamic Smagorinsky model is used. Full details of this simulation, which has an acoustic emphasis, are given in Chap. 6. Leschziner et al. (2009) repeat this trailing edge case. This time a two-layer model is used. The switch over to LES takes place at $y^+ = 40$ and a substantially coarser grid with $N = 1.5 \times 10^6$ is used. When the pressure gradient is included, in the two-layer model, encouraging agreement is found with the benchmark solution. Leschziner and Tessicini (2009) report further studies for this case.

Different codes and hybrid RANS-LES strategies are used. These include, SA and Menter SST based DES along with an approach that utilizes the $v'^2 - f$ RANS model of Durbin (1995). Substantial differences in the location of the separation point and turbulence levels is found between the different studies. Most trailing edge computations are in a noise context. Hence LES and hybrids for this type of flow are discussed more deeply in Chap. 6.

5.3.3 Multi-component Aerofoils

Spalart et al. (2003) explore the use of DES for active flow control of a V22 type tilt rotor system wing. The angle of attack is $-85°$. An idealized flap is also modelled. The $Re_c = 3 \times 10^5$ and $N = 0.6 \times 10^6$. The flow configuration considered is where the rotors are generating vertical lift. Hence presenting a flow quasi-normal to the wing surfaces. Notably, in the simulation of Spalart et al., better agreement is found with measurements if the near wall RANS layer is omitted. A similar simulation is performed by Tucker (2006) in a hybrid RANS-NLES framework. This is shown in Fig. 5.25. However, this time without flow control. Spalart et al. found DES to probably be inaccurate when modelling this. Figure 5.25a gives instantaneous vorticity contours. Figure 5.25b gives eddy viscosity contours. A striking feature is that the only area where there is substantial viscosity is on the right hand side lowered flap zone. This is largely because this is the only area with classical boundary layer content. Indeed, interestingly, for the NACA0012 DES, the RANS layer is of little importance (Tucker 2006). The key need for the RANS layer is to ensure plausible behavior at lower angles of attack. Kjellgren et al. (2000) perform Smagorinsky based LES for the tilt rotor system explored by Spalart et al. with $N \sim 2.5 \times 10^5$, just 10 spanwise nodes, and $Re_c = 130,000$. The domain depth is $0.3c$. A finite element procedure is used with a RK temporal integration and a lumped mass matrix. The pressure field is computed by a fractional step method. In accordance with the measurements, the LES captures the beneficial influence of the oscillatory blowing, active flow control.

This need for a RANS layer in attached boundary layer zones is illustrated in Figs. 5.26a–5.26c. This gives instantaneous vorticity contours for a wing-flap configuration. Simulations are for $Re_c = 2.2 \times 10^6$ and $M = 0.18$. Frame (a) gives Menter SST based DES results, Frame (b) hybrid RANS-NLES and (c) NLES. The spanwise extent of the domain is 2.5 times the flap chord. Clearly the NLES results show a non-physical flow separation on the suction surface. This is primarily because the cross-stream grid is too coarse for LES. Hence, a RANS zone is needed to model the fine near wall turbulence scales, which the grid is too coarse to resolve. Hence, this a key point in relation to MILES type simulations. There is no special exception with MILES. Some form of near wall modelling is needed just as with ordinary LES.

Cummings et al. (2004) apply the unstructured COBALT code to a three-dimensional slat-wing-flap geometry in DES mode. Broadly, encouraging agreement is found with surface pressure measurements. However, at certain spanwise

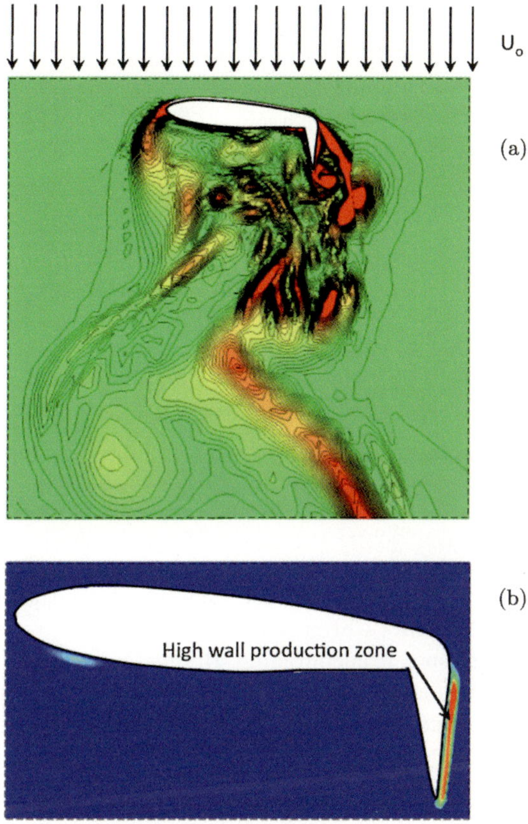

Fig. 5.25 Hybrid RANS-NLES vorticity and turbulence viscosity contours for generic tilt-rotor wing geometry: (**a**) vorticity contours; and (**b**) viscosity contours (Tucker 2006). Published with kind permission from Wiley and Sons

locations, the RANS models has incorrectly captured the separation location. This has strongly impacted on the DES. The complex unsteady flow physics that arises when considering high-lift wings, for a three-dimensional geometry is evident from this work. Deck (2005b, 2009b) apply zonalized DES to a three-element aerofoil. The slat and flap deflection angles are 30 and 40°, respectively. Also, the main element is at 9° and $Re_c = 1.7 \times 10^6$, based on main element c. For the structured grid $N = 5.5 \times 10^6$. The spanwise domain extent is $6c$. The DES zones are localized to the slat and flap cove zones, the rest of the domain being treated using URANS. Comparison is made with surface pressure measurements and turbulence fluctuations. The complexity of the flow for slat-wing-flap configurations is illustrated in this work. Hence, the need for such flows to be modelled using eddy resolving approaches is demonstrated. For example, as the detached eddies pass through inter-element zones, they are subjected to a complex external strain field. Reasonable agreement is found with surface pressure, mean velocity and turbulent fluctuation measurements.

Fu et al. (2007) apply k–ω based DES and DDES to two wing-body junction flows. One is idealized, and NACA0020 profile based. The other involves a more complex swept wing. The simulations use modest grids with $N \sim 1$–2×10^6. A third

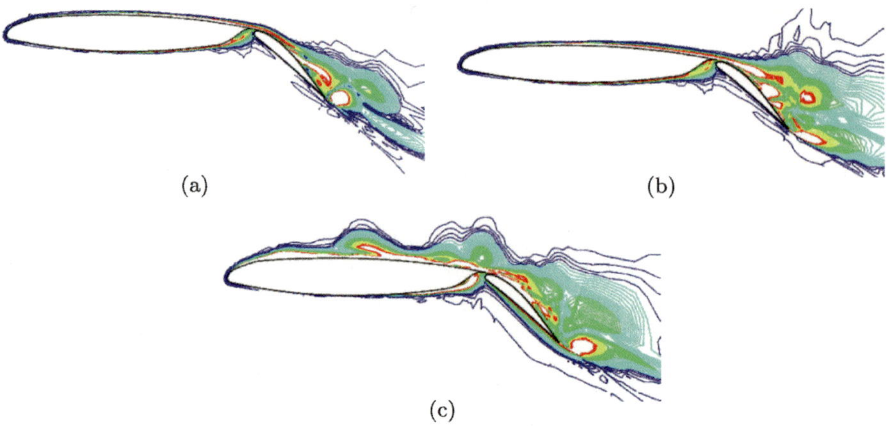

Fig. 5.26 Wing-flap configuration, vorticity contours: (a) Menter SST based DES; (b) hybrid RANS-NLES; and (c) NLES (Tucker 2006). Published with kind permission from Wiley and Sons

order upwind based Roe scheme is used to discretize the convective terms, in a multiblock, structured code. Standard DES showed particularly poor performance for both flows—even worse than RANS in many instances. However, the DDES performance is encouraging. Both eddy resolving simulations used a modified k–ω RANS model. A key advantage of the eddy resolving calculations is found to be in sensibly capturing the turbulence kinetic energy levels.

5.3.4 Swept and Delta Wings

Li and Leschziner (2007), Leschziner et al. (2009) model flow over a highly swept wing. The flow incidence is 8.8° at a $Re_c = 21000$ (root chord based) with $N = 23.6 \times 10^6$. Simulations make use of a 2nd order BD in time with spatial CDs, in a collocated, curvilinear flow solver. LES are performed with a two layer model. Unlike the turbulence kinetic energy, mid-span velocity profiles, at different axial locations, are found to be in close agreement with measurements. Both the position and size of leading edge vortices are well predicted. However, there are also differences to be observed between the simulations and measurements/flow visualization. For example, the simulations show secondary flows near the trailing edge region, inboard on the wings. Also, near the wing tips, the general flow structure is substantially different to the measurements.

Hahn and Drikakis (2009) repeat the above case using MILES with a grid density of around half. Comparison is also made with hybrid RANS-LES. In attached boundary layer zones, the hybrid RANS-LES is found to do well. In other zones, the MILES results are superior. Comparisons are made with mean velocity, Reynolds stress profiles and also surface flow visualization and vortex extent data. Broadly encouraging results are found.

Ceresola (2009) report a range of DES results for a delta wing (at TU Munich) at an angle of attack of 35°. Structured grids are used with $N = 5.2 \times 10^6$ and unstructured for $N = 13 \times 10^6$. Comparison is made with surface mean pressure and pressure fluctuation measurements along with velocity fluctuations away from surfaces. Standard DES, an EASM based DES and Menter et al.'s SAS model are tested. These hybrid methods are used in different flow solvers. Mean properties and the general flow structure (vortex location) are generally well captured. Some considerable discrepancies are found for fluctuating properties. Lack of grid isotropy and resolution are identified as a potential error sources.

5.3.5 Full Aircraft Configurations

Forsythe et al. (2004) perform DES of an F15. The angle of attack is 65°, $Re_c = 13.6 \times 10^6$ and $M = 0.3$. The unstructured, commercial COBALT solver with TVD flux limiters is used for $N = 2.85 \times 10^6$–10×10^6. Both the DES and RANS results are found to be within 10 % of integrated forces (lift, drag and pitching moments). The DES C_L and C_D values are within the error bars of the measurements. The accuracy of the DES is clearly better than the RANS. Nonetheless, the RANS does surprisingly well for such a massively separated flow. Also, the cost of DES is found to be seven times higher than RANS.

Morton et al. (2007) perform DES of an F/A–18 fighter configuration. Tail buffet is explored and comparison made with real flight data. The unstructured commercial COBALT flow solver is again used. The simulations are performed for $N = 3.9 \times 10^6$ and $Re = 13 \times 10^6$. Prismatic cells are used in the boundary layers and tetrahedral elsewhere. Agreement is found with measurements of vortex location and surface pressure spectra. However, for the spectra, only frequency range and shape can be compared. No reference pressure is measured for the flight data.

Leicher (2009) reports the performance of a range of DES related methods (DES97, DDES and EASM based DES, SAS etc.) for a FA–5, generic, fighter configuration. The angle of attack is 15°, the $M = 0.125$ and the $Re = 2.8 \times 10^6$. Both structured and unstructured grids are used with $N = 11.4 \times 10^6$ and 13×10^6, respectively. Comparison is made with mean velocity, turbulence fluctuations and axial velocity spectra. As would be expected, the hybrid RANS-LES methods showed superior performance to URANS. However, even with the hybrid RANS-LES methods, strong sensitivity of the turbulence fluctuations to the chosen RANS model is found. The Menter SST SAS model appeared to capture the velocity spectra slightly better than the other approaches.

In a transonic buffet study, Brunet and Deck (2008) make zonal DES of flow over a scaled civil aircraft with an angle of attack of 4.2°. The full DES zone is restricted to the suction surface away from the leading edge—thus reducing grid requirements. The $Re = 2.8 \times 10^6$, $M = 0.82$ and $N = 1 \times 10^7$. The measurement data consists of, surface pressures, their spectra and r.m.s. values. Relative to RANS, the eddy resolving methods show improvements. This is especially so away from the outboard region. Here, the wing deflection is identified as a potential error in the wind tunnel measurements. The need for validation data in the fluid zone is identified.

Gand (2012) uses DDES and zonal DES for a spoiler mounted to a complete airframe. A Chimera code is used with a 2nd order backwards difference temporal scheme. The Chimera grid allows the key LES zone to have an especially refined grid around the spoiler zone. A large part of the airframe is globally treated in RANS mode. Spectra either side of Chimera interfaces are checked for contamination from spurious reflections. Results from these tests are encouraging. The zonal DES uses a volume based filter scale. Comparisons are made with time average surface pressures, downstream total pressure profiles and wing surface pressure spectra. Generally the agreement is encouraging. The zonal DES is richer in scales than the DES. The topology of the separated flow behind the spoiler is considerably improved relative to RANS with the DES approaches.

5.3.6 Base Flows

Supersonic base flows have detached shear layers and a recirculation zone with expansion waves. The latter are located at the start of the separated shear layer. A recompression region follows them. Obviously, the base drag can be a large fraction of the total drag. This type of flow is relevant to missiles and rockets. The base pressure coefficient should be constant. RANS models fail to capture this having a wavy nature. Hence, a substantial number of relatively successful applications of DES and LES for this type of flow have arisen. Fureby et al. (1999) perform MILES along with Smagorinsky and Yoshizawa based LES for $N = 1.3\text{–}2.1 \times 10^6$ with $Re \cong 2.2 \times 10^6$ (based on the diameter). Little impact of the LES modelling is found, the MILES and classical LES being in close accord. Encouraging agreement is found with velocity profile and turbulence measurements. Successful ENO based MILES for this case are reported by Baurle et al. (2003) for $N = 2.2 \times 10^6$. Further successful MILES (AUSM based) with $N = 1.26\text{–}5.84 \times 10^6$, along with Smagorisnky LES and hybrid RANS-LES are performed by Kawai and Fujii (2005). A series of DNS, at lower Reynolds numbers are reported by Sandberg and Fasel (2005, 2006b,a) and Sandberg (2012). These involve high order finite differences with a spectral discretization in the tangential direction.

Forsythe et al. (2002) who made $N \sim 0.3\text{–}2.6 \times 10^6$ (both structured and tetrahedral with near wall prism layers), $Re = 2.86 \times 10^6$ simulations with the unstructured commercial COBALT flow solver. Both Menter SST and SA DES are considered. Comparison is made with measurements of mean velocity profile, base pressure profiles and turbulence kinetic energy levels in the wake. Forsythe et al. note the need to accurately capture the velocity profile coming into the wake.

Simon et al. (2006) make standard DES, zonal DES , hybrid RANS-MILES and LES for the supersonic axisymmetric base flow considered above. Comparison is made with mean velocity and turbulence kinetic energy profiles. The inflow mean velocity profile leading into the separation zone is found to be important. Also, sensitivity to numerical method is observed, both the Roe and AUSM based schemes being considered with $N = 5\text{–}13.5 \times 10^6$. A 2nd order BD temporal scheme is used.

Simon et al. (2007b) extend this work using a 2nd order AUSM based convective scheme for $N = 20.3 \times 10^6$. Complex flow physics with a wide range of time scales is observed. A low frequency component is found in the separation bubble region and some form of excitation feedback loop. Two-point—two-time correlations are used to estimate eddy convection velocities. Encouraging agreement is found with Reynolds stress profiles, the evolution of shear layer thickness and the number of structures in the azimuthal direction. The simulations have a deep level of flow analysis.

Simon et al. (2007a) perform zonal DES of a projectile in the subsonic and transonic flow regimes. The standard RANS modelling is found to be accurate on the attached surface of the body. However, as would be expected, in the base flow zone substantial discrepancies are observed. These are largely rectified through use of the ZDES. Simulations involved $N = 7 \times 10^6$. For the spatial discretization the Roe scheme is used with a BD temporal integration.

Deck and Thorigny (2007) perform zonal DES of a base type flow. This is representative of a space launcher vehicle 1st stage. The simulations are made for $Re = 1.1 \times 10^6$ (based on body diameter) and $N = 8.3 \times 10^6$. A multiblock, structured flow solver is used with a BD temporal scheme. Roe's scheme is used for the convective fluxes. This is extended to 3rd order accuracy through use of the MUSCL approach. Reasonable agreement is found with pressure spectra, for this complex flow. For example, it has multiple shear layers with very different time scales and large-scale coherent structures.

Garbaruk et al. (2009b) report a range of eddy resolving simulations for a supersonic base flow. The Reynolds number is 3.2×10^6 and the Mach number 2.46. Both unstructured and structured grid solutions are considered. These are for a range of numerical schemes and DES variants including the zonal DES approach of Deck. A wide range of grid densities are considered with $N = 0.9\text{–}14 \times 10^6$.

As would be expected, all the eddy resolving approaches did considerably better than RANS. A key issue is the initial development of the Reynolds stresses in the free shear layer. As pointed out in the work of Deck, this is probably due to the inappropriate grid based length scale used in standard DES.

Forsythe et al. (2002) note the need for further validation data. They point out that, although Reynolds stress data is available for such flows, it is for limited Mach number, Reynolds number and geometry ranges.

5.3.7 Landing Gear

Hedges et al. (2002) carry out a DES performance study, primarily focusing on landing gear aerodynamics for a simplified Boeing 757. However, the strong deriver for such predictions is acoustics. Hence, such flows are also considered further in Chap. 6. Hedges et al. perform multiblock simulations with $N = 2.5 \times 10^6$ and $Re = 0.6 \times 10^6$. The NTS code is used with a blend of 5th order upwinding and 4th order CD. The latter is used in wakes zones/away from boundary layers. The pressure field is computed using the Rogers and Kwak artificial compressibility method (see Sect. 2.20.3). The time averaged surface pressure from the DES is shown to be in considerably better agreement with measurements than URANS. However, the DES lift and drag coefficients do not agree well with the measurements. Further application of the NTS code to landing gear is reported in Spalart et al. (2010) for $N = 18 \times 10^6$. This work has a far field sound emphasis and is also discussed in Chap. 6. Lockard et al. (2004) make simplified (but with considerably more geometric detail than considered by Hedges et al.) landing gear DES predictions. For these simulations, the structured NASA CFL3D code is used with $N = 13 \times 10^6$. Soemarwoto et al. (2009) report on 4th order hybrid RANS-LES simulations for the same landing gear as studied by Hedges et al. The hybrid approach, with $N = 3 \times 10^6$, uses a near wall k–ω model. A turbulence kinetic energy based subgrid scale model is used away from walls. Comparison is made with surface pressure measurements. Broadly, encouraging agreement is found. However, the flow on the ground side, at the back of the front wheel, is found to strongly control the flow downstream.

5.3.8 Cavity Flows

An early example of the successful application of DES to a three-dimensional, supersonic, cavity flow is given by Shieh and Morris (2001). These simulations are for $Re = 0.2 \times 10^6$ and $N = 1.7 \times 10^6$. They involve a 4th order DRP spatial discretization. Close agreement is found with measured cavity surface pressures. Hamed et al. (2003) also find encouraging agreement with Menter SST based DES for an N of just 0.77×10^6. In this work, essentially a RANS based flow solver is used.

Rizzetta and Visbal (2003) make MILES type computations with a 4th order compact scheme that includes a 6th order smoother. Cavity doors are included in the simulations. Also, there is mass injection upstream of the shear layer. A precursor, flat plate simulation, is used to produce turbulence inflow. The mass injection is found effective in reducing the acoustic excitation. Encouraging agreement is found the spectra of cavity wall pressures. Thomber and Dfikakis (2008) look at MILES modelling for a deep cavity. Simulations are made for $Re = 86000$ (based on the cavity length) with $0.8 \times 10^6 < N < 3 \times 10^6$. A structured compressible flow solver is used with a 3rd order RK temporal scheme. The spatial discretization is MUSCL based and 5th order, away from discontinuities. Comparisons are made with measured mean velocity profiles, Reynolds stress and surface pressure spectra. Broadly, encouraging agreement is found. However, the maxima in spectra, corresponding to tonal elements are over-predicted. Notably, the Rossiter based frequency estimate is around 30 % in error for this case. Hence, eddy resolving approaches appear to be valuable. There is also close accord with the LES predictions of Larchevêque et al. (2003) for the same case. These make use of a selective mixed scale model with

$N = 2.3 \times 10^6$ (this is the maximum grid density used, a range of densities being considered). For some simulations, Larchevêque et al. (2003) use a two-dimensional grid embedded into a three-dimensional zone. The former is located where the experiments show there are no large-scale turbulent structures. Larcheveque et al. also perform MILES simulations. For the LES, 2nd order CDs are used. Where excessive dispersion is detected dissipation is added. This is achieved using the approach of Mary and Sagaut (2002)—see Sect. 2.16. Interestingly, the turbulence levels in the inflow, boundary layer, did not seem to greatly influence the cavity flows. However, it is found important to have the correct mean flow velocity profile.

Allen et al. (2005) perform a range of DES based computations (SA, k–ω and k–ϵ) for the M219 cavity with and without doors. Trimmed cell Cartesian grids with local shear layer refinements are used ($N = 1.1 \times 10^6$). The grids give large first off-wall y^+ values. However, even so, since the flow has little boundary layer content, there is encouraging agreement with surface pressure spectra measurements. Second order CDs are adaptively blended with a 2nd order UP scheme. The intention being to have the predominately centred scheme active in the LES target zone. The pressure field is computed using the PISO (see Sect. 2.20) scheme in a collocated grid flow solver. Little difference in results is found for the three chosen near wall RANS closures. This, again, is probably because of the low level of classical boundary layer content.

Peng (2009) report on the performances of a range of DES methodologies for the M219 cavity. These simulations use several flow solvers, with typically $1 \times 10^6 < N < 6 \times 10^6$ and $Re = 13.5 \times 10^6$. The studies encompass both structured and unstructured solvers, with DES, DDES, SST SAS and a range of other hybrid RANS-LES variants. Comparisons are made with cavity surface pressures including their spectra, mean velocity and Reynolds stress profiles. For the finer grids, broadly the agreement with the measurements is encouraging. This is perhaps not surprising. The flow is basically dominated by large scale detached eddies. This work is extended by Lawson et al. (2009) for the same cavity but with idealized doors. Both SA DES and LES are performed for $N = 4.5 \times 10^6$ and 8.5×10^6, respectively. Again agreement with measurements is broadly encouraging.

Jefferson-Loveday et al. (2012) model a cavity flow using NLES. For the grid, $N = 7 \times 10^6$ and $Re = 50{,}000$ (based on the cavity height). Encouraging agreement is found with mean velocity and Reynolds stress profiles. The numerical scheme used is 6th order with a 7th order smoother. The level of smoothing is controlled, based on the dispersion levels, using the approach of Mary and Sagaut (2002).

A DES of a nose cone cavity, with landing gear, is performed by Langtry and Spalart (2009). Since this has more of an acoustic emphasis, this case is discussed in Chap. 5.

5.3.9 Miscellaneous Flows and Zones

Deck (2009a) makes DDES computations for a rocket nozzle with $N \cong 11 \times 10^6$ and $Re = 21 \times 10^6$ based on throat diameter. At ground conditions, the ambient pressure is higher than the nozzle exit pressure (over expanded flow). This can give rise to nozzle flow separation, which can have a strong impact on the nozzle unsteady loading levels. It can result in nozzle damage. The ONERA, multiblock, FLU3M code is used. This again uses Roe's flux difference splitting scheme. It is extended to 3rd order using the MUSCL approach. A 2nd order BD time scheme is used. Wall pressure fluctuation spectra are in close accord with measurements. Pressure fluctuation amplitudes reach 20 % of the ambient pressure.

Le Chuiton et al. (2009) report on a range of hybrid RANS-LES type computations for an EC145 helicopter fuselage. The Reynolds number is 2.48×10^6 and $M = 0.12$. Several structured and unstructured flow solvers are considered for $N = 2.2$–14.3×10^6. The finer grid is used for DES and the coarser for a Menter SST based SAS computation.

Hadjadj (2012) perform dynamic Smagorinsky based LES for a zero pressure gradient boundary layer with oblique shock impingement. A complex, low frequency (relative to the boundary layer time scales ($O(\delta/U_\infty)$), unsteady shock movement arises. This takes place in the region of the reflected shock with its local separation bubble. This suggests a coupled dynamic shock-bubble system. The simulations are for $Re = 5350$ (based on momentum thickness), a free stream Mach number of over 2 and $N = 8.5 \times 10^6$. Span-wise domain extents of one and five times the incoming boundary layer thickness are considered. Both velocity and pressure are found to correctly spanwise de-correlate. The WENO scheme is used at the shock locality with a 4th order CD scheme used away from shocks. In the shock vicinity, the Ducros switch (see Chap. 2) is used to activate the WENO scheme. The temporal integration involves a 3 stage explicit RK scheme. The turbulent inflow is produced through digital filtering (see Chap. 3). The LES results are in agreement with measurements, displaying an extreme level of flow complexity.

Bisek et al. (2013) use MILES to explore the use of plasma flow control to reduce separation induced by a shock boundary layer interaction. A high order compact scheme is used with $N = 1.3$–95×10^6, For inflow, a body force based trip, with an extended upstream domain, is used to develop a turbulent boundary layer. The shock location is identified using the Swanson and Turkel (1992) pressure gradient detector. In shock zones, the compact scheme is switched to a flux limited Roe scheme. No direct comparison is made with measurements. However, the plasma flow control is found to be effective in reducing the extent of separation.

Sreedhar and Ragab (1994) perform 4th order in space and 2nd order in time LES for the development of a vortex. Such simulations are important for understanding of wing trailing edge vortices. The dynamic Smagorinsky model is used. The vortex is temporally evolved, based on various initial disturbances. Elongated vortices aligned with the vortex core direction of rotation are found. These are reminiscent of those found in Taylor-Coutte flow. Reasonable agreement is found with the measured radial distributions of circulation. Duraisamy and Lele (2008) perform pseudo-spectral DNS. The governing equations are solved in the velocity-vorticity form. Solutions elucidate the complex, multi-component flow physics and the Reynolds stress generation process. Uzun et al. (2006) make ILES type computations (with an explicit high order smoother) of a NACA0012 wing tip vortex. The spatial discretization involves a 4th order centred compact scheme applied to overset grids. The ILES is run at a much lower Reynolds number than the experiments. However, qualitative agreement with the experiments is found.

You et al. (2006a) perform dynamic Smagorisnky LES, with Lagrangian averaging, of vortex generators (wall mounted half-delta wings) on a flat plate. An immersed boundary method is used to represent the vortex generators. Lund's recycling generates a turbulent inflow. The CN scheme is used with 2nd order CDs. Encouraging agreement is found with mean velocity and Reynolds stress measurements for two vortex generator configurations. It is proposed that the simulations can be used to prescribe endwall region inflow data, for turbomachinery blade simulations downstream of a stator tip. Evidently the vortex generators will replicate the intense turbulent fluctuations in the near wall region generated by the stator tip.

Table 5.9 gives samples of typical aerodynamic validation data and validation grades. As noted by Zhou and Wygnanski (1993), wall jets can be found on airplane wings like the F–104, Starfighter and A–6 Crusader and on the tail boom of the NOTAR helicopter. Hence wall jet data is included in Table 5.9. Some of the data sources in Table 5.3, on cut-back trailing edges, is relevant to wall jets in general and filling in wakes of aerofoils for low drag and noise.

5.4 General Discussion

5.4.1 Summary of Validation Data Sets and Levels

Table 5.10 summarizes the validation level in the propulsion and airframe zones. The tabulated data is just a distillation of that in earlier tables. The key point is that for airframes, there is more availability of higher-level data and hence more in depth LES validation. The lower level, for propulsion validation data, perhaps partly relates to the fact that measurements in rotating machinery are more challenging. Measurement systems can be subjected to large centrifugal forces. Also, slip rings or telemetry can be needed to monitor instrumentation and phase locking needed for optical measurements etc. Also, downstream of the combustor the environment is more hostile.

Notably, for the low-pressure turbine, there is generally detailed data; understanding the particularly complex flow physics in this zone yielding potentially high gas turbine performance benefits. However, there is considerably less data for both the compressor and other turbine zones. Data is lacking at blade extremities (end wall and rim seal localities). What is more, much of the data sets, for commercial reasons, do not define geometry and other potentially key parameters. This restricts the practical extent of their use to a very limited number of stakeholders. The crucial need for validation data, in combustion modelling, is discussed by Swaminathan and Bray (2011). The lack of basic validation data for combustion models is also noted by Pitsch (2006). Forsythe et al. (2002) identify the need for extended validation data for base flows.

Table 5.9 Sample of typical available aerodynamic validation data and validation grades

Airframe zone	Nature of data	Data source	Comments	Potential validation level
Simplified landing gear	PIV with turbulence statistics, surface flow visualization	Lazos (2002b,a, 2004)	Flow transitional and hence challenging to model and also physically unrealistic	1, 2, 3
Rudimentary landing gear	Surface flow visualization, force coefficients, unsteady surface pressures	Karthikeyan and Venkatakrishnan (2011), Venkatakrishnan et al. (2011)	Has sharp edged geometrical features, larger scale geometry and trips to ensure turbulent flow	1, 2, 3
Basic aerofoils	NACA0012: surface pressures, surface normal profiles of mean velocity and LDA turbulence kinetic energy	Revell et al. (2011)		1, 2, 3, 4
	Also, surface pressures, mean velocity and Reynolds stress profiles, surface pressure spectra and correlations between this and wake turbulence	Garcia-Sagrado (2007), Garcia-Sagrado and Hynes (2011)	Garcia-Sagrado work focuses on acoustic sources	
	NACA4412: extensive flying hot-wire measurements for surface and wake profiles	Coles and Wadcock (1979)		
	A–Airoil ONERA tunnel measurements—extensive LDA measurements	Mary and Sagaut (2002)	Substantial laminar flow content and separated flow transition	
Trailing edge flow	NACA0021—surface pressure and lift coefficient spectra (60° angle of attack)	Swalwell et al. (2003)	–	2, 3
	Mean velocity and Reynolds stress profiles, surface pressure distributions	Blake and Gershfeld (1989), Blake (1975)		

Table 5.9 (Continued)

Airframe zone	Nature of data	Data source	Comments	Potential validation level
Wingtip vortices	NACA0012: Hot-wire, surface flow visualization, pressure probes	Dacles-Mariani et al. (1995), Devenport et al. (1996), Chow et al. (1997a,b)	$Re_c = 0.53\text{–}4.6 \times 10^6$	1, 2, 3, 4
	Lifting hydrofoil: LDA	Baker et al. (1974)	$Re_c = 2.2\text{–}0.75 \times 10^4$ Explores unsteady vortex wandering	2
	NACA66–209	Green and Acosta (1991)	$Re_c = 0.3\text{–}1.3 \times 10^6$ Explores unsteady vortex wandering	2
Vortex generators	Flat plate with two vortex generator configurations (different orientations of half delta wing). LDA and hot-wire measurements of mean velocity and Reynolds stress profiles.	Several data sources reported in You et al. (2006a) including Pauley and Eaton (1988)	–	2, 3
Wing-body junction	NACA0020: LDA—flow physics analysed in detail	Devenport and Simpson (1990)	$Re_c = 0.11 \times 10^6$	2, 3, 4
	NASA TN–712 junction—time averaged surface pressures	Reported in Fu et al. (2007)	$Re_c = 7.5 \times 10^6$	2
Slat aerodynamics	PIV and LDA, surface flow visualization and pressures	Takeda et al. (2001), Paschal et al. (2000)	Most data qualitative/flow visualization Several other data sources given in Khorrami et al. (2002)	1, 2
Wing, slat flap. Based on an RA16SC1 profile	Surface pressures, PIV, includes turbulence fluctuations	Reported in Deck (2005a)	Low Speed Wind Tunnel of Airbus Bremen. Part of the EUROPIV2 Project G4RD–CT–2000–00190	1, 2, 3

Table 5.9 (Continued)

Airframe zone	Nature of data	Data source	Comments	Potential validation level
Wing, slat flap. Based on an NACA 63_2–215 profile.	Surface pressure and flow visualization	Reported in Cummings et al. (2004)	NASA Ames 7 × 10 foot wind tunnel—three-dimensional geometry but no wing sweep	1, 2
Deflected spoiler	Wake total pressures, mean surface pressure distributions and their spectra	Reported in Gand (2012)	Complete aircraft geometry included	2, 4
Shock buffet	LDA, Mean velocity profiles, surface pressures, Reynolds stresses, surface pressure spectra, skewness factor	Jacquin et al. (2005)	Comprehensive data set	1, 2, 3, 4
Shock buffet—half wing model	Surface pressures, their spectra and rms values	Reported in Brunet and Deck (2008)—Measurements by Depre	Wing-body CAT3D model studied in ONERA S2MA tunnel. Wing profile mostly OAT15A	1, 2, 3
Deep dynamic stall of plunging aerofoil (SD7003 section)	PIV from two different water tunnel facilities (University of Michigan and US Air force Research Laboratory), Mean velocities and turbulent fluctuations	Kang et al. (2009), Baik et al. (2009), Ol et al. (2009)	Significant discrepancies between two measurement sets are found at later stages of flow development Low Reynolds numbers—$Re_c = 1 \times 10^4$–6×10^4	1, 2, 3
Pitching and plunging flat plate	PIV from water tunnel facility (University of Michigan)	Baik et al. (2009)	Low Reynolds numbers—$Re_c = 1 \times 10^4$–6×10^4. However, flow found to be *Re* independent	1, 2, 3
F/A–18 Tail buffet	Vortex breakdown location, surface pressure spectra—Kulites used.	Flight report data given in Morton et al. (2007)	Real flight data from HARV—High Angle of Attack Research Vehicle. Unfortunately, the reference pressure is not know and hence amplitudes of pressure traces	1, 2, 3

Table 5.9 (Continued)

Airframe zone	Nature of data	Data source	Comments	Potential validation level
Wing in down wash of rotor (V22 type configu-ration)	Drag, surface pressure measurements, flow visualization	Kjellgren et al. (2000)	Evidence that aerofoil flow is laminar (Spalart et al. 2003)	1, 2
Wall jets	Hot-wire and LDA data, flat and curved surfaces See also entries in Table 5.6	Bradshaw and Gee (1960), Eriksson et al. (1998), Tachie et al. (2004)	Wall jets are often associated with the need for high lift and increased recirculation (Zhou and Wygnanski 1993)	1, 2, 3
Highly swept, high incidence wing	Surface flow visualization and mean velocity profiles. LDA and PIV based data	Manchester University data reported in Leschziner et al. (2009), Hahn and Drikakis (2009)	Data not published at the time of writting	1, 2
Sharp leading edge delta wing	Surface pressure measurements, surface pressure fluctuations, hot-wire measurements of velocity and velocity fluctuations	Breitsamter (1997)	Tests performed for the following angles of attack: 12.5°, 25°, 30°, and 35°	1, 2, 3
FA–5 generic fighter	Mean velocity, Reynolds stresses and velocity spectra	Breitsamter (1997)	–	1, 2, 3, 4
Wing-slat-flap	Surface pressures, PIV of mean velocities and fluctuating velocities	Arnott et al. (2003)	Low speed wind tunnel, Airbus, Bremen	1, 2, 3
Supersonic base flow	Schlieren, shadowgraphs and Mie scattering, surface pressures, velocity and Reynolds stresses—LDA based	Axisymmetric blunt base—Bourdon and Dutton (1999, 2000), Cannon et al. (2005), Herrin and Dutton (1994a, 1995), Janssen and Dutton (2004)	Blow down tunnel—University of Illinois at Urbana-Champaign	1, 2, 3
		Boat tailed base—Herrin and Dutton (1994b, 1995, 1997), Bourdon and Dutton (2001)		

Table 5.9 (Continued)

Airframe zone	Nature of data	Data source	Comments	Potential validation level
		Planar case—Smith and Dutton (1996), Messersmith and Dutton (1996), Smith and Dutton (1999, 2001)		
EC145 helicopter fuselage	Force and moment coefficients, surface pressures and surface flow visualization	Le Chuiton et al. (2009)	Support strut needs to be accounted for and Mach number is low.	1, 2
Deep cavity	Schlieren and two-component LDA, surface pressure spectra via Kulite sensor	Forestier et al. (2003)	$M = 0.8$, phased average data and deep level of flow physics analysis	2, 3, 4
Cavity	Mean velocity profiles, Reynolds stress, spectra from hot-wire and microphone measurements in cavity/mainstream shear layer	Grace et al. (2004)	–	2, 3, 4
M219 cavity flow with and without doors	Mean velocity profiles and Reynolds stresses, cavity surface pressure and their spectra	de Henshaw (2002)—report by Peng (2009)	–	2, 3, 4
Cavity flow control	Surface pressures	Shaw (1998)	Leading edge flap, pulsed fluidic actuator, tone generator	4

With regards to validation data, the need to accurately measure inflow conditions is often critical. Frequently data sets do not even give mean inflow quantities.

LES and Hybrids

As with the field of RANS modelling, LES best practices are needed. However, the structure of these is a matter of debate. They could be general, broken down to consider generic building block flow types or be customized to, for example,

Table 5.10 Summary of validation levels in the propulsion and airframe zones for aerodynamics

Level	Available validation data		LES and hybrids validation	
	Propulsion (%)	Airframe (%)	Propulsion (%)	Airframe (%)
1	62	81	11	59
2	100	100	75	71
3	17	81	14	46
4	7	28	0	24
4+	1	0	0	0

particular propulsion or airframe components. These guidance types are already available for different fields of numerical modelling but not LES.

Few of the simulations discussed use resolved inflow turbulence. However, for the flows governed by intense shear layers this is not an issue. Evidence suggests that capturing the mean inflow profile is important. As noted in Chap. 3, many of the popular inflow modelling techniques, in their raw form, can have limited relevance to many practical configurations.

Figure 5.27 broadly summarizes the popularity of different eddy resolving strategies for propulsion. The vertical axis is frequency of use. Akin to RANS, even though there are a wide range of LES model variants, most simulations make use of either the Smagorinsky model in its standard or dynamic forms. For practical flows, the former is most popular. Limitations of the Smagorinsky model are listed in Chap. 3.

For the work reviewed on airframe aerodynamics, Fig. 5.28 summarizes the frequency of use of different eddy resolving strategies. As would be expected, for the higher Reynolds numbers found in airframes, the use of hybrid approaches is much more prevalent.

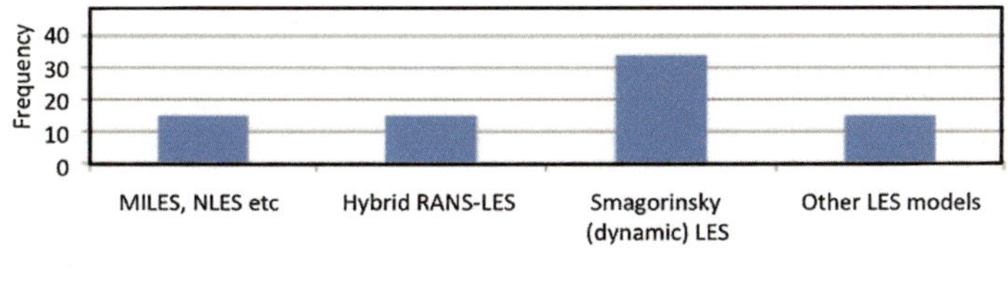

Fig. 5.27 Frequency of use of different modelling strategies in turbomachinery

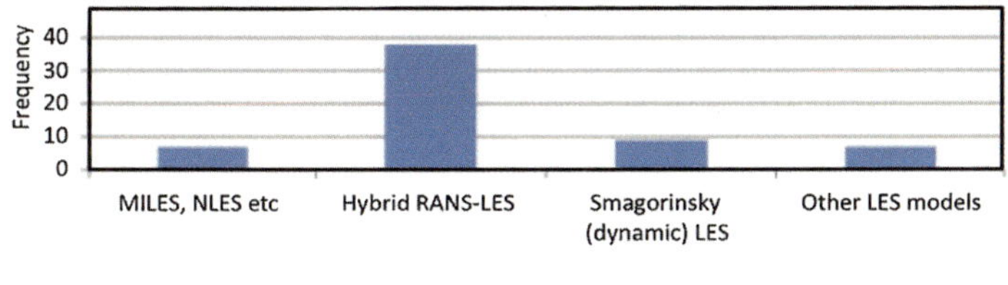

Fig. 5.28 Frequency of use of different modelling strategies for airframes

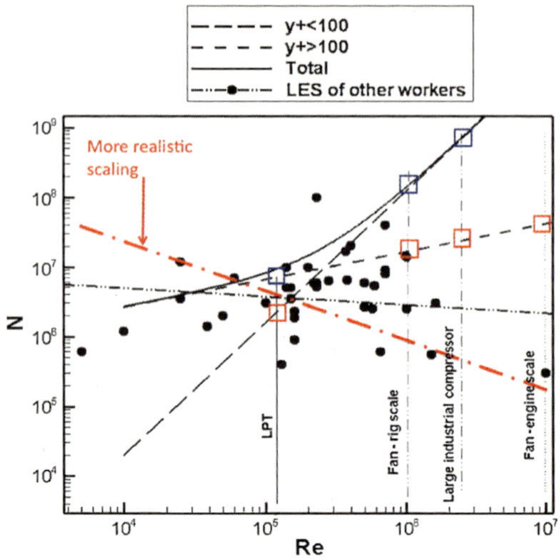

Fig. 5.29 Grid requirements for LES and hybrid RANS-LES combined with typical grid densities actually used in turbomachinery LES studies (Tucker 2011b). Published with kind permission from Elsevier

Figure 5.29 takes values of Re_c and N from the different propulsion LES, discussed above. These Re_c and N pairs are represented by the full symbols. The dashed-dot-dot line is the best fit to this data. The striking thing from this best fit is that its gradient is nearly -0.2. This is inconsistent with the high positive power for Re_c that we expect for correctly performed LES. This is represented by the full line. This line, in turn, being the sum of the long and short dashed lines, representing the inner and outer layer grid requirements for LES. Notably, most of the LES at higher Re_c involve complex geometries. Hence, they include end wall and tip effects and even multiple blade rows. The symbols relating to the lower *Re* tend to involve just blade sections. Hence, the actual lack of resolution with increasing *Re* is probably worse than suggested in the figure. It is more like that represented by the chain dashed line (Tucker 2011a). A key aspect is perhaps that the higher *Re* calculations tend to be linked to computational studies having a greater practical emphasis. Hence, there may well be tighter project time frames associated with such simulations. However, the lack of *Re* scaling does suggest that the use of hybrid approaches in turbomachinery warrants serious consideration. LES computations seem, for whatever reason, not to assimilate the need for grids scaling with *Re* to a high power. Figure 5.30 is the equivalent to 5.29 but this time for airframes. The sample size for LES is much smaller. There are two plausible reasons for this. These being that the review in this area is less extensive and also that the Reynolds numbers are much higher. Hence, the use of hybrid RANS-LES is much more compelling in airframes. The small sample, unlike for turbomachinery, correctly shows a scaling of grid count that has Reynolds number to a positive power.

The benefits of LES, versus computational costs are a key question. Clearly for flows where transition processes are important there are improvements in predictive accuracy. Nonetheless, the transition modelling capability of LES models is an area of open debate. As noted in Chap. 3, various averaging procedures are used for

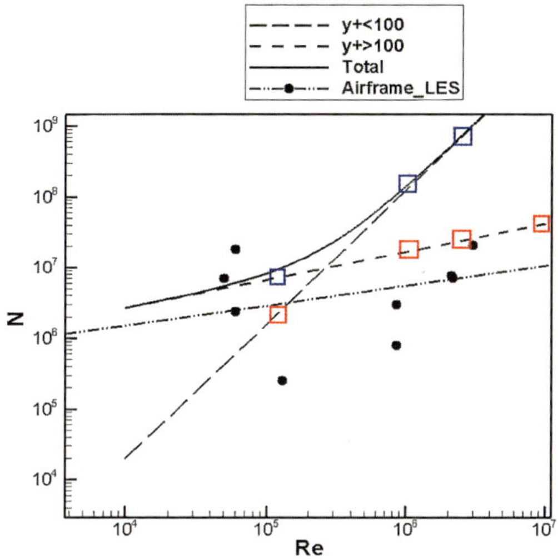

Fig. 5.30 Grid requirements for LES and hybrid RANS-LES combined with typical grid densities actually used in various airframe LES studies

the coefficient in the dynamic Smagorinsky model. The procedure chosen will certainly alter the transition trajectory. Hence, for LES involving transition the use of grids tending to quasi-DNS seems sensible. For regions where the flow is subjected to large unsteady disturbances, such as blades downstream of combustors, cutback trailing edges and ribbed roughened passages (inside turbine blades) there are clear LES accuracy advantages. Also, where a broadband turbulent acoustic source description is needed there are obvious benefits (see Chap. 6). However, for geometries that have attached, turbulent, boundary layers with relatively standard flow physics, that will be in the range of calibration of most RANS models, the potentially orders of magnitude extra LES cost is hard to justify. For ribbed passage, jets at high *Re* and cutback trailing edge flows, large-scale unsteady, Reynolds number independent, structures are generated. These are hard for RANS to rationally capture. However, these large-scale structures, where much turbulence energy is concentrated, are relatively easy to resolve on modest grids. Hence, LES becomes especially attractive for such flows. The cut back trailing edge (for sufficiently high blowing ratios) and ribbed passage flows can be viewed as top down flows. By this it is meant that large-scale structures are generated outside the boundary layer. These then strongly interact with the boundary layer, potentially overwhelming the finer structures that extend to the buffer layer. It is these latter structures that give rise to the extreme cost of LES.

5.5 Conclusions and Closing Remarks

LES and, to an extent, hybrid methods alleviate the serious area of modelling uncertainty associated with RANS simulations. However, this comes at an extreme increase in computational cost. There seems a lack of detailed validation data to explore in depth the performance of LES and thus refine it. This is especially so for turbomachinery. LES best practices are needed. Although LES is much less model dependent than RANS, grids currently used in propulsion are clearly insufficiently fine for the LES model (and numerical schemes) not to be playing an excessively strong role. The accurate convection of moving wakes also presents challenges. Most of the practical LES studies reviewed, essentially made use of RANS solvers where robustness features more greatly than accuracy.

Few simulations make use of properly constructed, correlated turbulence inflow. Even if this is attempted, most measurement sets are incomplete and lack an adequate basis for defining this inflow. The use of idealized synthetic and basic outflow boundary conditions seems excessively simplistic in such a complex coupled system as a gas turbine engine.

Despite the limitations of the excessively dissipative Smagorinsky model, it still sees wide spread use, especially with dissipative flow solvers. As with RANS, the modelling of transition is still a key issue, especially where LES grids tend to be too coarse. For propulsion systems, MILES related approaches, hybrid RANS-LES and more advanced LES models seem to have an equal frequency of use, none of these approaches being especially avoided. At the higher Reynolds numbers found in airframes hybrid RANS-LES methods dominate. Clearly the introduction of a RANS layer can have a substantial accuracy penalty, but does allow LES to be rationally used albeit in a diluted sense. Most airframe related LES studies are for more canonical flows.

Despite the excessively coarse grids used, in many of the propulsion system simulations reviewed, meaningful results are often gained. This can perhaps be attributed to the choice of cases, these being ones for which RANS modelling gives poor performance. Certain flows are highly amenable to LES and are forgiving with regards to modelling defects. Others need substantial expertise and computational resource to even gain results of comparable quality to RANS. Flows where wall modelling is not critical, such as those in combustors, cutback trailing edges, wake flow related systems (base flows, cavities, ribbed passages) and free shear flows in jets are notable areas of LES success. Transitional flows present a stronger LES challenge.

References

S. Acharya, S. Dutta, T.A. Myrum, R.S. Baker, Periodically developed flow and heat transfer in a ribbed duct. Int. J. Heat Mass Transf. **36**(8), 2069–2082 (1993)

D. Ahlman, G. Brethouwer, A.V. Johansson, Direct numerical simulation of a plane turbulent wall-jet including scalar mixing. Phys. Fluids **19**(6), 065102 (2007)

R. Allen, F. Mendonca, D. Kirkham, RANS and DES turbulence model predictions of noise on the M219 cavity at $M = 0.85$. Int. J. Aeroacoust. **4**(1), 135–152 (2005)

H.I. Andersson, M. Lygren, LES of open rotor-stator flow. Int. J. Heat Fluid Flow **27**(4), 551–557 (2006)

A. Arnott, G. Schneider, K.P. Neitzke, J. Agocs, A. Schröder, B. Sammler, J. Kompenhans, Detailed characterisation, using piv, of flow around an airfoil in high-lift configuration, in *Particle Image Velocimetry: Recent Improvements Proceedings of the EUROPIV*, vol. 2 (2003)

L.U. Axelsson, W.K. George, Spectral analysis of the flow in an intermediate turbine duct, in *Proc. of ASME Turbo Expo 2008*, Berlin, Germany, 9–13 June 2008. ASME Paper Number GT2008–51340

Y.S. Baik, J.M. Rausch, L.P. Bernal, M.V. Ol, Experimental investigation of pitching and plunging airfoils at Reynolds. number between 1×10^4 and 6×10^4. AIAA Paper Number AIAA–2009–4030 (2009)

G.R. Baker, S.J. Barker, K.K. Bofah, P.G. Saffman, Laser anemometer measurements of trailing vortices in water. J. Fluid Mech. **65**(2), 325–336 (1974)

W. Balzer, H.F. Fasel, Direct numerical simulations of laminar separation bubbles on a curved plate: Part 1—simulation setup and uncontrolled flow, in *Proc. of ASME Turbo Expo*, San Antonio, Texas, USA, 3–7 June 2013a. ASME Paper Number GT2013–95277

W. Balzer, H.F. Fasel, Direct numerical simulations of laminar separation bubbles on a curved plate: Part 2—flow control using pulsed vortex generator jets, in *Proc. of ASME Turbo Expo*, San Antonio, Texas, USA, 3–7 June 2013b. ASME Paper Number GT2013–95278

R.S. Barlow, J.H. Frank, Effects of turbulence on species mass fractions in methane/air jet flames, in *Symposium (International) on Combustion*, vol. 27 (1998), pp. 1087–1095

J.W. Baughn, X.J. Yan, M. Mesbah, The effect of Reynolds number on the heat transfer distribution from a flat plate to a turbulent impinging jet, in *Proc. of ASME Winter Annual Meeting 1992*, ASME HTD-vol. 226 (1992), pp. 1–7

R.A. Baurle, C.J. Tam, J.R. Edwards, H.A. Hassan, Hybrid simulation approach for cavity flows: blending, algorithm, and boundary treatment issues. AIAA J. **41**(8), 1463–1480 (2003)

N.J. Bisek, D.P. Rizzetta, J. Poggie, Plasma control of a turbulent shock boundary-layer, interaction. AIAA J. (2013). doi:10.2514/1.J052248

W.K. Blake, A statistical description of pressure and velocity fields at the trailing edges of a flat strut. Technical Report, David Taylor Naval Ship Research and Development Center, Rept. 4241, Bethesda, MD (1975)

W.K. Blake, J.L. Gershfeld, The aeroacoustics of trailing edges. Front. Exp. Fluid. Mech. **1**, 457–532 (1989)

D.J. Bodony, S.K. Lele, Review of the current status of jet noise predictions using large-eddy simulation (invited), in *44th AIAA Aerospace Sciences Meeting and Exhibit*, Reno, Nevada, USA (2006). AIAA Paper Number AIAA–2006–0468

D.J. Bodony, S.K. Leley, Jet noise prediction of cold and hot subsonic jets using large-eddy simulation, in *10th AIAA/CEAS Aeroacoustics Conference* (2004). AIAA Paper Number AIAA–2004–3022

D.E. Bohn, A. Decker, H. Ma, M. Wolff, Influence of sealing air mass flow on the velocity distribution in and inside the rim seal of the upstream cavity of a 1.5-stage turbine, in *Proc. of ASME Turbo Expo 2003*, Atlanta, Georgia, USA, 16–19 June 2003. ASME Paper Number GT2003–38459

D.E. Bohn, A. Decker, N. Ohlendorf, R. Jakoby, Influence of an axial and radial rim seal geometry on hot gas ingestion into the upstream cavity of a 1.5-stage turbine, in *Proc. of ASME Turbo Expo 2006*, Barcelona, Spain, 8–11 May 2006. ASME Paper Number GT2006–90454

G. Boudier, L.Y.M. Gicquel, T. Poinsot, D. Bissieres, C. Bérat, Comparison of LES, RANS and experiments in an aeronautical gas turbine combustion chamber. Proc. Combust. Inst. **31**(2), 3075–3082 (2007)

C.J. Bourdon, J.C. Dutton, Planar visualizations of large-scale turbulent structures in axisymmetric supersonic separated flows. Phys. Fluids **11**, 201–213 (1999)

C.J. Bourdon, J.C. Dutton, Shear layer flapping and interface convolution in a separated supersonic flow. AIAA J. **38**(10), 1907–1915 (2000)

C.J. Bourdon, J.C. Dutton, Effects of boattailing on the turbulence structure of a compressible base flow. J. Spacecr. Rockets **38**, 534–541 (2001)

P. Bradshaw, M.T. Gee, Turbulent wall jets with and without an external stream. Aeronaut. Res. Counc. R & M 3252 (1960)

C. Breitsamter, Turbulente Strömungsstrukturen an Flugzeugkonfigurationen mit Vorderkantenwirbeln. PhD thesis, Technical University, Munich, January 1997

J. Bridges, C.A. Brown, Parametric testing of chevrons on single flow hot jets, in *10th AIAA/CEAS Aeroacoustics Conference*, Manchester, UK, May 2004. AIAA Paper Number AIAA–2004–2824

J. Bridges, C.A. Brown, Validation of the small hot jet acoustic rig for aeroacoustic research, in *11th AIAA Aeroacoustics Conference*, Monterey, California, 23–25 May 2005. AIAA Paper Number AIAA–2005–0363

J. Bridges, M.P. Wernet, Measurements of the aeroacoustic sound source in hot jets. Technical Report NASA–TM–221508, NASA Glenn Research Centre (2004)

J. Bridges, M.P. Wernet, Effect of temperature on jet velocity spectra, in *13th AIAA/CEAS Aeroacoustics Conference (28th AIAA Aeroacoustics Conference)*, Rome, Italy, 21–23 May 2007. AIAA Paper Number AIAA–2007–3628

K.M. Britchford, A.P. Manners, J.J. McGuirk, S.J. Stevens, Measurement and prediction of low in annular s-shaped ducts. Exp. Therm. Fluid Sci. **9**(2), 197–205 (1994)

V. Brunet, S. Deck, Zonal-detached eddy simulation of transonic buffet on a civil aircraft type configuration. AIAA Paper Number AIAA–2008–4152 (2008)

R.K. Byskov, Large eddy simulation of flow structures in a centrifugal pump impeller. PhD thesis, Aalborg University, Denmark (2000)

P.M. Cannon, G.S. Elliott, J.C. Dutton, Time-series axisymmetric base-pressure measurements with simultaneous near-wake visualizations. AIAA Paper Number AIAA–2005–5285 (2005)

N. Ceresola, TU Munich, delta wing, in *DESider—A European Effort on Hybrid RANS-LES Modelling*, ed. by W. Haase, M. Braza, A. Revell. Notes on Numerical Fluid Mechanics and Multidisciplinary Design, vol. 103 (Springer, Berlin, 2009)

N. Chauvet, S. Deck, L. Jacquin, Zonal detached eddy simulation of a controlled propulsive jet. AIAA J. **45**(10), 2458–2473 (2007)

J.S. Chow, G.G. Zilliac, P. Bradshaw, Mean and turbulence measurements in the near field of a wingtip vortex. AIAA J. **35**(10), 1561–1567 (1997a)

J. Chow, G. Zilliac, P. Bradshaw, Turbulence measurements in the near field of a wingtip vortex. Technical Report NASA–TM–110418, NASA (1997b)

D. Coles, A.J. Wadcock, Flying-hot-wire study of flow past an naca 4412 airfoil at maximum lift. AIAA J. **17**(4), 321–329 (1979)

Y. Colin, B. Aupoix, J. Boussuge, P. Chanez, Numerical simulation of the distortion generated by crosswind inlet flows, in *Proceedings of the 18th ISABE Conference*, Beijing, China (2007). Paper Number ISABE–2007–1210

S. Connell, M. Braaten, L. Zori, R. Steed, B. Hutchinson, G. Cox, A comparison of advanced numerical techniques to model transient flow in turbomachinery blade rows, in *Proc. of ASME Turbo Expo 2011*, Vancouver, BC, Canada, 6–11 June 2011. ASME Paper Number GT2011–45820

S. Connell, B. Hutchinson, P. Galpin, R. Campregher, P. Godin, The efficient computation of transient flow in turbine blade rows using transformation methods, in *Proc. of ASME Turbo Expo*, Copenhagen, Denmark, 11–15 June 2012. ASME Paper Number GT2012–69019

D. Cooper, D.C. Jackson, B.E. Launder, G.X. Liao, Impinging jet studies for turbulence model assessment—I. Flow-field experiments. Int. J. Heat Mass Transf. **36**(10), 2675–2684 (1993)

J.D. Coull, Wake induced transition in low pressure turbines. PhD thesis, Engineering Department, Cambridge University (2009)

R.M. Cummings, S.A. Morton, J.R. Forsythe, Detached-Eddy simulation of slat and flap aerodynamics for a high-lift wing, in *42nd AIAA Aerospace Sciences Meeting and Exhibit*, Reno, Nevada, USA, 5–8 January 2004. AIAA Paper Number AIAA–2004–1233

J. Dacles-Mariani, G.G. Zilliac, J.S. Chow, P. Bradshaw, Numerical/experimental study of a wingtip vortex in the near field. AIAA J. **33**(9) (1995)

S. Dahlström, L. Davidson, Large eddy simulation of the flow around an airfoil. AIAA Paper Number AIAA–2001–0425 (2001)

J.W. Daily, R.E. Nece, Chamber dimension effects on induced flow and frictional resistance of enclosed rotating disks. J. Basic Eng. **82**, 217–232 (1960)

B.B. Dally, A.R. Masri, R.S. Barlow, G.J. Fiechtner, Instantaneous and mean compositional structure of bluff-body stabilized nonpremixed flames. Combust. Flame **114**(1–2), 119–148 (1998)

M.J.C. de Henshaw, M2129 cavity case: verification and validation data for computational unsteady aerodynamics. Technical Report RTO–TR–26, AC/323(AVT)TP/19, QinetiQ, UK, pp. 453–472 (2002)

C.C. De Wiart, K. Hillewaert, P. Geuzaine, DNS of a low pressure turbine blade computed with the discontinuous Galerkin method, in *Proc. of ASME Turbo Expo*, Copenhagen, Denmark, 11–15 June 2012

J.R. DeBonis, Progress toward large-eddy simulations for prediction of realistic nozzle systems, in *44th AIAA Aerospace Sciences Meeting and Exhibit* (2006). AIAA Paper Number AIAA–2006–487

J.R. Debonis, J.N. Scot, Large-eddy simulation of a turbulent compressible round jet. AIAA J. **40**(7), 1346–1354 (2002)

S. Deck, Numerical simulation of transonic buffet over a supercritical airfoil. AIAA J. **43**(7), 1556–1566 (2005a)

S. Deck, Zonal-Detached-Eddy simulation of the flow around a high-lift configuration. AIAA J. **43**(11), 2372–2384 (2005b)

S. Deck, Delayed detached eddy simulation of the end-effect regime and side-loads in an overexpanded nozzle flow. Shock Waves **19**(3), 239–249 (2009a)

S. Deck, TU Munich, three element airfoil, in *DESider—A European Effort on Hybrid RANS-LES Modelling*, ed. by W. Haase, M. Braza, A. Revell. Notes on Numerical Fluid Mechanics and Multidisciplinary Design, vol. 103 (Springer, Berlin, 2009b)

S. Deck, P. Thorigny, Unsteadiness of an axisymmetric separating-reattaching flow: numerical investigation. Phys. Fluids **19**, 65–103 (2007)

J. Denecke, K. Dullenkopf, S. Wittig, H.J. Bauer, Experimental investigation of the total temperature increase and swirl development in rotating labyrinth seals, in *Proc. of ASME Turbo Expo 2005* (2005). ASME Paper Number GT2005–47268

W. Devenport, R. Simpson, Time-dependent and time-averaged turbulence structure near the nose of a wing-body junction. J. Fluid Mech. **210**(1), 23–55 (1990)

W.J. Devenport, M.C. Rife, S.I. Liapis, G.J. Follin, The structure and development of a wing-tip vortex. J. Fluid Mech. **312**, 67–106 (1996)

G. Dufour, N. Gourdain, F. Duchaine, O. Vermorel, L.Y.M. Gicquel, J.F. Boussuge, T. Poinsot, *Large Eddy Simulation Applications*. VKI Lecture Series Numerical Investigations in Turbomachinery: the State of the Art (2009)

K. Duraisamy, S.K. Lele, Evolution of isolated turbulent trailing vortices. Phys. Fluids **20**, 035102 (2008)

P.A. Durbin, Separated flow computations with the k–ϵ–v-squared model. AIAA J. **33**(4) (1995)

S.J. Eastwood, P.G. Tucker, H. Xia, P. Carpenter, P. Dunkley, Hybrid LES computations and measurements of a small scale high speed coflowing jet. AIAA J. **48**(5), 963–974 (2010)

Y. Elazar, R.P. Shreeve, Viscous flow in a controlled diffusion compressor cascade with increasing incidence. J. Turbomach. **112**(2), 256–266 (1990)

J.G. Eriksson, R.I. Karlsson, J. Persson, An experimental study of a two-dimensional plane turbulent wall jet. Exp. Fluids **25**(1), 50–60 (1998)

P.R. Farthing, C.A. Long, J.M. Owen, J.R. Pincombe, Rotating cavity with axial throughflow of cooling air: flow structure. J. Turbomach. **114**(1), 237–246 (1992a)

P.R. Farthing, C.A. Long, J.M. Owen, J.R. Pincombe, Rotating cavity with axial throughflow of cooling air-heat transfer. J. Turbomach. **114**(1), 229 (1992b)

M. Fenot, Etude du refroidissement par impact de jets. Application aux aubes de turbines. PhD thesis, Universite de Poitiers, Portiers, France (2004)

N. Forestier, L. Jacquin, P. Geffroy, The mixing layer over a deep cavity at high-subsonic speed. J. Fluid Mech. **475**(1), 101–145 (2003)

J.R. Forsythe, K.A. Hoffmann, R.M. Cummings, K.D. Squires, Detached-eddy simulation with compressibility corrections applied to a supersonic axisymmetric base flow. J. Fluids Eng. **124**(4), 911–923 (2002)

J.R. Forsythe, K.D. Squires, K.E. Wurtzler, P.R. Spalart, Detached-eddy simulation of the f–15E at high alpha. J. Aircr. **41**(2), 193–200 (2004)

S. Fu, Z. Xiao, H. Chen, Y. Zhang, J. Huang, Simulation of wing-body junction flows with hybrid RANS/LES methods. Int. J. Heat Fluid Flow **28**(6), 1379–1390 (2007)

S. Fujimoto, Large eddy simulation of film cooling flows using octree hexahedral meshes, in *Proc. of ASME Turbo Expo*, Copenhagen, Denmark, 11–15 June 2012. ASME Paper Number GT2012–70090

C. Fureby, Large eddy simulation modelling of combustion for propulsion applications. Philos. Trans. R. Soc. A, Math. Phys. Eng. Sci. **367**(1899), 2957–2969 (2009)

C. Fureby, Y. Nilsson, K. Andersson, Large eddy simulation of supersonic base flow. AIAA Paper Number AIAA–99–0426 (1999)

C. Fureby, N. Alin, N. Wikström, S. Menon, N. Svanstedt, L. Persson, D. Knight, R. Friedrich, W. Rodi, M. Lesieur et al., Large eddy simulation of high Reynolds number wall bounded flows. AIAA J. **42**(3), 457–468 (2004)

A.J.M. Gamal, J.M. Vance, Labyrinth seal leakage tests: tooth profile, tooth thickness, and eccentricity effects. J. Eng. Gas Turbines Power **130**(1), 012510 (2008)

F. Gand, Zonal detached eddy simulation of a civil aircraft with a deflected spoiler. AIAA J. 1–11 (2012)

U. Ganz, S.A.L. Dlegg, P. Joppa, Measurement and prediction of broadband fan noise. AIAA Paper Number AIAA–98–2316 (1998a)

U.W. Ganz, P.D. Joppa, T.J. Patten, D.F. Scharpf, Boeing 18–inch fan rig broadband noise test. Technical Report NASA–CR–1998–208704, NASA (1998b)

A. Garbaruk, M. Shur, M. Strelets, A. Travin, NACA0021 at 60 degrees incidence, in *DESider—A European Effort on Hybrid RANS-LES Modelling*, ed. by W. Haase, M. Braza, A. Revell. Notes on Numerical Fluid Mechanics and Multidisciplinary Design, vol. 103 (Springer, Berlin, 2009a)

A. Garbaruk, M. Shur, M. Strelets, A. Travin, Supersonic base flow, in *DESider—A European Effort on Hybrid RANS-LES Modelling*, ed. by W. Haase, M. Braza, A. Revell. Notes on Numerical Fluid Mechanics and Multidisciplinary Design, vol. 103 (Springer, Berlin, 2009b)

A. Garcia-Sagrado, Boundary layer and trailing edge noise sources. PhD thesis, Engineering Department, Cambridge University (2007)

A. Garcia-Sagrado, T. Hynes, Wall-pressure sources near an airfoil trailing edge under separated laminar boundary layers. AIAA J. **49**(9), 1841–1856 (2011)

B. Gardarin, L. Jacquin, P. Geffroy, Experimental studies on a rudimentary four wheel landing gear. AIAA Paper Number AIAA–2008–3773 (2008)

E. Garnier, S. Deck, Large-eddy simulation of transonic buffet over a supercritical airfoil. ERCOFTAC Ser. (2008). doi:10.1007/978-90-481-3652-0_81

S. Gdadebo, Three dimensional separation in compressors. PhD thesis, Engineering Department, Cambridge University (2003)

L.F.G. Geers, M.J. Tummers, K. Hanjalić, Experimental investigation of impinging jet arrays. Exp. Fluids **36**(6), 946–958 (2004)

O. Gentilhomme, Turbine rim seal ingestion. PhD thesis, School of Engineering, University of Sussex (2004)

O. Gentilhomme, N.J. Hills, A.B. Turner, J.W. Chew, Measurement and analysis of ingestion through a turbine rim seal. J. Turbomach. **125**, 505–512 (2010)

B. Geurts, D. Holm, Leray and LANS—a modeling of turbulent mixing. J. Turbul. **7**(10), 1–33 (2006)

M. Goodhand, Compressor leading edges. PhD thesis, Engineering Department, Cambridge University (2010)

M.N. Goodhand, R.J. Miller, Compressor leading edge spikes: a new performance criterion. J. Turbomach. **133**(2), 021006 (2010)

E. Göttlich, A. Marn, R. Pecnik, F.J. Malzacher, O. Schennach, H.P. Pirker, The influence of blade tip gap variation on the flow through an aggressive S-shaped intermediate turbine duct downstream a transonic turbine stage: Part II—time-resolved results and surface flow, in *Proc. of*

GT2007 ASME Turbo Expo 2007, Montreal, Canada, 14–17 May 2007. ASME Paper Number GT2007–28069

N. Gourdain, L.Y.M. Gicquel, E. Collado, Comparison of RANS and LES for prediction of wall heat transfer in a highly loaded turbine guide vane. AIAA J. Propuls. Power **28**(2) (2012)

S.M. Grace, W.G. Dewar, D.E. Wroblewski, Experimental investigation of the flow characteristics within a shallow wall cavity for both laminar and turbulent upstream boundary layers. Exp. Fluids **36**(5), 791–804 (2004)

S.I. Green, A.J. Acosta, Unsteady flow in trailing vortices. J. Fluid Mech. **227**(1), 107–134 (1991)

A. Hadjadj, Large-eddy simulation of shock/boundary-layer interaction. AIAA J. (2012, in press)

C. Hah, Large eddy simulation of transonic flow field in NASA rotor 37, in *47th AIAA Aerospace Sciences Meeting* (2009). AIAA Paper Number AIAA–2009–1061

C. Hah, M. Voges, M. Mueller, H.P. Schiffer, Investigation of unsteady flow behaviour in transonic compressor rotors with LES and PIV measurements. ISABE. 2009 Paper No. Technical Report, ISABE–2009–02 (2009)

M. Hahn, D. Drikakis, Implicit large-eddy simulation of swept-wing flow using high-resolution methods. AIAA J. **47**(3), 618–630 (2009)

E.E. Halila, D.T. Lenahan, T.T. Thomas, Energy efficient engine high pressure turbine test hardware detailed design report. Technical Report NASA–CR–167955, NASA (1982)

C.A. Hall, T.P. Hynes, Measurements of intake separation hysteresis in a model fan and nacelle rig. J. Propuls. Power **22**(4), 872–879 (2006)

D.E. Halstead, D.C. Wisler, T.H. Okiishi, G.J. Walker, H.P. Hodson, H.W. Shin, Boundary layer development in axial compressors and turbines: Part 1 of 4—composite picture. J. Turbomach. **119**(1), 114–127 (1997)

A. Hamed, D. Basu, K. Das, Detached eddy simulations of supersonic flow over cavity. AIAA Paper Number AIAA–2003–0549 (2003)

J.C. Han, P.R. Chandra, S.C. Lau, Local heat/mass transfer distributions around sharp 180 deg turns in two-pass smooth and rib-roughened channels. ASME Trans. J. Heat Transf. **110**, 91–98 (1988)

M. Harper-Bourne, Some observations on the noise of heated jets, in *13th AIAA/CEAS Aeroacoustics Conf.*, Rome, Italy, 21–23 May 2007. AIAA Paper Number AIAA–2007–3632

L. He, Efficient computational model for nonaxisymmetric flow and heat transfer in rotating cavity. ASME J. Turbomach. **133**, 021018 (2011)

L.S. Hedges, A.K. Travin, P.R. Spalart, Detached-eddy simulations over a simplified landing gear. J. Fluids Eng. **124**(2), 413–423 (2002)

I. Helms, Effect of non-uniform inlet turbulence distributions on high-pressure turbine heat transfer. Master's thesis, Oxford University (2011)

J.L. Herrin, J.C. Dutton, Supersonic base flow experiments in the near wake of a cylindrical afterbody. AIAA J. **32**(1) (1994a)

J.L. Herrin, J.C. Dutton, Supersonic near-wake afterbody boattailing effects on axisymmetric bodies. J. Spacecr. Rockets **31**(6), 1021–1028 (1994b)

J.L. Herrin, J.C. Dutton, Effect of a rapid expansion on the development of compressible free shear layers. Phys. Fluids **7**, 159–171 (1995)

J.L. Herrin, J.C. Dutton, The turbulence structure of a reattaching axisymmetric compressible free shear layer. Phys. Fluids **9**, 3502–3512 (1997)

C.G. Himmel, Ultra-high lift blades for low pressure turbines. PhD thesis, Engineering Department, Cambridge University (2010)

G.V. Hobson, D.J. Hansen, D.G. Schnorenberg, D.V. Grove, Effect of Reynolds number on separation bubbles on compressor blades in cascade. J. Propuls. Power **17**(1), 154–162 (2001)

B.M. Holley, S. Becz, L.S. Langston, Measurement and calculation of turbine cascade endwall pressure and shear stress. J. Turbomach. **128**(2), 232–239 (2006)

D.S. Holloway, J.H. Leylek, F.A. Buck, Pressure-side bleed film cooling: Part I—steady framework for experimental and computational results, in *Proc. of ASME Turbo Expo 2002* (2002a). ASME Paper Number GT2002–30471

D.S. Holloway, J.H. Leylek, F.A. Buck, Pressure-side bleed film cooling: Part II—unsteady framework for experimental and computational results, in *Proc. of ASME Turbo Expo 2002* (2002b). ASME Paper Number GT2002–30472

R.J. Howell, Wake-separation bubble interactions in low Reynolds number turbomachinery. PhD thesis, Engineering Department, Cambridge University (1999)

M. Inagaki, T. Kondoh, Y. Nagano, A mixed-time-scale sgs model with fixed model-parameters for practical LES. Eng. Turbul. Model. Exp. 257–266 (2002)

I.V. Iourokina, S.K. Lele, Towards large eddy simulation of film-cooling flows on a model turbine blade leading edge, in *43rd Aerospace Sciences Exhibit* (2005). AIAA Paper Number AIAA–2005–0670

M. Itoh, Experiments on the turbulent flow in the narrow clearance between a rotating and a stationary disk. Turbul. Flows **1995**, 27–32 (1995)

M. Itoh, Y. Yamada, S. Imao, M. Gonda, Experiments on turbulent flow due to an enclosed rotating disk. Exp. Therm. Fluid Sci. **5**(3), 359–368 (1992)

M.C. Jacob, J. Boudet, D. Casalino, M. Michard, A rod-airfoil experiment as a benchmark for broadband noise modeling. Theor. Comput. Fluid Dyn. **19**(3), 171–196 (2005)

L. Jacquin, P. Molton, S. Deck, B. Maury, D. Soulevant, An experimental study of shock oscillation over a transonic supercritical profile. AIAA Paper Number AIAA–2005–4902 (2005)

J.R. Janssen, J.C. Dutton, Time-series analysis of supersonic base-pressure fluctuations. AIAA J. **42**(3), 605–613 (2004)

R.J. Jefferson-Loveday, P.G. Tucker, LES of impingement heat transfer on a concave surface. Numer. Heat Transf., Part A, Appl. **58**(4), 247–271 (2010)

R.J. Jefferson-Loveday, P.G. Tucker, Wall-resolved LES and zonal LES of round jet impingement heat transfer on a flat plate. Numer. Heat Transf., Part B, Fundam. **59**(3), 190–208 (2011)

R.J. Jefferson-Loveday, P.G. Tucker, N.R. Vadlamani, J.D. Northall, Differential equation specification of integral turbulence length scales, in *Proc. of ASME Turbo Expo*, Copenhagen, Denmark, 11–15 June 2012. ASME Paper Number GT2012–68091

T. Jimbo, D. Biswas, Y. Yokono, Y. Niizeki, A high-order LES turbulent model to study unsteady flow characteristics in a high pressure turbine cascade, in *Proc. of ASME Turbo Expo 2008* (2008). ASME Paper Number GT2008–51458

P.B.V. Johansson, L.U. Axelsson, Numerical and experimental analysis of the flow in an aggressive intermediate turbine duct, in *Proc. of ASME Turbo Expo 2009*, Orlando, Florida, USA, 8–12 June 2009. ASME Paper Number GT2009–59299

J. Joo, P. Durbin, Simulation of turbine blade trailing edge cooling. J. Fluids Eng. **131**, 021102 (2009)

J.C. Jouhaud, L.Y.M. Gicquel, B. Enaux, M.J. Esteve, Large-eddy-simulation modeling for aerothermal predictions behind a jet in crossflow. AIAA J. **45**(10), 2438–2447 (2007)

S.C. Kacker, J.H. Whitelaw, The effect of slot height and slot-turbulence intensity on the effectiveness of the uniform density, two-dimensional wall jet (impervious wall effectiveness of two dimensional wall jet, discussing effectiveness dependence on slot height and turbulence intensity). ASME Trans., Ser. C, J. Heat Transf. **90**, 469–475 (1968a)

S.C. Kacker, J.H. Whitelaw, Some properties of the two-dimensional, turbulent wall jet in a moving stream (open circuit wind tunnel experiment to measure properties of two dimensional plane turbulent wall jet in moving stream). ASME Trans., Ser. E, J. Appl. Mech. **35**, 641–651 (1968b)

S.C. Kacker, J.H. Whitelaw, An experimental investigation of the influence of slot-lip-thickness on the impervious-wall effectiveness of the uniform-density, two-dimensional wall jet. Int. J. Heat Mass Transf. **12**(9), 1196–1201 (1969)

S.C. Kacker, J.H. Whitelaw, The turbulence characteristics of two-dimensional wall-jet and wall-wake flows. J. Appl. Mech. **38**, 239 (1971)

C.K. Kang, Y. Baik, L. Bernal, M.V. Ol, W. Shyy, Fluid dynamics of pitching and plunging airfoils of Reynolds number between 1 × 104 and 6 × 104. AIAA–2009–0536 (2009)

M.K. Karakasis, E.M.J. Naylor, R.J. Miller, H.P. Hodson, The effect of an upstream compressor on a non-axisymmetric s-duct. Glasgow, UK, June 2010. ASME Paper Number GT–2010–2340414–18

N. Karthikeyan, L. Venkatakrishnan, Application of photogrammetry to surface flow visualization. Exp. Fluids **50**(3), 689–700 (2011)

C. Kato, M. Kaiho, A. Manabe, An overset finite-element large-eddy simulation method with applications to turbomachinery and aeroacoustics: flow simulation and modeling. J. Appl. Mech. **70**(1), 32–43 (2003a)

C. Kato, H. Mukai, A. Manabe, Large-eddy simulation of unsteady flow in a mixed-flow pump. Int. J. Rotating Mach. **9**(5), 345–351 (2003b)

S. Kawai, K. Fujii, Computational study of supersonic base flow using hybrid turbulence methodology. AIAA J. **43**(6), 1265–1275 (2005)

M.R. Khorrami, M.E. Berkman, F. Li, B.A. Singer, Computational simulations of a three-dimensional high-lift wing. AIAA Paper Number AIAA–2002–2804 (2002)

N.S. Kim, Analyse expérimentale d'un jet turbulent impactant sur une plaque plane et sur un obstacle de section carrée. PhD thesis, Universite Paul Sabatier Toulouse III, Toulouse, France (2005)

W.W. Kim, S. Menon, An unsteady incompressible Navier-Stokes solver for large eddy simulation of turbulent flows. Int. J. Numer. Methods Fluids **31**(6), 983–1017 (1999)

P. Kjellgren, N. Anderberg, I. Wygnanski, Download alleviation by periodic excitation on a typical tilt-rotor configuration-computation and experiment. AIAA Paper Number AIAA–2000–2697 (2000)

C. Klostermeier, Investigation into the capability of large eddy simulation for turbomachinery design. PhD thesis, Engineering Department, University of Cambridge (2008)

T. Knopp, K. Weinman, D. Schwamborn, Oscillating airfoil NACA0012 at 15 degrees—a basic case for aero-elasticity, in *DESider—A European Effort on Hybrid RANS-LES Modelling*, ed. by W. Haase, M. Braza, A. Revell. Notes on Numerical Fluid Mechanics and Multidisciplinary Design, vol. 103 (Springer, Berlin, 2009)

U. Köller, R. Mönig, B. Kösters, H.A. Schreiber, Development of advanced compressor airfoils for heavy-duty gas turbines—Part I: design and optimization. J. Turbomach. **122**(3), 397–405 (2000)

M. Konopka, M. Meinke, W. Schröder, Large-eddy simulation of shock/cooling-film interaction. AIAA J. **50**(10), 2102–2114 (2012)

J.E. Kopriva, G.M. Laskowski, M. Reza, H. Shiekhi, Assessment of scale resolution on high pressure turbine blade wakes via URANS and LES at engine scale conditions, in *Proc. of ASME Turbo. Expo., San Antonio*, Texas, USA (2013), pp. 3–7. ASME Paper Number GT2013–94285

B. Kosovic, Subgrid-scale modelling for the large-eddy simulation of high-Reynolds-number boundary layers. J. Fluid Mech. **336**(1), 151–182 (1997)

B. Küsters, H.A. Schreiber, U. Köller, R. Mönig, N.A. Cumpsty, Development of advanced compressor airfoils for heavy-duty gas turbines: Part II: experimental and theoretical analysis. Discussion. J. Turbomach. **122**(3), 406–415 (2000)

G. Lalizel, Q. Sultan, M. Fenot, E. Dorignac, Aerothermal study of periodic pulsed jet issuing into a crossflow: application to temperature mixture near wall, in *7th Int. Symposium on Turbulence, Heat and Mass Transfer*, Palmero, Sicily, Italy, 24–27 September 2012

L.S. Langston, Crossflows in a turbine cascade passage. ASME J. Eng. Power **102**(4), 866–874 (1980)

L.S. Langston, M.L. Nice, R.M. Hooper, Three-dimensional flow within a turbine cascade passage. J. Eng. Power **99**, 21–28 (1977)

R. Langtry, P. Spalart, Detached eddy simulation of a nose landing-gear cavity, in *IUTAM Symposium on Unsteady Separated Flows and Their Control*, ed. by M. Braza, K. Hourigan (2009), pp. 357–366

R. Laraufie, S. Deck, Zonal detached eddy simulation (zdes) study of a 3d curved duct, in *7th Int. Symposium on Turbulence, Heat and Mass Transfer*, Palmero, Sicily, Italy, 24–27 September 2012

L. Larchevêque, P. Sagaut, I. Mary, O. Labbé, P. Comte, Large-eddy simulation of a compressible flow past a deep cavity. Phys. Fluids **15**, 193 (2003)

S. Lardeau, M. Leschziner, T. Zaki, Large eddy simulation of transitional separated flow over a flat plate and a compressor blade. Flow Turbul. Combust. **88**(1–2), 19–44 (2012)

S.J. Lawson, G.N. Barakos, P. Nayyar, Appendix for M219 cavity flow: computations with and without bay doors, in *DESider—A European Effort on Hybrid RANS-LES Modelling*, ed. by W. Haase, M. Braza, A. Revell. Notes on Numerical Fluid Mechanics and Multidisciplinary Design, vol. 103 (Springer, Berlin, 2009)

B. Lazos, Surface topology on the wheels of a generic four-wheel landing gear. AIAA J. **40**(12), 2402–2411 (2002a)

B.S. Lazos, Mean flow features around the inline wheels of four-wheel landing gear. AIAA J. **40**(2), 193–198 (2002b)

B.S. Lazos, Reynolds stresses around the wheels of a simplified four-wheel landing gear. AIAA J. **42**(1), 196–198 (2004)

F. Le Chuiton, A.D. Alascio, G. Barakos, R. Steijl, D. Schwamborn, H. Lu-deke, EC145 helicopter fuselage—an industrial case, in *DESider—A European Effort on Hybrid RANS-LES Modelling*, ed. by W. Haase, M. Braza, A. Revell. Notes on Numerical Fluid Mechanics and Multidisciplinary Design, vol. 103 (Springer, Berlin, 2009)

B.H.K. Lee, Transonic buffet of a supercritical airfoil. Aeronaut. J. **94**(935), 143–152 (1990)

D.H. Lee, Y.S. Chung, S.Y. Won, The effect of concave surface curvature on heat transfer from a fully developed round impinging jet. Int. J. Heat Mass Transf. **42**(13), 2489–2497 (1999)

K. Lehmann, Heat Transfer and aerodynamics of over-shroud leakage flows in a high pressure turbine. PhD thesis, Engineering Department, University of Cambridge (2009)

S. Leicher, FA–5 configuration, in *DESider—A European Effort on Hybrid RANS-LES Modelling*, ed. by W. Haase, M. Braza, A. Revell. Notes on Numerical Fluid Mechanics and Multidisciplinary Design, vol. 103 (Springer, Berlin, 2009)

J. Leray, Sur les movements d'un fluide visqueux remplaisant l'espace. Acta Math. **64**, 193 (1934)

M.A. Leschziner, F. Tessicini, Separated flow behind an aerofoil trailing edge without camber, in *DESider—A European Effort on Hybrid RANS-LES Modelling*, ed. by W. Haase, M. Braza, A. Revell. Notes on Numerical Fluid Mechanics and Multidisciplinary Design, vol. 103 (Springer, Berlin, 2009)

M. Leschziner, N. Li, F. Tessicini, M. Leschziner, N. Li, F. Tessicini, Simulating flow separation from continuous surfaces: routes to overcoming the Reynolds number barrier. Philos. Trans. R. Soc. A, Math. Phys. Eng. Sci. **367**(1899), 2885–2903 (2009)

J. Li, Hybrid RANS/LES modelling of OGV/prediffuser flow. PhD thesis, Department of Aeronautical and Automotive Engineering, Loughborough University (2012)

N. Li, M.A. Leschziner, Large-eddy simulation of separated flow over a swept wing with approximate near-wall modelling. Aeronaut. J. **111**(1125), 689–697 (2007)

Y. Liu, P.G. Tucker, G. Lo Iacono, Comparison of zonal RANS and LES for a non-isothermal ribbed channel flow. Int. J. Heat Fluid Flow **27**(3), 391–401 (2006)

D.P. Lockard, M.R. Khorrami, F. Li, High resolution calculation of a simplified landing gear. AIAA Paper Number AIAA–2004–2887 (2004)

C.A. Long, P.G. Tucker, Shroud heat transfer measurements from a rotating cavity with an axial throughflow of air. J. Turbomach. **116**(3) (1994)

M. Lygren, H.I. Andersson, Turbulent flow between a rotating and a stationary disk. J. Fluid Mech. **426**(1), 297–326 (2001)

M. Lygren, H.I. Andersson, Turbulence statistics in an open rotor-stator configuration. Phys. Fluids **14**, 1137–1145 (2002)

D. Lytle, B.W. Webb, Air jet impingement heat transfer at low nozzle-plate spacings. Int. J. Heat Mass Transf. **37**(12), 1687–1697 (1994)

W. Ma, X. Ottavy, L. Lu, F. Leboeuf, F. Gao, Experimental study of corner stall in a linear compressor cascade. Chin. J. Aeronaut. **24**(3), 235–242 (2011)

P. Martini, A. Schulz, Experimental and numerical investigation of trailing edge film cooling by circular coolant wall jets ejected from a slot with internal rib arrays. J. Turbomach. **126**(2), 229–236 (2004)

P. Martini, A. Schulz, C.F. Whitney, E. Lutum, Experimental and numerical investigation of trailing edge film cooling downstream of a slot with internal rib arrays. Proc. Inst. Mech. Eng. A, J. Power Energy **217**(4), 393–401 (2003)

P. Martini, A. Schulz, H.J. Bauer, C.F. Whitney, Detached eddy simulation of film cooling performance on the trailing edge cutback of gas turbine airfoils. J. Turbomach. **128**(2), 292–299 (2006a)

P. Martini, A. Schulz, H.J. Bauer, C.F. Whitney, Film cooling effectiveness and heat transfer on the trailing edge cutback of gas turbine airfoils with various internal cooling designs. J. Turbomach. **128**(1), 196–205 (2006b)

I. Mary, P. Sagaut, Large eddy simulation of flow around an airfoil near stall. AIAA J. **40**(6), 1139–1145 (2002)

K. Matsuura, C. Kato, Large-eddy simulation of compressible transitional cascade flows with and without incoming free-stream turbulence. JSME Int. J., Ser. B **49**(3), 660–669 (2006)

N.E. May, C.A. McHugh, A.J. Peace, An investigation of two intake/S-bend diffuser geometries using the Sauna CFD system—phase 1. Aircraft Research Association, MEMO 386, 1993

J.J. McGuirk, Development, validation and application of large eddy simulation for gas-turbine combustion system fluid mechanics, in *Proc. 2nd ISJPPE, 2nd Int. Symposium on Jet Propulsion and Power Engineering* Gulin, China, 22–26 September 2008. Paper Number 2008–ISJPPE–0006

J.J. McGuirk, A. Taylor, Numerical simulation of gas turbine combustion processes-present status and future trends. J. Gas Turbine Soc. Jpn. **30**(5), 347–361 (2002)

W.A. McMullan, G.J. Page, Large eddy simulation of a compressor cascade and the influence of spanwise domain. J. Power Energy **225**(6), 817–831 (2011)

G. Medic, O. Sharma, Large-eddy simulation of flow in a low-pressure turbine cascade, in *Proc. of ASME Turbo Expo*, Copenhagen, Denmark, 11–15 June 2012. ASME Paper Number GT2012–68878

Y. Mei, A. Guha, Implicit numerical simulation of transonic flow through turbine cascades on unstructured grids. Proc. Inst. Mech. Eng. A, J. Power Energy **219**(1), 35–47 (2005)

C.P. Mellen, J. Fröhlich, W. Rodi, Lessons from LESFOIL project on large-eddy simulation of flow around an airfoil. AIAA J. **41**(4), 573–581 (2003)

F. Mendonca, O. Baris, G. Capon, Simulation of radial compressor aeroacoustics using CFD, in *Proc. of ASME Turbo Expo*, Copenhagen, Denmark, 11–15 June 2012. ASME Paper Number GT2012–70028

K. Menzies, Large eddy simulation applications in gas turbines. Philos. Trans. R. Soc. A, Math. Phys. Eng. Sci. **367**(1899), 2827–2838 (2009)

N.L. Messersmith, J.C. Dutton, Characteristic features of large structures in compressible mixing layers. AIAA J. **34**(9), 1814–1821 (1996)

V. Michelassi, J.G. Wissink, J. Fröhlich, W. Rodi, Large-eddy simulation of flow around low-pressure turbine blade with incoming wakes. AIAA J. **41**(11), 2143–2156 (2003a)

V. Michelassi, J.G. Wissink, W. Rodi, Direct numerical simulation, large eddy simulation and unsteady Reynolds-averaged Navier-Stokes simulations of periodic unsteady flow in a low-pressure turbine cascade: a comparison. J. Power Energy **217**(4), 403–411 (2003b)

R. Mittal, S. Venkatasubramanian, F.M. Najjar, Large-eddy simulation of flow through a low-pressure turbine cascade. AIAA Paper Number AIAA–2001–2560 (2001)

M.V. Morales, The effect of compressibility on unsteady boundary layers in high lift low pressure turbines. PhD thesis, Engineering Department, University of Cambridge (2004)

S.A. Morton, R.M. Cummings, D.B. Kholodar, High resolution turbulence treatment of f/a–18 tail buffet. J. Aircr. **44**(6) (2007)

S. Murari, S. Sunnam, J.S. Liu, Steady state and transient cfd studies on aerodynamic performance validation of a high pressure turbine, in *Proc. of ASME Turbo Expo*, Copenhagen, Denmark, 11–15 June 2012. ASME Paper Number GT2012–68853

A. Murata, S. Mochizuki, Large eddy simulation with a dynamic subgrid-scale model of turbulent heat transfer in an orthogonally rotating rectangular duct with transverse rib turbulators. Int. J. Heat Mass Transf. **43**(7), 1243–1259 (2000)

C. Muthanna, Flowfield downstream of a compressor cascade with tip leakage. PhD thesis, Virginia Polytechnic Institute and State University (1998)

Y. Nagano, H. Hattori, T. Houra, DNS of velocity and thermal fields in turbulent channel flow with transverse-rib roughness. Int. J. Heat Fluid Flow **25**(3), 393–403 (2004)

W.B. Nicoll, J.H. Whitelaw, The effectiveness of the uniform density, two-dimensional wall jet. Int. J. Heat Mass Transf. **10**(5), 623–639 (1967)

M.V. Ol, L. Bernal, C.K. Kang, W. Shyy, Shallow and deep dynamic stall for flapping low Reynolds number airfoils. Exp. Fluids **46**(5), 883–901 (2009)

T.S.D. O'Mahoney, Large-eddy simulation of turbine rim seals. PhD thesis, School of Engineering, University of Surrey (2011)

T.S.D. O'Mahoney, N.J. Hills, J.W. Chew, T. Scanlon, Large-eddy simulation of rim seal ingestion, in *Proc. of ASME Turbo Expo 2010*, Glasgow, UK, 14–18 June 2010. ASME Paper Number GT2010–22962

M. Opiela, R. Keinke, W. Schroder, LES of wake-blade interaction, in *Proc. of the 4th European Conference on Turbomachinery Fluid Dynamics and Thermodynamics*, Firenze, Italy, 20–23 March 2001. AIAA Paper Number AIAA–2011–2881

M.M. Opoka, The effect of upstream and downstream blade rows on transition in low pressure turbines. PhD thesis, Engineering Department, University of Cambridge (2007)

L.R. Owens, B.G. Allan, S.A. Gorton, Boundary-layer-ingesting inlet flow control. J. Aircr. **45**(4), 1431–1440 (2008)

J. Panda, Two point space-time correlation of density fluctuations measured in high velocity free jets. Technical Report NASA–CR–2006–214222, NASA (2006)

J. Panda, R.G. Seasholtz, Velocity and temperature measurement in supersonic free jets using spectrally resolved Rayleigh scattering. Technical Report NASA–TM–2004–212391, NASA (2004)

J. Panda, R.G. Seasholtz, K.A. Elam, A.F. Mielke, D.G. Eck, Effect of heating on turbulent density fluctuations and noise generation from high speed jets, in *10th AIAA Aeroacoustics Conference* (2004). AIAA Paper Number AIAA–2004–3016

K. Paschal, L. Jenkins, C. Yao, Unsteady slat-wake characteristics of a high-lift configuration. AIAA Paper Number AIAA–2000–0139 (2000)

W.R. Pauley, J.K. Eaton, Experimental study of the development of longitudinal vortex pairs embedded in a turbulent boundary layer. AIAA J. **26**(7), 816–823 (1988)

S.-H. Peng, M219 cavity flow, in *DESider—A European Effort on Hybrid RANS-LES Modelling*, ed. by W. Haase, M. Braza, A. Revell. Notes on Numerical Fluid Mechanics and Multidisciplinary Design, vol. 103 (Springer, Berlin, 2009)

G. Petit, M. Mallet, Contribution of Dassault to DESider, in *DESider—A European Effort on Hybrid RANS-LES Modelling*, ed. by W. Haase, M. Braza, A. Revell. Notes on Numerical Fluid Mechanics and Multidisciplinary Design, vol. 103 (Springer, Berlin, 2009)

A. Pfau, M. Treiber, M. Sell, G. Gyarmathy, Flow interaction from the exit cavity of an axial turbine blade row labyrinth seal. J. Turbomach. **123**(2), 342–352 (2001)

A. Pfau, A.I. Kalfas, R.S. Abhari, Making use of labyrinth interaction flow. ASME Trans. J. Turbomach. **129**(1), 164–174 (2007)

J.R. Pietrzyk, D.G. Bogard, M.E. Crawford, Effects of density ratio on the hydrodynamics of film cooling. J. Turbomach. **112**, 437–443 (1990)

H. Pitsch, Large-eddy simulation of turbulent combustion. Annu. Rev. Fluid Mech. **38**, 453–482 (2006)

I. Popovic, Aerothermal investigation of hub leakage flows in high-pressure turbines. PhD thesis, Engineering Department, University of Cambridge (2010)

G. Rau, M. Cakan, D. Moeller, T. Arts, The effect of periodic ribs on the local aerodynamic and heat transfer performance of a straight cooling channel. J. Turbomach. **120**(2), 368–375 (1988)

B. Raverdy, I. Mary, P. Sagaut, N. Liamis, High-resolution large-eddy simulation of flow around low-pressure turbine blade. AIAA J. **41**(3), 390–397 (2003)

H.J. Rehder, A. Dannhauer, Experimental investigation of turbine leakage flows on the three-dimensional flow field and endwall heat transfer. J. Turbomach. **129**(3), 608–618 (2007)

A.J. Revell, T.J. Craft, D.R. Laurence, Turbulence modelling of unsteady turbulent flows using the stress strain lag model. Flow Turbul. Combust. **86**(1), 129–151 (2011)

C.M. Rhie, W.L. Chow, Numerical study of the turbulent flow past an airfoil with trailing edge separation. AIAA J. **21**(11), 1525–1532 (1983)

D.P. Rizzetta, M.R. Visbal, Large-eddy simulation of supersonic cavity flowfields including flow control. AIAA J. **41**(8), 1452–1462 (2003)

D.P. Rizzetta, M.R. Visbal, P.E. Morgan, A high-order compact finite-difference scheme for large-eddy simulation of active flow control. Prog. Aerosp. Sci. **44**(6), 397–426 (2008)

B. Rosic, The control of shroud leakage flows in low aspect ratio multistage turbines. PhD thesis, Engineering Department, University of Cambridge (2006)

S. Roy, S. Kapadia, J.D. Heidmann, Film cooling analysis using des turbulence model, in *Proceedings of ASME Turbo Expo*, Citeseer (2003), pp. 16–19

A. Rozati, D.K. Tafti, Large eddy simulation of leading edge film cooling: Part I—computational domain and effect of coolant pipe inlet condition, in *Proc. of ASME Turbo Expo 2007*, Montreal, Canada, 14–17 May 2007. ASME Paper Number GT2007–27689

A. Rozati, D.K. Tafti, Large eddy simulation of leading edge film cooling—Part II: heat transfer and effect of blowing ratio. J. Turbomach. **130**, 041015 (2008)

P. Sagaut, LES and DNS simulation of turbomachinery flows, in *Recent Developments in Numerical Methods for Turbomachinery Flows*. VKI Lecture (2001)

P. Sagaut, S. Deck, Large eddy simulation for aerodynamics: status and perspectives. Philos. Trans. R. Soc. A, Math. Phys. Eng. Sci. **367**(1899), 2849–2860 (2009)

A.K. Saha, S. Acharya, Flow and heat transfer in an internally ribbed duct with rotation: an assessment of large eddy simulations and unsteady Reynolds-averaged Navier-Stokes simulations. J. Turbomach. **127**(2), 306–320 (2005)

R.D. Sandberg, Numerical investigation of turbulent supersonic axisymmetric wakes. J. Fluid Mech. **1**(1), 1–33 (2012)

R.D. Sandberg, H.F. Fasel, Direct numerical simulations of transitional supersonic base flows. AIAA Paper Number AIAA–2005–98 (2005)

R.D. Sandberg, H.F. Fasel, Numerical investigation of transitional supersonic axisymmetric wakes. J. Fluid Mech. **563**(1), 1–41 (2006a)

R.D. Sandberg, H.F. Fasel, Direct numerical simulations of transitional supersonic base flows. AIAA J. **44**(4), 848–858 (2006b)

S. Sarkar, Effects of passing wakes on a separating boundary layer along a low-pressure turbine blade through large-eddy simulation. J. Power Energy **221**(4), 551–564 (2007)

S. Sarkar, Identification of flow structures on a lp turbine blade due to periodic passing wakes. J. Fluids Eng. **130**(6), 061103 (2008)

S. Sarkar, Influence of wake structure on unsteady flow in a low pressure turbine blade passage. J. Turbomach. **131**(4) (2009)

S. Sarkar, P.R. Voke, *LES of Passing Wakes Influencing Transition on Turbine Blades*. Number ERCOFTAC Series, Direct and Large-Eddy Simulation V (Kluwer Academic, Dordrecht, 2003)

S. Sarkar, P.R. Voke, Large-eddy simulation of unsteady surface pressure over a low-pressure turbine blade due to interactions of passing wakes and inflexional boundary layer. J. Turbomach. **128**(2), 221–231 (2006)

J.U. Schlüter, Large eddy simulations of flow and mixing in jets and swirl flows: application to a gas turbine. PhD thesis, INPT, Toulouse, France (2000)

H. Schneider, D. von Terzi, H.J. Bauer, Large-eddy simulations of trailing-edge cutback film cooling at low blowing ratio. Int. J. Heat Fluid Flow **31**(5), 767–775 (2010)

P. Schuler, K. Dullenkopf, H.J. Bauer, Investigation of the influence of different rim seal geometries in a low-pressure turbine, in *Proc. of ASME Turbo Expo* (2011). ASME Paper Number GT2011–45682

V. Schulte, Unsteady separated boundary layers in axial-flow turbomachinery. PhD thesis, Engineering Department, University of Cambridge (1995)

A.X. Sengissen, J.F. Van Kampen, R.A. Huls, G.G.M. Stoffels, J.B.W. Kok, T.J. Poinsot, LES and experimental studies of cold and reacting flow in a swirled partially premixed burner with and without fuel modulation. Combust. Flame **150**(1), 40–53 (2007)

E. Séverac, E. Serre, A spectral vanishing viscosity for the LES of turbulent flows within rotating cavities. J. Comput. Phys. **226**(2), 1234–1255 (2007)

E.A. Sewall, D.K. Tafti, Large eddy simulations of the developing region of a rotating ribbed internal turbine blade cooling channel, in *Proc. of ASME Turbo Expo 2004*, Vienna, Austria (2004). ASME Paper Number GT2004–53833

L. Shaw, Active control for cavity acoustics. AIAA Paper Number AIAA–98–2347 (1998)

C.M. Shieh, P.J. Morris, Comparison of two- and three-dimensional cavity flows, in *39th AIAA, Aerospace Sciences Meeting and Exhibit*, Reno, Nevada, USA (2001). AIAA Paper Number AIAA–2001–0511

M. Shur, P.R. Spalart, M. Strelets, A. Travin, Detached-eddy simulation of an airfoil at high angle of attack. Eng. Turbul. Model. Exp. **4**, 669–678 (1999)

C.H. Sieverding, D. Ottolia, C. Bagnera, A. Comadoro, J.F. Brouckaert, J.M. Desse, Unsteady turbine blade wake characteristics. J. Turbomach. **126**(4), 551–559 (2004)

F. Simon, S. Deck, P. Guillen, P. Sagaut, Reynolds averaged Navier-Stokes/Large-Eddy simulations of supersonic base flow. AIAA J. **44**(11), 2578–2590 (2006)

F. Simon, S. Deck, P. Guillen, R. Cayzac, A. Merlen, Zonal-detached-eddy simulation of projectiles in the subsonic and transonic regimes. AIAA J. **45**(7), 1606–1619 (2007a)

F. Simon, S. Deck, P. Guillen, P. Sagaut, A. Merlen et al., Numerical simulation of the compressible mixing layer past an axisymmetric trailing edge. J. Fluid Mech. **591**, 215–254 (2007b)

A.K. Sinha, D.G. Bogard, M.E. Crawford, Film cooling effectiveness downstream of a single row of holes with variable density ratio. J. Turbomach. **113**, 442–449 (1991)

K.M. Smith, J.C. Dutton, Investigation of large-scale structures in supersonic planar base flows. AIAA J. **34**(6), 1146–1152 (1996)

K.M. Smith, J.C. Dutton, Evolution and convection of large-scale structures in supersonic reattaching shear flows. Phys. Fluids **11**, 2127–2138 (1999)

K.M. Smith, J.C. Dutton, The effects of expansion strength on large-scale structures in compressible free shear layers. Phys. Fluids **13**, 2076–2086 (2001)

B.I. Soemarwoto, J.C. Kwok, H. van der Ven, Contribution of Dassault to DESider, in *DESider—A European Effort on Hybrid RANS-LES Modelling*, ed. by W. Haase, M. Braza, A. Revell. Notes on Numerical Fluid Mechanics and Multidisciplinary Design, vol. 103 (Springer, Berlin, 2009)

P. Spalart, L. Hedges, M. Shur, A. Travin, Simulation of active flow control on a stalled airfoil. Flow Turbul. Combust. **71**(1), 361–373 (2003)

P.R. Spalart, M.L. Shur, M.K. Strelets, A.K. Travin, Towards noise prediction for rudimentary landing gear, in *IUTAM Symposium on Computational Aero-Acoustics for Aircraft Noise Prediction Procedia IUTAM*, vol. 1 (2010), pp. 283–292

M. Sreedhar, S. Ragab, Large eddy simulation of a longitudinal vortex, in *32nd AIAA, Aerospace Sciences Meeting and Exhibit*, Reno, Nevada, USA, 10–13 January 1994. AIAA Paper Number AIAA–94–0529

R.D. Stieger, The effects of wakes on separating boundary layers in low pressure turbines. PhD thesis, Engineering Department, University of Cambridge (2002)

R.D. Stieger, H.P. Hodson, The transition mechanism of highly loaded low-pressure turbine blades, in *Proc. of ASME Turbo Expo 2003*, Atlanta, Georgia, USA, 16–19 June 2003. ASME Paper Number GT2003–38304

M. Strelets, Detached eddy simulation of massively separated flows. AIAA Paper Number AIAA–2001–0879 (2001)

K.L. Suder, Ml. Celestina, Experimental and computational investigation of the tip clearance flow in a transonic axial compressor rotor. Technical Report NASA–TM–106711, NASA (1994)

K.L. Suder, R.V. Chima, A.J. Strazisar, W.B. Roberts, The effect of adding roughness and thickness to a transonic axial compressor rotor. Technical Report NASA–TR–95–30524, NASA STI/Recon Technical Report N (1995)

K.E. Swalwell, J. Sheridan, W.H. Melbourne, Frequency analysis of surface pressure on an airfoil after stall, in *21st AIAA Aerodynamics Conference* (2003). AIAA Paper Number AIAA–2003–3416

N. Swaminathan, K. Bray, *Turbulent Premixed Flames* (Cambridge University Press, Cambridge, 2011)

R.C. Swanson, E. Turkel, On central-difference and upwind schemes. J. Comput. Phys. **101**(2), 292–306 (1992)

M.F. Tachie, R. Balachandar, D.J. Bergstrom, Roughness effects on turbulent plane wall jets in an open channel. Exp. Fluids **37**(2), 281–292 (2004)

D.K. Tafti, Evaluating the role of subgrid stress modeling in a ribbed duct for the internal cooling of turbine blades. Int. J. Heat Fluid Flow **26**(1), 92–104 (2005)

K. Takeda, G.B. Ashcroft, X. Zhang, P.A. Nelson, Unsteady aerodynamics of slat cove flow in a high-lift device configuration. AIAA Paper Number AIAA–2001–0706 (2001)

H.K. Tanna, An experimental study of jet noise. Part I: turbulent mixing noise. J. Sound Vib. **50**(3), 405–428 (1977)

N. Tauveron, Simulation of a compressor cascade with stalled flow using large eddy simulation with two-layer approximate boundary conditions. Nucl. Eng. Des. **240**(2), 321–335 (2010)

B. Thomber, D. Dfikakis, Implicit large-eddy simulation of a deep cavity using high-resolution methods. AIAA J. **46**(10), 2634–2645 (2008)

S. Trapier, P. Duveau, S. Deck, Experimental study of supersonic inlet buzz. AIAA J. **44**(10), 2354–2365 (2006)

S. Trapier, S. Deck, P. Duveau, P. Sagaut, Time-frequency analysis and detection of supersonic inlet buzz. AIAA J. **45**(9), 2273–2284 (2007)

S. Trapier, S. Deck, P. Duveau, Delayed detached-eddy simulation and analysis of supersonic inlet buzz. AIAA J. **46**(1), 118–131 (2008)

P.G. Tucker, Temporal behavior of flow in rotating cavities. Numer. Heat Transf., Part A, Appl. **41**(6–7), 611–627 (2002)

P.G. Tucker, Turbulence modelling of problem aerospace flows. Int. J. Numer. Methods Fluids **51**(3), 261–283 (2006)

P.G. Tucker, Computation of unsteady turbomachinery flows: Part 1—progress and challenges. Prog. Aerosp. Sci. **47**(7), 522–545 (2011a)

P.G. Tucker, Hybrid Hamilton-Jacobi-Poisson wall distance function model. Comput. Fluids **44**(1), 130–142 (2011b)

P. Tucker, S. Eastwood, C. Klostermeier, R. Jefferson-Loveday, J. Tyacke, Y. Liu, Hybrid les approach for practical turbomachinery flows—Part I: hierarchy and example simulations. J. Turbomach., Trans. ASME **134**(2) (2012)

J. Tyacke, R.J. Jefferson-Loveday, P.G. Tucker, Numerical modelling of seal type geometries, in *Proc. of ASME Turbo Expo*, Copenhagen, Denmark, 11–15 June 2012a. ASME Paper Number GT2012–68840

J.C. Tyacke, P.G. Tucker, R. Jefferson-Loveday, N.R. Vadlamani, R. Watson, I. Naqavi, LES of aero engine turbines, in *ECCOMAS 2012, 6th European Congress on Computational Methods in Applied Sciences and Engineering*, Vienna, Austria, 10–14 September 2012b

M. Tyagi, S. Acharya, Large eddy simulation of turbulent flows in complex and moving rigid geometries using the immersed boundary method. Int. J. Numer. Methods Fluids **48**(7), 691–722 (2005)

A. Uzun, M.Y. Hussaini, C.L. Streett, Large-eddy simulation of a wing-tip vortex on overset grids. AIAA J. **44**(6), 1229–1242 (2006)

N.R. Vadlamani, High fidelity large eddy simulation of turbines: current status and future outlook. PhD thesis, Engineering Department, Cambridge University (2013)

L. Venkatakrishnan, N. Karthikeyan, K. Mejia, Experimental studies on a rudimentary four wheel landing gear. AIAA Paper Number AIAA–2011–0354 (2011)

M.R. Visbal, Numerical investigation of deep dynamic stall of a plunging airfoil. AIAA J. **49**(10), 2152–2170 (2011)

K. Viswanathan, Aeroacoustics of hot jets. J. Fluid Mech. **516**, 39–82 (2004)

A.K. Viswanathan, D.K. Tafti, Detached eddy simulation of turbulent flow and heat transfer in a two-pass internal cooling duct. Int. J. Heat Fluid Flow **27**(1), 1–20 (2006)

Y. Wang, W.J. Devenport, Wake of a compressor cascade with tip gap, Part 2: effects of endwall motion. AIAA J. **42**(11), 2332–2340 (2004)

M. Wang, P. Moin, Computation of trailing-edge flow and noise using large-eddy simulation. AIAA J. **38**(12), 2201–2209 (2000)

Z.-N. Wang, X. Yuan, Unsteady mechanisms of compressor corner separation over a range of incidences based on hybrid LES/RANS, in *Proc. of ASME Turbo Expo*, San Antonio, Texas, USA, 3–7 June 2013

K. Watanabe, T. Takahashi, LES simulation and experimental measurement of fully developed ribbed channel flow and heat transfer, in *Proc. of ASME Turbo. Expo. 2002*, Amsterdam, The Netherlands, 3–6 June 2002. ASME Paper Number GT–2002–30203

S.R. Wellborn, T.H. Okiishi, B.A. Reichert, A study of the compressible flow through a diffusing S-duct. Technical Report NASA–TM–106411, NASA (1993)

A. Wheeler, Effect of unsteady flows on compressor performance. PhD thesis, Engineering Department, Cambridge University (2007)

J. Winkler, S. Moreau, LES of the trailing-edge flow and noise of a naca6512–63 airfoil at zero angle of attack, in *Center for Turbulence Research, Proceedings of the Summer Program* (2008), pp. 331–342

J.G. Wissink, DNS of separating, low Reynolds number flow in a turbine cascade with incoming wakes. Int. J. Heat Fluid Flow **24**(4), 626–635 (2003)

J. Wissink, W. Rodi, DNS of a laminar separation bubble affected by free-stream disturbances. ERCOFTAC Ser. **9**, 213–220 (2004)

J.G. Wissink, W. Rodi, H.P. Hodson, The influence of disturbances carried by periodically incoming wakes on the separating flow around a turbine blade. Int. J. Heat Fluid Flow **27**(4), 721–729 (2006)

X. Wu, P.A. Durbin, Evidence of longitudinal vortices evolved from distorted wakes in a turbine passage. J. Fluid Mech. **446**(1), 199–228 (2001)

X. Wu, J.P. Hickey, Visualization of continuous stream of grid turbulence past the Langston turbine cascade. AIAA J. **50**(1), 215–225 (2012)

X. Wu, K.D. Squires, Prediction and investigation of the turbulent flow over a rotating disk. J. Fluid Mech. **418**(1), 231–264 (2000)

X. Wu, R.G. Jacobs, J.C.R. Hunt, P.A. Durbin, Simulation of boundary layer transition induced by periodically passing wakes. J. Fluid Mech. **398**, 109–153 (1999)

X. Wu, L.T. Li, M.S. Hilaire, Migration of a turbulent patch through a high-pressure turbine cascade. Phys. Fluids **21**, 025110 (2009)

X. Yan, A preheated-wall transient method using liquid crystals for the measurement of heat transfer on external surfaces and in ducts. PhD thesis, University of California, Davis (1993)

A. Yoshizawa, Bridging between eddy-viscosity-type and second-order models using a two-scale dia, in *Ninth Symposium on Turbulent Shear Flows*, vol. 18, Kyoto, Japan, 16 August 1993, pp. 1–23

D. You, M. Wang, R. Mittal, P. Moin, Study of rotor tip-clearance flow using large eddy simulation. AIAA J. **838**.(2003)

D. You, R. Mittal, M. Wang, P. Moin, Computational methodology for large-eddy simulation of tip-clearance flows. AIAA J. **42**(2), 271–279 (2004)

D. You, M. Wang, R. Mittal, P. Moin, Large-eddy simulations of longitudinal vortices embedded in a turbulent boundary layer. AIAA J. **44**(12), 3032 (2006a)

D. You, M. Wang, P. Moin, R. Mittal, Effects of tip-gap size on the tip-leakage flow in a turbomachinery cascade. Phys. Fluids **18**, 105102 (2006b)

L.L. Yuan, R.L. Street, J.H. Ferziger, Large-eddy simulations of a round jet in crossflow. J. Fluid Mech. **379**(1), 71–104 (1999)

T. Zaki, J. Wissink, W. Rodi, P. Durbin, Direct numerical simulations of transition in a compressor cascade: the influence of free-stream turbulence. J. Fluid Mech. **1**, 1–42 (2006)

T.A. Zaki, J.G. Wissink, P.A. Durbin, W. Rodi, Direct computations of boundary layers distorted by migrating wakes in a linear compressor cascade. Flow Turbul. Combust. **83**(3), 307–322 (2009)

X.F. Zhang, Separation and transition control on ultra-high-lift low pressure turbine blades in unsteady flow. PhD thesis, Engineering Department, Cambridge University (2005)

M.D. Zhou, I. Wygnanski, Parameters governing the turbulent wall jet in an external stream. AIAA J. **31**(5), 848–853 (1993)

Chapter 6
Computational Aeroacoustics

6.1 Introduction

Aeroacoustics can involve the modelling of tonal or broadband noise. For tonal noise, (U)RANS can be adequate. Broadband noise needs the detailed modelling of unsteady source structures. This need largely invalidates (U)RANS modelling which can be grossly wrong in such situations. Critically, (U)RANS lacks the necessary temporal information needed for accurately recovering sound. For simpler, tonal noise prediction, analytical approaches have shown some success in understanding and predicting key elements of acoustics behaviour (Tyler and Sofrin 1962; Rossiter 1964). However, perhaps their true value, in aeroacoustics, is as a means of converting unsteady turbulent flow and near field sound information, from CFD, into far field sound data. Nowadays, key problems of interest are in the area of broadband noise. Analytically based approaches, that allow this near field unsteady data conversion process to far field sound, are due to Lighthill (1952), Ffowcs-Williams and Hawkings (1969)—FWH—and also the method of Kirchhoff (Lyrintzis 2003).

Most far field sound prediction approaches have a sound wave propagation operator

$$L(U,\phi') = S_{\phi'} \tag{6.1}$$

In this, U, is the mean flow velocity field vector. This field can refract the acoustic field. This component is defined by ϕ', an acoustic perturbation quantity. This could be, for example, fluctuating density, ρ', or pressure, p'. The input to the operator, $L(U,\phi')$, is a representation of say the consequences of vortical disturbances supplied by turbulence. This is in the form of a source term, $S_{\phi'}$. If necessary, this can also be formulated to account for refraction. As shown by Ewert and Schroder (2003), vortical structures generate acoustic waves with a length scale of $1/M$, (M is Mach number) bigger. For aircraft at take off and landing M is relatively low. Hence the acoustic wavelengths are much longer than the turbulence scales. Notably, in contrast, the acoustic amplitudes are orders of magnitude lower than the fluctuating amplitudes of the turbulent sources. Turbulence is extremely inefficient at generating sound (Crighton 1975, p. 31). Hence, when solving for such small relative

P.G. Tucker, *Unsteady Computational Fluid Dynamics in Aeronautics*,
Fluid Mechanics and Its Applications 104, DOI 10.1007/978-94-007-7049-2_6,

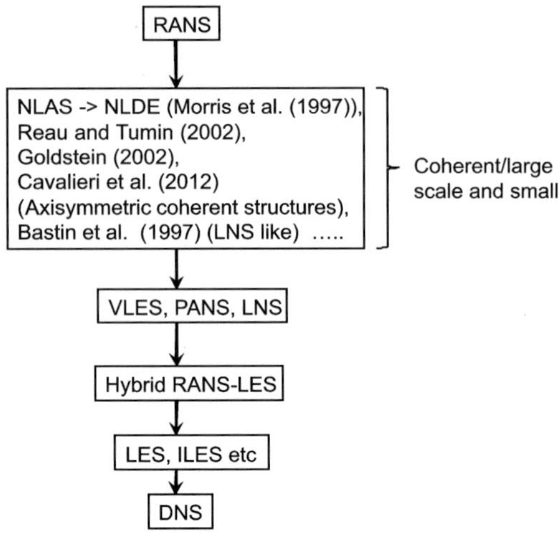

Fig. 6.1 Turbulent noise source/flow field modelling hierarchy

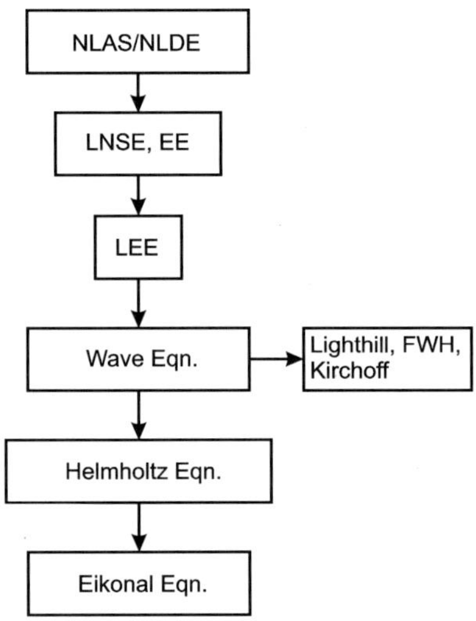

Fig. 6.2 Wave propagation hierarchy

acoustic perturbations in a non-linear flow solver, rounding errors can contaminate solutions. Also, as noted in Chap. 2, to deal with the extreme disparities in wavelengths and amplitudes specialised numerical schemes are beneficial for acoustics, such as dispersion relation preserving approaches.

6.1.1 Noise Prediction Hierarchy and Methods

Figure 6.1 gives a noise source, $S_{\phi'}$, prediction hierarchy. PANS and LNS (see Chap. 3) have not, to the author's knowledge, explicitly been directly used for noise prediction. They are included in the Fig. 6.1 hierarchy to see how the other methods

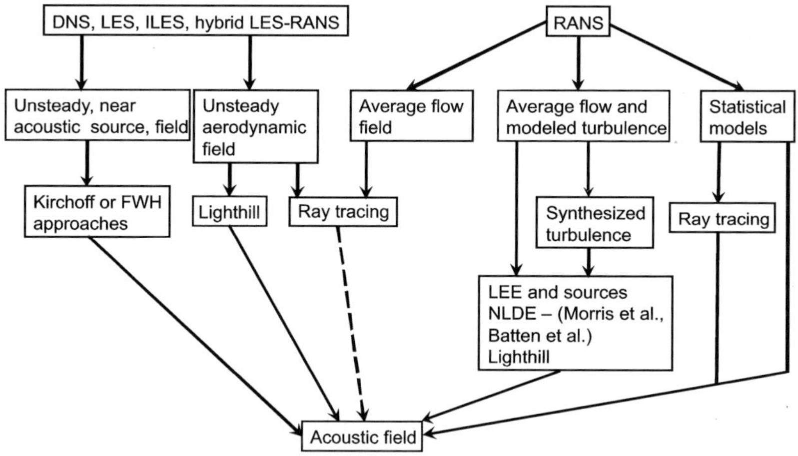

Fig. 6.3 Some potential combinations of CFD and CAA for far-field noise prediction

fit in the wider turbulence-modelling context. The hierarchy is not that different to that presented in the aerodynamics methods, Chap. 4. However, a notable difference is between the RANS and VLES fidelities. In this zone, there sits a substantial amount of work that attempts to partition the turbulence flow field into larger coherent and smaller scales. This reflects the flow dynamics of shear layers and most strongly relates to the area of jet noise modelling. Figure 6.2 gives a hierarchy of wave propagation operators, $L(U, \phi')$. These can have some source characterization information implicit in them.

Figure 6.3 summarizes some of the potential paths from the Fig. 6.1 flow field evaluation procedures and Fig. 6.2 wave propagation techniques to far field sound predictions. These will be considered in this chapter. Steady RANS based methods are only briefly discussed here, being outside the scope of this text. Next forms of $L(U, \phi')$ are discussed. The Fig. 6.2 hierarchy is approximately followed.

6.2 Disturbance Equation Based Approaches

Note, in this section the term disturbance is interpreted broadly, encompassing both non-linear and the linear disturbances found in the linearized Euler equations (LEE).

6.2.1 Nonlinear Disturbance Equation

First we will consider a non-linear acoustic solver (NLAS). This makes use of non-linear disturbance equations (NLDEs). Restating the flux equations given in Chap. 2

we have

$$\frac{\partial \phi}{\partial t} + \frac{\partial E}{\partial x} + \frac{\partial F}{\partial y} = \text{Viscous terms} \tag{6.2}$$

where $\phi = [\rho, \rho u, \rho v, e]^T$, $E = [\rho u, \rho uu + p, \rho uv, eu + pu]^T$, $F = [\rho v, \rho uv, \rho vv + p, ev + pv]^T$.

Just to give the general idea, we restrict ourselves to considering the NLDE in two-dimensions. Clearly the extension to three-dimensions is straightforward. The equations are fully written in Morris et al. (1997). To derive the equations, the vector ϕ is decomposed into time mean, $\bar{\phi}$, and fluctuating, ϕ', components as below

$$\phi = \bar{\phi} + \phi' \tag{6.3}$$

Substituting the decomposition Eq. (6.3) in to Eq. (6.2) yields, after some rearrangement, and ignoring the viscous perturbation terms (large-scale fluctuations are inviscid in nature), yields

$$\frac{\partial \phi'}{\partial t} + \frac{\partial E'}{\partial x} + \frac{\partial F'}{\partial y} + \frac{\partial \tilde{E}'}{\partial x} + \frac{\partial \tilde{F}'}{\partial y} = S_\phi \tag{6.4}$$

The primed terms (Φ') relate to linear disturbances. The primed terms with tildes as overbars ($\tilde{\Phi}'$) represent nonlinear disturbances. The solution vector for the disturbance is

$$\phi' = \begin{bmatrix} \rho' \\ \bar{\rho} u' + \rho' \bar{u} + \rho' u' \\ \bar{\rho} v' + \rho' \bar{v} + \rho' v' \\ e' \end{bmatrix} \tag{6.5}$$

The source term takes the form

$$S_\phi = -\left(\frac{\partial \bar{E}}{\partial x} + \frac{\partial \bar{F}}{\partial y} \right) + \text{Mean viscous stresses} \tag{6.6}$$

The relevant flux terms are given in Table 6.1 below. The form of the mean viscous fluxes is not outlined here. If the nonlinear disturbance equation is time averaged, it yields the usual Reynolds averaged Navier-Stokes equations. The left hand side of the disturbance equation would give the Reynolds stress terms. The source gives the remaining elements of the Reynolds averaged equations. For laminar flow, the left-hand side would be zero.

In the NLAS work of Morris et al., a RANS solution provides the mean flow field. The perturbation equation provides non-linear perturbations around this. Also, viscous terms are ignored in actual simulations. In the NLDE work of Batten et al. (2004) viscous terms are fully accounted for. Their inclusion or not is flow physics dependent. In the approach of Batten et al., subgrid scale turbulence fluctuations can be imposed on solutions. This is achieved through using the synthetic turbulence construction procedure of Batten et al. outlined in Sect. 3.10.2. The turbulence input

Table 6.1 Flux like terms for non-linear disturbance equation

	Φ'	$\tilde{\Phi}'$	$\bar{\Phi}$
E	$\bar{\rho}u' + \rho'\bar{u}$	$\rho'u'$	$\bar{\rho}\bar{u}$
	$\rho'\bar{u}^2 + 2\bar{\rho}\bar{u}u' + p'$	$2\rho'u'\bar{u} + \bar{\rho}u'^2 + \rho'u'^2$	$\bar{\rho}\bar{u}^2 + \bar{p}$
	$\bar{\rho}\bar{u}v' + \bar{\rho}\bar{v}u' + \rho'\bar{u}\bar{v}$	$\bar{\rho}u'v' + \rho'v'\bar{u} + \rho'u'\bar{v} + \rho'u'v'$	$\bar{\rho}\bar{u}\bar{v}$
	$u'(\bar{e} + \bar{p}) + \bar{u}(e' + p')$	$u'(e' + p')$	$\bar{u}(\bar{p} + \bar{e})$
F	$\bar{\rho}v' + \rho'\bar{v}$	$\rho'v'$	$\bar{\rho}\bar{v}$
	$\bar{\rho}\bar{v}u' + \bar{\rho}\bar{u}v' + \rho'\bar{u}\bar{v}$	$\bar{\rho}u'v' + \rho'v'\bar{u} + \rho'u'\bar{v} + \rho'u'v'$	$\bar{\rho}\bar{u}\bar{v}$
	$\rho'\bar{v}^2 + 2\bar{\rho}\bar{v}v' + p'$	$2\rho'v'\bar{v} + \bar{\rho}v'^2 + \rho'v'^2$	$\bar{\rho}\bar{v}^2 + \bar{p}$
	$v'(\bar{e} + \bar{p}) + \bar{v}(e' + p')$	$v'(e' + p')$	$\bar{v}(\bar{p} + \bar{e})$

to this is based on the RANS data. Hence, the decomposition of the flow involves both subgrid scale and non-linear disturbance fluctuations. The approach therefore decomposes the flow field into three components. Hence, the approach allows the extraction of acoustic source terms from the subgrid scale structures.

Terracol et al. (2005) apply the NLDE approach to a three-element, high-lift wing. The key NLDE zone was in the slat cove-wing leading edge vicinity—this being a key noise source zone with wake impingement, shear layers, additional erratic vortex shedding due to a larger vortex 'breathing' process.

The NLDE approach displays less sensitivity to numerical errors. However, when the disturbance zone is locally embedded into *steady* RANS zones avoiding interface reflections is a key challenge.

6.2.2 Other NLDE Decompositions

Jet noise can be viewed in some senses at quite a specific class a problem. There is much experimental evidence to suggest that the flow from jets can be decomposed into three quite distinct levels. These being the usual mean flow, turbulent stochastic fluctuations and also large coherent unsteady flow structures. The latter are especially key elements in supersonic jet noise. When these larger eddies move at the speed of sound they become efficient noise radiators. Bastin et al. (1997) apply the three-level decomposition noted above to jet noise. However, in essence a basic URANS approach is taken. The eddy viscosity coefficient (C_μ-directly scaling eddy viscosity, see Appendix B) in the k–ϵ model is reduced, allowing the solution to support larger coherent strictures. Then, the k–ϵ model is left as a type of 'subgrid scale model' to account for the more 'stochastic' fluctuations. The method is reminiscent, in a crude sense, of the LNS approach (see Chap. 3). However, the partitioning of the coherent and stochastic scales is ad hoc. Reau and Tumin (2002) outline a more rigid procedure. Disturbance equations for the coherent scales are specifically formulated, closure being achieved through Boussinesq type modelling

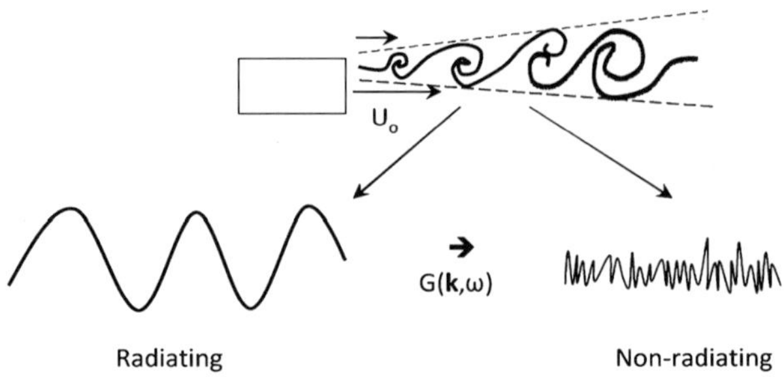

Fig. 6.4 Decomposition of jet into coherent and incoherent fields

terms, as used in RANS. Further related techniques are the Coherent Vortex Simulation (CVS) (Farge and Schneider 2001) and semi-deterministic method (SDM) of Ha Minh (reported in Druault et al. 2004). These both isolate coherent structures in differing ways. Farge and Schneider use wavelets to enforce the scale separation considering a mixing layer and other flows.

6.2.3 Linearized Navier-Stokes Equations

Linearized Navier-Stokes Equations (LNSE) method of Goldstein (2003) takes the Navier-Stokes equations and makes the usual decomposition of variables $\phi = \bar{\phi} + \phi'$. This is substituted into the Navier-Stokes equations. The variable $\bar{\phi}$ is furnished by a RANS like equation system that has unknown fluctuating source terms (These can be later modelled using Boussinesq type approximations). The RANS type equation is subtracted from that resulting from the combination of the Navier-Stokes equations and the decomposition $\phi = \bar{\phi} + \phi'$. This subtracted equation set is then rearranged to yield a linearized, left hand side, Euler operator with an exact source term. The resulting equations are an exact consequence of the Navier-Stokes equations. The mean flow field that goes into the resulting system can come from RANS, URANS, LES or DNS. For the latter three, the instantaneous flow would be time averaged. If working totally with mean target quantities the source specification reduces to working out forth order space-time correlations of velocity fluctuations and, for heated flows, velocity-temperature fluctuations. In the pioneering work of Harper-Bourne (2003) the 4th order, space-time, velocity correlations are measured.

When URANS is used, then the application of Goldstein's approach is more complex. A time average of the URANS field can be made and the fluctuations resolved by the URANS expressed as additional 'turbulent' stresses to be characterized by source terms. In an LES context typically the grid will be sufficiently fine that any subgrid scale contributions can be ignored.

When using the general LNSE approach, for flows with large high speed coherent structures, such as jets (see Fig. 6.4) Fourier partitioning can be used (see later). This allows the non-radiating source element (perhaps characterized via Boussinesq type modelling) to be segregated from the more coherent jet structures.

6.2.4 *Linearized Euler Equations (LEE)*

When considering the linearized Euler equation, Eq. (6.5) reduces to

$$\phi' = \begin{bmatrix} \rho' \\ \bar{\rho}u' \\ \bar{\rho}v' \\ p' \end{bmatrix} \tag{6.7}$$

Also, Eq. (6.4) can be expressed as (Bailly and Juve 1999)

$$\frac{\partial \phi'}{\partial t} + \frac{\partial E'}{\partial x} + \frac{\partial F'}{\partial y} = S_\phi \tag{6.8}$$

In the above, the primed terms again involve mean quantities and disturbances—but this time the latter are linear. Also, the source term, as formulated by Bailly and Juve (1999) contains both sound sources and refraction terms. The simple linearized equations only capture first order influences. Hence, the scattering of sound by turbulence is ignored.

The noise prediction procedure used by Bailly and Juve is similar in spirit, in a sense, to that of Morris et al. and Batten et al. There is a three level decomposition with a mean flow field, turbulent fluctuations and acoustic fluctuations. The former is obtained from a RANS solution. The turbulence energy and length scales from this are fed into the synthetic turbulence generation procedure of Kraichnan (1969)—see Sect. 3.10.2. The RANS solution is interpolated onto a separate acoustics mesh and the LEE with synthetic turbulence source used to compute far field sound.

6.3 Solution of Linearized Equations

The LEE and LNSE can be cast into the frequency domain. Then the acoustic wave propagation can be solved at specified frequencies. This presents and efficient solution process. However, it can have the following drawbacks:

1. As sound propagates (for example between rotating blades rows in turbomachinery) the dominant acoustic frequency can shift, requiring solution of the Euler equations;
2. If say a RANS predicted base flow is interpolated onto an acoustics propagation mesh, since the base flow solution will not be compatible with the mesh, convergence of the LEE can be difficult to secure. Strong solvers (e.g. Krylov) can be used but this process, itself, can lead to physically incorrect solutions
3. The LEE/LNSE can support hydrodynamic instabilities and this can contaminate the predicted sound.

The sound fluctuations are typically orders of magnitude smaller than those of the turbulence. Also, the acoustic fluctuations are orders of magnitude smaller then the numerical errors associated with computing the turbulence (Goldstein 2002). Hence, it is important when solving the LNSE/LEE that the acoustic perturbations are not corrupted by numerical errors. According to Goldstein, in this context it is helpful to have the base flow non-radiating. If the base flow is radiating then the source will be incorrect, having a mix of both acoustic and turbulence fluctuations. Even though the former is low, it can corrupt the far field predicted sound (Ewert and Schroder 2003). For RANS, a non-radiating field naturally occurs. The alternative is to have an incompressible base flow or the requirement that the time derivative of density in the base flow is small. However, to enforce this, source terms need to be added to ensure the equations can be satisfied and hence driven to iterative convergence. Hardin and Pope (1994) ensure that the base flow is non-radiating by solving the Navier-Stokes equations in their incompressible form. The approach is applied to various unsteady acoustic problems including an unsteady cavity flow. A difficulty with this rigid segregation of the acoustic and hydrodynamic information is that acoustic feedback can profoundly influence the actual turbulent flow field. Another approach, suggested by Goldstein (2002), to make an *unsteady base flow* non-radiating, is to use filtering. The filter function can take the broad formulation outlined in Sect. 3.2.2 but this time extended as a space-time filter

$$\bar{\phi} = \int_{-\infty}^{\infty} \int_V \phi(\mathbf{r}, \tau) \tilde{G}(\mathbf{x} - \mathbf{r}, t - \tau) d\mathbf{r} d\tau \tag{6.9}$$

where this time $\tilde{G}(\mathbf{x} - \mathbf{r}, t - \tau)$ is a space-time filter function. Note, such a filter has also been proposed by Pruett (2000) and other workers in a purely LES context. As suggested by Goldstein the filtering operation can be passed into Fourier space

$$\bar{\Phi}(\mathbf{k}, \omega) = (2\pi) G(\mathbf{k}, \omega) \Phi(\mathbf{k}, \omega) \tag{6.10}$$

In the above the capital letters now represent the Fourier transform. The Fourier transform of the filter $G(\mathbf{x} - \mathbf{r}, t - \tau)$, $G(\mathbf{k}, \omega)$ that only passes subsonically convecting eddies, is given by

$$G(\mathbf{k}, \omega) = \begin{cases} 1/2\pi^3 & \text{for } \omega^2 < c_o^2 k^2 \\ 0 & \text{otherwise} \end{cases} \tag{6.11}$$

In the above c_o—is the ambient speed of sound. Naturally, $\tilde{G}(\mathbf{x} - \mathbf{r}, t - \tau)$ can be constructed by taking the inverse Fourier transform of $G(\mathbf{k}, \omega)$. If using a spectral solver this separation of scales can be more naturally imposed by selection of the base functions used in the computation.

As noted above the LEE and LNSE can, where there are inflectional velocity profiles, support hydrodynamic instabilities (these equations support acoustic, entropy and vorticity waves). Hence, at the noise propagation stage results can be compromised by the generation of hydrodynamic instabilities. The situation is worsened by the fact that for wave propagation, the high order schemes typically used, will readily tend to pick up such instabilities. In the example of engine intakes, the background turbulence levels will typically be low. This will again facilitate instability growth (Ozyoruk and Long 1997). These can be suppressed by specialised numerical treatments, such as strong solvers, time-step selection procedures (Karabasov et al. 2010), or the use of specialized equations. An example of the latter are the, linearized Euler based, Acoustic Perturbation Equations (APE) of Ewert and Schroder (2003). These are designed to prevent the growth of hydrodynamic instabilities. There are various APE formulations. Koh et al. (2012) apply a revised APE-4 source formulation to modelling of jet noise from hot co-flowing jets. The flow field is computed using LES. The standard source formulation produced an overestimate of low frequency sound. In the above work, the APE equations are solved using a 7-point DRP scheme with an alternating 5–6 stage LDRK scheme. Birkefeld and Munz (2012) solve the APE-4 formulation, using it to explore noise propagation around a wing-flap-slat system. The noise source description is based on RANS data and converted into a time dependent noise source using the RPM (see later) method. In the near wing-surface zone, the APE equations are solved using the discontinuous Galerkin approach (see Chap. 2). This is coupled to a high order, strictly Cartesian approach away from the surfaces. The solution order is 4th order in both space and time but, long period, time integration instability is observed.

6.4 Wave Equation and Simplifications

6.4.1 Wave Equation

The wave equation

$$\nabla^2\phi - \frac{1}{c^2}\frac{\partial^2\phi}{\partial t^2} = 0 \tag{6.12}$$

is the obvious choice of operator for propagating source information but does not account for refraction. Note, c is the speed of sound.

6.4.2 Helmholtz Equation

The Helmholtz is a simplified wave propagation equation. If we assume $\phi' = e^{i\omega t}$ where ω is the sound wave frequency, in radians, the wave equation can be directly re-expressed as $-(\omega^2/c^2)\phi' - \nabla^2\phi'$. Defining the wave number, w, of the sound field as $w = \omega/c$ allows the Helmholtz equation below to be written. Then through use of Fourier transforms the Helmholtz equation can be expressed as below

$$\nabla^2\phi' = -w^2\phi' \tag{6.13}$$

In the above w is a chosen wave number. The above equation is relatively easy to solve and is amenable to Boundary Element Method (BEM) solution.

6.4.3 Eikonal Equation

The eikonal equation is the high frequency limit to the wave equation. Since it is the high frequencies that can be most annoying to the human ear, eikonal equation solutions are potentially useful. The human ear is most sensitive to frequencies in the range of 2000–4000 Hz. Hence, since wavelength, $\lambda = c/\omega$, where ω is frequency, this corresponds to λ in the range of 0.08–0.16 m. The eikonal equation is based on the ray-tracing theory which is valid for $\lambda \ll H$ (where, here, H, is now a characteristic length scale) or rather $\omega H/c \gg 1$. The latter, for example, makes the eikonal approach suitable for long range propagation such as in the design of noise shielding for aero engines. The general limitation of the eikonal approach, however, is that it is not accurate for sound propagation through thin shear layers. In these, the mean velocity gradients are of the order of the high frequency range of audible sound. Freund (2003) solved the eikonal equation in an unsteady form for turbulent jets with the mean flow coefficients based on the results of DNS. Khritov et al. (2005) and Secundov et al. (2007) directly solved the eikonal for unsteady jet flows representative of those found in aero engines. Apart from more theoretical studies, the eikonal equation can be a valuable engineering tool in the context of studying acoustic shielding designs. For example, with aircraft, the engines can be located so that the airframe shields observer noise, thus reducing environmental impact (Agarwal et al. 2007, Moore 2004, Moore and Mead 2003).

The eikonal equation can be derived in a variety of ways. For example, by performing Fourier transform in space and time of the linearised Euler equations and keeping just the highest order terms in frequency of the acoustic part of the equations (e.g., Colonius et al. 1994) leads to

$$\left(1 - u_n H \frac{\partial \tilde{\phi}}{\partial x_n}\right)^2 - H^2 c^2 \left(\frac{\partial \tilde{\phi}}{\partial x_n}\right)^2 = 0 \tag{6.14}$$

In the above $\tilde{\phi}$ is the dimensionless eikonal where $\phi = \tilde{\phi} H$, which is proportional to the propagation time of the sound ray. Note, H is a system length scale used to assist with non-dimensionalization.

6.4.4 Eikonal Equation Applications Contrasting with Euler Equation Solutions

Plane Wave Propagating through a Vortex In aerospace engineering vortices can be frequently found, for example, in the wake of aircraft. Studying the interaction of acoustic rays with vortices is also necessary for understanding the passage of noise through turbulent flows. Following Colonius et al. acoustic ray deflection for a plane wave passing through a vortex (Georges 1972) is first considered through solution of both the Euler and eikonal equations. Four rotational Mach numbers (M_{max})

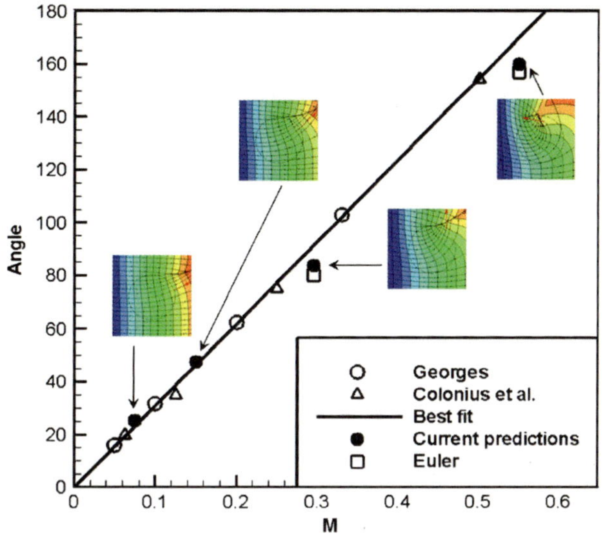

Fig. 6.5 Plane wave front propagating though a viscous vortex for $M_{max} = 0.075$, 0.15, 0.3 and 0.55 (Copyright, Multiscience Publishing, used with permission

of 0.075, 0.15, 0.33 and 0.55 are considered. These are based on the maximum vortex tangential velocity. Figure 6.5 plots maximum ray deflections against M_{max}. The open circles represent the predictions of Georges (1972). The triangles give those of Colonius et al. (1994). The line is a best fit to their data. The square symbols are for Euler simulations. These use the high resolution CABARET scheme, discussed in Sect. 2.12.2. The closed circles are for eikonal solutions. For the high frequency considered there is close agreement with the Euler solutions. Note, the insets show wave fronts and rays for eikonal solutions. The ray bending with increasing M_{max} is evident from the ray traces.

The problem is next considered further with the Euler method. Figure 6.6 gives normalized r.m.s. directivity of the scattered pressure field. Frame (a) shows comparison with the 6th-order compact scheme, Navier-Stokes, solution of Colonius et al. This is for the dimensionless acoustic frequency $\widetilde{f} = 2\pi R/\lambda = 2.5$, where R is the vortex radius and λ is the wavelength of the incoming acoustic wave. Frame (b) compares the Euler results with the ray-theory characteristic solution (dotted lines shows the caustics locations) of Colonius et al. In this comparison the frequency is higher with $\widetilde{f} = 82.5$. Note, since the CABARET scheme is high resolution, just, 5 points per vortex radius for, $\widetilde{f} = 2.5$, and 7 points per wavelength for $\widetilde{f} = 82.5$ are needed for accurate solutions.

Acoustic Rays in Jets As noted earlier, the eikonal equation has multiple uses. Hence, as well as understanding fundamental physics, it can be utilized in system design. For example the solid lines, in Fig. 6.7a, represent acoustic rays passing through a time averaged flow field from a hybrid RANS-(N)LES type computation (Eastwood et al. (2012) for full details). This simulation is for a heated coflowing jet. Acoustic rays are shown propagating from the centre of the jet at different locations in the x–z plane. The background contour plot shows the mean streamwise velocity.

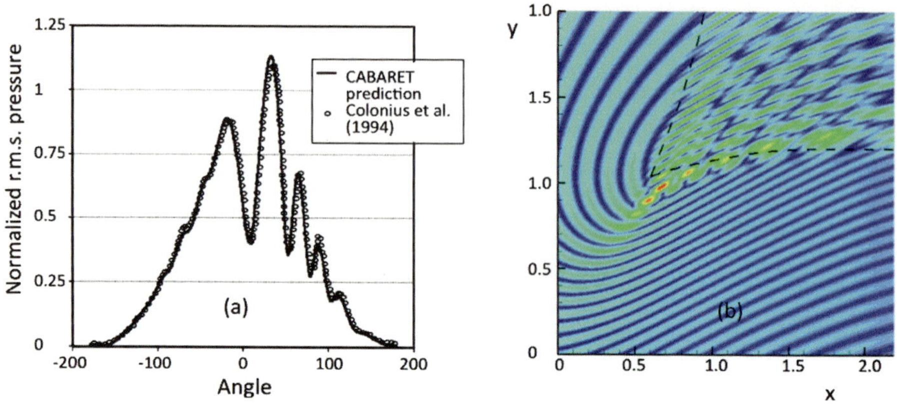

Fig. 6.6 Validation of Euler solution for the r.m.s. directivity of the scattered pressure field for $M_{max} = 0.25$ against results of Colonius at al.: (**a**) comparison with the reference 6th-order compact scheme Navier-Stokes solution for $\widetilde{f} = 2.5$, (**b**) comparison with the reference ray-theory characteristic solution (*dotted lines* shows the caustics locations) for the dimensionless acoustic frequency $\widetilde{f} = 82.5$ (Copyright, Multiscience Publishing, used with permission)

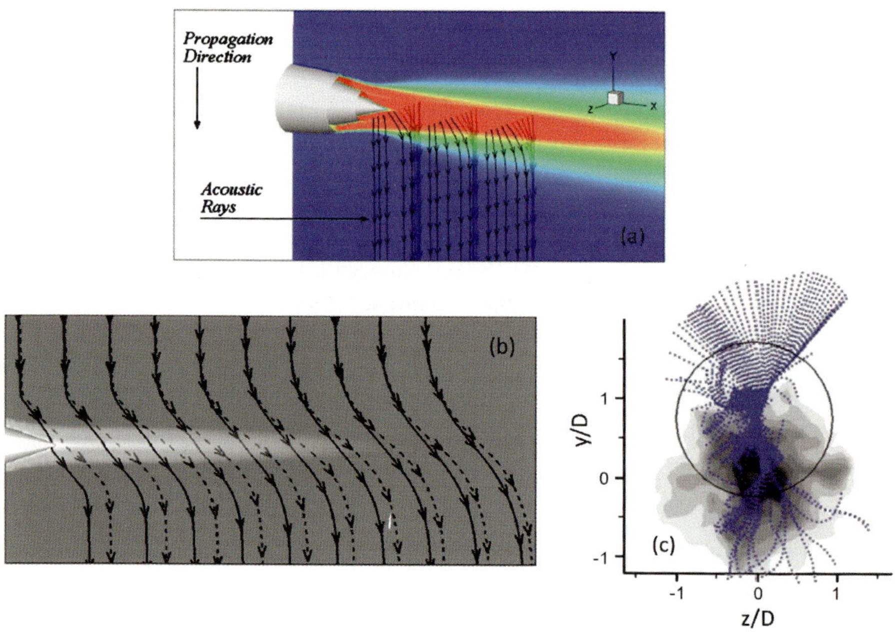

Fig. 6.7 Acoustic rays passing through coflowing jets with (N)LES flow fields: (**a**) three-dimensional ray field in time averaged flow; (**b**) two-dimensional planar view of acoustic rays (for heated and unheated flow) and (**c**) rays in the instantaneous flow field (adapted from Eastwood et al. 2012 and Secundov et al. 2007)

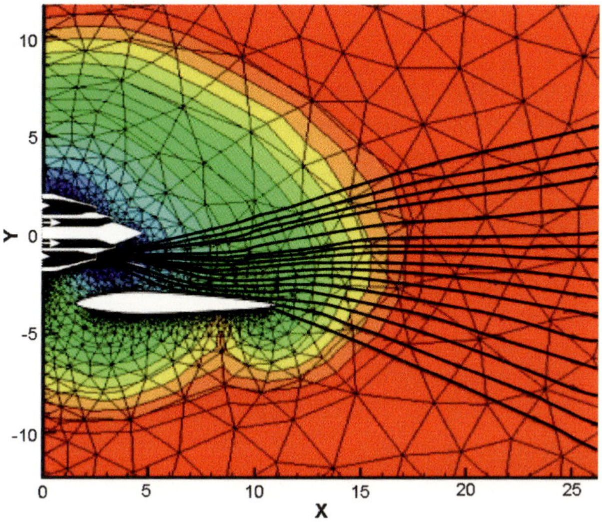

Fig. 6.8 Wing-jet schematic and eikonal solutions: (**a**) schematic of geometry; (**b**) eikonal solution (Copyright, Multiscience Publishing, used with permission)

As can be seen, the acoustic rays are strongly deflected (in the streamwise direction) by the flow. The greatest deflection is on the centerline of the jet. By varying c, Fig. 6.7b explores the effect of temperature on the acoustic rays. In an adjoint sense the rays are released from the far field and their subsequent deflection as they pass through the jet explored. The solid and dashed lines plot rays for c based on hot and cold fields. The background contour plot shows c for the heated jet. For conditions representing heated flow the rays are deflected least.

Figure 6.7c is from Secundov et al. (2007). The instantaneous flow field from an LES of a coflowing jet is internally seeded with numerous rays. From comparison with frames (a, b) it can be seen that the rays are actually deflected more by the instantaneous turbulent flow events than by the refraction due to mean flow field and temperate gradients. Notably, also rays are unable to pass from one side of the jet to the other—the jet being 'opaque' to high frequency sound. Information from ray tracing can be used to refine statistical RANS based far field noise prediction procedures (see Secundov et al.) and even be used as an integral part of them. For example, Ilario et al. (2010, 2011a,b), use Lighthill's acoustic analogy in a RANS based predictive framework to get far field sound. Ray tracing is used to predict the influence of refraction and hence correct the Lighthill based predictions. The analysis was applied to three-dimensional propulsive jet nozzles.

Airframe-Engine Aeroacoustic Integration Cleary it is not enough to model jet noise in isolation. The sound source and its propagation can potentially be modified by the presence of the airframe. Hence, finally, Fig. 6.8 gives a schematic of an idealized wing-jet geometry along with an eikonal solution. Frame (a) gives (for proof of concept) the two-dimensional geometry. This consists of a coflowing jet nozzle, with a wing beneath it. Note, the flow field is steady. The intention is that the 'wing's' position is optimized for acoustics shielding. There is an acoustics source placed in the vicinity of the lower co-flow stream. Frame (b), gives eikonal equation computed acoustic wave fronts and acoustic rays emanating from the sound source. These are all superimposed on the triangulated background mesh used to solve the eikonal equation in a Eulerian framework. The acoustic shielding influence of the wing is clear from the Frame (b) simulation. Since, here, the eikonal equation is solved in a Eulerian frame, any reflected rays would need to be computed as a separate solution, the ray starting points being placed on the wing surface. The application of a Lagrangian ray tracing approach to a blended wing-body aircraft, to explore airframe-engine integration noise aspects, is outlined by Agarwal et al. (2007).

6.5 Integral Equation Based Approaches

As noted above, near field sound/source information can be extrapolated to the far field through post processing procedures based on analytical expressions. These are sometimes called integral methods and are reviewed by Lyrintzis (2003). The key integral methods, that are most widely used, with eddy resolving simulations, are the Kirchhoff and FWH.

6.5.1 The Lighthill Equation

The founding work, relating sound sources to wave propagation is due to Lighthill (1952). Lighthill's analogy rearranged the Navier-Stokes equations to give the wave equation operator, $L(\rho')$ [1] with a quadrupole type source term $S_{\rho'}$. Hence, with Lighthill's approach

$$L(\rho') = \frac{\partial^2 \rho'}{\partial t^2} - c^2 \nabla^2 \rho' \quad \text{and} \quad S_{\rho'} = \frac{\partial^2 T_{ij}}{\partial x_i \partial x_j} \tag{6.15}$$

If c and ρ can be considered constant

$$T_{i,j} = \rho u_i' u_j' \tag{6.16}$$

Therefore, we can see turbulence is a key sound source. If the unsteady turbulence field is known, Eq. (6.15) source term could be directly specified. Equation (6.15)

[1] Note, the wave operator can be expressed in the Helmholtz (harmonic) form—see later.

is exact. Even though the wave equation is used, for L, refraction and convection is accounted for via $S_{\rho'}$ (Ewert and Schroder 2003). For the calculation of far field sound, Lighthill derived the following (Green's function) based analytical solution to Eq. (6.15)

$$4\pi p'(\mathbf{x}, t) = \frac{x_i x_j}{|\mathbf{x}|^3} \int_V \frac{\partial^2 [T_{ij}(\mathbf{y}, \tau_r)]}{c^2 \partial t^2} dV \tag{6.17}$$

In the above (which cannot model refraction influences) volume (V) integral, $\mathbf{y}$ is the location of the quadrupole sound source (i.e. turbulence). The observer location is $\mathbf{x}$. The sound generated at $\mathbf{y}$ arrives at $\mathbf{x}$ at the later time $|\mathbf{x} - \mathbf{y}|/c$. Hence, the quantities in square brackets are computed at a retarded time, which is the moment when sound was generated and where $\tau_r = t - |\mathbf{x} - \mathbf{y}|/c$. Note, the observer location is assumed relatively far from the source. As can be seen from Eq. (6.17), the effect of the sound source diminishes inversely with distance from the source i.e. as $1/|\mathbf{x} - \mathbf{y}|$. Also, simple scaling analysis can, for example, be used to calculate the power from a jet of diameter, *D*. The characteristic time scale of the turbulence will be approximately *D/U* where *U* is the jet velocity. In relation to Eq. (6.17) we can apply the following scalings $t \to D/U$, $\int dV \to D^3$, $T_{ij} \to \rho U^2$ to give $4\pi p'(\mathbf{x}, t) \sim U^4 \rho D / x c^2$. Then the sound intensity is proportional to $\overline{p'^2} \propto U^8$ giving Lighthill's famous result that jet noise goes as U^8.

6.5.2 Ffowcs Williams and Hawkings Equation

There are other alternatives to Lighthill's volume integral expression. For example, the equation of Curle (1955), to predict far field sound from a flow region, V, adjacent to a fixed surface, s. Ffowcs-Williams and Hawkings (1969) extended this to when s is moving. Hence, solid moving surfaces can be considered such as rotors. For the FWH equation, L is the same as Lighthill's. Indeed, the FWH theory generalizes Lighthill's acoustic analogy. However, now

$$S_{\rho'} = \underbrace{\frac{\partial^2 T_{ij}}{\partial x_i \partial x_j}}_{\text{Volume integral}} \underbrace{- \frac{\partial}{\partial x_i}\left[p'_{ij} n_j \delta(\varphi)\right] + \frac{\partial}{\partial t}\left[\rho_o u_n \delta(\varphi)\right]}_{\text{Surface integrals}} \tag{6.18}$$

where $p'_{ij} = p_{ij} - p_o \delta_{ij}$ is a perturbation stress tensor and δ_{ij}, the Kronecker delta. Also, φ is a level-set variable, and δ the Dirac delta function. It takes a value of zero on a surface, is negative inside s and positive outside it. On analytical solution of the FWH equation, the first source term gives rise to a volume integral. The latter two source terms give surface integrals. Equation (6.18) solution yields the following relation for the far field acoustic pressure

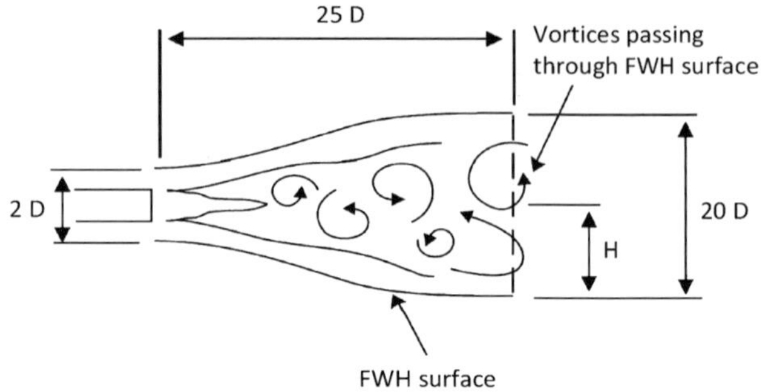

Fig. 6.9 FWH surface surrounding a jet with vortical structures passing through the downstream closing disk (shown by the *vertical dashed line*)—from (Eastwood et al. 2009)

$$4\pi p'(\mathbf{x},t) = \frac{x_i x_j}{|\mathbf{x}|^3 c^2}\frac{\partial^2}{\partial t^2}\left[\int_V T_{ij} dV\right] + \frac{x_j}{|\mathbf{x}|^2 c}\frac{\partial}{\partial t}\left[\int_s p' n_j + \rho u_j u_n ds\right] + \frac{1}{|\mathbf{x}| c}\frac{\partial}{\partial t}\left[\int_S \rho u_n ds\right] \tag{6.19}$$

In the above, V, is the volume outside s. Also, n_j are the projections of the vector of outer normal to s. $T_{ij} = \rho u_i u_j + (p' - c_0^2 \rho')\delta_{ij}$ is the high Reynolds number form of the Lighthill tensor.[2] Also, u_n is the velocity component normal to s.

If all noise sources are inside s, then the first term on the right hand side of Eq. (6.19) can be omitted. Since it is a volume integral, over the whole FWH volume, this term is expensive to compute. For an LES, it would require the storage of the time history of the data in V. Hence, the volume integral omission simplifies the solution. Its presence detracts from the use of the Lighthill equation.

There are various FWH formulations intended for different purposes. The one used here is called Farassat's formulation 1. For a wing or landing gear, the surfaces could define s. There could also be a volumetric turbulent source zone outside of this. Use of the equation in this way is not made use of here. A principle difficulty with the FWH approach is enclosing all the noise sources within s. For example, with jets, eddies convect downstream and pass through s. Figure 6.9 illustrates the problem. It shows typical dimensions of the FWH surface used in jet LES work discussed here. Note, D is the jet diameter. For results given here, during an LES type computation, flow variables are saved on s. The solution of the FWH equation is made in the time domain. Saving is done at time intervals of $\Delta\tau$, that can be multiples of Δt, the flow field solution time step. A postprocessor then later reads in the saved FWH surface data. It calculates the sound pressure level (SPL) at user specified locations, $\mathbf{x}$, from the jet. It is also possible to evaluate the FWH integral 'on the fly.' Shur et al. (2003) solve the FWH equation in both the frequency and time domains.

[2] Note, this is still not the full form of the Lighthill stress tensor.

6.5.3 Kirchhoff Approach

A popular alternative to the FWH approach is the Kirchhoff. This is applicable to any phenomena governed by the linear wave equation e.g. optics, electromagnetics, etc. It involves specification of a single field property and the integral surface. For acoustics this would be the pressure fluctuation. This is different to the FWH approach where the pressure, velocity and density fluctuations need to be specified at the surface. Notably, Francescantonio (1997) uses the FWH equations to try and form a Kirchhoff like formulation. This is intended to make the best of the pros and cons of these different methods (see later). Andersson et al. (2005) and Bodony and Lele (2004) use a Kirchhoff surface method obtaining encouraging results for far field sound from LES of axisymmetric jets. However, even at 30° there is variation of up to 8 dB between different predictions.

6.5.4 Advantages and Disadvantages

As noted above, the volume integral in Lighthill's approach is problematic for LES. A key issue with the Kirchhoff approach is the especially high sensitivity of solutions to surface location. Although the Kirchhoff approach has the simplification of having less variables that need to be specified at the surface it does need the normal gradient of pressure. This is more complicated to specify and could become singular. However, advantageously, for rotating systems a stationary Kirchhoff surface can be used. This avoids problems associated with supersonic surfaces in motion (Francescantonio 1997).

As noted above, there are different FWH formulations. Care is required that the most appropriate formulation is being used. Lew et al. (2005), Shur et al. (2003) and Eastwood et al. (2009) use the FWH method to predict noise from LES type computations of axisymmetric jets. Shur et al. (2003) considered an $M = 0.9$ cold jet. Placement of the FWH surface in the optimal position yielded sound predictions within 2 dB, for overall far field sound intensity. However, there is a sensitivity of up to 10 dB. This depends on the shape and size of the FWH surface. This is explored next.

Figure 6.10 shows the location of the FWH surface for a typical jet noise LES. Contours of instantaneous jet axial velocity are shown along pressure fluctuations, p'.

Figure 6.11a plots the overall sound pressure level (OASPL) against directivity for NLES computations of an axisymmetric jet. The OASPL is plotted at $100D$ from the nozzle. The angle 0° is the downstream direction and 90° is the sideline. The symbols show the measurements of Tanna (1977). The lines show results for various extents of the FWH closing disk, i.e. values of H between 0 and 1 (illustrated in Fig. 6.9). The variation of OASPL for different values of H is up to 8 dB. The trends for different H are similar, except for $0H$ (solid line), where there is a rise in OASPL for angles greater than 80°. Integration of the OASPL curves, in Fig. 6.11a allows a

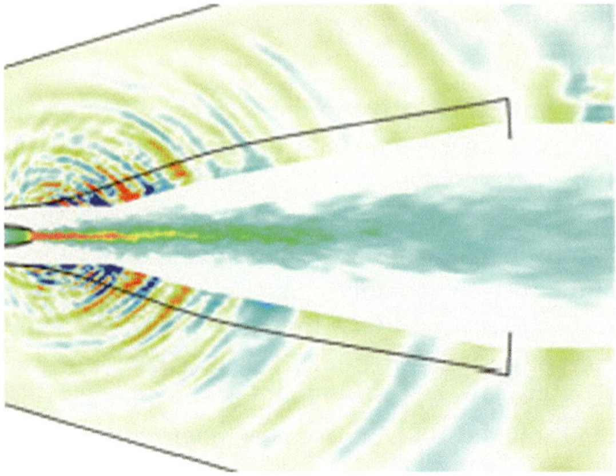

Fig. 6.10 Near field pressure fluctuation, p', snapshot, overlaid with central instantaneous jet velocity contours (the open FWH surface also shown) for eddy resolving jet simulation

single value to be plotted against H. This is shown in Fig. 6.11b. The straight dashed line shows the integral of the measurements. The results show that the FWH surface configuration can have a significant influence on the predicted sound.

A range of remedies have been attempted to allow $H = 1$. The simplest is to have multiple disc surfaces (of extent $H = 1$). These are located at different axial locations. An average is then performed of the different results (Shur et al. 2005). The idea behind this is that the pseudo-sound, related to the turbulent eddies, passing through the closing disk, corrupts the far field sound calculation. The pseudo-sound travels at a very different speed to the sound. Averaging is expected to result in a cancellation of pseudo-sound errors (Spalart et al. 2010). If the lateral surfaces are too far away, acoustic wave amplitudes will be diminished. Hence far field sound can be underestimated. If the surface is too close, it will no-longer be in the irrotational flow zone. Then the accuracy of the solution is highly compromised.

The mesh density at the FWH surface and time resolution of the data at it will obviously limit the maximum frequency of the sound that can be resolved. To just increase the frequency resolution by a factor of two equates to grid doubling in all three coordinate directions. Hence, an approximate 10-fold increase in computational cost will only increase the maximum acoustic Strouhal number that can be resolved by a factor of two. Hence, this is a harsh frequency resolution limit. This can be overcome through use of an acoustic subgrid scale model (see, for example, Bodony and Lele 2002, Spalart et al. 2010 and Xia et al. 2012). The other problem is that to adequately resolve low frequencies long LES runs are needed. To overcome this, Xia et al. (2011) proposed a frequency relay approach. With this, the low frequency field, associated with larger structures is resolved on a relatively coarse LES grid. This is amenable to simulating a sufficient physical period at tolerable computational cost. Then this coarser mesh flow field is interpolated onto a finer mesh and the run continued. From the preliminary results of Xia et al., the two spectra (coarse mesh—low frequency and fine mesh—high frequency) appear relatively easy to match together. What is more, acoustic subgrid scale type modelling can also be added. This creates a three-layer approach that addresses both the low and high frequency resolution barriers.

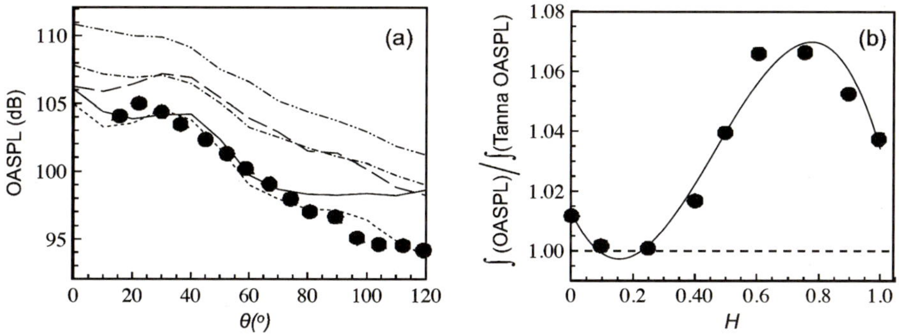

Fig. 6.11 Variation of overall sound pressure level (OASPL) with different FWH closing disk surfaces and integral of OASPL curves: (**a**) OASPL at 100D. *Solid curve*, $H = 0.0$; *dashed curve*, $H = 0.25$; *dot-dashed curve*, $H = 0.5$; *double dot-dashed curve*, $H = 0.75$; *dotted curve*, $H = 1.0$; *triangles*, measurements of Tanna (1977). (**b**) Integral of OASPL at 100*D*. *Squares*, NLES results; *solid curve*, best-fit line of NLES results; *dashed line*, integral of Tanna (1977) measurements (from Eastwood et al. 2009)

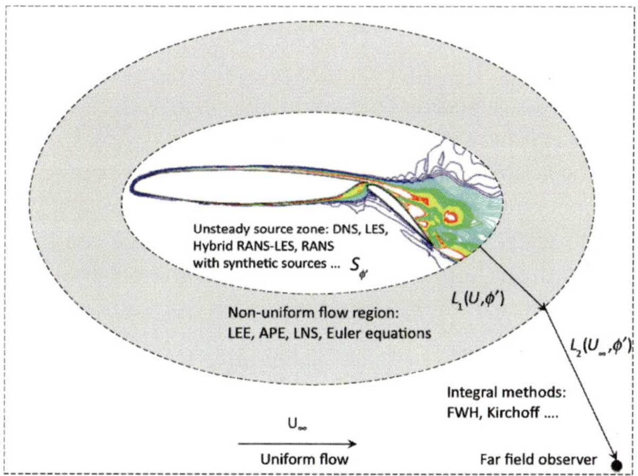

Fig. 6.12 Two component sound propagation

6.6 Hybridization of Wave Propagation Approaches

As shown by Terracol et al. (2005), the far field fluctuations from solution of say a higher fidelity operator $L_1(U, \phi')$, on a grid, can be also be input into integral and other approaches—$L_2(\phi')$. This allows data to be obtained at an even more distant observer location. Figure 6.12 shows potential options for most economical sound propagating from a source region, $S_{\phi'}$, through a zone of non-uniform flow and then finally through a uniform flow zone. For fan noise, issuing from an engine intake, Ozyoruk and Long (1997) solve the Euler equations in the near field. This is then coupled to a Kirchhoff surface. Similarly, for a rotorcraft fuselage, Agarwal and Morris (2000) couple the FWH approach to a near field linearized Euler solution.

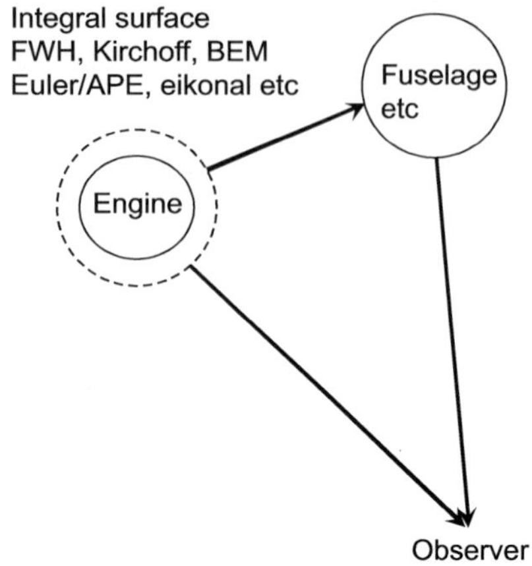

Fig. 6.13 Hybrid superposition approach for modelling far field sound to include engine-airframe interaction

As noted by Mincu and Manoha (2012), the most practical way to explore the sound propagation from an aircraft engine is to use a superposition approach. This is summarized in Fig. 6.13. With this, for example, there could be the following stages:

1. An integral surface can be placed around the engine and the unsteady acoustic emissions stored. This could also involve acoustic propagation modelling inside the engine.
2. The acoustic information at these surfaces can be propagated to an observer and the airframe.
3. The scattered field from the aircraft surface can be propagated to the observer.
4. The direct sound form the engine to the observer and that from the airframe can be summed.

This currently seems to be the most suitable approach for practical design.

6.7 Source Descriptions

6.7.1 Input of Data for Disturbance Equation Methods

Typically, the noise source is input to the propagation equation through source terms. In practical terms, for LES the unsteady flow in a source zone must be stored. This can be volumetrically, as for the Lighthill approach. Alternatively, it can be stored at surfaces. The latter would typically happen for integral methods and is used by Terracol et al. (2005) with the Euler equation. Interpolation can be used between the stored time-levels to produce data at desired points in time for input to the $L(U, \phi')$ operator. Under certain circumstances, if the time sample is insufficiently long, data recycling can be used. The process is repeated until sufficient data is gained for the low frequency part of the far field noise spectrum to be adequately resolved. Typically, when using a field based wave operator, the acoustics grid will have a substantially different topology to that used for the flow field. Hence, spatial interpolation is also required. LES based source information can be used to inform the source description in RANS based methods. This is discussed next.

6.7.2 Turbulence Source Descriptions

The statistical properties of turbulence have a great impact on noise. Far field sound prediction approaches can need a description space-time correlations to characterise the source region. Hence, they are an important validation parameter. Such correlations can also be useful for understanding turbulence physics and for example estimating eddy scales and convection speeds. These again can be key inputs to source models. The 2nd order space-time correlation of turbulence can be expressed as

$$R(\mathbf{x}, \bar{\Delta}, \tau) = \frac{\overline{\phi'(\mathbf{x}, t) \cdot \phi'(\mathbf{x} + \bar{\Delta}, t + \tau)}}{\overline{\phi'(\mathbf{x}, t) \cdot \phi'(\mathbf{x}, t)}} \tag{6.20}$$

where ϕ' is now a fluctuating turbulence quantity. Also, $\bar{\Delta}$ represents a spatial shift of a 'probe' and τ a time shift. The correlation is normalised by the auto-correlation of the signal for zero time delay.

Figure 6.14 takes results from Xia and Tucker (2012) for hybrid RANS-NLES (see Chap. 3) predicted flow for a serrated nozzle. The reference probe point is fixed at $x/D = 2.0$ and a radius of $0.5D$. This is identified in the Fig. 6.14 insets. The results for the R_{11} correlation (corresponding to an axial probe shift involving the fluctuations in axial velocity) are plotted in the upper frames of Fig. 6.14. Each point on the space correlation, representing a spatial separation, becomes a curve in the $R_{11} - \tau$ space. The envelopes, shown as dashed lines, of these curves are the auto-correlation in a frame moving with the energy-containing eddies at the convection speed U_c. The integral of the envelope is the moving-axis time-scale of the turbulence. Figure 6.14 shows correlations on both tip and notch planes in columns (a) and (b), respectively. The envelope curve for the tip plane is replicated to the upper frame of Column (b). The difference between the envelopes indicates a time-scale disparity. The correlations can also be presented as iso-correlation contours in the $\bar{\Delta} - \tau$ space. This is shown in Fig. 6.15. At each contour level, $\max(\tau)$ can be located in the $\bar{\Delta} - \tau$ space. The dashed lines are linear fits of these locations. The slope of the dashed lines give the convection speed U_c. The difference in the slopes indicates that there is nearly a 20 % difference in the convection velocity between the tip and root zones. Such information is useful for RANS based acoustics models and to understand flow physics at a basic level.

As Goldstein's work shows, it is the 4th order-space time correlations that are key to characterizing turbulent flow noise sources. The fourth order correlation coefficient intrinsically has the form

$$R(\mathbf{x}, \bar{\Delta}, \tau) = \overline{\phi_i'(\mathbf{x} + \bar{\Delta}, t + \tau)\phi_j'(\mathbf{x} + \bar{\Delta}, t + \tau)\phi_k'(\mathbf{x}, t)\phi_l'(\mathbf{x}, t)} \tag{6.21}$$

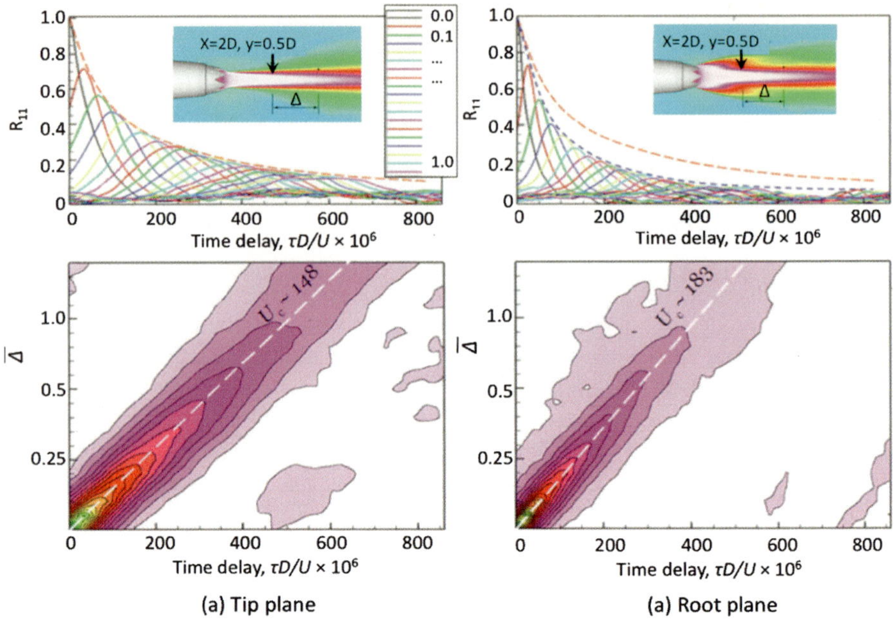

Fig. 6.14 Cross correlation of axial velocity fluctuations with downstream separation (*upper frames*) and iso-correlation contours of cross-correlations (*lower frames*). Different $\bar{\Delta}$ values are listed in a top-down order in the *upper frames* (from Xia and Tucker 2012 published with kind permission from Springer Science and Business Media B. V.)

Fig. 6.15 Fourth order correlation coefficients at (*solid*: $\frac{\Delta x}{D} = 0$; *dashed*: $\frac{\Delta x}{D} = 0.2$; *dash-dot*: $\frac{\Delta x}{D} = 0.4$; *dash-dot-dot*: $\frac{\Delta x}{D} = 0.6$; Δ: measurements (From Eastwood et al. 2012)

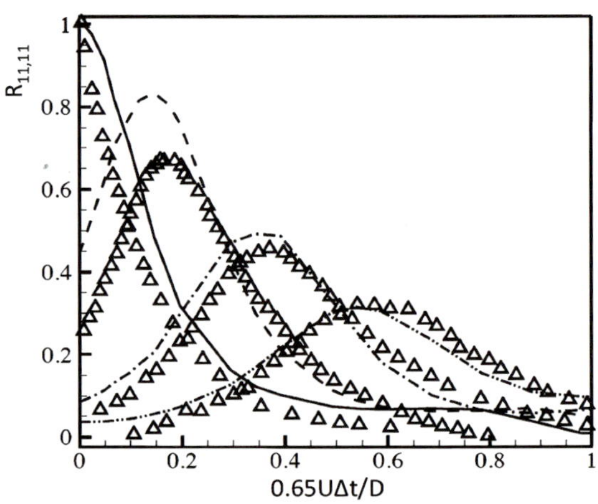

Again, this can be normalised and i, j, k and l can be 1, 2 or 3, giving 3 components of velocity. Hence $R_{11;11}$ gives $R_{uu;uu}$, the fourth order correlation of the streamwise velocity. This is plotted in Fig. 6.15, which presents results from Eastwood et al. (2012). Lines plot the predictions and symbols the measurements of

Harper-Bourne (2003). The LES grids used are modest at around 5 million cells but nonetheless the results are encouraging.

6.7.3 Models for Turbulence Correlations

A popular model is due to Tam and Auriault (1999). This is also used by Morris and Farassat (2002). It is based on Gaussian and exponential functions. The former expresses the spatial correlation and the latter the temporal. The equation for this is given below

$$R_{ijkl}(\mathbf{x}, \Delta, \tau) = \hat{A}_{ijkl}(\mathbf{x}) \exp\left[\frac{-\Delta_1}{(U_c \cdot \tau(\mathbf{x}))} - \frac{\ln 2((\Delta_1 - U_c \cdot \tau)^2 + \Delta_2^2 + \Delta_3^2)}{l^2(\mathbf{x})}\right] \tag{6.22}$$

In the above l and τ represent the correlation length and time scales. Also, $\hat{A}$ represents the root-mean-square of the corresponding quantity for zero time and spatial, $\bar{\Delta}$, separations. Ewert and Munsch (2006) use this correlation with the APE procedure to make slat noise predictions. In this work, based on RANS predicted turbulence length and energy scales, broad-band sound sources are synthetically generated using what is termed the Random Particle Mesh (RPM) method. Alternatively the approach of Batten et al. (2004) or Kraichnan (1969), among others could be used. These are described in Sect. 3.10.2. The Goldstein based round jet, noise prediction work of Karabasov et al. (2010) with RANS input makes use of 4th order space-time correlations. Again the equation above is used to express the form of the 4th order space time correlations. However, the turbulence length scale, l_i, is made tensorial as below

$$\begin{aligned} R_{ijkl}(\mathbf{x}, \Delta, \tau) = \hat{A}_{ijkl}(\mathbf{x}) \exp\Bigg(&\frac{-\Delta_1}{(U_c \cdot \tau(\mathbf{x}))} \\ &- \ln 2\left[\frac{(\Delta_1 - U_c \cdot \tau_s)^2}{l_1^2(\mathbf{x})} + \frac{\Delta_2^2}{l_2^2(\mathbf{x})} + \frac{\Delta_3^2}{l_3^2(\mathbf{x})}\right]\Bigg) \end{aligned} \tag{6.23}$$

As noted above, LES data can be used to calibrate RANS predicted turbulence time and length scales. For a RANS (k–ε) solution the following expressions can be used:

$$A_{RANS} = (2\rho k)^2, \qquad l_{RANS} = k^{3/2}/\varepsilon, \qquad \tau_{RANS} = k/\varepsilon \tag{6.24}$$

Also,

$$\hat{A}_{ijkl}(\mathbf{x}) = \hat{C}_{ijkl} A_{RANS}, \qquad l_\alpha(\mathbf{x}) = c_\alpha l_{RANS}, \qquad \tau(\mathbf{x}) = c_\tau \tau_{RANS} \tag{6.25}$$

where the constants $\hat{C}_{ijkl}$, c_α (where $\alpha = 1, 2, 3$ to allow for turbulence anisotrophy) and c_τ can be unity or defined by calibration of the RANS at fixed locations from LES or other higher fidelity data.

6.7.4 Noise Sources in Hot Turbulent Fluids

Noise sources in hot turbulent fluids are not well understood. However, it is likely that a key element in the modelling of them will be incorporating the enthalpy fluctuations. Indeed a key aspect will be characterizing their spatial and temporal nature through the two-time, two-space, velocity temperature, correlation coefficient—H_{ij}, where

$$
\begin{aligned}
H_{\mathrm{ij}}(\mathbf{x}, \bar{\Delta}, \tau) &= \overline{\rho u_{\mathrm{i}}' h_o'(\mathbf{x}, t) \cdot \rho u_{\mathrm{j}}' h_o'(\mathbf{x} + \bar{\Delta}, t + \tau)} \\
&\quad - \overline{\rho u_i' h_o'(\mathbf{x}, t)} \cdot \overline{\rho u_{\mathrm{j}}' h_o'(\mathbf{x} + \bar{\Delta}, t + \tau)}
\end{aligned}
\tag{6.26}
$$

In the above h_0 is a modified form of the stagnation enthalpy. The prime indicates a perturbation based on Favre averaging. Hence, for extension to hot jets the key issue is estimation of H_{ij} and indirect products with R_{ijkl}. The effect of temperature (enthalpy fluctuations) on the jet noise is considered by several researchers (see, for example, Bridges and Brown 2004, Tanna 1977, Viswanathan 2004 and Birch et al. 2006). All this work shows the great challenges of RANS modelling jet noise. There is a complex relationship between velocity and thermal fluctuations. This presents great challenges for modelling both R_{ijkl} and H_{ij}. RANS-based thermal time and length scales can be expressed as

$$
\tau_\theta(\mathbf{x}) = \overline{\theta^2}/\varepsilon_\theta \quad \text{and} \quad l_\theta(\mathbf{x}) = \left(\overline{\theta^2}\right)^{3/2}/\varepsilon_\theta \tag{6.27}
$$

where $\overline{\theta^2}$ is the mean square temperature fluctuation and ε_θ is its rate of dissipation. As with k and ε, model transport equations are available for $\overline{\theta^2}$ and ε_θ (Gibson and Launder 1978; Newman et al. 1981; Karcz and Badur 2005; Khavaran and Kenzakowski 2007).

Combustion Noise Combustion is emerging as an important source of noise from aero engines (Flemming et al. 2005). For noise from combustion correlations of the rate of change of heat release are important for characterizing sound. Liu et al. (2011, 2012), using DNS, found this correlation is best expressed as using a combination of Hermite-Gaussian functions as given below

$$
R = \left(1 - C_0 \overline{\Delta}^2\right) \exp\left(-C_1 \overline{\Delta}^2\right) \tag{6.28}
$$

where C_0 and C_1 are constants. Liu et al. later show that although the above is adequate to capture SPL levels, if the interest is in the spectral content of the far field sound a two-time correlation is better. This is because the flame zone gives a discontinuity in spatial properties. This results in problematic behaviour for spatial correlations.

6.8 (U)RANS Predictions

6.8.1 RANS Based Procedures

Since the focus here is on unsteady flows, minimal attention is paid to RANS based noise prediction. These, like LES based methods, can make use of the Lighthill or the Ffowcs Williams and Hawkings equations, discussed earlier. As indicated above, in the RANS procedure, the RANS equations supply the mean flow field and turbulent source information. Then, typically, the $L(U, \phi') = S_{\phi'}$ system, which has a time varying source to model turbulence, is cast into the frequency domain to yield a steady equation. Using the Green's function, this equation can be converted into one that directly yields acoustic fluctuations ϕ'. Then ϕ' can be converted into, for example, a noise spectrum. The key aspect, where there is considerable variation in techniques, is the characterization of $S_{\phi'}$—the turbulent source. Notably, the analytical Green's function solution is only tractable with simplifications such as axisymmetric flow and high frequency components being considered etc. The acoustic equations of Lilley (1974) (effectively a moving media inhomogeneous wave equation—see Goldstein 2002) can account for refraction and convection of acoustic rays. Hence, it is popular in leading RANS based noise predictions codes such as MGBK of General Electric (Secundov et al. 2007) and the NASA Langley JET3D code (Hunter and Thomas 2003).

6.8.2 URANS Tonal Noise Predictions

Khorrami et al. (2002) apply the FWH approach to a slat wing-flap configuration. Just surface integral terms are used. The flow is simulated using URANS with the Menter SST turbulence model. The slat's free shear layer is modeled with and without forcing. Excessively rapid dissipation of the shear layer vortices is observed and attributed to the turbulence model. The predicted far field sound levels are found to be sensitive to the forcing frequency. Hence, in these simulations the key far field sound modelling uncertainty seems to be more with the source characterization than with the extrapolation of sound to the far field.

Defoe et al. (2009) look at airframe-acoustic integration modelling with blended wing-body type aircraft in mind. With these, distorted flow will enter the engine. As well as an aerodynamic, there will be an acoustic impact from this distortion. Defoe et al. apply a modified version of Gong's body force approach to this problem exploring fan rotor shock noise. Gong's model is outlined in Sect. 4.6. Defoe et al. use this model to predict tonal noise from a fan. As shown by Gong, the simple body force model can correctly account for distortion transfer. Propagation to the far field achieved using the FWH approach. Defoe et al. use a body force perturbation, rotating at the blade speed superimposed on the usual steady force field to generate the tonal noise. The perturbation is estimated by making two-dimensional blade section

CFD simulations. The blade stagger angles are perturbed and Fourier analysis used to characterize the force field perturbations.

Schlinker et al. explore RANS based modelling of noise from a centrifugal impeller. This is part of an aircraft auxiliary power unit. The noise source is rotor locked shocks. These arise from blades operating at transonic relative Mach numbers near their tips (there are a multitude of other potential more broad band and tonal sources but these are ignored). The near field acoustic pressure field is used as a boundary condition for a finite element based code. This solves a wave equation. The casing pressure amplitudes are in reasonable accord with measured values. The far field acoustic power is evidently again in close agreement with the measurement.

Muhlbauer et al. (2008) applied URANS to study entropy noise through a convergent nozzle, deriving acoustic sources. The convergent nozzle is intended to represent the flow physics of combustion noise passing through a turbine. This is discussed further later.

6.8.3 LES Informed RANS

The jet noise procedure of Karabasov et al. (2010) is a hybrid one. It uses a LEE wave operator. The acoustic sources are characterized using 4th order correlation coefficients of velocity R_{ijkl}. The sources are based on RANS turbulence length and time scale estimates. However, these are informed by the LES, in terms of turbulence correlation coefficient shape and the level of turbulence anisotropy. Xia et al. (2011) apply this approach to chevron nozzles. The procedure involves using LES to find the best values of parameters (in Eqs. (6.25)) to allow Eq. (6.23) to fit the turbulence field from a chevron nozzle. A key challenge is that the noise source description involves a vast amount of potential correlations—we have just focused above on the R_{1111} correlation. Fortunately the results of Xia et al. (2011) show that there are just six large components of the source with indices: 1111, 2222, 3333, 1212, 1313 and 2323 (and terms equal to them from symmetry, e.g., 2112, 1221, etc.). It is fortunate that for the chevron jet, the number of significant source components does not increase relative to an axisymmetric jet (Karabasov et al. 2010). Figure 6.16 shows the azimuthally filtered spatial distribution of the autocorrelation coefficient R_{1111} for a chevron nozzle at various axial locations. Clear azimuthal variations can be seen.

Figure 6.17 relates to the hybrid RANS-(N)LES type computation discussed above. It shows comparison of the far-field sound pressure power spectral density (PSD) predictions at the sideline to a jet. Hybrid RANS-(N)LES + FWH results are represented by the solid line. The line with square symbols gives results from the Goldstein based RANS analysis where the 4th order space-time correlations, used in source terms, are informed by NLES. Comparison is made with measurements (line with round symbols). Clearly the hybrid RANS-NLES + FWH prediction exhibits quicker spectrum decay at high frequencies. This is mainly be-

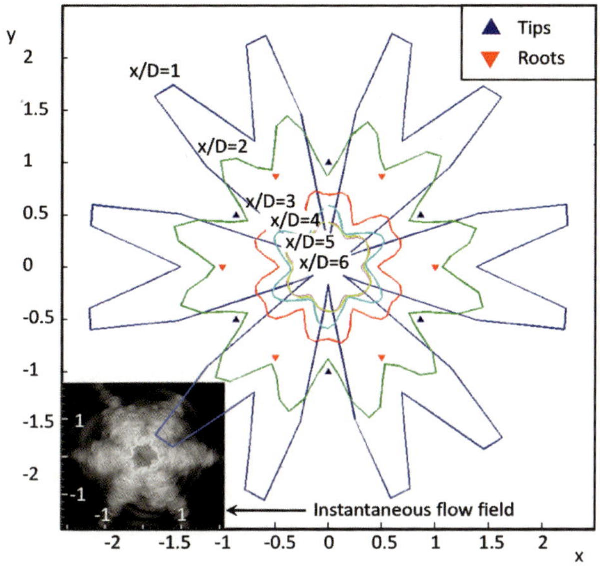

Fig. 6.16 Azimuthally filtered spatial distribution of the autocorrelation coefficient R_{1111} for a chevron nozzle at various axial locations

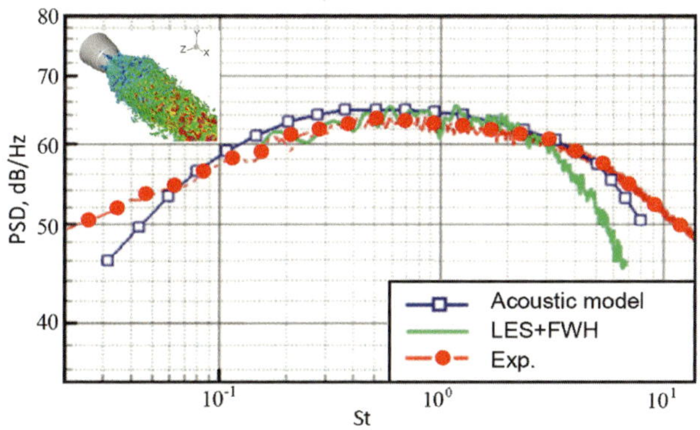

Fig. 6.17 Power spectral densities of the sound pressure in the far field at 90° for chevron nozzle (adapted from Xia and Tucker (2012) published with kind permission from Springer Science and Business Media B. V.)

cause of the mesh resolution at the FWH surface. For the NLES informed, RANS acoustic model, the agreement with the measurements is encouraging. The high frequency limit of spectral resolution is within the 4 dB error up to $St \approx 8$. Components of this RANS based acoustic model could be used as an acoustic subgrid scale model.

6.9 DNS, LES and Hybrid Acoustic Related Computations

Table 6.2 summarizes the Reynolds numbers (generally, chord based), grid counts, *N* and any far field sound prediction methods utilized for some cases to be discussed. Also, the validation level is summarized in the table. The latter is broken down into two elements. The first is the flow field validation level. These are as defined in Table 5.2. Then, additional acoustic levels are included. These indicate if an integral quantity for far field sound is validated against or comparison extended to the spectral level. Also, a grade is included to indicate if flow field measurements are explicitly tailored to explore sound source information. The additional acoustic validation levels are summarized in Table 6.3. Note, jet noise studies are omitted from Table 6.2. This is because, their number is much more extensive and hence would tend to dominate the table.

6.9.1 Compressor and Fan Noise

Kato et al. (2007) perform finite element based dynamic Smagorinsky LES of a five stage centrifugal pump. LES casing pressures are used as a boundary conditions in a structural finite element model of the pump casing. The casing response is then fed into a FWH related acoustic analogy to estimate far field sound. The discrepancy with far field sound measurements is typically less than 10 dB.

Riou et al. (2007) study LES modelling for an aerofoil downstream of a rod and also a fan with outlet guide vanes (OGVs). The object of the work is to reduce broadband fan noise. Due to improvements in the management of noise, this is now, for subsonic flows, greater than discrete tone noise. The broadband noise is mainly of the dipole type. It arises from ambient turbulence interacting with the rotor and also the OGVs (stators) being exposed to the upstream rotor wakes. This all results in surface pressure fluctuations. The rod-aerofoil is a good prototype for studying the flow physics of a fan section interacting with a downstream wake. Riou et al. use the WALE LES model in a fairly standard compressible flow solver. The blade counts are substantially altered to allow the modelling of single blades. This will influence tonal noise. However, this is a legitimate practice for exploring the broadband noise. No measurements are available for validation. Rulik et al. (2010) also carry our rod-aerofoil simulations using DES and the commercial CFX program. A separate LEE solver is used to predict far field sound. Broadly encouraging agreement is found with near field turbulence statistics. The predicted far field sound is nearly 20 dB away from measurements. However, the peak sound level in the noise spectrum is well captured. Berland et al. (2010, 2011) and Daude et al. (2012) study the rod aerofoil problem with a 6th order overset code comparing with the measurements of Jacob et al. (2005). A 6-stage LDRK temporal scheme is used. For subgrid scale modelling LES-RF is used—see Sect. 3.5.5. A key benefit of the high order scheme is for the acoustic propagation element. Indeed most of the grid is heavily focused in the acoustic source region. The overset grid methodology provides the flexibility

Table 6.2 Summary of eddy resolving simulations in acoustics

Component	Author(s)	Method	$Re_c/10^5$	$N/10^6$	Far field sound prediction method	Validation level	
						Aero	Acoustic
Rod-aerofoil	Ashcroft and Nuernberger (2001)	LES, DES	5	3	FWH	0	1, 2
Rod-aerofoil	Riou et al. (2007)	LES	5	2.6	–	0	0
Rod-aerofoil	Rulik et al. (2010)	Hybrid RANS-LES	4.6	2–3.4	LE	2, 3	1, 2
Rod-aerofoil	Berland et al. (2010, 2011), Daude et al. (2012)	ILES	5	20	NS (high order)	2, 3	1, 2
Fan with OGVs	Reboul et al. (2008)	LES	2.3	6	FWH	0	1
Fan with OGVs	Riou et al. (2007)	LES	2.2	12	–	0	0
Fan with OGVs	Ashcroft and Nuernberger (2001)	LES and some DES	2.3	1000	FWH	2, 3	0
Fan	Li et al. (2009)	LES	7	9	–	0	0
Fan	Tucker et al. (2012)	Hybrid RANS-NLES	30	8	–	0	1
Multistage centrifugal pump	Kato et al. (2007)	LES	100	0.3	Hybrid, LES, structures, FWH	4	1, 2
Compressor representa-tive profile	Winkler and Moreau (2008)	LES	1.9	3.2	Lighthill and FWH	2, 3	1
NACA0018	Lee et al. (2004)	LES	1.6	0.9	FWH	0	1
Compressor linear cascade	"	LES	–	1.4	FWH	0	1
Cooling fan	"	LES	1.3	0.4	FWH	0	1
Radial compressor	Mendonça et al. (2012)	DES	–	9	Near field pressure traces	1	0

Table 6.2 (Continued)

Component	Author(s)	Method	$Re_c/10^5$	$N/10^6$	Far field sound prediction method	Validation level	
						Aero	Acoustic
NACA0012	Kato et al. (2007)	LES	2	–	BEM based procedure	2, 3	2
NACA0012—trailing edge noise	Sandberg and Jones (2011, 2012)	DNS	0.5–1	1600–2000	–	–	–
Flat plate—trailing edge noise	Manoha et al. (2000)	LES	0.01 (based on edge thickness)	1.65	–	–	2 (NACA0012—different geometry)
Trailing edge (hydrofoil)	Marsden et al. (2007)	LES	19	7	Lighthill equation solution	0	0
Asymmetrically beveled trailing edge	Wang and Moin (2000)	LES	21.5	7.1	Lighthill equation solution	2, 3, 4	0
Nose landing gear cavity	Langtry and Spalart (2009)	DES	–	6	–	4	0
Landing gear	Spalart et al. (2010)	DES	–	18	FWH	0	0
Simplified landing gear	Lockard et al. (2004)	DES	–	13.3	FWH	0	0

Table 6.3 Validation level definitions for aeroacoustics

Grade	Validation data
1	Integral far field sound quantity
2	Far field sound spectra
3	Space-time source correlations or source spectral information linked to far field sound

to coarsen the grid away from the source zones(s). Encouraging agreement is found with the velocity and Reynolds stress measurements. The agreement with the far field sound levels is not as encouraging. Again, there is a nearly 20 dB disparity at low frequencies. However, there is a notably spurious asymmetry in the measurements.

Ashcroft and Nuernberger (2001) model the DLR low speed fan with OGVs using LES. Generally, LES gave greatly improved rotor wake predictions. Reboul et al. (2008) perform induct LES of the DLR low speed fan rig. The blade count is modified to allow the modelling of single blade passages. A compressible multi-

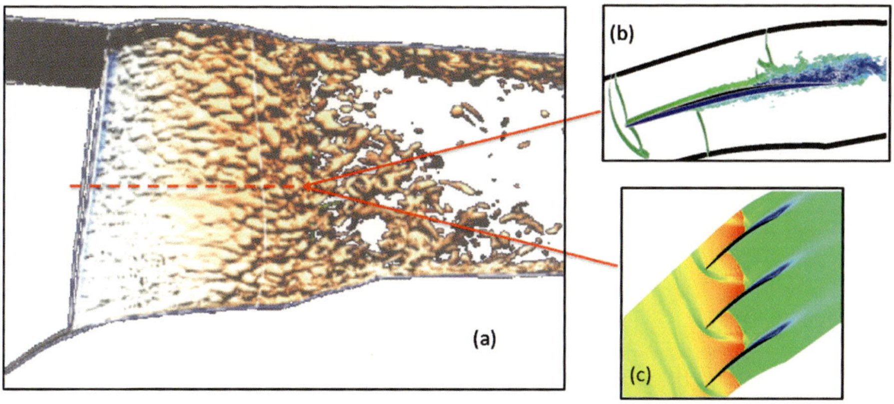

Fig. 6.18 Hybrid type LES simulations for fans: (**a**) Instantaneous vorticity iso-surfaces coloured by instantaneous pressure (**b**) Instantaneous Mach number of unity isosurface coloured by vorticity magnitude and (**c**) Contours of time-averaged M (from Tucker 2011 published with kind permission from Elsevier)

block flow solver is used. The form of the subgrid scale modelling is not reported. Duct noise propagation is modeled using a modified FWH formulation. The agreement of sound pressure and power levels with measurements is encouraging. Li et al. (2009) perform WALE model based LES of a fan blade (the Boeing 18″ fan experiment—see Ganz et al. (1998a,b)). No comparison is made with acoustic or source measurements. Figure 6.18a gives a DES computed instantaneous flow field for this rig taken from Tucker et al. (2012). Again, there is no comparison with measurements.

Ray and Dawes (2009) perform SA DES and Tucker et al. (2012) RANS-NLES of a transonic fan blade section. The flow conditions approximate the mid-span region of a blade rotating close to full speed. The motivation for this work is broad band acoustics. A relatively standard, 2nd order, compressible RANS based solver is used. There are no measurements to compare with. Figures 6.18a, b indicates the flow complexity (Note, these are *not* for the Boeing fan section). It includes multiple shocks and shock induced separation. For the flow, there is qualitative agreement with the experiments of Lepicovsky et al. (2002). At high frequencies, empirical models (Goody 2004) suggest turbulent boundary layers exhibit a f^{-5} scaling while at intermediate frequencies a f^{-1} scaling. Figure 6.19 shows pressure-side surface spectra at 80 % chord. The curves show the expected scalings. Hence, broadly the results are encouraging. Such data could be used as a source description in an impermeable FWH formulation or other types of wave propagation operator. Figure 6.19 does not validate the perturbation amplitudes and how the RANS layer might attenuate these. This is an area of open research.

Lee et al. (2004) explore aerodynamic and aeroacoustic LES prediction for a NACA0018, an axial compressor cascade and a cooling fan. The far field sound is predicted using the FWH approach. A variant on the dynamic Smagorinsky model

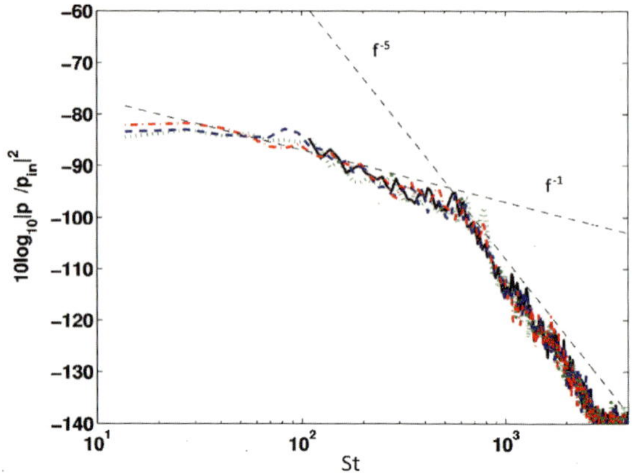

Fig. 6.19 Blade surface pressure spectra at around 80 % chord. (Published with kind permission from ASME.)

is used. Also, a structured, incompressible flow solver is used with a 2nd order CD scheme. The agreement with acoustics measurements is broadly encouraging.

Mendonça et al. (2012) make centrifugal fan, Menter SST based DES computations exploring acoustics. The unstructured, commercial, pressure based STAR_CCM+ code is used. The acoustic field is studied through near field pressure traces. Some of these are placed at the compressor inlet and outlet pipes and others in the compressor casing. No acoustic validation is performed but encouraging agreement found with system performance measurements. At certain conditions a tonal noise component is found that corresponds to rotating stall.

6.9.2 Trailing Edge Noise

Manoha et al. (2000) make trailing edge noise LES. A flat plate with a blunt trailing edge is considered. The equations are solved using a hybrid finite difference/element method of 2nd order accuracy. The turbulence inflow had two components. These are turbulent boundary layer DNS in precursor simulation form. A three-dimensional perturbation is imposed on this. It is designed to promote three-dimensional wake instabilities. Lighthill's analogy and other approaches are used to compute the far field sound. Even though the LES is run at a considerably lower Reynolds number to experiments encouraging agreement is found with far field sound measurements.

Wang and Moin (2000) make LES of trailing edge noise from a asymmetrically beveled trailing edge of a strut. The dynamic Smagorinsky model is used. A 2nd order spatial discretization is utilized with a Fourier method applied in the spanwise direction. For temporal integration, the CN scheme is used for viscous terms. A 3rd order RK scheme is used for the inviscid. Lund's recycling is used to generate the turbulent inflow. The far field sound is gained via Lighthill's analogy. The need for a wider domain, to capture low frequency source components is identified. However, broadly, encouraging agreement is found with mean velocity and Reynolds stress profiles along with surface pressure spectra.

Winkler and Moreau (2008) make LES of the trailing-edge flow and noise for a NACA6512-63 aerofoil at zero angle of attack. This aerofoil is most representative of a highly loaded, high-speed compressor blade. The commercial FLUENT CFD code is used. To correctly trip the flow, as in the experiments, a serrated trip zone is found necessary. Second order CDs are used with the dynamic Smagorinsky model. The span-wise domain extent is just under 0.75 % of chord. There is reasonable agreement between wake mean velocity and turbulent statistic profiles along with spectral information. To get far field sound, both the Lighthill acoustic analogy and the FWH approaches are used. Again reasonable agreement is found with measurements.

Key DNS work on aerofoil (NACA0012) acoustics is that of Sandberg and Jones (2011, 2012). They look at trailing edge noise scattering and the feedback of this to suction surface instabilities. This work involved both clean and serrated trailing edges. Simulations used 4th order CDs with a 4th order RK scheme. Stability is maintained in these compressible flow simulations though the use of the flux entropy splitting scheme (Sandham et al. 2002). The serrations are modeled using the immersed boundary method and are found not to greatly influence the flow field. The spanwise domain is set at 20 % of the chord.

Marsden et al. (2007) explore the use of dynamic Smagorinsky model based LES for the design optimization of a trailing edge. The cost function, in the non-gradient based optimization, is the trailing edge noise. This is evaluated based on a solution to Lighthill's analogy. A nearly 90 % reduction in noise power is demonstrated. This is achieved by the suppression of trailing edge vortex shedding. The approach is hybridized with RANS solutions being used to enforce design constraints on lift and drag.

6.9.3 General Airframe Studies

Langtry and Spalart (2009) perform DES of an aircraft nose landing gear cavity. A 2nd order, upwind (UP) biased TVD scheme is used for the convective flux discretization with a 2nd order BD temporal discretization. Hexahedral cells are used. The simulations captured the key tone within reasonable accuracy. Also, the SPL of this is within 4 dB of measurements.

Hedges et al. (2002) carry out a DES performance study primarily focusing on landing gear aerodynamics. Multiblock simulations are performed. A blend of 5th order UP with 4th order CD is used. The latter is applied in wake zones, away from boundary layers. Further simulations, with the same CFD program, but this time in compressible mode, for landing gear is reported in Spalart et al. (2010). Here, to gain far field sound the FWH equation is solved in the frequency domain. No validation measurements were available at the time of writing. However, a substantial difference is observed between the computed far field sound when the permeable and impermeable variants of the FWH approach are used. Lockard et al. (2004) make a similar observation when making DES simulations for landing gear. Discrepancies are attributed to the large turbulent eddies from wakes passing through the porous FWH surface. The resulting error is observed to be as large in magnitude as that of the radiated sound.

6.9.4 Jet Noise

Much jet LES work relates to single stream jets. The number of jet noise related LES computations is extensive. Hence, just selected examples are discussed here. Liu et al. (2008) explore the performance of the mixed Leray, Kosovic, α, Yoshizawa and Smagorinsky models for plane jets. These models are outlined in Chap. 3. Little LES model influence is observed. Hybrid RANS-(N)LES type studies for co-flowing jets are discussed in Shur et al. (2006), Eastwood et al. (2010, 2011, 2012). The latter involves heating. Even with modest grids ($N = 6 \times 10^6$), Eastwood et al. find encouraging agreement with PIV (Particle Image Velocimetry) and LDA mean flow field and Reynolds stress measurements. Grids considered had the following density range $6 \times 10^6 < N < 50 \times 10^6$.

To reduce perceived sound levels, serrated nozzles can be used and are now seen in operation on aircraft. For these nozzles, hybrid RANS-NLES type simulations for 12×10^6–20×10^6 cell grids are reported by Xia et al. (2009, 2011), Xia and Tucker (2012). Accord is again found with mean velocity and Reynolds stress measurements. The far-field sound level predictions also show agreement with measurements. With regards to power spectral densities agreement is also encouraging. However, as discussed earlier, capturing low and high frequencies is found to be challenging. Hence, the use of a sub-grid-scale parameterization of acoustic output is proposed by Xia et al. and Spalart (2011). Notable work on serrated nozzles is that of Uzun and Hussaini (2009) who makes 6th order ILES type computations for $N \cong 300 \times 10^6$. Paliath et al. (2011) also makes 6th order chevron, ILES, for $N = 24 \times 10^6$ using a compact scheme.

Notably, a key influence on jet development is the state of incoming boundary layers and this information is seldom measured. Also, as shown by Birch et al. (2008) and Secundov et al. (2007), jets have a tendency to stray from the axisymmetric state and this complicates modelling and validation. Hybrid RANS-NLES type computations ($N = 7 \times 10^6$) exploring the impact of eccentricity are reported in Xia et al. (2012). Figure 6.20 summarizes these results. Frame (a) gives time averaged axial velocity and (b) instantaneous axial velocity. Frame (c) gives jet deflection with eccentricity for both hot and cold jets. The symbols represent each simulation and the lines are best fits. The lines suggest a roughly linear relationship between nozzle eccentricity and jet deflection. Note, that the core nozzle deflection is expressed as a fraction of the outer nozzle diameter. The dashed lines take experimentally observed flow deflection levels and uses the simulations to infer potential

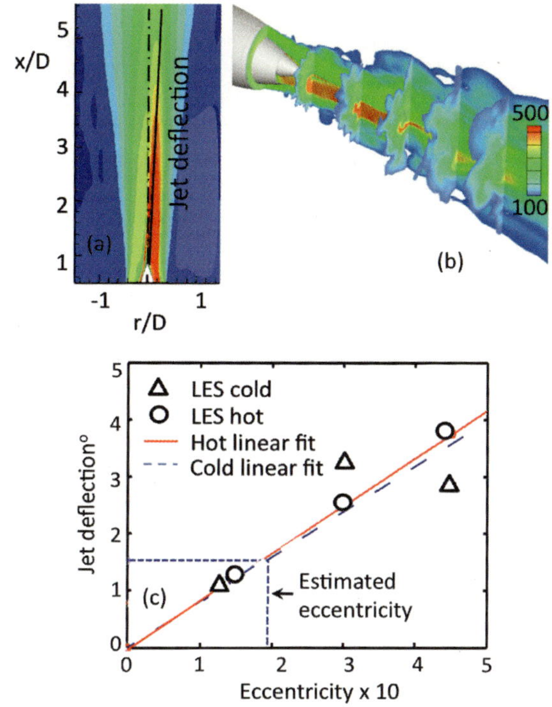

Fig. 6.20 Hybrid RANS-NLES exploring jet deflection with core nozzle eccentricity: (**a**) time averaged axial velocity (**b**) instantaneous axial velocity and (**c**) jet deflection with eccentricity

eccentricity levels. These estimates suggest that the hot jet has a nozzle eccentricity of around 2 %. Eccentricity could be related to structural deflections caused by heating.

Another area of uncertainty, outlined in Chap. 3, is that in real engines, upstream of the jet there is a tremendous amount of geometric complexity. The impact of this on the general state of the inflow and nozzle boundary layers is still not understood. Also, little is understood regarding the impact of external geometry such as the proximity of wings and flaps and the influence of this on the flow and acoustic fields. Figure 6.21 shows the influence of an engine mounting pylon and internal struts on jet plume development. Frame (a) gives the pylon influence and (b) the strut influence. Frame (a) plots circumferentially averaged turbulent kinetic energy profiles at different axial locations. The dashed lines are for a nozzle with no pylon—the base line—and the full for a nozzle with a pylon. As can be seen the pylon significantly influences the turbulence field and hence will influence the noise. The nozzle internal geometry also influences the shear layer turbulence levels by nearly 20 % in certain regions. This is shown in Frame (b). To account for the internal geometry body forces, discussed in Chap. 4, are used. These impose wake profiles, consistent with measured expectations (Eastwood 2009) for two blade rows located in the core. The differences between a nozzle with (full line) and without (dashed line) internal geometry are clear from Frame (b). Hence, clearly, modelling the real engine geometry and the turbulence structures generated in the upstream flow is of some importance there influences most likely being additive.

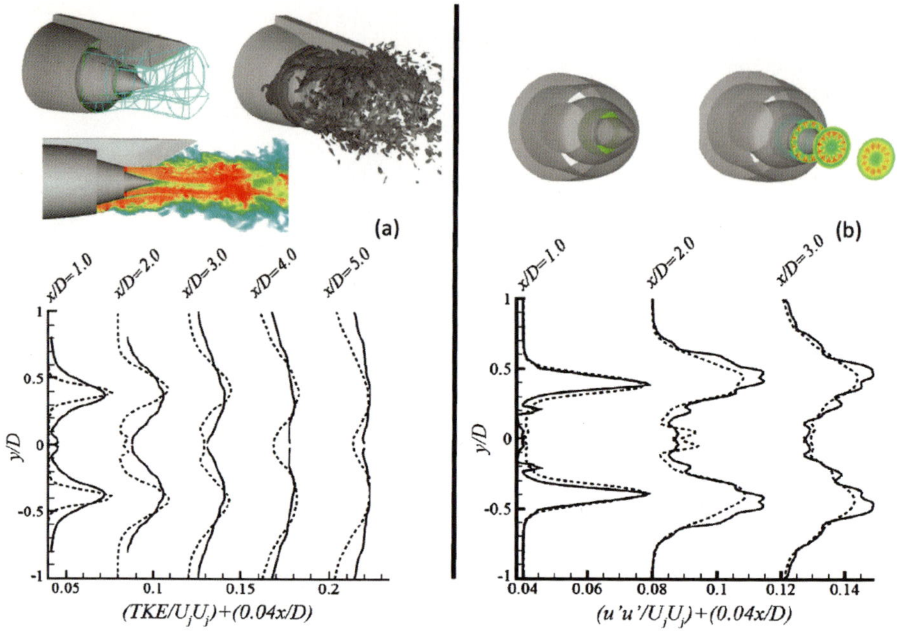

Fig. 6.21 Influence of engine mounting pylon and internal struts on jet plume development: (**a**) pylon influence and (**b**) strut influence (Copyright, Multiscience Publishing, used with permission)

Figure 6.22 plots contours of $\partial p/\partial t$ in the near nozzle region. Frame (a) plots contours in the pylon plane. Frame (b) plots contours out of the pylon plane at 90° to it. The propagating acoustic waves are weaker in Frame (a), where the pylon has blocked their escape from the turbulent jet (even though the pylon has potentially created more quadrupole noise). The turbulent region, which generates the noise, is more spread out surrounding the pylon at $x/D \sim 4$ and $y/D \sim 0.5$. Along the dashed directivity line in the figure the overall acoustic wave amplitude is reduced by over 50 % (Eastwood and Tucker 2011).

Jets are extremely sensitive to their surrounding environment (Klein et al. 2003). Hence, accounting for wing proximity is a key new research area its interaction with the jet being a major source of noise. Figure 6.23 shows another hybrid RANS-(N)LES type computation of a jet-pylon-wing-flap geometry, taken from Eastwood et al. (2012). Frame (a) shows the geometry and mesh blocks of the multi-block hexahedral system used. Frame (b) gives isosurfaces of axial velocity. As outlined by Eastwood et al. (2012), the pylon-wing-flap influence on the jet development is strong. The total of the geometrical influences discussed above could potentially be extremely high, making the value of more idealized jet simulations questionable.

Influence of Reynolds Number with Jets Beyond any incoming boundary layers jets are *Re* independent. However, with increasing *Re* more energy will be contained in the small scales. This will impact LES solution in three ways:

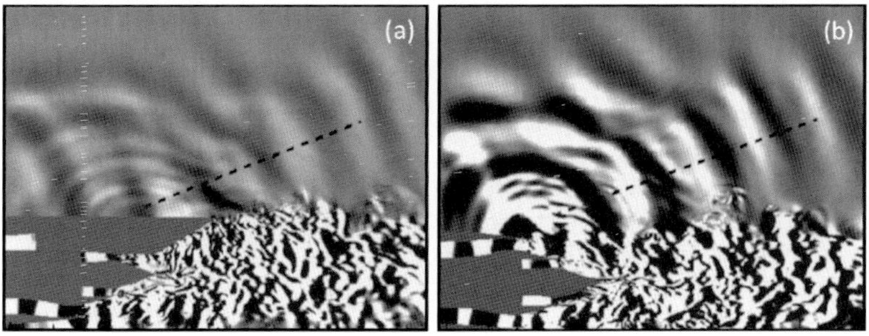

Fig. 6.22 Contours of $\partial p/\partial t$ in different planes: (**a**) contours in the pylon plane and (**b**) contours out of the pylon plane at 90° to it (from Eastwood and Tucker 2011) (Copyright, Multiscience Publishing, used with permission)

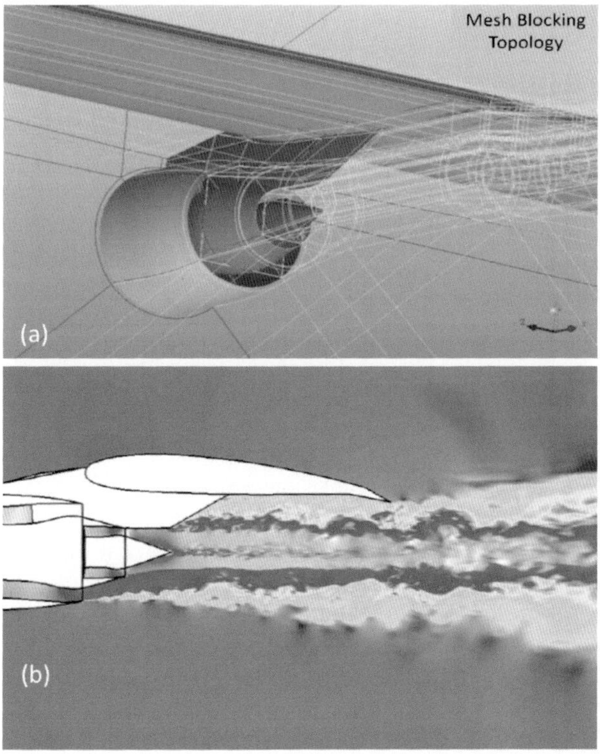

Fig. 6.23 Hybrid RANS-(N)LES of a jet-pylon-wing-flap geometry (Xia et al. 2012, published with kind permission from Elsevier): (**a**) geometry and mesh blocks and (**b**) isosurfaces of axial velocity

1. For a fixed filter, the numerical discretization will become more stressed and hence more critical, since greater energy will be interacting at the filter scale;
2. Any LES modelling at the filter scale will play a greater role and

3. The missing acoustic characterization at the subgrid scale will have a greater impact on the predicted far field sound levels.

For a jet, the size of the large scales increases downstream (and their energy levels). Hence, for a fixed filter width, moving downstream the characterization of energy in the smaller scales becomes less crtitical. For example, from DNS of a jet Bodony and Lele (2002) find that the ratio of the energy in the small to large scales is 42 %, 12 % and 5 % for $x/\delta_{\omega o} = 40$, 60 and 80, where $\delta_{\omega o}$ is the initial vorticity thickness. If the LES sits in the inertial range, then the turbulence energy spectrum can be expressed as

$$E(w) \sim \frac{\varepsilon^{2/3}}{k^{5/3}} \tag{6.29}$$

where here, w, is used to represent the wave number. Then it follows that the energy under the filter E_{under} is given by

$$E_{under} = \int_{w_c}^{\infty} E(w)dw \sim \frac{\varepsilon^{2/3}}{k_c^{2/3}} \tag{6.30}$$

where w_c is the filter cutoff frequency. Now $\varepsilon \simeq u'^3/l$ (where u' represents the turbulence fluctuation levels). Also, l will increase linearly with downstream distance while u' will remain relatively constant. Hence, $E_{under} \propto x^{-1}$ and the energy contained with the small scales will decrease with downstream distance. Figure 6.24 shows a hybrid RANS-NLES instantaneous flow field (Frame (a)) and RANS based estimate to the energy spectrum for this jet (Frame (b)). The latter is for two Reynolds numbers. The spectrum is evaluated near the jet exit. The greater energy at the smaller scales is evident for the higher Reynolds number. The above scaling argument is from Bodony and Lele. It implies that the greatest area of numerical and LES model sensitivity will be near the jet outlet for LES at high Reynolds numbers. The RANS estimates are consistent with the scaling argument. Hence, in this zone, near the jet outlet, an acoustic subgrid scale model could be helpful.

Mahak et al. (2012) make hybrid RANS-NLES for a jet in a large industrial facility. The instantaneous flow field is shown Fig. 6.24. The measurements are performed at a relatively high Reynolds number ($Re = 2 \times 10^6$). Hence, sensitivity to near nozzle grid resolution is observed. This is because, as noted above, there is a substantial amount of energy at the small scales. To deal with this, the jet Reynolds number is reduced. Since, jets are Reynolds number independent (above $Re = 10^5$) this has minimal accuracy implications. However, the incoming boundary layers are Re dependent. To deal with this, the eddy viscosity in the RANS layer is scaled. It is scaled to ensure a boundary layer thickness consistent with the higher Re. The scaling zones are shown in Fig. 6.25a. This approach results in little grid sensitivity. Simulations then show encouraging agreement with measurements. For example, Fig. 6.25b contrasts the predicted centre line velocity decay with measurements. The full lines are for simulations run at the rig Re and the dashed at the scaled down Re. Lines are shown for $N = 18 \times 10^6$ and 50×10^6. Clearly there is much greater

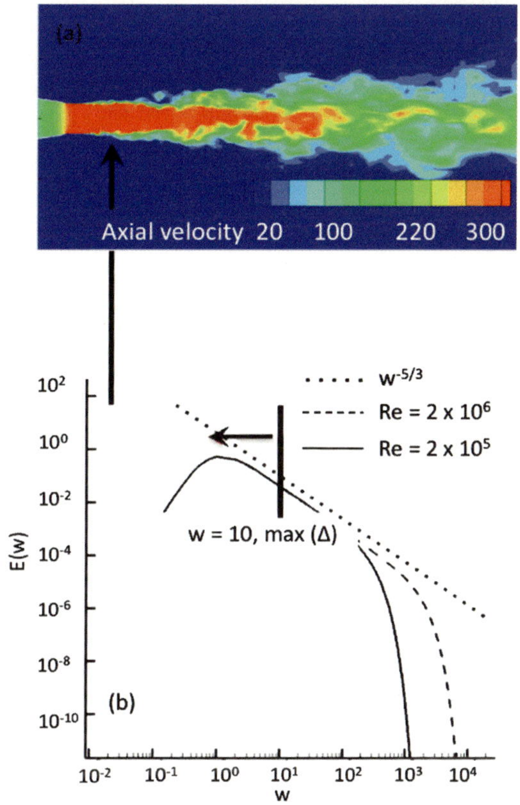

Fig. 6.24 Extent of energy spectrum in the initial shear layer of a jet: (**a**) instantaneous axial velocity field and (**b**) RANS estimate of energy spectrum

consistency for the lower *Re* jet. Also, as can be seen from Fig. 6.25c, the far field sound is close to the measurements.

6.9.5 Combustion Noise

The acoustic field generated within the combustor is modified as it propagates through the turbine. The noise production and transmission processes are coupled. Hence, ideally fully compressible flow LES are required for the combustor and turbine. However, obviously this is tremendously expensive. Hence, lower-order models are generally be used. For example, to model the acoustic, entropic and vortical interactions through rotating and stationery blades in turbines, actuator discs (Cooper and Peake 2000) with transmission and reflection coefficients can be used. There are several methods available to predict the propagation of sound through exit ducts, its scattering at the jet lip and radiation to the far field. One possibility is to use specialized analytical expressions. Alternatively the propagation can be addressed via solution of the LEE. Schemel et al. (2004) solved the LEEs for propagation of

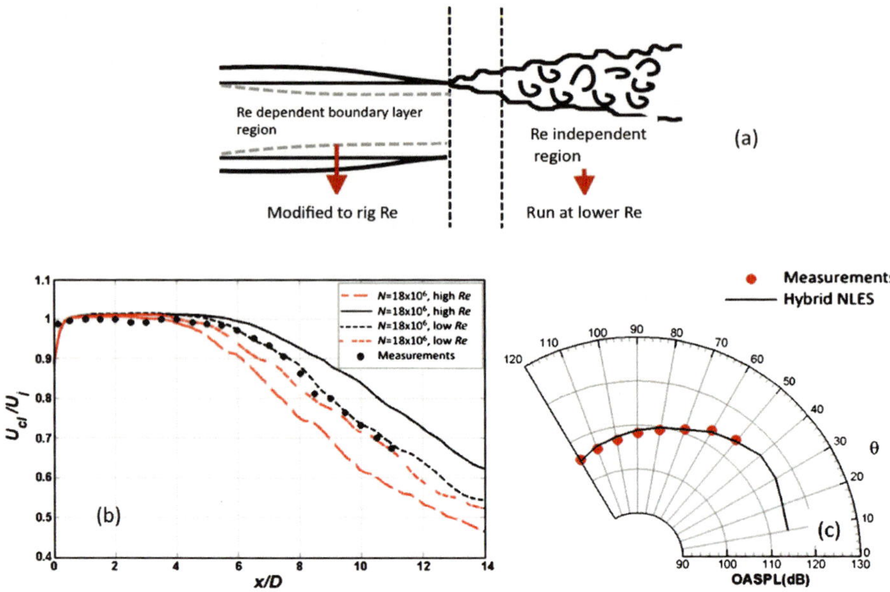

Fig. 6.25 Use of Reynolds number scaling for high *Re* jet simulations: (**a**) schematic of zones scaled; (**b**) centre line velocity decay and (**c**) far field sound

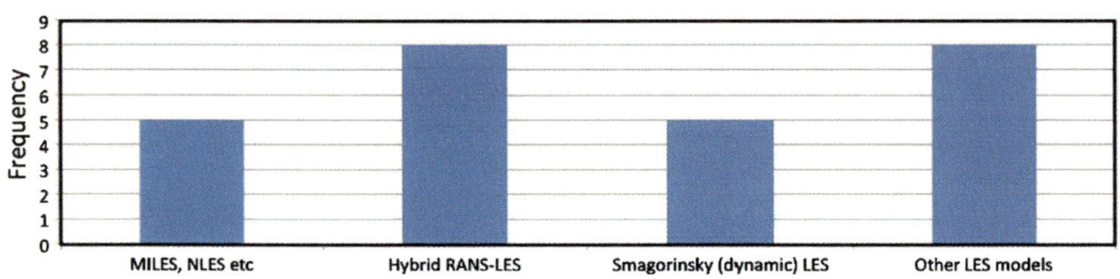

Fig. 6.26 Approximate frequency of use of different eddy resolving modelling strategies in aeroacoustics

entropy and acoustic waves through a convergent-divergent nozzle and exhaust duct (aimed at studying combustion noise).

For an open flame, Bui et al. (2007) take LES source data and feed it into an acoustic perturbation equation to gain far field sound. Again for an open flame, and using LES source data, Kaltenbacher et al. (2007) use Lighthill's acoustic analogy/wave equation along with the more complete Phillip's equation to get far field sound. This has greater fidelity than pure acoustic perturbation equations, containing acoustic, vorticity and entropy wave propagation modes.

Based on work reviewed in this chapter, Fig. 6.26 summarizes the approximate frequency of use of different eddy resolving modelling strategies. The vertical axis is the frequency of use. The use of different approaches seems relatively uniform,

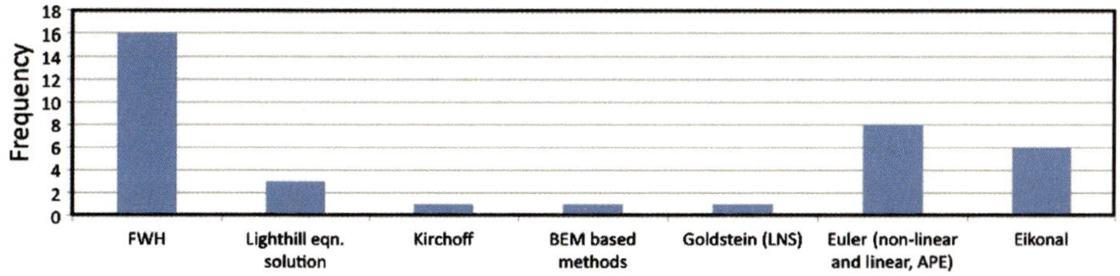

Fig. 6.27 Approximate frequency of use of different far field sound modelling strategies

but this could be due to the small sample size. Similarly, Fig. 6.27 summarizes the frequency of use of different modelling approaches for taking source data to yield far field sound. It is clear from this table that the FWH approach is popular.

6.10 Validation Data

A few experimental acoustics validation data sources related to aero engines and airframes are summarized in Tables 6.4 and 6.5, respectively. The tables are far from extensive. For acoustics propagation there are numerous potential analytical validation sources. These are not given here. The most extensive summary of propulsive jet measurements and their accuracy is given by Bridges and Wernet (2010). Note, not all of the data summarized is purely acoustics related. This is because, with aeroacoustics, to give a complete validation, it is necessary to characterize both the aerodynamic and noise propagation information. For landing gear, aerodynamic related validation data is also tabulated in Chap. 4.

Table 6.4 indicates that for jets, although more data is needed, relative to other areas this has the most extensive validation data. The data is relatively complete in terms of both flow field and acoustics measurements. Perhaps the key missing areas are temperature velocity correlations, characterizing noise sources for hot jets. Also, more often than not, the inflow boundary conditions are not measured. Frequently, even the most basic information, such as the boundary layer thickness at inflow is lacking. Also, there seems room for explorations into how real geometry influences the development of the jet shear layers. Fortunately, for single stream jets, as noted by Bridges and Wernet (2010), the precise clean nozzle geometry does not have a substantial aerodynamic impact. Hence, unlike many turbomachinery data sets, proprietary data issues, relating to gaining geometry are not a so restrictive. Ever increasing bypass ratios are bringing the propulsive jet in close proximity to wings. Hence, new data is needed to understand this interaction and also jet flap interaction. Some data is noted in Table 6.5. Both fan and jet noise are considered the high aerospace priorities. Hence, it is perhaps not surprising that fans come out as second highest in terms of validation grading level. However, data available for fans is far less extensive than for jets. This is not surprising. Making measurements in high speed rotating systems is challenging. The rod-aerofoil case is often cited in the literature as a case relevant to fans. However, there seems a large disconnect between

Table 6.4 Acoustic modelling validation data sources for aero engines

Engine zone	Nature of data	Data sources	General comments	Validation level	
				Aero	Acou
Intake	JT15D inlet—far field sound pressure level against observer angle	Ozyoruk and Long (1997)	–	0	1
Fan blade (aerofoil section)	Trailing edge surface pressures and far field broad band sound	Roger and Moreau (2004)	Open jet wind tunnel		1, 2
Fan	Characteristic curve data, total pressure radial profiles, wake profiles, some blade tip data, spectral information for noise and turbulence, mean velocities, Reynolds stresses. Both hot wire and LDA data	Boeing 18″ fan rig–Ganz et al. (1998a), Ganz et al. (1998b), DLR low speed fan–Ashcroft and Nuernberger (2001), NASA rotor 37–Suder and Celestina (1994), Suder et al. (1995)	Fan section type data would be useful but compromises flow physics in terms of shock dynamics	1, 2, 3, 4	1, 2
Compressor representative section NACA 6512-63	Surface pressure measurements, far field sound spectra, mean velocities and Reynolds stresses in wake along with turbulence spectral information	Winkler and Moreau (2008)	Most representative of a highly loaded, high-speed compressor blade. Has acoustic data	1, 2, 3, 4	1, 2
Rod-aerofoil	PIV and hot wire measurements. Far field pressure spectra, velocity profiles, Reynolds stresses	Jacob et al. (2005)	This case is a very generic representation of interactions of a blade with an upstream wake and has been used, for example, in the context of fan noise. Note—some evidence of asymmetry in measurements	1, 2, 3	1, 2

Table 6.4 (Continued)

Engine zone	Nature of data	Data sources	General comments	Validation level	
				Aero	Acou
Turbine	Far field sound data (spectral)	GE—Oerlemans et al. (2009)	Work Oerlemans et al. (2009) is for wind turbines and explores the impact of serrated trailing edges. The tests are at full scale	0	1, 2
Combustor noise with propagation in turbine	Pressure and temperature traces with time	Muhlbauer et al. (2008)	Idealized rig that places a hot spot in a nozzle. The latter is intended to represent flow acceleration in turbine	0	1, 2
Propulsive jet	Spectral acoustic information, 4th order space-time turbulence velocity correlations, limited velocity-temperature correlations	Harper-Bourne (2003), Tanna (1977), Bridges and Wernet (2004), Bridges and Brown (2005), Bridges and Wernet (2004), Bridges and Wernet (2007), Panda et al. (2004), Panda and Seasholtz (2004), Panda (2006), Viswanathan (2004), Eastwood et al. (2010)	Extensive range of studies for a wide range of geometries. Lack of characterization of inflow conditions. Need more data to explore interaction with wings and flap, flight stream, Mach number and heating influences	1–7	1, 2, 3

this and the tabulated, relatively high-speed fan, rig data. For the latter fan wake—OGV interaction data is perhaps needed.

Noise sources, such as those from the propulsive jet and fan have been reduced through improved understanding and advanced technologies. Novel, lean premixed combustors, are necessary to reduce pollutants. But, these burn more unsteadily. Hence, due to the reduction in other noise sources and the increase in combustion noise more data is perhaps needed in this area. Especially data that explores how this noise is modified as it passes through the turbine.

For basic aerofoil noise studies there seems conflicting evidence on noise sources. To avoid problems with noise contamination open jet wind tunnels are generally used. For example, Nash et al. (1999) carried out experiments in a closed section wind tunnel. On finding erroneous tonal components they replaced the upper and lower wind tunnel walls with boxes filled with foam. This approximated an anechoic environment. The potential for the interaction of acoustic waves with wind

Table 6.5 Acoustic modelling validation data sources for airframe zone

Airframe zone	Nature of data	Data sources	General comments	Validation level	
				Aero	Acou
Wide range of rudimentary landing gear configurations	Phased array measurements with 48 microphones	Quayle et al. (2007)	Parametric type study for 1/12th scale models that also include local fairings	0	1, 2, 3
Basic aerofoils—broad band and tonal noise	Surface pressures, mean velocity and Reynolds stress profiles, surface pressure spectra and correlations between this and wake turbulence	Garcia-Sagrado (2007), Garcia-Sagrado and Hynes (2011a) Garcia-Sagrado and Hynes (2011b)	Work Oerlemans et al. (2009) is for wind turbines and explores the impact of serrated trailing edges. The tests are at full scale	2, 3, 4, 5	1, 2
	Surface pressures, far field sound, spectral information, CEPRA 19 anechoic wind tunnel	Garcia (1989)			
	NASA quiet flow facility—surface pressure, spectra, far-field sound spectra, hot-wire measurements of flow around the aerofoil	Hodgson and Brooks (1981)			
	Far field sound data (spectral) GE	Oerlemans et al. (2009)			
	Farfield sound spectra (broad band—open jet facility)	Moreau and Roger (2009)	NACA 0112, wind tunnel noise interference issues		
	LDA and flow visualization and far field sound (mean velocity and Reynolds stress profiles)	Nash et al. (1999)			
	Far field noise spectra	Arbey and Bataille (1983)	Open jet wind tunnel-broad band noise content		

Table 6.5 (Continued)

Airframe zone	Nature of data	Data sources	General comments	Validation level	
				Aero	Acou
	PIV, Reynolds stress, mean velocity and far field sound spectra. Far field sound correlated to flow field	Nakano et al. (2006)	NACA0018, glass wool for upper and lower wind tunnel boundaries to avoid acoustic reflections		
	LDA (no useable profile data published) and far field sound spectra	McAlpine et al. (1999)	FX79 W151 and NACA0012 profiles		
	Far field sound and directivity	Hutcheson and Brooks (2004)	Beam forming used, focus on trailing edge noise		
	Surface pressures and far field sound	Hodgson and Brooks (1981)	Open jet facility, focus on trailing edge noise		
Trailing edge noise—flat strut	Mean velocity and Reynolds stress profiles, surface pressure distributions, far field sound spectra, wall pressure spectra	Blake and Gershfeld (1988), Blake (1975)		2, 3, 4	1, 2
Fan-airframe interaction	Far field circular array of 48 microphones. Kulite pressure transducers just upstream of exhaust	Data reported in Mincu and Manoha (2012)	NACRE fan noise experiment	3	1, 2
Jet-wing interaction	Far field sound spectra	Lawrence et al. (2012)	Wing idealized to a flat plate	–	2
Jet-flap interaction	Far field sound and source location	Mengle et al. (2007)	Serrated nozzles and serrated flaps	–	1

tunnels walls is evident from the Fig. 6.28(a) simulation. In this the acoustic waves emanate from the trailing edge zone. However, for open jet tunnels, the strong jet shear layers can have a contaminating influence in far field sound measurements. The shear layers are in themselves a noise source. Also, acoustic waves fronts will be refracted as they pass through the shear layers. What is more, the impact of the shear layer unsteadiness, for the free jet in the flow, over the aerofoil is uncertain.

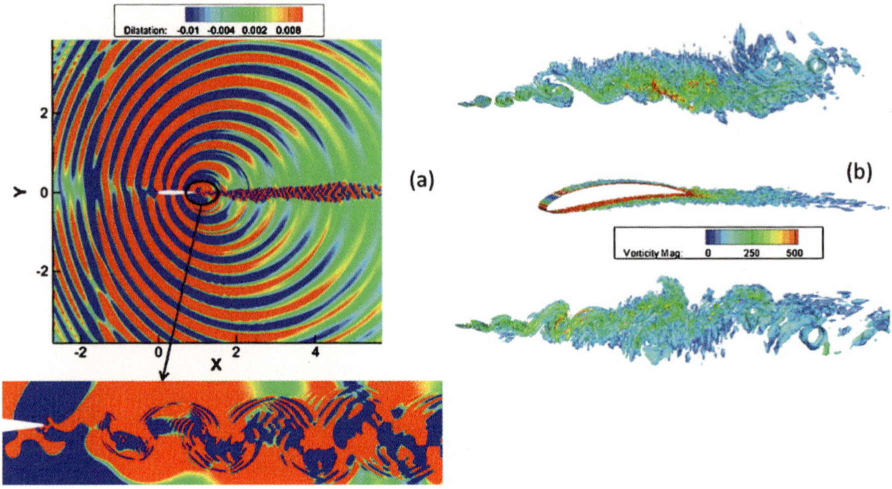

Fig. 6.28 Two-dimensional aerofoil simulations: (**a**) low Re NLES and (**b**) NLES for open jet configuration

Figure 6.28(b) shows a simulation for the case of Winkler and Moreau (2008). The shear layer flow in the open jet zone is close in proximity to the trailing edge noise zone. Many of the more fundamental experiments are at low Reynolds numbers. Those involving laminar separations are hard to numerically replicate.

6.11 The Need for Holistic Simulations

Aerodynamic and aeroacoustic activities in a design context are frequently not dealt with as holistically as they might be and in a multi-objective design sense. Epstein et al. (1986) explored the use of noise to control compressor stability and Collins and Zelenevitz (1975) separation control on aerofoils. Ffowcs-Williams (2002) also outlines the potential of noise to both reduce/cancel noise and also enhance aerodynamic performance. The latter are not unchartered waters and pose great research challenges. Nonetheless, the rich flow physics and acoustics insights provided by LES, along with the potential to precisely control excitations might enable us to make news advances in the area of using sound to reduce sound and also aerodynamic losses. Hence, there seems a need to treat aerodynamics and aeroacoustics more holistically. Extended explorations involving use of residual acoustic energy for modest positive flow control purposes are also perhaps needed. There is frequently an intrinsic link between the aeroacoustic and aerothermal behaviours. Hence, we should perhaps move to multi-objective, aero-thermo-acoustic design. Unfortunately, as noted earlier, acoustic fluctuation amplitudes are low relative to for example that of the turbulence. The acoustic energy levels needed to exert substantial flow control can be high—sound pressure levels of around 100 dB (Collins and Zelenevitz 1975) and there is frequency dependence. However, an area of worthwhile exploration would perhaps be to see if ambient noise could be used to produce some modest positive gain. Certainly, with aero engines fractions of a percent in efficiency are of significant economic importance. LES solutions readily reveal both flow structure and acoustic information. Hence, complex interactions can be more readily observed and understood. Techniques are available to correlate acoustic signatures with flow events. An adjoint process, although challenging, does not seem unreasonable where the influence of the acoustic events on the flow can be explored and adjustments made to use the acoustic signatures to positively influence the flow. At a more basic level, filling in rotor wakes can have both a positive aerodynamic and aeroacoustic impact. Hence, it makes sense to treat these areas holistically. Similarly, recent work suggests the holistic treatment of fan-OGV interactions can yield design breakthroughs.

Accurately predicting aerodynamic and aeroacoustic behaviour can need coupled fluid-structure calculations. For example, with open rotor engines to accurately capture aerodynamics, blade rotor untwist needs to be captured (Zachariadis and Hall 2011). In the hydroacoustics of marine propellers structural deflections are a key sound source component as is the multi-physics element of cavitation. Techniques for dealing with moving boundary problems are discussed in the next chapter which focuses on coupled simulations.

6.12 Conclusions

There is a wide range of far field sound prediction techniques. The FWH approach has considerable popularity. Many of the techniques can be hybridized with, for example, a linearized Euler approach in the near field linked with the lower fidelity FWH or BEM based approaches. In practical terms, the assembly of spatial and temporal unsteady time history fields, that have been created with high quality numerics and grids, using well characterized unsteady turbulent inflow, ready to be fed into far field acoustic methods, presents substantial computational process challenges. The solution of LEEs can present a range of numerical challenges. As would be expected, the acoustic related problems discussed in this chapter tend to use more specialized numerical schemes. For acoustics there seems a roughly equal split between the use of ILES, LES and hybrid RANS-(N)LES methods. However, the sample here is relatively small. With regards to subgrid scale modelling, the use of acoustic subgrid scale modelling appears a useful area for future exploration. The application of eddy resolving methods to acoustics yields the added benefit of providing reliable (relative to RANS) acoustic source information. For complex geometry systems such data can be used to refine RANS based approaches. The Reynolds number independence of jets makes them especially attractive to LES based noise explorations. However, LES studies are needed with a flight stream. This will rapidly weaken shear layers, placing much greater numerical modelling and turbulent inflow challenges. For jets, although the level of validation data has not reached the ideal, it appears high relative to other areas. With regards to aircraft propulsion fan validation data slightly lags. More validation data appears to be needed in the area of combustion noise and its propagation through the turbine. For jets, considerable scatter can be found between equivalent validation data sets and hence this scatter is quite probable for fans too. It appears difficult to find high quality acoustics data in relation to aerofoil noise. More, multi-objective studies are needed that optimize both aerodynamics and aeroacoustics concurrently. Further refinements in predictive accuracy can require coupled calculations that include, for example, characterizing of the structural behaviour.

References

A. Agarwal, P.J. Morris, Direct simulation of acoustic scattering on a rotorcraft fuselage. AIAA Paper Number AIAA-2000-2030 (2000)

A. Agarwal, A.P. Dowling, H.-C. Shin, W. Graham, S. Sefi, Ray-tracing approach to calculate acoustic shielding by a flying wing airframe. AIAA J. **45**(5), 1080–1090 (2007)

N. Andersson, L.-E. Eriksson, L. Davidson, Investigation of an isothermal Mach 0.75 jet and its radiated sound using large-eddy simulation and Kirchhoff surface integration. Int. J. Heat Fluid Flow **26**, 393–410 (2005)

H. Arbey, J. Bataille, Noise generated by airfoil profiles placed in a uniform laminar flow. J. Exp. Fluids **134**, 33–47 (1983)

G. Ashcroft, D. Nuernberger, A computational investigation of broadband noise generation in a low-speed axial fan. AIAA Paper Number AIAA-2001-0512 (2001)

C. Bailly, D. Juve, A stochastic approach to compute subsonic noise using linearized Euler equations. AIAA Paper Number AIAA-99-1872 (1999)

F. Bastin, P. Lafon, S. Candel, Computation of jet mixing noise due to coherent structures: the plane jet case. J. Fluid Mech. **335**, 261–304 (1997)

P. Batten, E. Ribaldone, M. Casella, S. Chakravarthy, Towards a generalized non-linear acoustics solver, in *Proceedings of the 10th AIAA/CEAS Aeroacoustics Conference*, Manchester, United Kingdom, 10–12 May 2004. AIAA Paper Number AIAA-2004-3001

J. Berland, P. Lafon, F. Crouzet, F. Daude, C. Bailly, Numerical insight into sound sources of a rod-airfoil flow configuration using direct noise calculation, in *16th AIAA/CEAS Aeroacoustics Conference* (2010). AIAA Paper Number AIAA-2010-3705

J. Berland, P. Lafon, F. Crouzet, F. Daude, C. Bailly, A parametric study of the noise radiated by the flow around multiple bodies: direct noise computation of the influence of the separating distance in rod-airfoil flow configurations, in *17th AIAA/CEAS Aeroacoustics Conference (32nd AIAA Aeroacoustics Conference)*, Portland, Oregon, 5–8 June 2011. AIAA Paper Number AIAA-2011-2819

S.F. Birch, D.A. Lyubimov, V.P. Maslov, A.N. Secundov, Noise prediction for chevron nozzle flows. AIAA Paper Number AIAA-2006-2600 (2006)

S.F. Birch, D.A. Lyubimov, V.P. Maslov, A.N. Secundov, K.Y. Yakubovsky Coaxial nozzles with deflected fan flows, in *14th AIAA/CEAS Aeroacoustics Conference (29th AIAA Aeroacoustics Conference)* (2008). AIAA Paper Number AIAA-2008-3066

A. Birkefeld, C.D. Munz, Simulations of airfoil noise with the discontinuous Galerkin solver NoisSol. ERCOFTAC Bull. **90**, 28–33 (2012)

W.K. Blake, A statistical description of pressure and velocity fields at the trailing edge of a flat strut. David Taylor Naval Ship Research and Development Center, Rept. 4241, Bethesda, MD (1975) NR

W.K. Blake, J.L. Gershfeld, The aeroacoustics of trailing edges, in *Frontiers in Experimental Fluid Mechanics*, ed. by M. Gad-el-Hak (Springer, New York, 1988), pp. 457–532. Chapter 10

D.J. Bodony, S.K. Lele, Spatial scale decomposition of shear layer turbulence and the sound sources associated with the missing scales in a large-eddy simulation, in *8th AIAA/CEAS Aeroacoustics Conference*, Breckenridge, Colorado (2002). AIAA Paper Number AIAA-2002-2454

D.J. Bodony, S.K. Lele, Jet noise prediction of cold and hot subsonic jets using large-eddy simulation, in *10th AIAA/CEAS Aeroacoustics Conference*, Manchester, UK, 10–12 May 2004. AIAA Paper Number AIAA-2004-3022

J. Bridges, C. Brown, Parametric testing of chevrons on single flow hot jets, in *Proceedings of the 10th AIAA/CEAS Aeroacoustics Conference*, Manchester, UK (2004). AIAA Paper Number AIAA-2004-2824

J. Bridges, C. Brown, Validation of the small hot jet acoustic rig for aeroacoustic research, in *Proceedings of the 11th AIAA Aeroacoustics Conference*, Monterey, California (2005). AIAA Paper Number AIAA 2005-363

J. Bridges, M. Wernet, Measurements of the aeroacoustic sound sources in hot jets. Technical Report, NASA Glenn Research Centre (NASA/TM-2004-221508) (2004)

J. Bridges, M. Wernet, Effect of temperature on jet velocity spectra, in *Proceedings of the 13th AIAA Aeroacoustics Conference*, Italy (2007). AIAA 2007-3628

J. Bridges, M.P. Wernet, Establishing consensus turbulence statistics for hot subsonic jets, in *16th AIAA/CEAS Aeroacoustics Conference*, Stockholm, Sweden, 7–9 June 2010. AIAA Paper No. AIAA 2010-3751

T.P. Bui, W. Schroder, M. Meinke, Acoustic perturbation equations for reacting flows to computer combustion noise. Int. J. Aeroacoust. **6**(4), 335–355 (2007)

F.G. Collins, J. Zelenevitz, Influence of sound upon separated flow over wings. AIAA J. **13**(3), 408–410 (1975)

T. Colonius, S.K. Lele, P. Moin, The scattering of sound waves by a vortex: numerical simulations and analytical solutions. J. Fluid Mech. **260**, 271–298 (1994)

A.J. Cooper, N. Peake, Trapped acoustic modes in aeroengine intakes with swirling flow. J. Fluid Mech. **419**, 151–175 (2000)

D.G. Crighton, Basic principles of aerodynamic noise generation. Prog. Aerosp. Sci. **16**(1), 31–96 (1975)

N. Curle, The influence of solid boundaries on aerodynamic sound. Philos. Trans. R. Soc., Ser. A, Math. Phys. Eng. Sci. **231**(1187), 505–514 (1955)

F. Daude, J. Berland, T. Emmert, P. Lafon, F. Crouzet, C. Bailly, A high-order finite-difference algorithm for direct computation of aerodynamic sound. Comput. Fluids **61**, 46–63 (2012)

J. Defoe, A. Narkaj, Z.S. Spakovszky, A novel MPT noise prediction methodology for highly-integrated propulsion systems with inlet flow distortion, in *15th AIAA/CEAS Aeroacoustics Conference (30th AIAA Aeroacoustics Conference)*, Miami, Florida, 11–13 May 2009. AIAA Paper Number AIAA-2009-3366

P. Druault, S. Lardeau, J.P. Bonnet, F. Coiffet, J. Delville, E. Lamballais, J.F. Largeau, L. Perret, Generation of three-dimensional turbulent inlet conditions for large-eddy simulation. AIAA J. **42**(3), 447–456 (2004)

S. Eastwood, Hybrid LES–RANS of complex geometry jets. PhD thesis, Department of Engineering, University of Cambridge (2009)

S. Eastwood, P.G. Tucker, Hybrid LES-RANS of complex geometry jets. Int. J. Aeroacoust. **10**(5–6), 661–686 (2011)

S. Eastwood, P.G. Tucker, H. Xia, C. Klostermeier, Developing large eddy simulation for turbomachinery applications. Philos. Trans. R. Soc., Ser. A, Math. Phys. Eng. Sci. **367**, 2999–3013 (2009)

S. Eastwood, P.G. Tucker, H. Xia, P. Carpenter, P. Dunkley, Hybrid LES computations and measurements of a small scale high speed coflowing jet. AIAA J. **48**(5), 963–974 (2010)

S. Eastwood, P.G. Tucker, H. Xia, Large eddy simulation of a hot high speed coflowing jet, in *Proceedings of the ASME Turbo. Expo.*, Vancouver (2011). ASME Paper Number GT-46837

S. Eastwood, H. Xia, P.G. Tucker, Large eddy simulation of complex geometry jets. AIAA J. Propuls. Power **28**(2), 235–245 (2012)

A.H. Epstein, J.E. Ffowcs-Williams, E.M. Greitzer, Active suppression of aerodynamic instabilites in turbomachines. AIAA J. Propuls. Power **5**(2), 204–211 (1986)

R. Ewert, M. Munsch, Slat noise predictions based on APE and stochastic sound sources from RANS, in *25th International Congress of the Aeronautical Sciences* (2006)

R. Ewert, W. Schroder, Acoustic perturbation equations based on flow decomposition via source filtering. J. Comput. Phys. **188**, 365–398 (2003)

M. Farge, K. Schneider, Coherent vortex simulation (CVS), a semi-deterministic turbulence model using wavelets. Flow Turbul. Combust. **66**(4), 393–426 (2001)

J.E. Ffowcs-Williams, Noise, anti-noise and fluid flow control. Philos. Trans. R. Soc. Lond., Ser. A **360**, 821–832 (2002)

J.E. Ffowcs-Williams, D.L. Hawkings, Sound generation by turbulence and surfaces in arbitrary motion. Philos. Trans. R. Soc. Lond., Ser. A **264**, 321–342 (1969)

F. Flemming, C. Olbricht, B. Wegner, A. Sadiki, J. Janicka, F. Bake, U. Michel, B. Lehmann, I. Rohle, Analysis of unsteady motion with respect to noise sources in a gas turbine combustor: isothermal flow case. Flow Turbul. Combust. **75**, 3–27 (2005)

P.D. Francescantonio, A new boundary integral formulation for the prediction of sound radiation. J. Sound Vib. **202**(4), 491–509 (1997)

J.F. Freund, Jet noise physics and modelling using first principles simulations NASA/CR–2003-212123 (2003)

U. Ganz, P.D. Joppa, T.J. Patten, D.F. Scharpf, Boeing 18-inch fan rig broadband noise test. NASA/CR-1998-208704 (1998a)

U. Ganz, P. Joppa, S.A.L. Clegg, Measurement and prediction of broadband fan noise. AIAA Paper Number (AIAA-98-2316) (1998b)

P. Garcia, Airfoil noise in a uniform flow. Rech. Aérosp. (Engl. Version) **3**(6), 1–7 (1989)

A. Garcia-Sagrado, Boundary layer and trailing edge noise sources. PhD thesis, University of Cambridge (2007)

A. Garcia-Sagrado, T. Hynes, Stochastic estimation of flow near the trailing edge of a naca0012 airfoil. Exp. Fluids, Exp. Methods Appl. Fluid Flow **51**, 1057–1071 (2011a)

A. Garcia-Sagrado, T. Hynes, Wall-pressure sources near an airfoil trailing edge under separated laminar boundary layers. AIAA J. **49**, 1841–1856 (2011b)

T.M. Georges, Acoustic ray paths through a model vortex with a viscous core. J. Acoust. Soc. Am. **51**(1), 206–209 (1972), Part 2

M.M. Gibson, B.E. Launder, Ground effects on pressure fluctuations in atmospheric boundary layer. J. Fluid Mech. **86**, 491–511 (1978)

M.E. Goldstein, A unified approach to some recent developments in jet noise theory. Int. J. Aeroacoust. **1**(1), 1–16 (2002)

M.E. Goldstein, A generalized acoustic analogy. J. Fluid Mech. **488**, 315–333 (2003)

M. Goody, Empirical spectral model of surface pressure fluctuations. AIAA J. **42**(9), 1788–1794 (2004)

J.C. Hardin, D.S. Pope, An acoustic/viscous splitting technique for computational aeroacoustics. Theor. Comput. Fluid Dyn. **6**, 323–340 (1994)

M. Harper-Bourne Jet noise turbulence measurements, in *9th AIAA Aeroacoustics Conference* (2003). AIAA Paper Number AIAA-2003-3214

L.S. Hedges, A. Travin, P.R. Spalart, Detached-eddy simulations over a simplified landing gear. J. Fluids Eng. **124**(2), 413–423 (2002)

T.H. Hodgson, T.F. Brooks, Trailing edge noise predicted from measured surface pressures. J. Sound Vib. **78**(1), 69–117 (1981)

C.A. Hunter, R.H. Thomas, Development of a jet noise prediction method for installed jet configurations. AIAA Paper Number AIAA-2003-3169 (2003)

F.V. Hutcheson, T.F. Brooks, Effects of angle of attack and velocity on trailing edge noise, in *42nd AIAA Aerospace Sciences Meeting*, Reno, NV (2004)

C.R.S. Ilario, J.R. Meneghini, M. Azarpeyvand, R.H. Self, A ray-tracing method applied to the propagation of jet noise, in *Proc. 17th International Congress on Sound and Vibration (ICSV17)*, Cairo, Egypt, 18–22 July 2010

C.R.S. Ilario, J.R. Meneghini, M. Azarpeyvand, R.H. Self, Refraction effects on far-field noise predictions and sources distribution of coplanar coaxial jet flows, in *18th AIAA/CEAS Aeroacoustics Conference*, Portland, USA, June 2011a. AIAA Paper Number AIAA-2011-2749

C.R.S. Ilario, J.R. Meneghini, M. Azarpeyvand, R.H. Self, Aeroacoustics investigation of the noise from an asymmetric nozzle, in *Proc. 18th International Congress on Sound and Vibration*, Rio de Janeiro, Brazil, 10–14 July 2011b

M.C. Jacob, J. Boudet, J. Casalino, M. Michard, A rod-airfoil experiment as benchmark for broadband noise modeling. Theor. Comput. Fluid Dyn. **19**(3), 171 (2005)

M. Kaltenbacher, M. Ihme, H. Pitsch, Enhanced hybrid LES/CAA approach for combustion-induced noise in complex geometries, in *13th AIAA/CEAS Aeroacoustics Conference (28th AIAA Aeroacoustics Conference)* (2007). AIAA Paper Number AIAA-2007-3403

S.A. Karabasov, M.Z. Afsar, T.P. Hynes, A.P. Dowling, W.A. McMullan, C.D. Pokora, G.J. Page, J.J. McGuirk, Jet noise—acoustic analogy informed by large eddy simulation. AIAA J. **48**, 1312–1325 (2010). ISSN 0001-1452

M. Karcz, J. Badur, An alternative two-equation turbulent diffusivity closure. Int. J. Heat Mass Transf. **48**, 2013–2022 (2005)

C. Kato, Y. Yamade, H. Wang, Y. Guo, M. Miyazawa, T. Takaishi, S. Yoshimura, Y. Takano, Numerical prediction of sound generated from flows with a low Mach number. Comput. Fluids **36**(1), 53–68 (2007)

A. Khavaran, D.C. Kenzakowski, Progress toward improving jet noise predictions in hot jets. NASA/CR–2007-214671 (2007)

M.R. Khorrami, B.A. Singer, M.E. Berkman, Time-accurate simulations and acoustic analysis of slat free shear layer. AIAA J. **40**(7), 1284–1291 (2002)

K.M. Khritov, V.Ye. Kozlov, S.Yu. Krasheninnikov, A.B. Lebedev, D.A. Lyubimov, V.P. Maslov, A.K. Mironov, K.S. Reent, A.N. Secundov, K.Ya. Yakubovsky, S.F. Birch, On the prediction of turbulent jet noise using traditional aeroacoustic methods. Int. J. Aeroacoust. **4**(3–4), 289–324 (2005)

M. Klein, A. Sadiki, J. Janicka, Investigation of the influence of the Reynolds number on a plane jet using direct numerical simulation. Int. J. Heat Fluid Flow **24**(6), 785–794 (2003)

S.R. Koh, G. Geiser, H. Foysi, W. Schroder, Impact of inhomogeneous density distribution on acoustic sources in turbulent jets. ERCOFTAC Bull. **90**, 15–20 (2012)

R.H. Kraichnan, Diffusion by a random velocity field. Phys. Fluids **13**(1), 22–31 (1969)

R. Langtry, P. Spalart, Detached eddy simulation of a nose landing-gear cavity, in *IUTAM Symposium on Unsteady Separated Flows and Their Control*, vol. 14, ed. by M. Braza, K. Hourigan (Springer, Netherlands, 2009), pp. 357–366

J.L.T. Lawrence, S. Nauri, M. Azarpeyvand, R.H. Self, Jet installations effects, in *18th AIAA/CEAS Aeroacoustics Conference (33rd AIAA Aeroacoustics Conference)* (2012)

S. Lee, H. Kim, A. Runchal, Large eddy simulation of unsteady flows in turbomachinery. Proc. Inst. Mech. Eng. A, J. Power Energy **218**(7), 463–475 (2004)

J. Lepicovsky, E. McFarland, R. Chima, V. Capece, J. Hayden, Intermittent flow regimes in a transonic fan airfoil cascade. NASA/TM–2002-211375 (2002)

P. Lew, G.A. Blaisdell, A.S. Lyrintzis, Recent progress of hot jet aeroacoustics using 3-D large-eddy simulation. AIAA Paper Number AIAA-2005-3084 (2005)

Q. Li, N. Peake, M. Savill, Grid-refined LES predictions for fan-OGV broadband noise, in *15th AIAA/CEAS Aeroacoustics Conference (30th AIAA Aeroacoustics Conference)*, Miami, Florida, 11–13 May 2009. AIAA Paper Number AIAA-2009-3147

M.J. Lighthill, On sound generated aerodynamically I. General theory. Proc. R. Soc., Ser. A **211**, 564–587 (1952)

G.M. Lilley, On the noise from jets, noise mechanisms. AGARD CP-131 (1974)

Y. Liu, P.G. Tucker, R.M. Kerr, Linear and non-linear large-eddy simulations of a plane jet. Comput. Fluids **37**, 439–449 (2008)

Y. Liu, A.P. Dowling, N. Swaminathan, T.D. Dunstan, R. Morvant, M.A. Macquisten, L.F. Caracciolo, Prediction of noise source for an aeroengine combustor, in *17th AIAA/CEAS Aeroacoustics Conference*, Portland, Oregon, USA, 5–8 June 2011. AIAA Paper Number AIAA-2011-2913

Y. Liu, A.P. Dowling, N. Swaminathan, T.D. Dunstan, Spatial correlation of heat release rate and sound emission from turbulent premixed flames. Combust. Flame **159**(7), 2430–2440 (2012)

D.P. Lockard, M.R. Khorrami, F. Li, High resolution calculation of a simplified landing gear, in *10th AIAA/CEAS Aeroacoustics Conference* (2004). AIAA Paper Number AIAA paper 04-2887

A.S. Lyrintzis, Integral acoustics methods: from the (CFD) near-field to the (acoustic) far-field. Int. J. Aeroacoust. **2**(2), 95–128 (2003)

M. Mahak, P.G. Tucker, P.K. Ray, Cost-effective hybrid RANS-NLES method for jet turbulence and noise prediction, in *Proc. ASME 2012 Gas Turbine India Conference*, Mumbai, India (2012). ASME Paper Number GT India 2012-9648

E. Manoha, B. Troff, P. Sagaut, Trailing edge noise prediction using large-eddy simulation and acoustic analogy. AIAA J. **38**(4), 575–583 (2000)

A.L. Marsden, M. Wang, J.E. Dennis, P. Moin, Trailing-edge noise reduction using derivative-free optimization and large-eddy simulation. J. Fluid Mech. **572**, 13–36 (2007)

A. McAlpine, E.C. Nash, M.V. Lowson, On the generation of discrete frequency tones by the flow around an aerofoil. J. Sound Vib. **222**, 753–779 (1999)

F. Mendonça, O. Baris, G. Capon, Simulation of radial compressor aeroacoustics using CFD, in *Proceedings of ASME Turbo. Expo. GT2012*, 11–15 June 2012, Copenhagen, Denmark (2012). Paper No. GT2012-70028

V.G. Mengle, R.W. Stoker, L. Brusniak, R. Elkob, Flaperon modification effect on jet-flap interaction noise reduction for chevron nozzles, in *13th AIAA/CEAS Aeroacoustics Conference (28th AIAA Aeroacoustics Conference)* (2007). AIAA 2007-3666

D.C. Mincu, E. Manoha, Numerical and experimental characterization of fan noise installation effects. ERCOFTAC Bull. **90**, 21–27 (2012)

A. Moore, A 3D prediction of the wing reflection of aero engine noise, in *10th AIAA/CEAS Aeroacoustics Conference* (2004). AIAA Paper Number AIAA-2004-2865

A. Moore, C. Mead, Reflection of noise from aero engines installed under an aircraft wing, in *9th AIAA/CEAS Aeroacoustics Conference and Exhibit*, Hilton Head, South Carolina, 12–14 May 2003. AIAA Paper Number AIAA-2003-3151

S. Moreau, M. Roger, Back-scattering correction and further extensions of Amiet's trailing-edge noise model. Part ii: application. J. Sound Vib. **323**, 397–425 (2009)

P.J. Morris, F. Farassat, Acoustic analogy and alternative theories for jet noise prediction. AIAA J. **40**(4), 671–680 (2002)

P.J. Morris, L. Long, A. Bangalore, Q. Wang, A parallel three-dimensional computational aeroacoustics method using nonlinear disturbance equations. J. Comput. Phys. **133**(1), 56–74 (1997)

B. Muhlbauer, B. Noll, M. Aigner, Numerical investigation of entropy noise and its acoustic sources in aero-engines, in *Proceedings of ASME Turbo Expo 2008: Power for Land, Sea and Air GT2008*, Berlin, Germany, 9–13 June 2008. ASME Paper Number GT2008-50321

T. Nakano, N. Fujisawa, S. Lee, Measurement of tonal-noise characteristics and periodic flow structure around naca0018 airfoil. Exp. Fluids **40**(3), 482–490 (2006)

E.C. Nash, M.V. Lowson, A. McAlpine, Boundary-layer instability noise on aerofoils. J. Fluid Mech. **382**, 27–61 (1999)

G.R. Newman, B.E. Launder, J.L. Lumley, Modelling the behaviour of homogeneous scalar turbulence. J. Fluid Mech. **111**, 217–232 (1981)

S. Oerlemans, M. Fisher, T. Maeder, K. Kögle, Reduction of wind turbine noise using optimized airfoils and trailing-edge serrations. AIAA J. **47**(6) (2009)

Y. Ozyoruk, L.N. Long, Multigrid acceleration of a high-resolution. AIAA J. **35**(3), 428–433 (1997)

U. Paliath, H. Shen, R. Avancha, C. Shieh, Large eddy simulation for jets from chevron and dual nozzles, in *Proceedings of the 17th AIAA/CEAS Aeroacoustics Conference (32nd AIAA Aeroacoustics Conference)*, Portland, Oregon (2011). AIAA Paper Number AIAA-2011-2881

J. Panda, Two point space–time correlation of density fluctuations measured in high velocity free jets, in *Proceedings of the NASA* (2006). CR–2006-214222

J. Panda, R. Seasholtz, Velocity and temperature measurement in supersonic free jets using spectrally resolved Rayleigh scattering, in *Proceedings of the NASA* (2004). TM–2004-212391

J. Panda, R. Seasholtz, K. Elam, A. Mielke, D. Eck, Effect of heating on turbulent density fluctuations and noise generation from high speed jets, in *Proceedings of the 10th AIAA Aeroacoustics Conference*, United Kingdom (2004). AIAA Paper No. AIAA 2004-3016

C.D. Pruett, Eulerian time-domain filtering for spatial large-eddy simulation. AIAA J. **38**(9), 1634–1642 (2000)

A.R. Quayle, A.P. Dowling, H. Babinsky, W.R. Graham, Y. Liu, Phased array measurements from landing gear models, in *13th AIAA/CEAS Aeroacoustics Conference (28th AIAA Aeroacoustics Conference)* (2007). AIAA 2007-3463

P.K. Ray, W.N. Dawes, Detached-eddy simulation of transonic flow past a fan-blade section, in *15th AIAA/CEAS Aeroacoustics Conference*, Miami, FL, 11–13 May 2009. AIAA Paper Number AIAA-2009-3221

N. Reau, A. Tumin, On harmonic perturbations in a turbulent mixing layer. Eur. J. Mech. B, Fluids **21**, 143–155 (2002)

G. Reboul, C. Polacsek, S. Lewy, S. Heib, Ducted fan broadband noise simulations using unsteady or averaged data, in *Inter-noise 2008, 37th Int' Congress and Exposition on Noise Control Engineering* (2008), pp. 2460–2475

J. Riou, S. Lewy, S. Heib, Large eddy simulation for predicting rotor-stator broadband interaction fan noise, in *36th International Congress and Exhibition on Noise Control Engineering*, Istanbul, Turkey, August 2007, pp. 28–31

M. Roger, S. Moreau, Broadband self-noise from loaded fan blades. AIAA J. **42**(3), 536–544 (2004)

J.E. Rossiter, Wind tunnel experiments on the flow over rectangular cavities at subsonic and transonic speeds. Technical Report RAE-ARC-3438R&M, Aeronautical Research Council Reports and Memoranda (1964)

Ş. Rulik, S. Dykas, W. Wroblewski, Modeling of aerodynamic noise using hybrid SAS and DES methods, in *Proc. ASME Turbo Expo 2010: Power for Land, Sea and Air GT2010*, Glasgow, 14–18 June 2010, pp. 1–9. ASME GT2010-22696

R.D. Sandberg, L.E. Jones, Direct numerical simulations of low Reynolds number flow over airfoils with trailing-edge serrations. J. Sound Vib. **330**, 3818–3831 (2011)

R.D. Sandberg, L.E. Jones, Direct numerical simulations of airfoil self-noise. ERCOFTAC Bull. **90**, 4–9 (2012)

N.D. Sandham, Q. Li, H.C. Yee, Entropy splitting for high-order numerical simulation of compressible turbulence. J. Comput. Phys. **178**, 307–322 (2002)

C. Schemel, F. Thiele, F. Bake, B. Lehmann, U. Michel, Sound generation in the outlet section of a gas turbine combustion chamber, in *10th AIAA/CEAS Aeroacoustics Conference* (2004). AIAA Paper Number AIAA-2004-2929

R.H. Schlinker, A. Blanc, E. Lurie, J. Joo, A. Reimann, D. Shannon, Simulation and validation of centrifugal impeller shock wave and acoustic power prediction, in *Proceedings of ASME Turbo Expo 2012 GT2012*, Copenhagen, Denmark, 11–15 June 2012. ASME Paper Number GT2012-70076

A.N. Secundov, S.F. Birch, P.G. Tucker, Propulsive jets and their acoustics. Philos. Trans. R. Soc., Ser. A, Math. Phys. Eng. Sci. **365**(1859), 2443–2467 (2007)

M.L. Shur, P.R. Spalart, M.Kh. Strelets, A.K. Travin, Towards the prediction of noise from jet engines. Int. J. Heat Fluid Flow **24**, 551–561 (2003)

M.L. Shur, P.R. Spalart, M.L.Kh. Strelets, Noise prediction for increasingly complex jets. Part I: methods and tests. Int. J. Aeroacoust. **4**(34), 247–266 (2005)

M. Shur, P. Spalart, M. Strelets, A. Garbaruk, Further steps in LES-based noise prediction for complex jets, in *Proceedings of the 44th AIAA Aerospace Sciences Meeting and Exhibit*, Nevada (2006). AIAA Paper Number AIAA-2006-485

P.R. Spalart, Jet noise from simulations: from path-finding to accuracy to industrial value, in *Proceedings of the 49th AIAA Aerospace Sciences Meeting Including the New Horizons Forum and Aerospace Exposition*, Florida (2011). Oral Presentation

P.R. Spalart, M.L. Shur, M.K. Strelets, A.K. Travin, Towards noise prediction for rudimentary landing gear, in *IUTAM Symposium on Computational Aero-Acoustics for Aircraft Noise Prediction, Procedia IUTAM 1* (2010), pp. 283–292

K. Suder, M.L. Celestina, Experimental and computational investigation of the tip clearance flow in a transonic axial compressor rotor. NASA, TM-106711 (1994)

K. Suder, M.L. Celestina, A.J. Strazisar, W.B. Roberts, The effect of adding roughness and thickness to a transonic axial compressor rotor. J. Turbomach. **117**, 491–505 (1995)

C.K.W. Tam, L. Auriault, Jet mixing noise from fine-scale turbulence. AIAA J. **37**(2), 145–153 (1999)

H.K. Tanna, An experimental study of jet noise. Part 1: turbulent mixing noise. J. Sound Vib. **50**(3), 405–428 (1977)

M. Terracol, E. Manoha, C. Herrero, E. Labourasse, S. Redonnet, P. Sagaut, Hybrid methods for airframe noise numerical prediction. Theor. Comput. Fluid Dyn. **19**, 197–227 (2005)

P.G. Tucker, Computation of unsteady turbomachinery flows: Part 2—LES and hybrids. Prog. Aerosp. Sci. **47**, 546–569 (2011)

P.G. Tucker, S. Eastwood, C. Klostermeier, H. Xia, P. Ray, J. Tyacke, W.N. Dawes, Hybrid LES approach for practical turbomachinery flows: Part 2—further applications. ASME J. Turbomach. **134**(2), 021024 (2012)

J. Tyler, T. Sofrin, Axial flow compressor noise studies. SAE Transact. **70**, 309–332 (1962)

A. Uzun, M. Hussaini, Simulation of noise generation in the near-nozzle region of a chevron nozzle jet. AIAA J. 1793–1810 (2009)

K. Viswanathan, Aeroacoustics of hot jets. J. Fluid Mech. **516**, 39–82 (2004)

M. Wang, P. Moin, Computation of trailing-edge flow and noise using large-eddy simulation. AIAA J. **38**(12), 2201–2209 (2000)

J. Winkler, S. Moreau, LES of the trailing-edge flow and noise of a NACA6512-63 airfoil at zero angle of attack, in *Proceedings of the Summer Program 2008*, Center for Turbulence Research (2008), pp. 331–342

H. Xia, P.G. Tucker, Numerical simulation of single-stream jets from a serrated nozzle. Flow Turbul. Combust. **88**(1), 3–18 (2012)

H. Xia, P. Tucker, S. Eastwood, Large-eddy simulations of chevron jet flows with noise predictions. Int. J. Heat Fluid Flow **30**(6), 1067–1079 (2009)

H. Xia, S. Karabasov, O. Graham, P.G. Tucker, A. Dowling, K. Depuru, T.P. Hynes, Hybrid RANS-LES modeling of chevron nozzles with prediction of far field sound, in *Proceedings of the 49th AIAA Aerospace Sciences Meeting and Exhibit*, Orlando, Florida (2011). AIAA Paper Number AIAA-2011-21

H. Xia, P.G. Tucker, S. Eastwood, M. Mahak, The influence of geometry on jet plume development. Prog. Aerosp. Sci. **52**, 56–66 (2012)

A. Zachariadis, C. Hall, Application of a Navier-Stokes solver to the study of open rotor aerodynamics. J. Turbomach. **133**, 031025 (2011)

Chapter 7
Coupled Computational Aerodynamics

7.1 Introduction

Many practical problems have an element of coupling. The aerodynamics can be coupled in the sense that a distinct component's flow field can be influenced by differing upstream and downstream aerodynamic elements. An example would be an engine intake duct that is upstream of a compressor and downstream of a rotor. The three elements are strongly aerodynamically coupled. The accurate prediction of heat transfer can also involve solving a coupled problem in resolving both heat conduction and heat transfer in an fluid. This presents another type of coupled problem. Aerodynamic behaviour can also be coupled to a structural response and the aerodynamics and structural behaviours interact. This coupled system in turn can produce noise and the noise can also be coupled to the aerodynamic and hence structural response. In the acoustics Chap. 6, typically just one-way coupling is considered i.e flow field linking to a noise propagation tool. Hence, naturally, care would be needed when using such a decoupled approach if the acoustic signature influenced flow fields when there are, for example, multi-component interactions. As indicated, by Spalart and Bogue (2003) the ultimate coupled aerospace model would include the modelling of the airframe aerodynamics, the propulsive system's aero-thermal behaviour with all this coupled to the pilot's system inputs. Coupled problems can involve moving meshes. Examples could include aero-elasticity, stores release and a vehicle piggybacked to another launch vehicle, groups of aircraft and their interaction (see Aftosmis et al. 1998 for a static configuration) to name but a few. In this chapter we outline some methods needed for such moving grid problems. Also, some examples of moving grid problems will be given. As with Chaps. 5 and 6 validation levels/grades will be defined. Also, examples of validation will be given and eddy resolving simulations reviewed.

P.G. Tucker, *Unsteady Computational Fluid Dynamics in Aeronautics*,
Fluid Mechanics and Its Applications 104, DOI 10.1007/978-94-007-7049-2_7,

Table 7.1 Mesh adaptation strategies and their advantages and disadvantages

Meshing strategy	Advantages	Disadvantages
Enrichment	Most practical for steady flows	Memory intensive in 3D; Distorted elements can be generated; De-refinement beyond initial mesh challenging
Movement	Efficient for modest boundary movements	Need to define sufficient initial points; Not easy to guarantee valid mesh
Re-meshing	Can deal extensive boundary movements	Time consuming in 3D

7.2 Moving Meshes

7.2.1 Mesh Adaptation Approaches

For unsteady flows, meshes can be either adapted to track/resolve flow features or geometrical deflections. Various mesh adaptation approaches and their advantages and disadvantages are summarized in Table 7.1.

Mesh Enrichment

As noted, an initial coarse mesh can be enriched to track flow features. This approach could be adapted to track moving flow features such as trailing edge vortices. Especially with LES in mind, it would be important to be sure that mesh movement does not introduce spurious motions. This aspect is to an extent dealt with later under the section on space conservation. Time averaging of results on a moving grid is also challenging. In addition, the continual changing of the LES filter width raises theoretical questions. A key, general, difficulty is that multiple grid refinements can generate distorted elements and mesh quality is especially important for LES. Also, in three-dimensions (3D) local mesh refinement—in a grid generation context—can be expensive. For example, if it is wished to de-refine the origins of the original refinement must be stored.

Boundary Movement

Rather than move the mesh, for small boundary deflections a transpiration boundary condition can be used. It involves imposing a surface velocity, correctly aligned, to account for the surface movement. Aero-elasticity problems can have moving aerodynamic surfaces in close proximity to stationary. Then, if one surface is moved, there is potentially strong grid shearing. Hence, in such zones, the transpiration boundary condition can be convenient. Examples of the use of such approaches, in

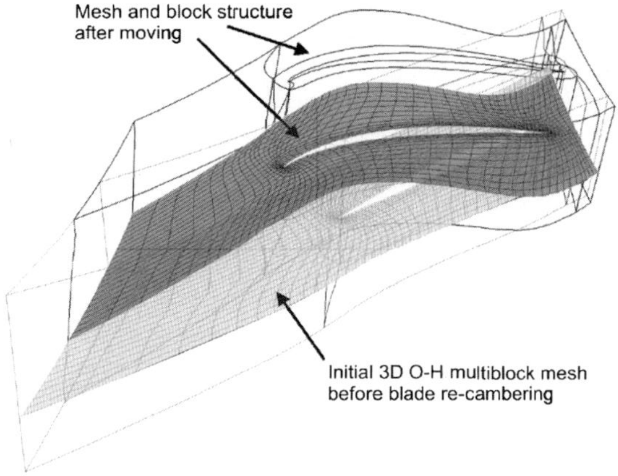

Fig. 7.1 Mesh movement using simple algebraic interpolations (from Page et al. 2013)

aero-elasticity calculations, are given by Stephens et al. (1998) and Cavagna et al. (2009).

Just moving the mesh to account for simple boundary movements is relatively straight forwards. This is especially so if structured meshes are involved. However, when there are large surface movements and complex mesh topologies it is not easy to guarantee a physically sensible mesh. If there are not enough initial points, the solution will need to be re-started. Full re-meshing is expensive but essential for problems with large boundary movements. Ideally, a moving boundary scheme needs to combine the advantages of the three methods noted above i.e. mesh movement, local enrichment and also zone marking and complete re-meshing. Such a strategy is discussed later. Next key mesh movement strategies are outlined.

7.2.2 Mesh Movement Algorithms

Transfinite Interpolation

For structured multiblock meshes transfinite interpolation is a fast efficient means of mesh generation and mesh movement (Gaitonde and Fiddes 1993). The most popular methods produce perturbations on existing grids. Algebraic shearing is widely used (Potsdam and Guruswamy 2001). With this surface displacements are distributed along grid lines of constant index. Grid line rotation operations can also be implemented. The displacement is weighted so that it tends to zero towards the outer boundary. Such a method is not readily extendible to multiblock grids where the outer boundary in likely to be another grid block. Figure 7.1 shows the use of a simple algebraic interpolation based technique for a fan blade. As can be seen reasonable grid quality can be (robustly) maintained for substantial grid displacements.

Delaunay Mapping

With the approach of Liu et al. (2006) the Delaunay graph of the boundary points is stored. There is a one to one mapping between this and interior grid points. Hence, once the boundary has moved and the boundary point Delaunay graph adjusted through mapping the mesh can be updated. With boundary rotation the approach (Delaunay graph) can become invalid. Since searches take place in the preprocessing stage the method is especially efficient.

Spring Analogy

The most widely used approach is to assume that the edges connecting nodes behave like springs that respond to small boundary displacements in either a linear or non-linear fashion. This is achieved by solving the following internal equilibrium equation

$$\sum_j k_{ij}\left[d_{ij}^n - d_{ij}^{n-1}\right] = 0 \tag{7.1}$$

In the above, the subscripts i and j are the indices for the two ends of a grid line. The spring coefficient, k_{ij}, can taken to be the inverse of the edge length or some adjacent volume based scale. Also, d_{ij}^n and d_{ij}^{n-1} give edge lengths before and after movement, respectively. For highly stretched near wall grids negative volumes can arise. In this zone the spring stiffness can be locally increased. This will result in the near wall cells moving as a solid. The spring analogy process can take around 10–15 % of total computing time (Hassan et al. 2007a).

Overset and Cut Cell Grids

For bodies with relative movement, the use of overset grids often proves robust and efficient. The overset grids can either be structured or unstructured. Components can slide through a background mesh or grids have relative movement. A the key issue with such grids is how to define the inter-grid boundaries. Nakahashi and Togashi (2000) use the local level set to determine the classification of vertices and the hole cutting. Notably, the eikonal equation (see later) is most suitable for this purpose.

It seems worth here briefly noting the Cartesian cut cell approach of Yang et al. (1997). This can naturally deal with moving objects by time dependently changing the cell cutting associated with an object moving through a Cartesian background mesh. The approach has a range of limitations. For example, like all cut cell techniques it is non-ideal for cases where accurate skin friction is critical. However, like many approaches it has its niche.

Table 7.2 summarizes advantages and disadvantages of different mesh movement approaches.

Table 7.2 Advantages and disadvantages of different mesh movement approaches

Approach	Advantages	Disadvantages
Transfinite interpolation	Fast	Lacks generality
Delaunay mapping	Can be applied to unstructured meshes, fast, allows large mesh movements	With boundary rotation the approach can become invalid
Spring analogy	Robust, general and provides good quality meshes	Computationally expensive, memory intensive
Overset grids with relative movement	Allows extreme mesh movements	Grid interface definition can be awkward, lack of interface conservation

Mesh-less methods are attractive. This is especially so for unsteady free surface flows. This type of flow is found in (Pelton) turbines. Marongiu et al. (2007, 2010) model this flow using the meshless, smooth particle hydrodynamics methods. For such free surface flows, volume of fluid (Sussman and Puckett 2000) with level set (Sethian 1999) methods can also be utilized.

7.2.3 Hybrid Mesh Movement Approach

Ideally a moving boundary scheme needs to combine advantages of the three methods noted above i.e. mesh movement, local enrichment and also zone marking and complete remeshing.

Figure 7.2 gives a flow chart showing how such meshing strategy, proposed by Hassan et al. (2007b), works. As can be seen from the chart, the initial mesh is generated and its quality assessed. The latter could be based on element angle or volume or both. In the time loop the following happens: surface coordinates are updated; based on these the volume mesh is updated and then the quality of this new volume mesh is tested. Based on the flow solution and the current grid spacings, grid spacings necessary to keep the solution error below a specified level are worked out. If the new grid is satisfactory the new time step is started. If it is not, the elements that failed tests are marked and grouped to form zone boundaries or holes. The elements in the zone boundary region are deleted and the hole remeshed. Then the solution is interpolated onto the new grid. The new time step is recomputed. Note, where possible, for viscous moving boundary cases the surface grid is held fixed.

An example of modelling an F16 stores release using the above hybrid simulation strategy is given in Hassan et al. (2007a,b). With this simulation, the payload trajectory is not prescribed but its 6 degrees of freedom solved for. Impressive agreement is found with measured trajectories. Hassan et al. (2007a,b) also show a shuttle booster simulation with a prescribed component motion. The shuttle moves back and then rotates through 15°. The surface mesh is fixed. This simulation also involves zonal remeshing. It is important to stress that the mesh generation

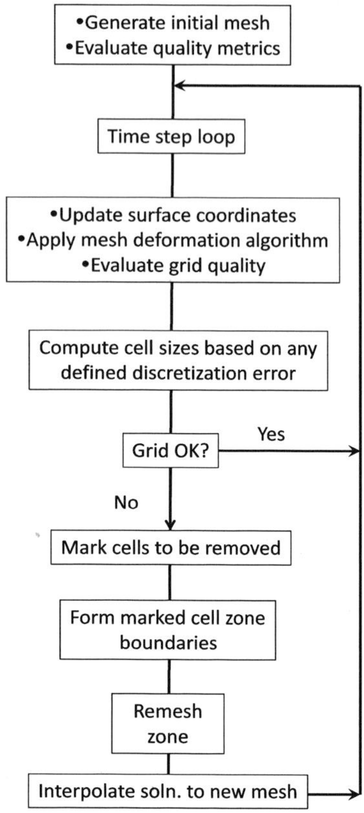

Fig. 7.2 Flow chart showing a typical moving meshing strategy (Procedure of Hassan et al. 2007b)

is these simulations is done in parallel. Dynamic parallel mesh partitioning takes place. Nonetheless, the volume meshing takes around half the computational time. Ideally any zones to be remeshed must be contained completely on one processor i.e. a zone marked for remeshing should not be spread across multiple processors.

7.3 Space Conservation Laws

For moving grids, in essence (either explicitly or implicitly), grid velocities are required in the discretized equations. If these velocities are not carefully specified, the grid movement can result in mass conservation errors. To avoid these, the grid velocities must be such that the Space Conservation Law (SCL) is obeyed. For an arbitrary shaped domain, containing a constant density fluid, the SCL can be expressed (Demirdžić and Perić 1988) in the following integral form

$$\frac{V^{new} - V^{old}}{\Delta t} = \oint v_g \cdot ds^n \tag{7.2}$$

where v_g is the grid velocities, V cell volumes and ds surface areas. For a region containing rectangular, indices notated, elements the following form of Eq. (7.2) is most convenient

$$\frac{V^{new} - V^{old}}{\Delta t} = \sum v_{g,NB} \cdot s_{NB}^{n} \tag{7.3}$$

In the above the subscript NB refers to a summation over neighbouring points. To enforce the SCL, grid velocities must be chosen so that Eqs. (7.2) or (7.3) are satisfied. This implies

$$\left(\frac{\Delta V}{\Delta t}\right)_{SCL} = \left(\frac{\Delta V}{\Delta t}\right)_{a} \tag{7.4}$$

the subscript a referring to an actual value. First it is instructive to consider the case where the temporal grid movement is in just one direction. This is shown in Fig. 7.3a. The grid is stretched vertically at a velocity $v_{g,j+1}$ such that

$$v_{gn} = \frac{\delta y_{j+1}}{\Delta t} \tag{7.5}$$

Equation (7.3) then gives (since $u_{g,i+1} = u_{g,i-1} = v_{g,j-1} = 0$)

$$\left(\frac{\Delta V}{\Delta t}\right)_{SCL} = v_{g,j+1}\Delta x \tag{7.6}$$

Also, it can be readily seen that for this simple case

$$\left(\frac{\Delta V}{\Delta t}\right)_{a} = \frac{\delta y_{j+1}\Delta x}{\Delta t} \tag{7.7}$$

Using Eq. (7.5), Eq. (7.7) can be re-expressed as

$$\left(\frac{\Delta V}{\Delta t}\right)_{a} = v_{g,j+1}\Delta x \tag{7.8}$$

Comparison of Eqs. (7.6) and (7.8) shows the SCL is automatically satisfied. It will now be considered for the simple unidirectional coordinate transformation. The moving control volumes relating to this are shown in Fig. 7.3b, L_y^{n-1} and L_y^n corresponding to the temporal height change of the cell. The SCL (Eq. (7.3)) is again as given by Eq. (7.6), where (assuming L_y is centrally located within the cell, which is not always the case)

$$v_{g,j+1} = \frac{\delta L_y}{\Delta t} \tag{7.9}$$

Also, it can be seen, since the volumes of the left and right hatched areas in the L_y^n frame will cancel out (this is again assuming L_y is centrally located in the cell)

$$\left(\frac{\Delta V}{\Delta t}\right)_{a} = \frac{\delta L_y \Delta x}{\Delta t} \tag{7.10}$$

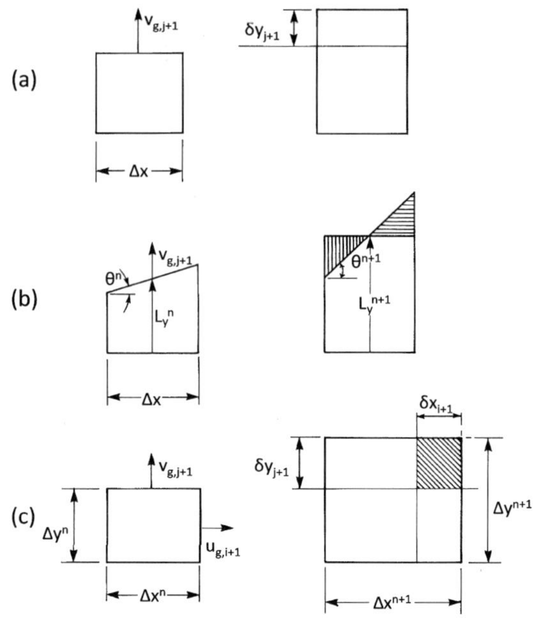

Fig. 7.3 Some basic temporal control volume deformations: (**a**) deformation just in one direction; (**b**) deformation for unidirectional coordinate transformation and (**c**) deformation in two directions

Substituting Eq. (7.9) into (7.10) gives Eq. (7.6), showing again that the SCL is satisfied. Finally, the more general case will be considered where two perpendicular surfaces move (the $j+1$ and $i+1$ control volume faces) as shown in Fig. 7.3c. For this case we have the additional velocity equation

$$u_{g,i+1} = \frac{\delta x_{i+1}}{\Delta t} \tag{7.11}$$

The space conservation law (Eq. (7.3)) then becomes

$$\left(\frac{\Delta V}{\Delta t}\right)_{SCL} = v_{g,j+1}\Delta x^n + u_{g,i+1}\Delta y^n \tag{7.12}$$

Using Eqs. (7.5) and (7.11) (the obvious grid velocity expressions) gives the following artificial continuity equation mass source error expression

$$\left(\frac{\Delta V}{\Delta t}\right)_{ERROR} = u_{g,i+1}(-v_{g,j+1})\Delta t \tag{7.13}$$

If Eqs. (7.5) and (7.11) are to be used in computations, it must be ensured that the error term $\varepsilon = (\Delta V)_{ERROR}/(\Delta V)_{SCL}$ (this assumes $(\Delta V)_a \cong (\Delta V)_{SCL}$) is small, i.e.

$$\epsilon \approx \frac{u_{g,i+1}(-v_{g,j+1})\Delta t}{u_{g,i+1}\Delta y^n + v_{g,j+1}\Delta x^n} \tag{7.14}$$

Equation (7.14) implies that to make the error negligible Δt must be small or the grid spacing large! On the other hand, (7.14) shows that keeping one set of grid lines fixed ensures $\varepsilon = 0$. The SCL time-step constraint (i.e. Eq. (7.14)) can be more severe than the Δt constraint for temporal accuracy, hence the importance of its consideration. A possible grid velocity specification that automatically satisfies the SCL, in the example above, is as follows

$$u_{g,i+1} = \frac{\Delta y^{n-1} + \Delta y^n}{2\Delta y^n} \frac{\delta x_{i+1}}{\Delta t} \tag{7.15}$$

$$v_{g,j+1} = \frac{\Delta x^{n-1} + \Delta x^n}{2\Delta x^n} \frac{\delta y_{j+1}}{\Delta t} \tag{7.16}$$

Further expressions, for arbitrary moving grids, are presented by Demirdžić and Perić (1988) and Hassan et al. (2007b). The latter applies to dual meshes constructed with assemblies of triangular facets

7.4 Wall Distance Computation

Wall distances, *d*, are still a key parameter in many key turbulence models (Spalart and Allmaras 1992; Menter 1993). Surprisingly, for highly optimised RANS solvers, the effort in calculating *d* can be a significant fraction of the total solution time. For example, even with a Cray C90 class computer it took 3 hours just to gain *d* (Wigton 1998). For flows with time dependent geometry (such as Computational Aeroelasticity and design optimization) or mesh refinement clearly this feature is exacerbated (Boger 2001). Because of *d* evaluation expense in some codes dangerous approximations are made (Spalart 2000).

Avoiding using nearest surface search (NSS) procedures Spalding (1994) proposed solving a Poisson differential equation for a wall distance related function ϕ. This equation is given below.

$$\nabla^2 \phi = -1 \tag{7.17}$$

The variable ϕ can be converted into a wall distance function, $\tilde{d}_P$, through the auxiliary equation below

$$\tilde{d}_P = \pm \sqrt{\sum_{j=1,3} \left(\frac{\partial \phi}{\partial x_j} \right)^2} + \sqrt{\sum_{j=1,3} \left(\frac{\partial \phi}{\partial x_j} \right)^2 + 2\phi} \tag{7.18}$$

The analytical derivation of (7.18) includes the assumption that surfaces are extensive in the non wall normal direction.

Fares and Schröder (2002) also derived a differential equation for $\tilde{d}$. This again is aimed at reflecting the traits of Eqs. (7.17), (7.18). However, an exact equation for *d*, that can be readily derived using coordinate geometry is the hyperbolic eikonal equation below

$$|\nabla d| = 1 + \Gamma \nabla^2 d \tag{7.19}$$

where $\Gamma \to 0$ giving viscosity solutions. This equation models a front propagating at unit velocity from surfaces. In fact, d is the first arrival time of the front. For a unit velocity this time is equivalent to wall distance.

Robust hybrid Poisson-eikonal approach is outlined in Tucker (2011b). This uses a hybrid front propagation based on a weighted average of that for the Poisson and eikonal equations.

Notably, the differential equations discussed above have the key advantage that they are naturally amenable to parallel post processing. Next the performance of these approaches are considered for some basic aeroelasticity problems. The space conservation law is used with curvilinear grids. Where there is grid movement transfinite interpolation is used.

7.5 Examples of Moving Mesh Related Calculations

7.5.1 Pitching Aerofoils

First a pitching NACA 0012 case is considered. Figure 7.4 compares NACA 0012 lift and moment coefficient (C_M) measurements, represented by symbols, with predictions. Figure 7.5 gives instantaneous pressure contours for angles of attack of 2° and 5°. Substantial temporal pressure variations can be seen. In Fig. 7.4 the full lines give an NSS procedure d solution with a non-deforming grid that rotates with the aerofoil. The long dashed and dotted lines are for deforming grid NSS and Poisson d distribution solutions, respectively. Importantly, the NSS solution uses a time invariant d distribution taken from the non-deforming grid case. Frames (a) and (b) give lift and moment coefficient data, respectively. Full test case details can be found in Tucker et al. (2003). The similarity of the lines suggests, as might be expected for this simple geometry, the grid redistribution scheme keeps the surface grid form quasi-constant with time. Hence it is very similar to that for the fixed mesh case. Therefore, the major differences between the fixed and deforming mesh results can most probably be attributed to discretization errors arising through the mesh movement and not wall distance differences. As might be expected, the Poisson equation approach needs little convergence effort. However, this is to a large extent due to the relatively low levels of grid deformation for this case. Nonetheless, the Poisson approach is found to require a factor of around 2.25 less computing time than the search. Also, the Gauss-Siedel solver iterations used for this case comprise about a factor of $1/100$ of the total Poisson cost. Therefore, with large grid movements the Poisson method is likely to be quite efficient.

7.5.2 Double-Delta Wing

Next a double-delta case is considered. For this the eikonal, NSS, and Poisson methods are used. The angle of attack is 6°, $M = 0.96$ and $Re = 2.2 \times 10^6$. A key

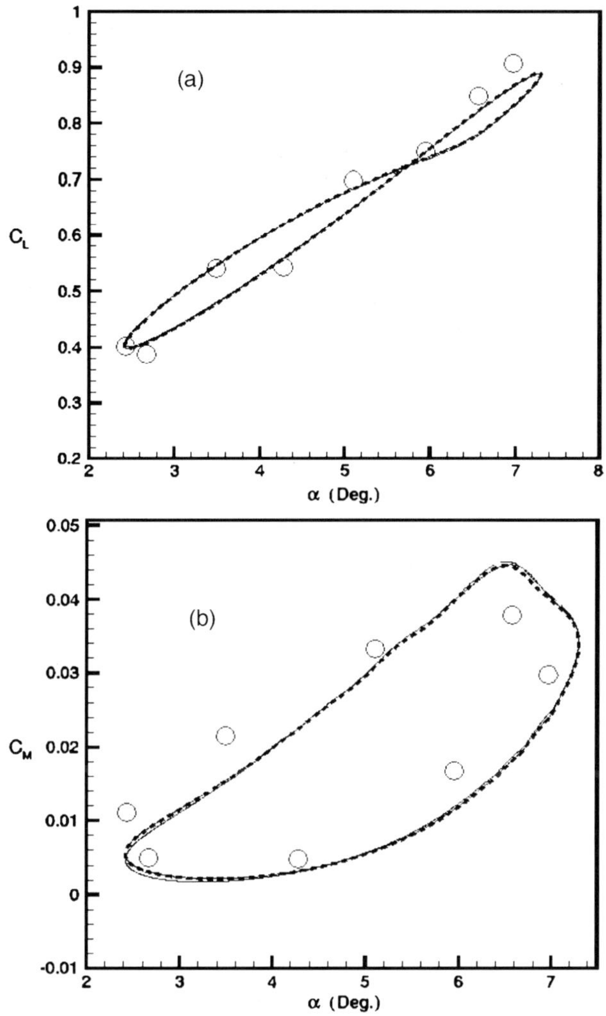

Fig. 7.4 Pitching NACA 0012 results: (**a**) lift coefficient variations and (**b**) moment coefficient variations. (o measurements, --- non-deforming grid, - - - deforming grid with NSS, ··· deforming grid with Poisson)

motivation for this case was aeroelasticity studies made as part of Boeing's sonic cruiser development. Figure 7.6 shows initial and severely deformed aeroelasticity calculation surface grids. These use a total number (N) of 1.2×10^6 cells. The meshes are far more severely deformed than for a realistic engineering calculation. Extreme plunging generates the deformations. They are intended to produce highly deformed cells to strongly test the robustness of differential equation based d approaches. Frames (a, b) of Fig. 7.6 are three-dimensional views. Frames (c, d) are two-dimensional y–z plane views. For moving mesh performance studies two approximately equi-spaced deflection increments between the Fig. 7.6 extremes are considered. Figure 7.7 gives eikonal equation d contours in three z–y planes for the undeflected geometry. Notably, if the search procedure is carried out every time step,

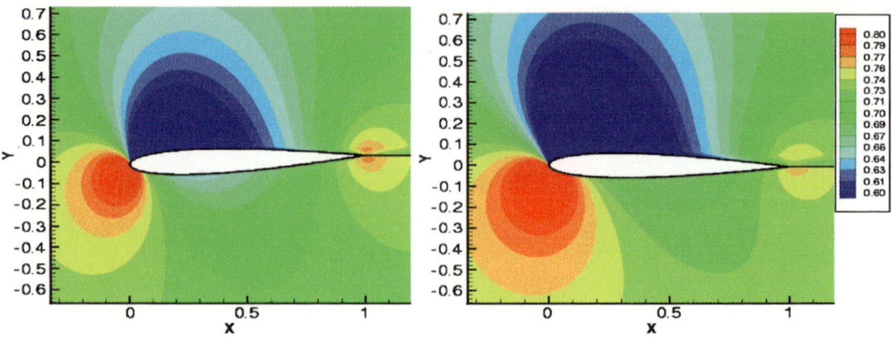

Fig. 7.5 NACA 0012 pressure contours for angles of attack of 2° and 5°

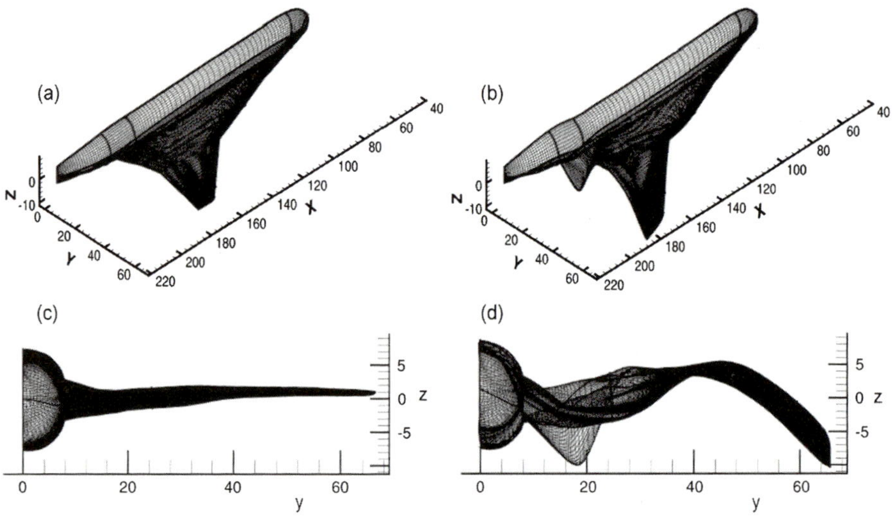

Fig. 7.6 Double delta initial and fully deformed surface grids:(**a**), (**b**) three-dimensional views and (**c**), (**d**) two-dimensional y–z plane views. Published with kind permission from Elsevier

this would constitute 20 % of the time step cost. The differential equation based time step methods constitute a fraction of a percent of the time step cost. This is mostly because of the effectiveness of the spring analogy surface grid redistribution procedure (near surface cell relative positions hardly change) and also because in a non-stationary grid mode the differential equations have a good initial guess for the iterative solution.

Figure 7.8 gives $y^+ < 400$ 'error' histograms where y^+ is dimensionless wall distance. Frame (a) is for the eikonal equation. Frame (b) is for the Poisson. The average Frame (a) 'error' is 3.27 %. The average Poisson error is 2.67 %. The lift and drag coefficients for the Poisson method are within 0.03 and 0.06 % of those

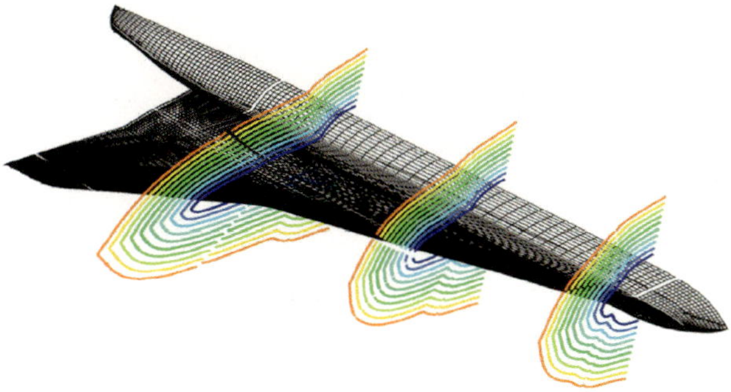

Fig. 7.7 Double delta eikonal equation *d* contours. Published with kind permission from Elsevier

Fig. 7.8 Double delta $y^+ < 400$, *d* error histogram: (**a**) eikonal equation and (**b**) Poisson. Published with kind permission from Elsevier

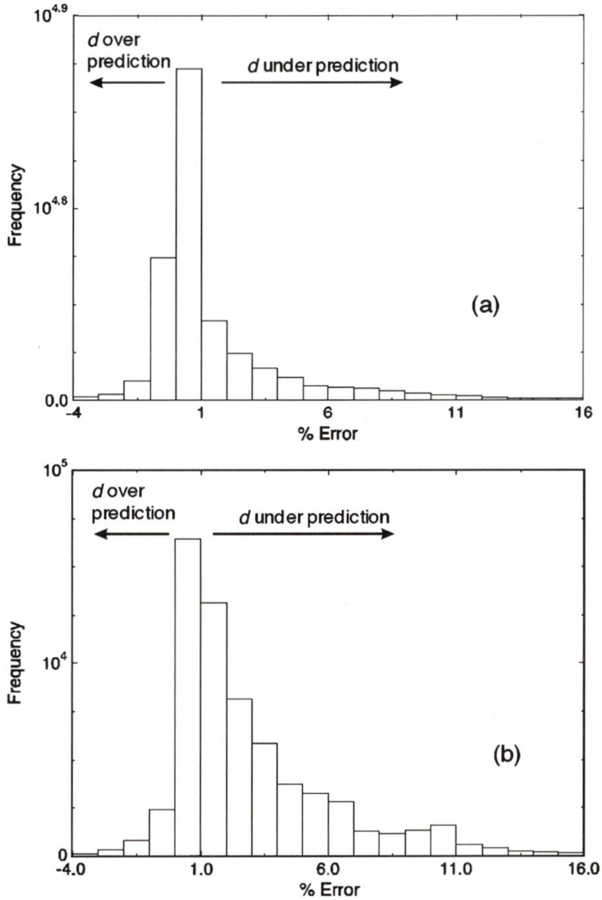

for the NSS procedure. Note, if the procedure of Hassan et al. is followed, where the near wall grid is frozen, the expense of making the wall distance update can be avoided.

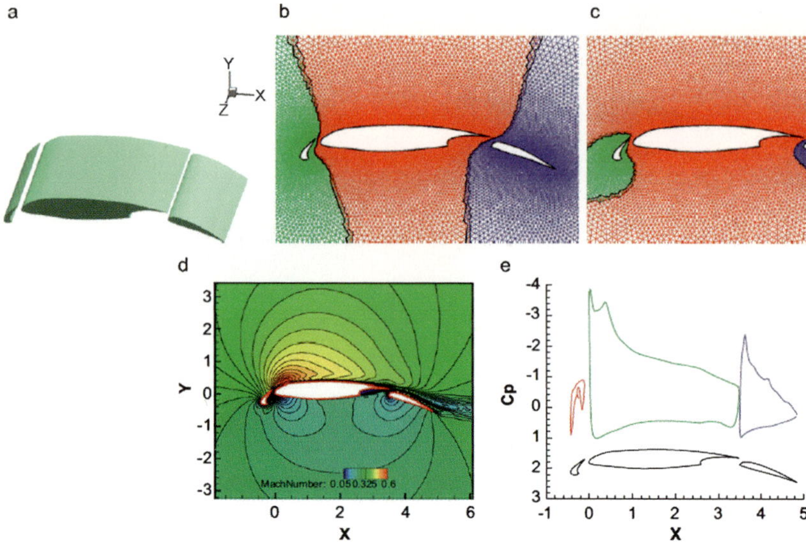

Fig. 7.9 Determination of the inter-grid boundary for unstructured overset grids: (**a**) three-dimensional wing section with slat and flap deployed, (**b**) inter-grid boundary from eikonal equation, (**c**) inter-grid boundary from biased eikonal equation, (**d**) Mach number contours and (**e**) C_p distribution (from Xia and Tucker 2010). Published with kind permission from Elsevier

Importantly, the eikonal equation is highly compatible with DES. Its front propagation can naturally be terminated when the wall distance reaches the LES modelled length scale ($C_{DES}\Delta$—see Sect. 3.8.1). Also, the eikonal equation can be readily extended to a Hamilton-Jacobi form. This can be used for palliating defects in RANS turbulence models (Tucker et al. 2005). It can also be used for setting optimal overset grid computational interfaces.

7.5.3 Overset Grid Computational Interfaces

As noted earlier, Nakahashi and Togashi (2000) make use of a local level set to determine the classification of vertices and the hole cutting for a moving overset mesh. Notably the eikonal equation is most suitable for this purpose. This is explored by Xia and Tucker (2010). Figure 7.9a, from Xia et al., shows a wing section with slat and flap. In Fig. 7.9b, c, two-dimensional unstructured overset meshes for the slat, wing and flap are shown. The interface between each domain is defined from a solution to the eikonal equation. Procedures for achieving this are outlined in Xia et al. (2010), Xia and Tucker (2011). Notably, the differences in the interfaces between frames (b) and (c) is achieved by setting different front propagation velocities in the eikonal equation. An inviscid flow solution (at $M = 0.3$ with an angle of attack of 5°) is obtained on the overset mesh shown in Frame (c). The Mach number contours and surface pressure coefficient are given in frames (d) and (e), respectively.

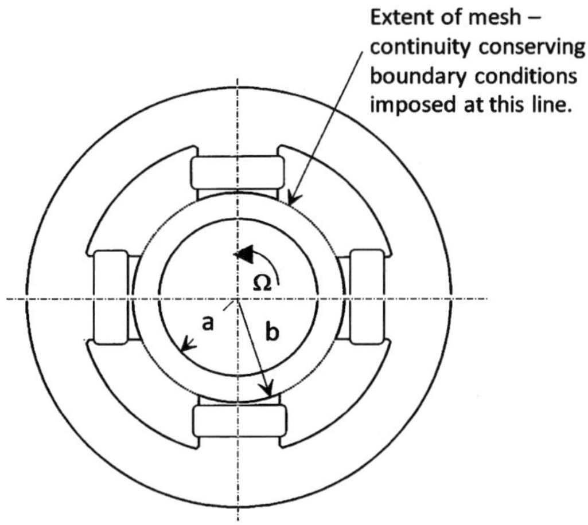

Fig. 7.10 Schematic diagram of magnetic bearing assembly. Published with kind permission from Elsevier

The above procedure is especially attractive for time moving grids where the use of search procedures for levels sets would be expensive.

7.5.4 Magnetic Bearings

As a final moving mesh example the aerodynamics of a magnetic bearing with prescribed shaft motion is considered. Full test case details can be found in Tucker (2001). A moving curvilinear grid is used. The geometry is shown in Fig. 7.10. Magnetic bearings offer a wide range of advantages over traditional oil lubricated bearings. For example, there is no need for a lubrication system and no wear. Also, active flow control can naturally implemented and extreme operating temperatures are possible. Bearing control offers the possibility of clearance and hence aero engine compressor stall control (Clark et al. 2004; Yoon et al. 2012). Figure 7.11 shows the variation of pressure in the z–θ plane when the shaft is in three different positions. The point x is used to indicate where the shaft is closest to pole surfaces. In frames (a) and (c) point x is at the leading edge of a pole face and so the pressure distributions are the same but shifted through 90°.

7.6 Coupled Simulations

In this section coupled simulation methods will be discussed in the following contexts: aeroelasticity, conjugate heat transfer and aerodynamically coupled components. Also, validation data is presented. Aeroelasticity is discussed first.

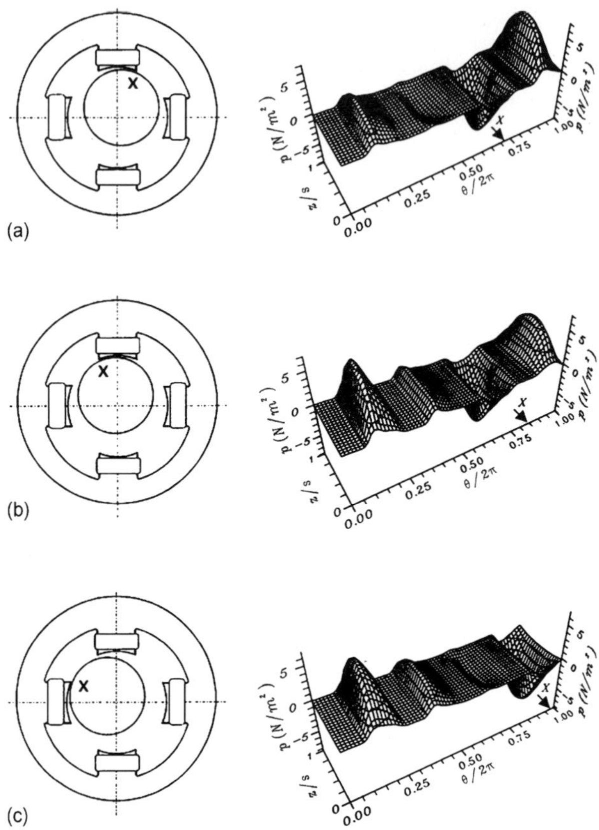

Fig. 7.11 Time variation of pressure in magnetic bearing (from Tucker 2001). Published with kind permission from Elsevier

7.6.1 Aeroelasticity

Typically, in an aeroelasticity process the FE (Finite Element) equation system given below is solved or modeled in some sense

$$[M_s]\{\ddot{U}\} + [K_s]\{U\} = \{F_s\} + [R]\{P\} \tag{7.20}$$

The, s, subscripts identify terms that involve the solid components. $[M_s]$ is the mass matrix and $[K_s]$ the stiffness matrix. The matrix $[F_s]$ is the force matrix. This involves both solid and fluid forces. $[R]$ is the coupling matrix giving surface areas at the solid-fluid interfaces. Finally, $[U]$ is the displacement matrix. The fluid system can be expressed in a similar matrix form

$$[M_f]\{\ddot{P}\} + [K_f]\{P\} = \{F_f\} - \rho_o[R]^T\{\ddot{U}\} \tag{7.21}$$

where ρ_0 is the fluid density. Rather than sequential solution, the above can be expressed as a coupled system (Dhopade et al. 2012)

$$\begin{bmatrix} M_s & 0 \\ \rho_o R^T & M_f \end{bmatrix} \begin{Bmatrix} \ddot{U} \\ \ddot{P} \end{Bmatrix} + \begin{bmatrix} K_s & -R \\ 0 & K_f \end{bmatrix} \begin{Bmatrix} U \\ P \end{Bmatrix} = \begin{Bmatrix} F_s \\ F_f \end{Bmatrix} \tag{7.22}$$

Bartels and Sayma (2007) identify over a dozen key European and US CFD codes that have an aerolasticity capability. Examples for the EU are: ENSOLV, EURANUS, FLOWer, AETHER, FASTFLO, TAU, ENFLOW, UES3D/UNS3D and elsA. For the US the following are identified: CFL3D, USM3D, ENSAER, ENS3DAE and AERO-F. There is also a range of commercial and academic codes exploiting a range of techniques. Some are quite specialized. Only the briefest overview of this extensive field is given here.

CFD codes can either be coupled to a reduced order model for the structure, modal equations or a full FE model (linear or non-linear). For aerodynamic loadings that lead to long-term structural degradations, the flow analysis can be linear. For more extreme non-linear loads that can give rise to more immediate catastrophic damage non-linear CFD equations must be solved. For endurance aircraft, with long wing spans, or concepts exploiting control surface flexibility for control purposes non-linearity must be modelled. Modelling of structural non-linearity can also be needed for turbomachinery problems, for example, due to viscoelastic materials inside blades. Such non-linearities are challenging to modally introduce but can be of critical importance (Garcia 2005). For the design of an aerodynamic system, numerous aeroelasticity calculations are necessary. Hence, reduced order models are also needed in the design process. For example, the fluid system can be represented in a modal fashion, like the structure′s, using proper orthogonal decomposition (Carlson et al. 2005). As shown in Fig. 7.12, Silva and Bartels (2004) generate reduced order models for both the fluid and structure. The former is based on exciting different modes, separately, in a non-linear CFD program. The superposition of responses approximates the total system response. If the CFD solver were linearized the superposition of responses, to individual excitations, would give the exact fluid system behaviour. Rather then exciting the system with a range of sinusoidal inputs (of low amplitude excitations) of different frequencies an impulse is used to excite a range of frequencies. The reduced order fluid and structure models are linked in a MATLAB/SIMULINK environment by Silva and Bartels. The fluid equations can also be represented using the harmonic balance method. This method is briefly outlined in Sect. 4.2. Thomas et al. (2012) develop the approach for aeroelasticity calculations, applying it to the front stage of a high-pressure compressor with transonic flow. For application to a F-16 configuration see Thomas et al. (2004).

As discussed in Sect. 4.4.1, specialized periodic boundary conditions, such as phase lagged or time inclined can be needed for turbomachinery blade row simulations. Unless all the blades in the circumferential direction are to be explicitly represented, these boundary conditions are also needed for aeroelasticity calculations. As noted in Chap. 4, depending on blade counts, the number of blade rows that can be modelled using such boundary conditions can be limited. The work of

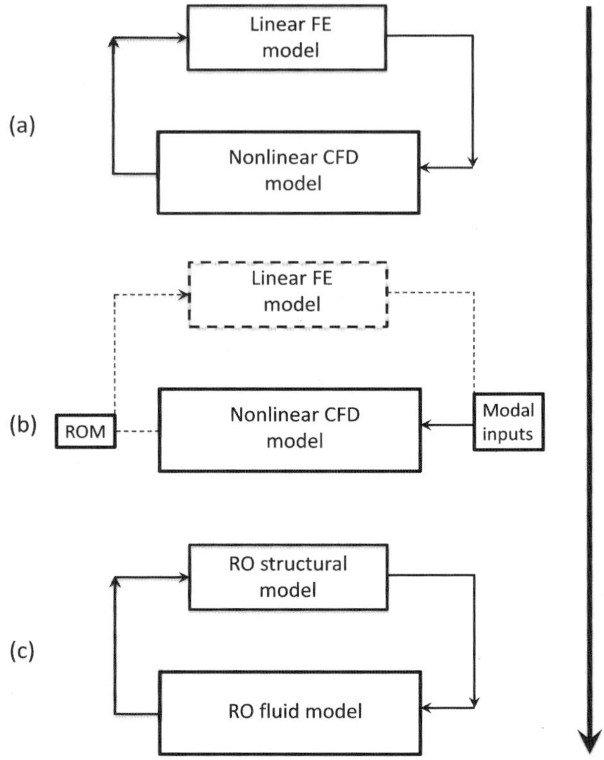

Fig. 7.12 Use of reduced order models for both fluid and structure: (**a**) analysis with full models; (**b**) generation of reduced order CFD model and (**c**) final reduced order model (Adapted from Silva and Bartels 2004)

Zhang et al. (2012) indicates that the assumption of a constant inter-blade phase angle, in the tangential direction, can be questionable. This would seem to have implications when using phase lagged boundary conditions, as would attempts to model structural deflections induced by azimuthally localized flow distortion.

Figure 7.13 shows the aeroelasticity process chain used by Kersken et al. (2012) for flutter prediction in turbomachinery. It involves a three-dimensional FE model. This is to capture mode shapes of the structure. The FE analysis involves two sub-stages. The 1st is static and captures centrifugal loading. It involves just steady aerodynamic loads. The 2nd stage involves modal analysis-studying the structure's response to a perturbation. It yields modal frequencies (Eigen values) and shapes (Eigen vectors). The surface deformations gained from the FE analysis are fed into modules that allow deformations of the CFD mesh. The nonlinear CFD solver is used to gain a steady base field around the static solid geometry. This is fed into a linearized CFD solver. This gives the flow field variation resulting from the blade vibration. The mode shape and blade surface pressure perturbations allows the blade surface displacements, aerodynamic work, stiffness and damping to be evaluated. The latter is calculated using the so called energy method (Micallef et al. 2012). This requires the integral of the blade surface pressure multiplied by area, over a vibration cycle to yield the aerodynamic work. In the process of Kersken et al. (2012),

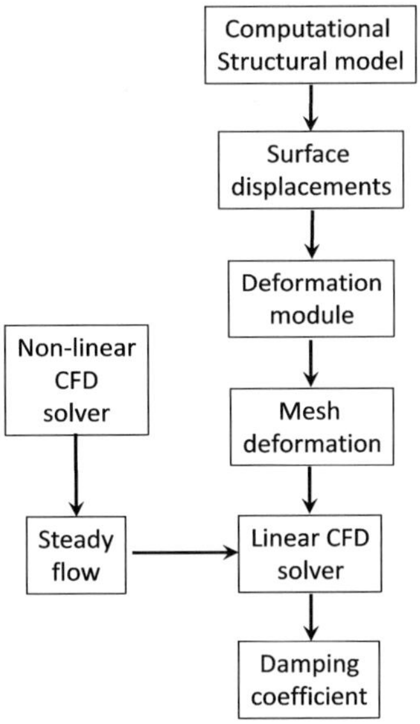

Fig. 7.13 Potential aeroelasticity process chain (from Kersken et al. 2012)

the non-linear CFD solver is commercial. The linear is in-house, as is the interfacing software between the different solution processes. It is found that care is needed to ensure a reasonable level of numerical consistency between the different CFD solver components. Little sensitivity to turbulence model choice is observed for the transonic turbine cases considered by Kersken et al. (2012). Again, for turbines, when evaluating aerodynamic damping coefficients, Micallef et al. (2012) find sensitivity to non-linear CFD solver choice.

Following Fig. 7.12, Doi (2002) directly couple the MSC/NASTRAN FE program to an unsteady CFD solver, successfully making flutter predictions for transonic fans. Figure 7.14 shows the code coupling strategy used. Rather than advance the fluid and the structure with the same time step, the fluid can be advanced with smaller time steps. The structural behaviour typically has a lower frequency than the fluid. Also, if an explicit temporal scheme is being used, for the fluid, a smaller time step can be required in this zone. Hence, the procedure of Doi (2002) is as follows:

1. The deformation of the structure, U, and its consequences (new cell volumes, normals and projected areas) are transferred to the fluid;
2. The fluid is integrated through time, potentially using the smaller time steps than the solid, to give the fluid solution ϕ;
3. The computed fluid pressures (the negligible fluid shear stresses are ignored) are converted to surface loads for the structure and
4. Based on new loads the structural calculation is made.

As can be seen from Fig. 7.14, there is a lag between the structural and fluid equations. Even though the fluid integration used by Doi (2002) is 2nd order, Taylor series analysis shows that this lag makes the temporal integration process 1st order.

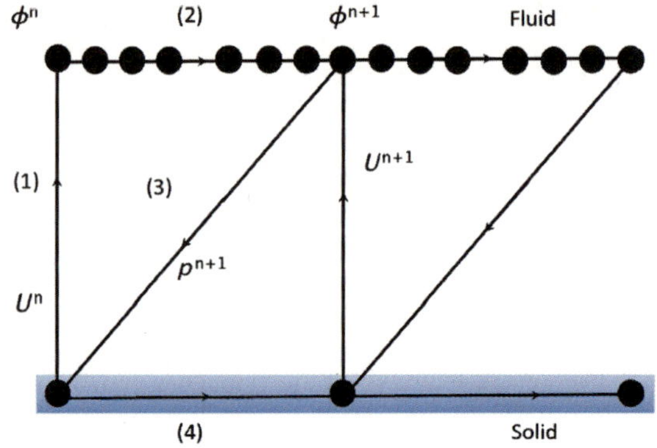

Fig. 7.14 Data transfer process between separate solid and fluid solvers in aeroelasticity calculations (Adapted from Doi 2002)

Geometry Alignment

The process of aligning the geometries for a fluid and structure robustly is not always trivial. In essence, Voigt et al. (2010) align the centres of area and principal axes of the two surfaces. The centre of area is evaluated from

$$x_0 = \frac{\sum_{i=1}^{N} ds_i r_i}{\sum_{i=1}^{N} ds_i} \tag{7.23}$$

where ds_i are element face areas and r_i the centres of them for an element i. Here, N gives the number of faces on the component surface being aligned. The axes of the system are the eigenvectors of a system of equations for the second moment of area tensor I.

$$I = \sum_{i=1}^{N} ds_i \left(\|r_i\| Id - r \otimes r \right) \tag{7.24}$$

Note, in the above Id is the identity matrix. The locations of the FE solid model's mesh points relative to the CFD's can be expressed as

$$x_{FE}^{CFD} = T(x_{FE} - x_0) \tag{7.25}$$

where T is a transformation matrix

$$T = I_{0,CFD} I_{0,FE} \tag{7.26}$$

This is the product of the principal axis systems, I_0, of the solid FE and CFD models. The above approach appears effective for simple topologies. However, for a shrouded turbine, and significantly different solid and fluid geometries (Voigt et al. 2010) for example, it will fail.

Interpolation

Loads and displacements need to be interpolated between the fluid and structure in a fashion that ensures the conservation of forces, momentum and energy. This will naturally occur if the surface elements of the fluid and structure match. Clearly, also the basic smoothness of the interpolation will impact on accuracy. The following range of spline based aeroelasticity interpolation procedures are cited by Bartels and Sayma (2007): infinite-plate, multi-quadratic biharmonic, thin-plate, finite-plate, non-uniform B-spline, non-uniform rational B-Spline (NURBS). Also, the following alternative approaches are given: inverse isoparametric mapping, interpolation-based algorithm (IBA), common refinement and quadrature projection. References for these procedures, which have different advantages and disadvantages, are given by Bartels and Sayma. As noted, the pressure force transferred from the CFD to the structure must satisfy force and energy conservation. There must also be compatibility of velocities, **v**, at location *i* of an interface relating the fluid, *f*, and solid, *s*, at the same point in space. These conditions are expressed below (Zhang et al. 2012).

$$\sum F_i = \int p dA; \qquad \sum F_i U_i = \int p dA dU; \qquad v_{f,i} = v_{s,i} \tag{7.27}$$

In the above dA is an elemental surface area, F_i is a nodal force and p a surface pressure. Also, U_i is a virtual nodal displacement. The 1st condition expresses that the sum of the nodal forces must give the same net force as the integrated pressure load. The 2nd expression indicates that the forces from the distributed surface pressures, moving through a virtual distance dU, must equate to the product of the nodal forces and their displacement. It seems worth noting here that computational aeroelasticity can involve extended couplings. For example, as shown by Vahdati et al. (2002), correctly incorporating the intake duct geometry and not just the fan can be key to the correct prediction of aero engine flutter. Coupling with acoustic modes in the duct can have a critical impact. The cut-on acoustic modes in the duct (see Appendix A) interact with the fan aerodynamics. Hence, the problem is holistic.

7.6.2 Conjugate Modelling

With gas turbines the aerodynamics is critical. However, also blades are exposed to extreme temperatures downstream of the combustor. Hence, to assess blade life, conjugate simulations of turbine blades, for example, are important. Understanding the thermal expansion of the drums to which the blades are mounted and their temporal response to engine accelerations and decelerations are also of importance. Furthermore, extreme temperatures are experienced in airframes in hypersonic vehicles. The conjugate fluid-solid problem can either be solved in one code, as one large system of equations (Rahman et al. 2005) or using coupled codes (Duchaine et al. 2009). The latter allows highly optimized bespoke programs to be used. However, in a parallel environment, the efficient connection of the codes and parallel load balancing becomes a challenge. The basic code interfacing requirements are in essence simple. They just need to impose compatibility of temperature, T, at the fluid—solid boundaries and continuity of the heat flux q. These conditions can be expressed as

$$\begin{aligned} T_{f,i} &= T_{s,i} \\ q_{f,i} &= q_{s,i} \end{aligned} \tag{7.28}$$

In the above, i, is the node number of the face being considered. These conditions are modified by Duchaine et al. (2009) to

$$\begin{aligned} T_{f,i} &= T_{s,i} \\ q_{f,i} + \epsilon_o T_{f,i} &= q_{s,i} + \epsilon_o T_{s,i} \end{aligned} \tag{7.29}$$

where ϵ_o is a positive relaxation type parameter. Following Giles (1997) and other workers, for stability it is recommend that the temperature from the solid is imposed on the fluid as a Dirichlet boundary condition. Then it follows that if Eqs. (7.29) are used we have

$$\begin{aligned} T^n_{f,i} &= T_{s,i} \quad \text{for } n = n_f \\ q^n_{s,i} &= q_{f,i} + \epsilon_o\left(T_{f,i} - T^n_{s,i}\right) \quad \text{for } n = n_s \end{aligned} \tag{7.30}$$

where n_f and n_s are the number of iterations, to converge to an acceptable level, the fluid and solid solvers, respectively. Care must be taken to ensure that the values of ϵ_o and solver iterations lead to a stable and physically meaningful solution. A stability analysis is presented by Duchaine et al. (2009).

As would be expected, the thermal time scales in solid components, τ_s, will be orders of magnitude larger than the fluid, τ_f. With this in mind, Duchaine et al. (2009) propose

$$n_f = \frac{\alpha_f \tau_f}{\Delta t_f}, \qquad n_s = \frac{\alpha_s \tau_s}{\Delta t_s} \tag{7.31}$$

where α is a dimensionless parameter.

If $\alpha_f t_f = \alpha_s t_s$ then the solution process is time accurate. For $\alpha_f = \alpha_s$ the steady state in the solid can be most optimally sought—if one exists. If one does not exist, to overcome the numerical stiffness, for turbine blades, He (2010) uses a Fourier based temporal discretization for the solid zone. A time-domain discretization is used for the fluid. The use of the former allows a small set of steady conduction related equations to be solved for in the solid domain. Details of this procedure are given in Chap. 4.

Sequential and parallel code coupling strategies proposed by Duchaine et al. (2009) are shown in Figs. 7.15 and 7.16, respectively. Note, the strategy, for the transfer of surface temperatures and heat fluxes, given in Fig. 7.15 can be run on a parallel machine. Each CFD and solid conduction run would utilize, P, processors—the total available number of processors. However, the interaction between the fluid and solid solvers is essentially sequential. Typically this sequential method is used with steady state codes (Duchaine et al. 2009). Note, n_{cpl}, in the figures, is the number of global solution cycles. For the parallel strategy, (Fig. 7.16) the two solvers

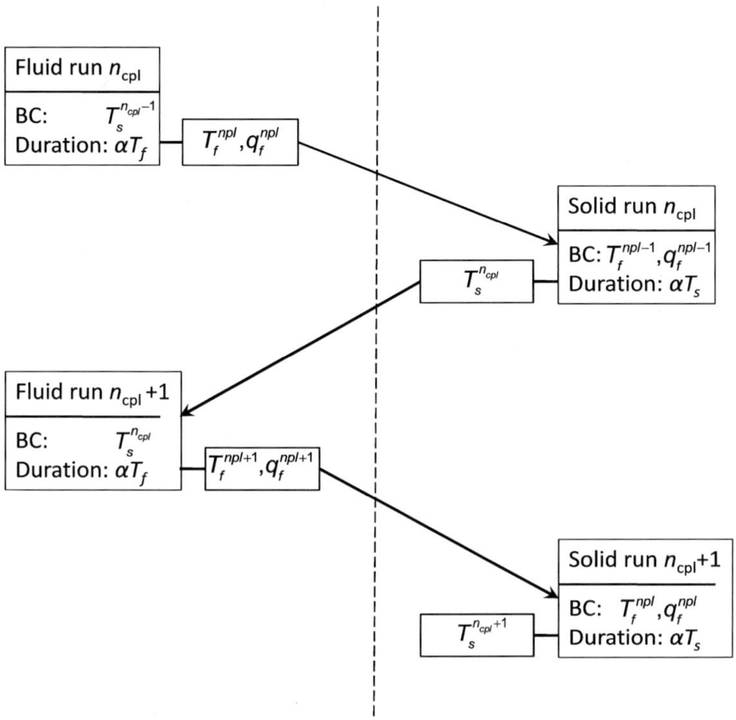

Fig. 7.15 Sequential coupling between fluid and solid codes (Adapted from Duchaine et al. 2009)

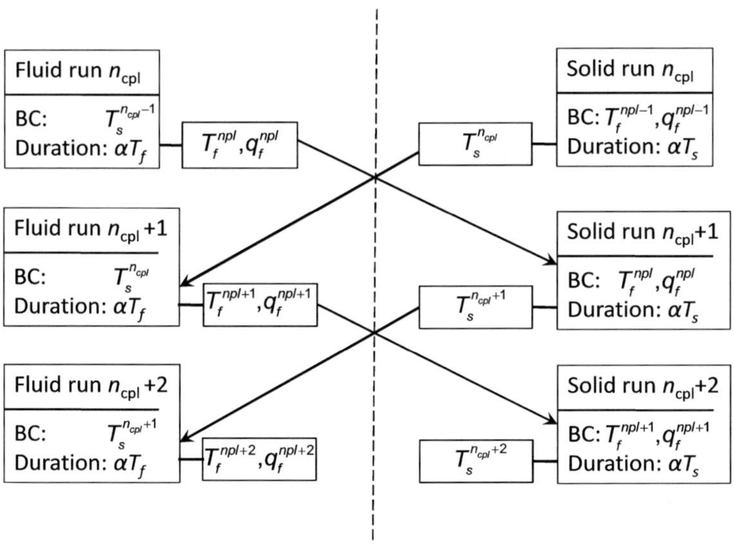

Fig. 7.16 Parallel coupling between fluid and solid codes (Adapted from Duchaine et al. 2009)

must share $P = P_f + P_s$ processors where P_f is the number of processors used for the fluid and P_s the number for the solid. Duchaine et al. (2009) state that for loading balancing the following approximation (it assumes, for example, perfect scaling of the solvers) can be used

$$\frac{P}{P_f} = \frac{1}{1 + Te_s/Te_f} \tag{7.32}$$

where Te_s and Te_f are the execution times for the solid and fluid solvers to compute the physical times $\alpha\tau_s$ and $\alpha\tau_f$, respectively.

Rather than exchange temperatures to the fluid and heat fluxes to the solid, Verdicchio et al. (2001) replace the latter exchange with a heat transfer coefficient and ambient temperature. Based on two CFD solutions (one perturbed by δ, hence $T_2 = T_1 + \delta$) the heat transfer coefficient, h, can be estimated as

$$h = \frac{\partial q}{\partial T} \simeq \frac{q_1 - q_2}{T_1 - T_2} \tag{7.33}$$

The subscripts in the above identify the two CFD solutions. The local nodal ambient temperatures are evaluated based on continuity of heat flux considerations.

Even with the overhead of making two CFD solutions, this procedure is found quicker than using the more usual process of exchanging surface temperatures and heat fluxes. Notably, the technique has not been tested for unsteady flows. Like Verdicchio et al. (2001), Ganine et al. (2012) consider the conjugate modelling of gas turbine cavity flows. They explore a range of techniques to improve numerical stability/convergence of the interface temperature update given below

$$T_k^{n+1} = T_{k-1}^{n+1} + QR_k^{n+1} \tag{7.34}$$

where k, is iteration number, n, the time step, R, residual and Q a function such as a Jacobian or under-relaxation parameter. Ganine et al. (2012) study approaches where Q is an Aitken dynamic relaxation parameters or an approximate inverse Jacobian (Broyden's, Anderson mixing or Newton-Krylov related method) or an inverse Jacobian in an interface GMRES (Generalized minimal residual) framework. The Anderson mixing method is found most effective. Michler et al. (2005) apply the Newton-Krylov approach to a fluid structure interaction problem involving flutter.

Conjugate Data Interpolation

When conjugate simulations are performed in a single solver, to ensure conservation of thermal energy, harmonic means are necessary. These need to be used for the diffusion coefficient, Γ, at the control volume face coincident with the solid-fluid interface. For a control volume face located exactly half way between nodes i, j, k and $i, j+1, k$ the diffusion coefficient at the control volume face takes the form

$$\Gamma = \frac{2\Gamma_{i,j,k}\Gamma_{i,j+1,k}}{\Gamma_{i,j,k} + \Gamma_{i,j+1,k}} \tag{7.35}$$

Similar expressions can be used for other faces. As with aeroelasticity problems, care can also be needed when interpolating data between separate solid and fluid meshes. If the meshes are of similar density and the variations of temperature across the surface relatively benign, fast nearest point based (Kd-Tree—see Shevtsov et al. 2007) interpolations can be used. However, if there is a large disparity in mesh sizes and relative fast changes in surface temperatures conservative interpolations are necessary.

Stability

Of course, with any numerical scheme care is needed with stability. Giles (1997) explores stability where the conjugate problem is performed by a single program and also coupled programs. The former involves jumps in material properties. The latter is explored for both explicit, implicit and hybrid temporal methods (i.e. implicit for one component and explicit for the other). For a single program no stability limit change is observed beyond the usual based on local material properties, etc. For the coupled codes, as long as the Dirichlet condition is applied to the fluid and the Neumann to the solid, no stability restriction is observed beyond the normal expected limits for the respective numerical schemes. Note, the analysis of Giles is based on the use of standard Fourier based analysis applied to one-dimensional diffusion equation(s). However, numerical experiments suggest the analysis is relevant to three-dimensional conjugate problems. The approach of Verdicchio et al. (2001) is stable for $|hL/k| < 1$ where L characterizes a geometrical length scale expressing the extent of the conduction path. Also, k is thermal conductivity.

Duchaine et al. (2009) consider stability when coupling separate programs with their mixed boundary condition. They consider global stability and also stability for conduction in just the solid (i.e. not a coupled problem) with the mixed condition. At the boundary opposite the one with the mixed boundary condition a Dirichlet condition is applied. The stability analysis is based on a one-dimensional problem—$\rho C \partial T/\partial x = k\partial^2 T/\partial x^2$, where C is heat capacity. It is also based on the problem being solved using 2nd order finite differences and a 1st order implicit time scheme. Hence, the discretized equation is $T_i^{n+1} = T_i^{n+1} + F(T_{i+1}^n - 2T_i^n + T_{i-1}^n)$. This discretization operator, written in matrix form with the boundary conditions included gives the following

$$M(F,\epsilon_o) = \begin{pmatrix} 1 & & & & & \\ F_1 & 1-2F_1 & F_1 & & & \\ & & \ddots & & & \\ & & F_1 & 1-F_1(1+D) & F_1 D & F_1 D/\epsilon_o \\ & & & & 1 & \\ & & & & & 1 \end{pmatrix} \tag{7.36}$$

In the above D is a Biot type number and F the Fourier

$$D = \frac{\epsilon_o \Delta x}{k}, \qquad F = \frac{k \Delta t}{\rho C \Delta x^2} \tag{7.37}$$

Also, Δx is the local grid spacing and Δt, the time step. Keeping the above matrix small (4 finite difference nodes) enables its spectral radius to be analytically evaluated. Numerical tests by Duchaine et al. (2009) indicate that the stability is only a weak function of the number of nodes. Based on keeping the spectral radius less then unity the stability limit is given by the equation below (Duchaine et al. 2009).

$$D \le \frac{F^3 - 12F^2 + 20F - 8}{-3F^3 + 8F^2 - 4F} \tag{7.38}$$

An extended stability analysis for the coupled solid-fluid system is outlined in Duchaine et al. (2009).

Unsteady Conjugate Simulation Examples

Next, the unsteady conjugate modelling of flow and heat transfer in a hydrodynamic journal bearing will be considered. The work is part of a study to explore a bearing failure mechanism in a (ground based) gas turbine engine (De Jongh et al. 1996). Orbiting of the bearing shaft potentially causes the failure. This gives rise to preferential shaft heating and hence excessive thermal distortion. Tucker and Keogh (1996) make, structured curvilinear grid, conjugate (solving heat conduction through the bearing shaft and housing), multiphase (to account for oil cavitation), unsteady, moving boundary (due to the shaft orbit) simulations. For these simulations a circular centered shaft orbit is chosen. Notably, the shaft orbit frequency is made equal to that of the shaft rotational frequency. This makes the shaft temperature field steady and hence removes the numerical stiffness. Equation (7.30) relationships are used but with $\epsilon_o = 0$.

Figure 7.17, Frame (a) shows the mesh. It also identifies the regions of metal surface where surface heat transfer coefficient boundary conditions are applied. Sliding grids are used. The shaft is solved in a rotating frame of reference. This minimizes the numerical precision and dissipation errors. The latter is especially strong, since

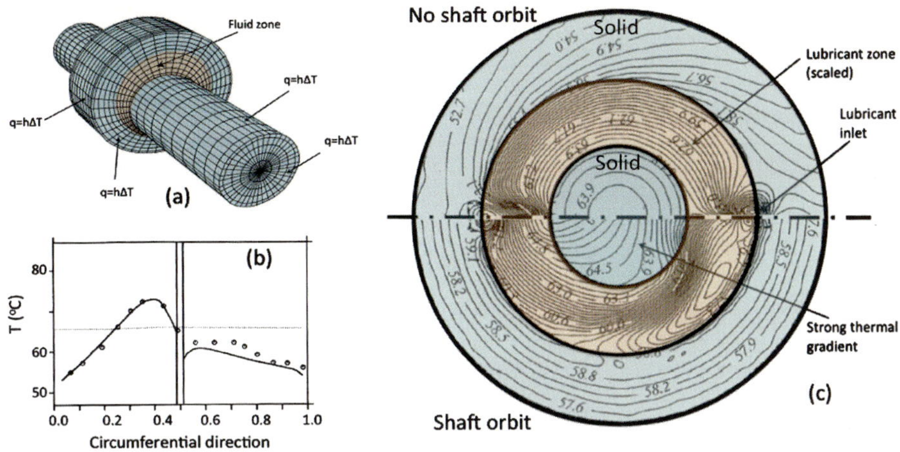

Fig. 7.17 Conjugate orbiting, hydrodynamic journal bearing: (**a**) mesh; (**b**) surface temperature validation when no shaft orbit and (**c**) temperature contours through, half, scaled, mid-axial planes for no orbit (*upper frame*) and circular centered orbit (*lower frame*). Published with kind permission from Elsevier

in the stationary frame of reference the shaft conduction equation includes a convective term. Such issues are discussed in Sect. 2.18. The lubricant and bearing solutions are made in the stationary frame. The conditions and geometry match the measurements of Lund and Tonnesen (1984). Frame (b) contrasts predicted bearing inner surface temperatures with the measurements of Lund and Tonnesen when there is no shaft orbit. The agreement is encouraging. The two vertical lines identify the lubricant inlet zone. The lower temperature zone corresponds to where there is cavitation. The upper frame of Fig. 7.17c shows a, half, bearing section of temperature contours for the static shaft. Note, for clarity the fluid region has been greatly scaled and displaced through use of a transformed coordinate. Correctly, in the upper half of Frame (c), the shaft temperature does not vary circumferentially. In the lower half, the shaft is moved in a circular centred orbit. As can be seen, a significant steady shaft temperature differential has developed. De Jongh and Morton postulated that the structural bending related to this caused bearing failure.

Amirante et al. (2012) make conjugate computations for the labyrinth type sealing zone (stator well) of a turbine blade. They explored how the seal clearance changes through a transient flight cycle. Separate CFD and FE based conduction codes are coupled. The latter is run in an axisymmetric mode. The importance of capturing the dynamic seal clearance and its strong impact on thermal response is shown. Because of the relatively fast fluid response time, the transient is represented by a series of steady state solutions with $N = 2.7 \times 10^6$. Notably, the mesh adaptation took 25 % of the total solution time. The process involves the FE code providing clearances for the CFD, which, in turn, provides information to the FE code. The CFD program runs with prescribed temperature boundary conditions and the FE with heat fluxes computed from the fluid. The simulations above are coupled but in a conjugate sense. Next aerodynamic coupling is explored.

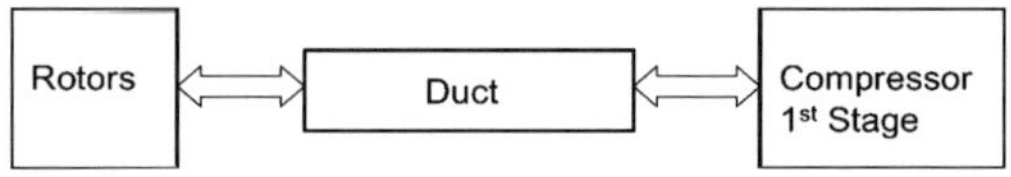

Fig. 7.18 Schematic of coupled aerodynamic simulation for an open rotor engine intake

Fig. 7.19 More detailed schematic of the rotor-duct-compressor configuration with location of sliding plane

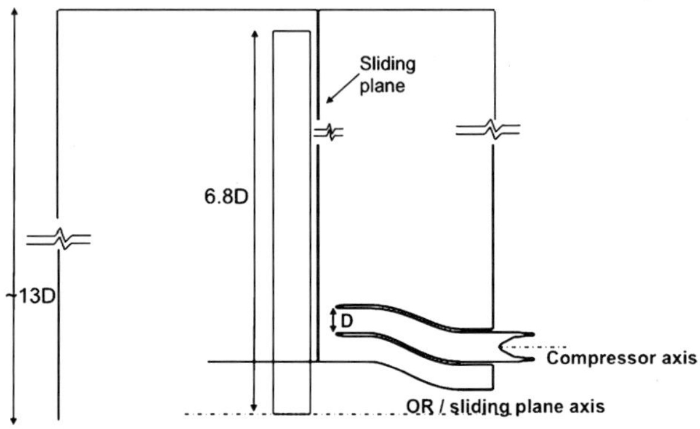

7.6.3 Coupled Aerodynamic Simulations

Chima et al. (2010b) perform coupled intake-fan simulations. An axisymmetric intake code is linked to a three-dimensional CFD code. The latter is used to perform fan simulations. A mixing plane type coupling is used at the code interface via a simple coupling code called SYNCEX. Perturbations about the axisymmetric intake solution are passed to the three-dimensional, fan, CFD solver. Blockage terms are used to model the presence of a gearbox placed in the bypass duct. Steady simulations are performed to explore coupled system performance over a complete envelope. Although found unnecessary for all parts of the flight envelope, the coupled simulations are found useful for gaining consistent solutions over the complete envelope. Chima et al. (2010a) make simulations of a serpentine inlet duct coupled to a fan. A reduced order, body force, model is used for the fan. This is coupled with a full three-dimensional Navier-Stokes solver and also one restricted to axisymmetric flow with blockage terms to account for struts. The SYNCEX coupling code is again used to link these differing elements.

A coupled aerodynamic simulation, outlined in Loiodice et al. (2010) is discussed next. This, as shown in Fig. 7.18, explores the interaction between the rotor, intake duct and compressor for a notional open rotor engine. For this simulation the rotors and the compressor are both modelled using the body force procedure outlined in Sect. 4.6. Figure 7.19 gives a more detailed schematic of the configuration with the location of the sliding plane. Figure 7.20 gives some mesh views. Frame (a) shows the full domain and Frame (b) the lip zone. Frame (c) gives the duct and compressor zones. As can be seen, a multiblock curvilinear mesh is used with $N = 5 \times 10^6$ and 0.6×10^6 nodes in the open rotor zone.

As discussed in Sect. 4.6, the body force is enhanced around the rotor blade zones to create wakes. The model also allows user control of the wake zone turbulence in-

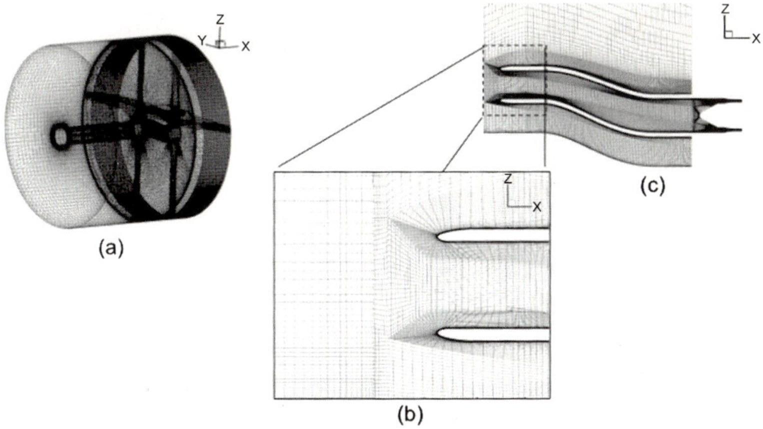

Fig. 7.20 Some mesh views: (**a**) full domain, (**b**) lip zone and (**c**) duct and compressor zones

Fig. 7.21 Isosurfaces of total pressure illustrating rotor zone and its proximity to the idealized intake (from Tucker 2011a). Published with kind permission from Elsevier

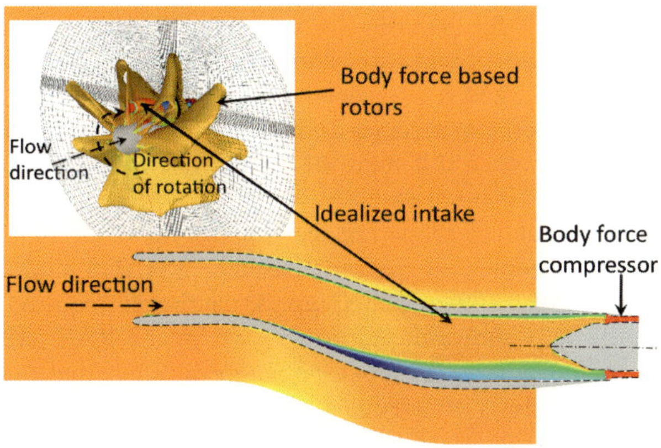

tensity and length scales. Figure 7.21 uses isosurfaces of total pressure to render the enhanced force zone, used to create wakes, visible. The figure also shows the idealized, RAE M2129, intake duct shape and its position relative to the rotor. The accuracy of the model at capturing rotor wakes is shown in Sect. 4.6. Note, the rotor is solved in a rotating frame of reference. The duct and the compressor are solved in the stationary frame. A sliding plane connects the rotating and stationary frames. The modelling of the dynamic consequences of wakes is found to substantially improve agreement with measurements of distortion at a compressor inlet. However, these results cannot be shown here due to commercial restrictions.

Figure 7.22, Frame(a) plots total pressure contours at the mid-plane of the RAE M2129 intake for the fully coupled rotor-compressor-duct simulations. Frame (b) plots the r.m.s. (root mean square) of the DC(60) (Note, the DC(60) parameter averages the distortion over the 60° sector that has the worst distortion, subtracts this from the average total pressure at the compressor inlet face and divides by the dynamic pressure) distortion parameter for different wake turbulence length scales,

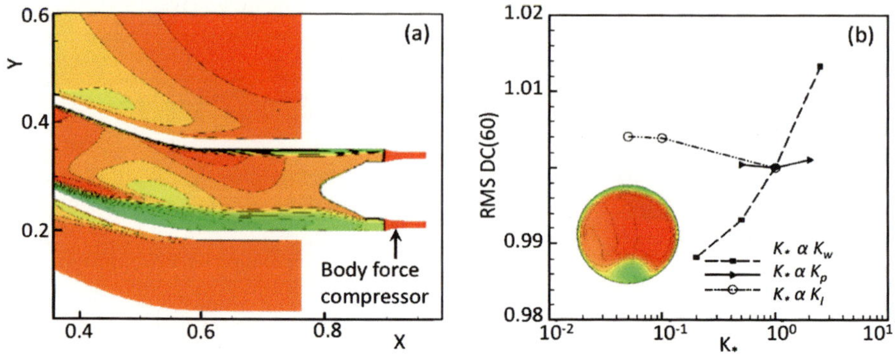

Fig. 7.22 The influence of unsteady wakes on flow in idealized intakes: (**a**) total pressure contours through mid-plane section of intake and (**b**) the variation of unsteady distortion with wake momentum deficit, length scale and turbulence intensity (inset total pressure distribution at compressor face for baseline conditions) (from Tucker 2011a). Published with kind permission from Elsevier

turbulence levels and wake momentum deficits. These are controlled by the parameters K_l, K_p and K_w, respectively—discussed in Sect. 4.6. As would be expected, the distortion experienced at the compressor face increases with wake momentum deficit. However, it is relatively insensitive to the wake turbulence length scale and intensity. This insensitivity is fortunate. Different RANS models can give very different turbulence properties in the wake. However, of course the global conclusions for this specific case are only based on URANS modelling and might not connect with reality. As noted in Chap. 3, caution is required with URANS simulations. Figure 7.23 shows the ratio of t_i/t_p for the RAE M2129 intake. The time scale t_p corresponds to the wake passing time scale and t_i to the modelled turbulence. Clearly there is no adequate spectral gap. Notably, this issue is more likely to be prevalent as the system scale grows. For example, with multistage turbomachinery, the greater the number of blade rows, the greater the number of harmonics present. Hence, there seems more potential for spectral gap violations with coupled systems. Also, as system complexity grows the potential for unsteady flow sources grow.

As another example of a coupled simulation, Fig. 7.24 relates to an unsteady simulation of flow field through the intake and bypass duct of an engine connected to a wing via a pylon. The wing has an idealized deployed flap. Frame (a) shows the geometry. Frame (b) gives a mesh view and Frame (c) a computed flow field. The fan is designed, with the help of a through flow model, to give the desired pressure rise and mass flow characteristics to link with the jet. The fan is then modeled in the CFD simulation using the body force with wake model also discussed above. A sliding plane is again used so that the fan flow field can be solved in a rotating frame of reference. Such simulations are important. As outlined in Eastwood (2009), in a real engine, upstream of the nozzle, there are numerous geometrical features.

A key gas turbine aero engine zone where code coupling is generally needed is the combustor turbine interface. For the combustor zone, where the Mach numbers are low, specialized (with a range of combustion models) pressure correction

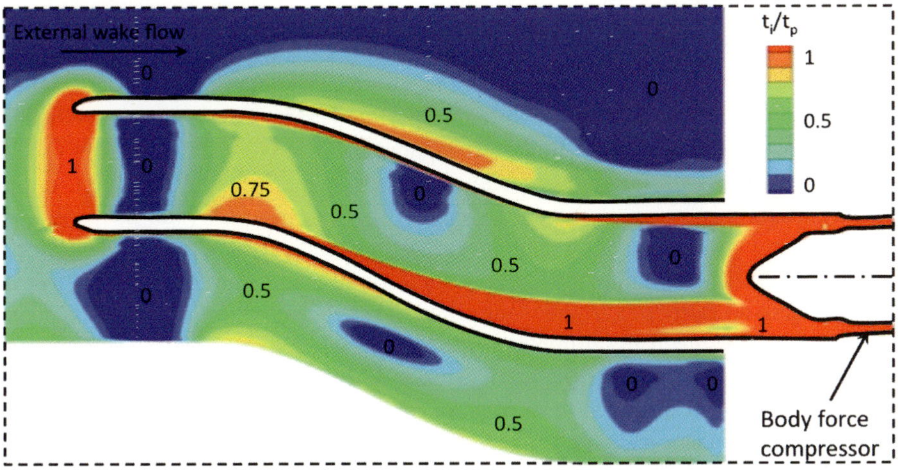

Fig. 7.23 The ratio on t_i/t_p for RAE M2129 intake (from Tucker 2011a)

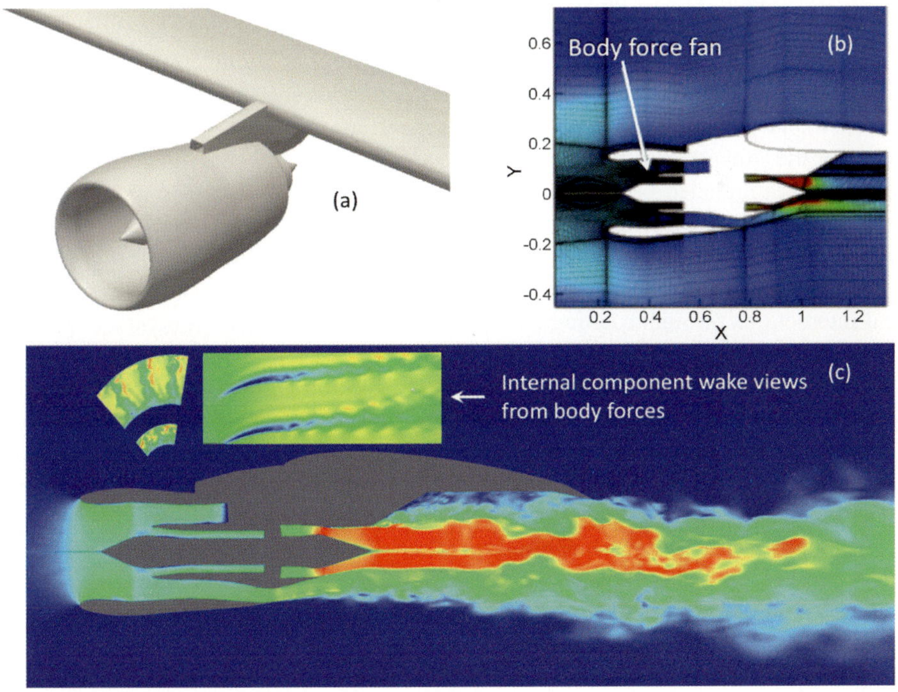

Fig. 7.24 Unsteady simulation of flow field through the intake and bypass duct of an engine through to a wing with a deployed, idealized, flap: (**a**) geometry; (**b**) mesh view and (**c**) flow field

solvers are generally regarded as most ideal. For the turbine zone, at realistic engine conditions, density based solvers are most efficient, the turbine blade passages having transonic or high subsonic flow. Salvadori et al. (2012) contrasted steady loosely coupled and uncoupled simulations around a combustor with a downstream turbine stator. The combustor flow is computed using a pressure correction based algorithm coupled with radiation modelling. The turbine flow is computed using a density-based solver. Simple bespoke coupling software is used. Total temperatures and pressures along with flow angles; turbulence intensity and length scales are transferred from the combustor outlet to the turbine stator inlet plane. Static pressures are passed from the turbine blade simulation to the combustor outlet. The data interpolations between the different grids are inverse distance weighted. The k–ϵ turbulence model is used for the combustor and Menter SST for the turbine. Over a 20 % variation in mid span heat transfer for the turbine stator is found between the coupled and decoupled simulations. Figure 7.25 shows a potential coupled code scenario centered around the combustor zone. It is inspired by the Stanford whole engine simulation work (Medic et al. 2008). In this work, RANS is used upstream of the combustor to characterize the compressor flow. Indeed, it is know that the accurate modelling of this upstream flow is critical for the correct characterization of the combustor flow. This presents one code coupling interface. However, the combustor flow is best modelled using LES where relative to cost, the accuracy impact can be high (see Sect. 6.9.5). Medic et al. (2008) use synthetic turbulence generation at the compressor-combustor interface. Details of this process are given in Sect. 3.10.2. Another code swap is needed at the combustor turbine interface. Here, in the work of Medic et al. (2008) the LES flow properties are averaged. Hence, they can be used in the downstream turbine RANS model zone. Figure 7.25 includes another element, not included in the Stanford work, this being acoustics. The impact of this on the shear layers emerging from the jet is uncertain. We might well, also wish to propagate the combustor/turbine noise down the turbine using some form of wave propagation method. As noted in Chap. 6, to model the acoustic, entropic and vortical interactions through rotating and stationary blades in turbines, actuator discs or body forces (see Chap. 5) can be used.

Like the combustor, the Mach number in film cooling holes will be low. Hence, again, it is most effectively modelled using an incompressible flow solver. This presents another code coupling interface zone. Therefore, Iourokina and Lele (2005) couple two different flow solvers to study film cooling, one for the main gas pass flow and one for the cooling hole.

For disparate Mach number zones, or where codes involving distinctly different flow physics are to be linked, the use of bespoke code coupling software can be attractive. However, MPI based commands can be used (Potsdam and Guruswamy 2001). The CHIMPS code coupling software is used in the pioneering Stanford who engine simulations work (see, Alonso et al. (2006), Medic et al. (2008)). It is designed for node based data with connectivity information. This allows direct interpolation to donor points. MpCCI and PALM (Buis et al. 2006) are readily available parallel code coupling programs. PALM was initially developed for ocean-

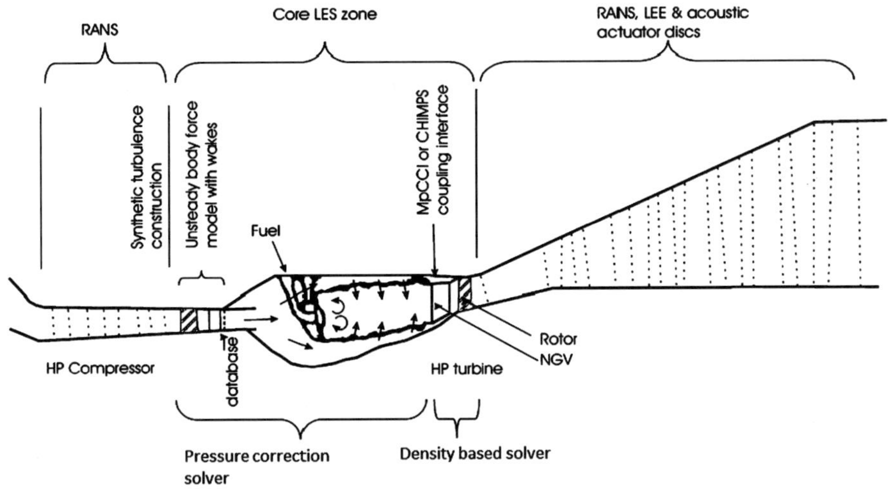

Fig. 7.25 Example of code coupling focusing around the combustor-turbine zone

atmosphere model coupling. It is free to academics and compatible with many platforms. It has seen successful use in combustor-turbine and conjugate turbine blade modelling. MDICE (MultiDisciplinary Computing Environment), see Kingsley and Siegel (1998), is another code coupling software variant.

Validation Data

Table 7.3 gives a very brief sample of validation data available for coupled simulations. There is a wide range of data sets in the area of aeroelasticity and no attempt is made to adequately cover these. Table 7.4 defines potential validation data levels. These are used in the far right hand column of Table 7.3.

Table 7.3 suggests that relative to pure aerodynamic and aeroacoustic cases, for coupled problems the general level of validation is lower. This is perhaps not surprising. The general complexity of an experiment for a coupled system is higher. Where this involves moving components, making aerodynamic measurements is more complex.

7.7 Examples of Eddy Resolving Coupled Simulations

Table 7.5 summarizes the eddy resolving simulations to be discussed giving typical grid densities, Reynolds numbers and validation levels.

Table 7.3 Sample of typical available validation data for coupled cases

Zone	Nature of data	Data source	Comments	Validation level
Conjugate heat transfer over an array of cubes—relevant to turbine blade internal cooling	LDA, infrared thermography and liquid crystals for surface temperatures and surface visualization	Meinders (1998) and Meinders and Hanjalić (1999)	Characterization of thermal boundary condition where cube joins plate surface complicates data comparisons	2, 3
Generic combustor	Laser vibrometry, surface pressure measurements	van Kampen (2006), Huls (2006)	Partially premixed flame. Square section chamber. Operates at 1.5 bar	1
Centrifugal pump	Casing displacement spectra, far field sound	Reported in Kato et al. (2007)	Non generic—multistage machine	4
Turbine blades	Time averaged surface heat transfer related properties	Reported in Duchaine et al. (2009)—some data in Technical reports	One blade has cooling holes	1
Serpentine duct coupled to a fan	Duct surface pressures and fan circumferential pressure distributions, fan characteristic curves	Reported in Chima et al. (2010a)	–	1
Aeroelasticity—Prismatic turbine profile, annular cascade facility at EPFL Lausanne	Real and imaginary parts of surface pressure perturbations	Reported in Kersken et al. (2012)	Part of a family of eleven standard international configurations (Micallef et al. 2012; Fransson et al. 1999)	3
Axial compressor blades	Blade mounted strain gauges and case mounted unsteady pressure measurements	Kielb et al. (2003)	Explores blade response to rotating instabilities	2
NACA 64A006 aerofoil	Lift and moment coefficients, surface pressures	Zwaan (1982a)	Oscillating flap	1
NACA 64A010 aerofoil	Lift and moment coefficients, surface pressures—instantaneous	Davis (1982a)	Oscillatory pitching	1

Table 7.3 (Continued)

Zone	Nature of data	Data source	Comments	Validation level
SD7003 aerofoil	Detailed PIV of mean flow and turbulence quantities	Kang et al., Baik et al. (2009), Ol et al. (2009)	Pitching and plunging aerofoil at low Reynolds numbers	1, 3
NACA 0012 aerofoil	Instantaneous pressures, force coefficients	Landon (1982)	Oscillatory and transient pitching	1
NLR 7301 supercritical aerofoil	Instantaneous surface pressures and their time histories	Zwaan (1982b), Davis (1982b)	Oscillatory pitching and oscillating flap	1
RAE wing	Instantaneous surface pressures and their spectra	Mabey (1982)	Oscillating flap	1
NORA model	Instantaneous surface pressures	Lambourne (1982)	Oscillating about swept axis	1
Helicopter sections	Surface pressures, hot films and wires, shadowgraphy and Pitot static tube wake traverses	McCroskey et al. (1982)	–	1

Table 7.4 Definitions of validation levels

Validation level	Data type
1	Unsteady integrated surface forces, time averaged surface properties
2	Surface force/pressure distribution histories
3	Second order statistics (r.m.s quantities)
4	One-point spectral analysis
5	Two-point space/time-correlations

7.7.1 Conjugate Large Eddy Simulations

Conjugate LES and hybrid LES-RANS are beginning to emerge. A relatively popular case for exploring the performance of RANS and eddy resolving simulations for conjugate problems is the flow over an array of cubes—see 8th ERCOFTAC/IAHR/COST workshop on refined turbulence modelling (Rautaheimo and Siikonen 1999; Mathey et al. 1999). In terms of flow physics, this conjugate cube case has relevance to the internal cooling found in turbine blades. Obviously, the

Table 7.5 Summary of eddy resolving simulations with typical grid densities, Reynolds numbers and validation levels for 'coupled' or moving boundary simulations

Component	Author(s)	Method	$Re_c/10^6$	$N/10^6$	Validation level
Oscillating NACA0012	Knopp et al. (2009)	Standard DES, Menter SST DES and SAS, EASM DES.	0.98	0.56–2.78	1
Plunging SD7003 aerofoil	Visbal (2011)—see Sect. 5.3.1	ILES	0.06	2.4–18.6	3
Combustion chamber vibration	Huls et al. (2007)	Fluid-structure, hybrid LES	–	2.7 (tetrahedral)	4
Combustion chamber vibration	Shahi et al. (2012)	Fluid-structure, hybrid LES-RANS (SAS)	–	1.2 (tetrahedral) for fluid, 2450 shell elements for solid	0
Multistage centrifugal pump	Kato et al. (2007)	LES	10	0.3	4
High-pressure turbine stator (cascade).	Dufour et al. (2009)	Conjugate LES	2.8	6.37	1
Turbine blade with film cooling holes (T120D cascade)	Dufour et al. (2009), Duchaine et al. (2009)	Conjugate LES	0.38	6.5	1
Conjugate flow over cubes	Mathey et al. (1999)	LES	0.013	–	2, 3
Conjugate flow over cubes	Ničeno et al. (2002)	Conjugate LES	0.013	0.428	2, 3
Conjugate flow over cubes	Zhong and Tucker (2004)	Conjugate LES	0.013	1.3	2, 3
Conjugate flow over cubes	Zhong and Tucker (2004)	Conjugate hybrid LES-RANS	0.013	0.421	2, 3
Turbine film cooling	Iourokina and Lele (2005)	Coupled code	–	1.9	0
Compliant surface	Carlson et al. (1995)	Imposed surface motion	0.002	0.075–0.17†	0
Compliant surface	Endo and Himeno (2002)	CFD coupled to spring supported plate	0.0046	0.9	0
Compliant surface	Xu et al. (2003)	CFD coupled to spring supported plate	0.003	0.2†	0

†Solutions have spectral content

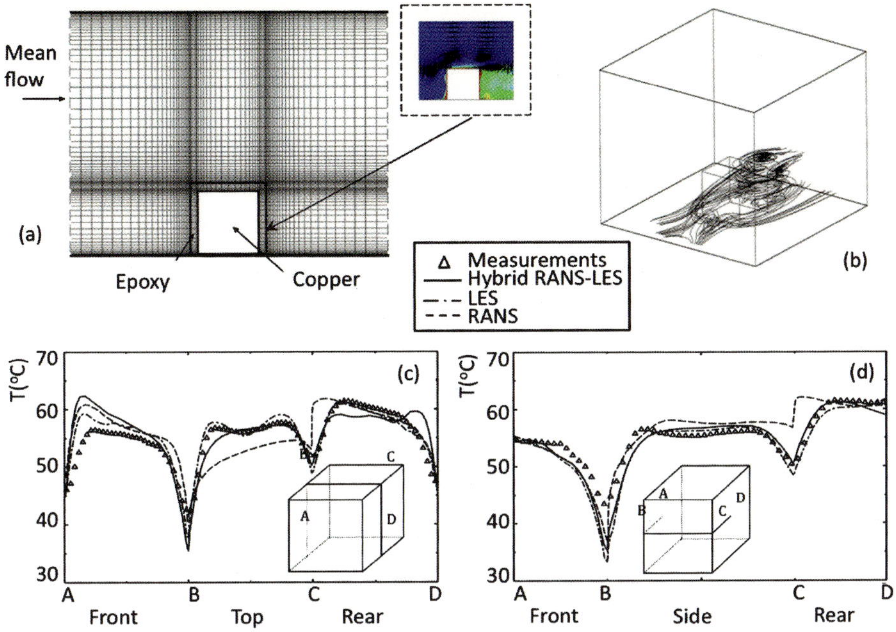

Fig. 7.26 Conjugate heat transfer over an array of cubes for LES and hybrid RANS-LES: (**a**) mesh and flow configuration (inset instantaneous temperature), (**b**) time averaged streamlines and (**c**), (**d**) heat transfer coefficients over surface of the cube

ribs create large wake flow zones and conjugate modelling, ideally, is needed. The cube case noted above has an internal copper core and an epoxy coating. The configuration allows the use of bi-periodic boundary conditions. Most simulations for this simple configuration make the conjugate simulations in a single solver. The Reynolds number, based on channel height, in which the cube sits is 13,000. Ničeno et al. (2002) and Mathey et al. (1999) perform LES for this configuration.

Figure 7.26 shows results for the hybrid LES-RANS and LES computations of Zhong and Tucker (2004). These are performed for $N = 421{,}875$ and $N = 1.3 \times 10^6$, respectively. An incompressible solver is used with 2nd order centered spatial differences and Crank-Nicolson temporal integration. Frame (a) gives a view of mesh and flow configuration. The inset is instantaneous temperature. Frame (b) gives time averaged streamlines. Frame (c) compares surface temperatures in the cube's vertical symmetry plane. Not surprisingly LES performs well. Compared to the k–ϵ RANS solution, the results from the hybrid LES-RANS approach are overall better. The improvement on the top surface (corner B to corner C) can be clearly identified. The discrepancies near the channel floor (front corner A and rear corner D) are due to uncertainties in connecting the simulations to the physical set-up (Ničeno et al. 2002). Frame (d) compares temperatures in a horizontal plane near the top surface of the cube. Again, LES provides accurate results. The results from the hybrid LES-RANS approach are quite close to those from the LES and much better than for the RANS.

As would be expected for this conjugate problem, due to the large ratio of the thermal diffusivities numerical stiffness is observed. To overcome this, Zhong and Tucker test three techniques. One approach is that the temperature equation is not solved until the velocity field has been established. The velocity field is then frozen and just the temperature equation solved until the temperature field has developed. After this, the velocity and temperature field are solved together. The second technique involves a gradual increase in the ratio of the thermal diffusivities. The third uses the fully implicit time scheme and a stable hybrid spatial scheme for the temperature equation. This allows larger initial time steps when solving the temperature equation. All these techniques are found effective in reducing the computing costs, especially the second.

Dufour et al. (2009) perform conjugate simulations for a high-pressure turbine stator (cascade). The Re_c is 2.8×10^6 and the WALE LES model used. Simulations are performed with $N = 6.37 \times 10^6$ and a compressible flow solver having a four-stage explicit Runge-Kutta scheme. The LES used a spanwise domain of 10 % of c. The LES is found to capture an important complex acoustic feedback mechanism. This substantially altered the energetics of the upstream separated shear layer. Duchaine et al. (2009) explore conjugate heat transfer from a high-pressure turbine blade with film cooling holes (T120D cascade). Again, a compressible flow solver is used (unstructured, cell vertex, FE) with the WALE LES model. The Re_c is 3.8×10^5 and $N = 6.5 \times 10^6$ (tetrahedral elements). The LES is found to substantially improve the prediction of the mixing process induced by the film cooling holes. The solid-fluid coupling is achieved using separate solid and fluid codes through a three-stage process. This is as follows:

1. An adiabatic surface CFD simulation is run;
2. The surface temperatures from (a) are used to produce a steady initial solid conduction solution and
3. The parallel, coupled simulation is run using the process summarized in Fig. 7.16.

The surface cooling effectiveness trends are in reasonable agreement with the measurements. However, due to lack of grid resolution, the magnitudes are over predicted.

7.7.2 Moving Surfaces

Kato et al. (2007) perform FE based Germano LES of a five stage centrifugal pump. These simulations are also discussed in Sect. 6.9.1, a key emphasis being noise prediction. The $Re = 1 \times 10^7$—based on the impeller disc outer radius and tip speed. For each blade passage $N = 0.3 \times 10^6$. LES casing pressures are used as a boundary condition in a structured FE model of the pump casing. The casing response is then fed into a FWH related acoustic analogy (see Sect. 6.5.2) to estimate far field sound.

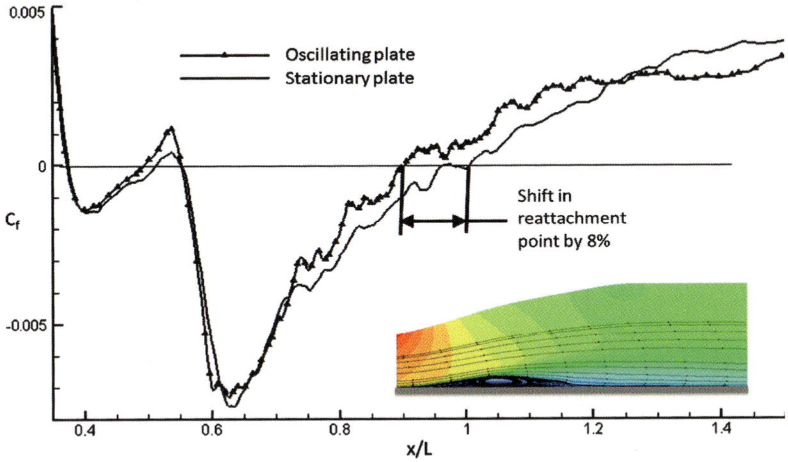

Fig. 7.27 Streamwise variation of skin friction coefficient for a stationary and oscillating surface subjected to a highly loaded low-pressure turbine blade pressure distribution Vadlamani (2013)

The discrepancy with far field sound measurements is typically less than 10 dB. The coupling processes used are one-way. Close agreement is found with pump casing displacement spectra.

Huls et al. (2007) make vibration predictions in combustion chambers by coupling FE and Smagorinsky based LES with wall functions. The surface pressure time histories from the, 3rd order in space and time, LES are stored and then fed into a FE vibration program. Hence, there is a one-way coupling between the flow field and structure. The calculated vibration levels/velocities agreed well with spectral test data. Huls (2006), using a reduced order model for the flow field, perform simulations that have a two-way coupling between the structure and the fluid. However, unlike the LES, these simulations are not able to capture the complex internal coupling between the flow field and acoustics. Shahi et al. (2012) make two-way coupled combustor simulations. CFX with the SAS hybrid LES-RANS approach is used for the fluid with 1.2×10^6 tetrahedral elements. For the solid 2450 shell elements are used. The tightly coupled simulations suggest minimal deflection ($\sim 10^{-9}$ m) of the combustor lining.

In Sect. 5.2.1, LES type results for the highly loaded low-pressure turbine case of Wissink and Rodi (2004) are briefly reported. Figure 7.27, plots the surface variation of skin friction coefficient C_f with streamwise distance. The full line and line with symbols represent skin friction distributions without and with plate vibrations, respectively. The figure inset gives the typical time averaged separation bubble structure. The case is intended to explore the influence of aeroelastic input on the transition process in a low-pressure turbine. The surface vibration amplitude is set to 0.01 % of chord (Sayma 2011). The frequency is equal to the 3rd harmonic of the shear layer frequency. The vibration is imposed using the method of Kato and Launder (1993). With this, a body force is used in the fluid domain. As can be seen from the figure, with vibration there is an 8 % reduction in the size of the separation bubble.

Compliant Walls

Compliant walls are a subset of fluid structure interaction problems. They have been considered as an attractive means of exerting passive flow control and hence to reduce drag and noise. They can potentially delay transition (Davies and Carpenter 1997). However, their ability to passively reduce drag in the turbulent regime is unclear. Carlson et al. (1995) perform DNS ($Re = 2000$) for imposed wall motion in a bi-periodic channel flow. The grid points considered are as follows: $48 \times 65 \times 24$ and $80 \times 65 \times 32$ (x, y, z). In the periodic directions the discretization is spectral. Chebyshev polynomials are used in the wall normal direction. Xu et al. (2003) apply the same numerical procedure as Carlson et al. to a channel flow with a compliant wall. The wall is modeled as a spring supported flexible plate. Hence, the simulations are fully coupled. Simulations involved $32 \times 65 \times 96$ (x, y, z) grid points and are for a Reynolds number of 3000. The turbulent flow field is found similar to that for a rigid wall with no skin friction decrease. Endo and Himeno (2002) perform a similar DNS, but find a skin friction reduction. The surface is represented by a spring, mass damper system. The phase delay in the system is suggested as being responsible for the suppression of streamwise streaks.

Vergne et al. (2007) consider aerovibroacoustics. With this, unsteady flow causes structural deflections and substantial noise. Vergne et al. treat the solid surface using a structural plate based model. This is coupled to a finite element based LES code. Use is made of the Smagorinsky LES model (with a type of wall function) and a second order, streamline upwind, Petrov-Galerkin, convective term treatment. A special LES treatment is used near walls to account for sound absorption from the un-resolved scales. For the fluid $N = 1 \times 10^6$ hexahedral elements. The structure is modelled using around 1,700 shell elements. The configuration is such that the vibrating plate is exposed to a turbulent boundary layer. The plate separates this layer from a cavity. The boundary layer flows over an upstream bluff strip. Comparison is made with mean velocity profiles, Reynolds stress and surface pressure spectra. Poor spanwise grid resolution is attributed to limiting the maximum frequencies that can be captured by the LES.

Oscillating Airfolds

Figure 7.28 gives the instantaneous flow field for an extreme circa $+/-45^\circ$ pitching aerofoil. With such high angle of attack variations massively separated flows arise. This can be seen from the figure which is for a hybrid RANS-(N)LES simulation. As noted in Chap. 3, eddy viscosity models cannot account for some of the affects of stress-strain misalignment for this type of unsteady flow. This can result in the prediction of too much turbulent viscosity in the separated wake regions. Hence, for this

Fig. 7.28 Instantaneous flow field for an extreme circa ±45° pitching on a aerofoil

type of flow hybrid RANS-LES methods are preferable. Knopp et al. (2009) report various eddy resolving simulations including DES and Menter SST based SAS for an oscillating NACA0012 aerofoil. The aerofoil has a 15° mean angle of attack and a 10° pitching amplitude. It has a reduced frequency $k_f = 0.1$ where $k_f = \omega c / 2U_\infty$ (U_∞ is the free stream velocity and ω frequency) and $Re_c = 1 \times 10^6$. A range of codes and grid topologies are used with $0.56 \times 10^6 < N < 2.78 \times 10^6$. On balance the eddy resolving results, although better than URANS appear disappointing. For, example, the negative pitching moment is over predicted by 100 %. The hybrid approaches capture the delay in stall, when the aerofoil passes the static stall angle better than URANS. Reasons identified for discrepancies with measurements are wind tunnel wall influences and the necessity to model transition. Also, the need for a deeper level of validation data that indentifies the state of the boundary layer is noted.

As discussed in Chap. 5, Visbal (2011) performs 6th order, compact scheme, ILES for deep dynamic stall of a plunging (SD7003) aerofoil with $Re_c = 6 \times 10^4$. Grids range from $2.4 \times 10^6 < N < 18.6 \times 10^6$ and the results found to be relatively insensitive to grid choice. The span-wise domain is 40 % of chord. Encouraging agreement is found with detailed measurements (phased averaged mean velocity, turbulence quantities and vorticity) for the initial stages of stall. However, at later stages, the level of agreement between the two sets of available measurements and also the LES breakdown. This appears to be due to inconsistencies in end wall boundary conditions.

Looking at the overall levels of validation the striking thing seems to be that validation tends to be either performed at level 1 or 4 but not at intermediate values. Hence, if the spectral information is wrongly predicted the reason will be hard to diagnose. Clearly the lower level of available data hampers the validation process.

7.8 Conclusions

As would be expected there are considerably less examples of eddy resolving simulations for more complex multifaceted coupled problems. However, for practical systems such simulations can be vital to gaining physically plausible solutions. Many practical problems are coupled and not amenable to being broken down into component forms. Hence, future modelling should encompass the following elements:

I Multi fidelity modelling that accounts for the coupled nature of aerospace systems and hence the challenges with regards to inflow and outflow boundary conditions—especially for LES;
II Multi-objective simulations that enable the tensions between aerodynamic, aeroacoustic and structural requirements etc to be explored and potential synergies between these elements and extended;
III Multi-physics simulations.

Static meshing can be a time consuming task. Hence it is perhaps not too surprising that when meshes undergo larger deflections mesh deformation becomes challenging. This is especially so in the parallel environment. There are other peripheral issues such as ensuring geometric conservation, conservation between differing domains and computation of the nearest wall distance.

Linking codes can be challenging in computational science and load balancing terms. Also, for numerical stability, care must also be taken to ensure solutions are valid. To remove numerical stiffness, specialized and rather problem specific approaches are needed for conjugate problems.

With regards to validation data, it seems to be true that detailed flow field data is generally lacking when assessing the performance of numerical predictions for coupled problems. It obviously follows that the validation level for more challenging coupled simulations is generally lower.

References

M.J. Aftosmis, M.J. Berger, J.E. Melton, Robust and efficient Cartesian mesh generation for component-based geometry. AIAA J. **36**(6), 952–960 (1998)

J.J. Alonso, S. Hahn, F. Ham, M. Herrmann, G. Iaccarino, G. Kalitzin, P. LeGresley, K. Mattsson, G. Medic, P. Moin et al., CHIMPS: a high-performance scalable module for multi-physics simulations. AIAA J. **5274**, 2006 (2006)

D. Amirante, N.J. Hills, C.J. Barnes, Use of dynamic meshes for transient metal temperature prediction, in *Proceedings of ASME Turbo Expo 2012* (2012). ASME Paper GT2012-68782

Y.S. Baik, J.M. Rausch, L.P. Bernal, M.V. Ol, Experimental investigation of pitching and plunging airfoils at Reynolds number between 1×10^4 and 6×10^4. AIAA J. **4030** (2009)

R.E. Bartels, A.I. Sayma, Computational aeroelastic modelling of airframes and turbomachinery: progress and challenges. Philos. Trans. R. Soc. A, Math. Phys. Eng. Sci. **365**(1859), 2469–2499 (2007)

D. Boger, Efficient method for calculating wall proximity. AIAA J. **39**(12), 2404–2406 (2001)

S. Buis, A. Piacentini, D. Déclat, PALM: a computational framework for assembling high-performance computing applications. Concurr. Comput., Pract. Exp. **18**(2), 231–245 (2006)

H.A. Carlson, G. Berkooz, J.L. Lumley, Direct numerical simulation of flow in a channel with complex, time-dependent wall geometries: a pseudospectral method. J. Comput. Phys. **121**(1), 155–175 (1995)

H.A. Carlson, J.Q. Feng, J.P. Thomas, R.E. Kielb, K.C. Hall, E.H. Dowell, Computational models for nonlinear aeroelasticity, in *43rd AIAA Aerospace Sciences Meeting and Exhibit*, Reno, NV (2005). AIAA Paper 2005-1085

L. Cavagna, P. Masarati, G. Quaranta, Simulation of maneuvering flexible aircraft by coupled multibody/CFD, in *Proceedings of ECCOMAS Thematic Conference, "Multibody Dynamics 2009"* (2009)

R.V. Chima, D.J. Arend, R.S. Castner, J.W. Slater, P.P. Truax, CFD models of a serpentine inlet, fan, and nozzle. AIAA J. **33** (2010a)
R.V. Chima, T.R. Conners, T.R. Wayman, Coupled analysis of an inlet and fan for a quiet supersonic jet. AIAA J. **479** (2010b)
D.J. Clark, M.J. Jansen, G.T. Montague, An overview of magnetic bearing technology for gas turbine engines. National Aeronautics and Space Administration. NASA/TM-20 4-213177 (2004)
C. Davies, P.W. Carpenter, Numerical simulation of the evolution of Tollmien-Schlichting waves over finite compliant panels. J. Fluid Mech. **335**, 361–392 (1997)
S.S. Davis, Data set 2, NACA 64A010 (NASA Ames model), Oscillatory pitching. Compendium of Unsteady Aerodynamics Measurements, AGARD (1982a), pp. 2.1–2.22. Report No. 702
S.S. Davis, Data set 5, NLR 7301 Supercritical airfoil oscillatory pitching. Compendium of Unsteady Aerodynamics Measurements, AGARD (1982b), pp. 5.1–5.27. Report No. 702
F.M. De Jongh, P.G. Morton, R. Holmes, The synchronous instability of a compressor rotor due to bearing journal differential heating. Discussion. J. Eng. Gas Turbines Power **118**(4), 816–824 (1996)
I. Demirdžić, M. Perić, Space conservation law in finite volume calculations of fluid flow. Int. J. Numer. Methods Fluids **8**(9), 1037–1050 (1988)
P. Dhopade, A.J. Neely, J. Young, K. Shankar, High-cycle fatigue of fan blades accounting for fluid-structure interaction, in *Proceedings of ASME Turbo Expo 2012* (2012). ASME Paper GT2012-68102
H. Doi, Fluid/structure coupled aeroelastic computations for transonic flows in turbomachinery. PhD thesis, Department of Aeronautics and Astronautics, Stanford University (2002)
F. Duchaine, A. Corpron, L. Pons, V. Moureau, F. Nicoud, T. Poinsot, Development and assessment of a coupled strategy for conjugate heat transfer with large eddy simulation: application to a cooled turbine blade. Int. J. Heat Fluid Flow **30**(6), 1129–1141 (2009)
G. Dufour, N. Gourdain, F. Duchaine, O. Vermorel, L.Y.M. Gicquel, J.F. Boussuge, T. Poinsot, *Large Eddy Simulation Applications*. VKI Lecture Series Numerical Investigations in Turbomachinery: The State of the Art (2009)
S. Eastwood, Hybrid RANS-LES of Complex Geometry Jets. PhD thesis, University of Cambridge (2009)
T. Endo, R. Himeno, Direct numerical simulation of turbulent flow over a compliant surface. J. Turbul. **3** (2002)
E. Fares, W. Schröder, A differential equation for approximate wall distance. Int. J. Numer. Methods Fluids **39**(8), 743–762 (2002)
T.H. Fransson, M. Jöcker, A. Bölcs, P. Ott, Viscous and inviscid linear/nonlinear calculations versus quasi-three-dimensional experimental cascade data for a new aeroelastic turbine standard configuration. J. Turbomach. **121**(4), 717–725 (1999)
A.L. Gaitonde, S.P. Fiddes, A three-dimensional moving mesh method for the calculation of unsteady transonic flows, in *Recent Developments and Applications in Aeronautical CFD* (1993), p. 13
V. Ganine, N.J. Hills, B.L. Lapworth, Nonlinear acceleration of coupled fluid-structure transient thermal problems by Anderson mixing. Int. J. Numer. Methods Fluids (2012)
J.A. Garcia, Numerical investigation of non-linear aeroelastic effects on flexible high aspect ratio wings. J. Aircr. **42**(4), 1025–1036 (2005)
M.B. Giles, Stability analysis of numerical interface conditions in fluid-structure thermal analysis. Int. J. Numer. Methods Fluids **25**(4), 421–436 (1997)
O. Hassan, K. Morgan, N. Weatherill, Unstructured mesh methods for the solution of the unsteady compressible flow equations with moving boundary components. Philos. Trans. R. Soc. A, Math. Phys. Eng. Sci. **365**(1859), 2531–2552 (2007a)
O. Hassan, K.A. Sørensen, K. Morgan, N.P. Weatherill, A method for time accurate turbulent compressible fluid flow simulation with moving boundary components employing local remeshing. Int. J. Numer. Methods Fluids **53**(8), 1243–1266 (2007b)
L. He, Fourier methods for turbomachinery applications. Prog. Aerosp. Sci. **46**(8), 329–341 (2010)

R.A. Huls, Acousto-elastic interaction in combustion chambers. PhD thesis, University of Twente (2006)

R.A. Huls, A.X. Sengissen, P.J.M. Van der Hoogt, J.B.W. Kok, T. Poinsot, A. de Boer, Vibration prediction in combustion chambers by coupling finite elements and large eddy simulations. J. Sound Vib. **304**(1), 224–229 (2007)

I.V. Iourokina, S.K. Lele, Towards large eddy simulation of film-cooling flows on a model turbine blade leading edge. AIAA J. **670** (2005)

C.K. Kang, Y. Baik, L. Bernal, M.V. Ol, W. Shyy, Fluid dynamics of pitching and plunging airfoils of Reynolds number between 1×10^4 and 6×10^4. AIAA Paper 2009-0536 (2009)

M. Kato, B.E. Launder, The modelling of turbulent flow around stationary and vibrating square cylinder, in *9th Symposium on Turbulent Shear Flows*, Kyoto, Japan (1993), pp. 10-4-1–10-4-6

C. Kato, Y. Yamade, H. Wang, Y. Guo, M. Miyazawa, T. Takaishi, S. Yoshimura, Y. Takano, Numerical prediction of sound generated from flows with a low Mach number. Comput. Fluids **36**(1), 53–68 (2007)

H.P. Kersken, G. Ashcroft, C. Frey, O. Putz, H. Stuer, S. Schmitt, Validation of a linearized Navier-Stokes based flutter prediction tool—Part 1: numerical methods, in *Proceedings of ASME Turbo Expo 2012* (2012). ASME Paper GT2012-68018

R.E. Kielb, J.W. Barter, J.P. Thomas, K.C. Hall, Blade excitation by aerodynamic instabilities: a compressor blade study, in *Proceedings of ASME Turbo Expo 2003* (2003). ASME Paper GT-2003-38634

G. Kingsley, J.M. Siegel, Development of a multi-disciplinary computing environment (MDICE), in *AIAA/USAF/NASA/ISSMO Symposium on Multidisciplinary Analysis and Optimization, 7th*, St. Louis, MO (1998). AIAA Paper 98-4738

T. Knopp, K. Weinman, D. Schwamborn, Oscillating airfoil NACA0012 at 15° a basic case for aero-elasticity, in *DESider a European Effort on Hybrid RANS-LES Modelling*. Notes on Numerical Fluid Mechanics and Multidisciplinary Design, vol. 103 (2009), pp. 261–269

N.C. Lambourne, Data set 7, NORA model, oscillation about a swept axis. Compendium of Unsteady Aerodynamics Measurements, AGARD (1982), pp. 7.1–7.35. Report No. 702

R. Landon, Data set 3, NACA 0012, Oscillatory and Transient Pitching. Compendium of Unsteady Aerodynamics Measurements, AGARD (1982), pp. 3.1–3.25. Report No. 702

R. Langtry, P. Spalart, Detached eddy simulation of a nose landing-gear cavity, in *IUTAM Symposium on Unsteady Separated Flows and Their Control* (Springer, Berlin, 2009), pp. 357–366

X. Liu, N. Qin, H. Xia, Fast dynamic grid deformation based on Delaunay graph mapping. J. Comput. Phys. **211**(2), 405–423 (2006)

S. Loiodice, P.G. Tucker, J. Watson, Coupled open rotor engine intake simulations, in *Proceedings of the 48th AIAA Aerospace Sciences Meeting and Exhibit*, Orlando, Florida (2010). AIAA Paper 2010-840

J.W. Lund, J. Tonnesen, An approximate analysis of the temperature conditions in a journal bearing. Part II: application. J. Tribol. **106**, 237 (1984)

D.G. Mabey, Data set 6, NLR 7301 RAE wing A, oscillating flap. Compendium of Unsteady Aerodynamics Measurements, AGARD (1982), pp. 6.1–6.15. Report No. 702

J.C. Marongiu, F. Leboeuf, E. Parkinson, Numerical simulation of the flow in a Pelton turbine using the meshless method smoothed particle hydrodynamics: a new simple solid boundary treatment. Proc. Inst. Mech. Eng. A, J. Power Energy **221**(6), 849–856 (2007)

J.C. Marongiu, F. Leboeuf, J.Ë. Caro, E. Parkinson, Free surface flows simulations in Pelton turbines using an hybrid SPH-ALE method. J. Hydraul. Res. **48**(S1), 40–49 (2010)

F. Mathey, J. Froehlich, W. Rodi, Flow in a matrix of surface-mounted cubes description of numerical methodology for test case 6.2, in *8th ERCOFTAC/IAHR/COST Workshop on Rened Turbulence Modeling*, Laboratory of Applied Thermodynamics, Espon, Helsinki University of Technology, Helsinki, Finland (1999), pp. 46–49. Report 127

W.J. McCroskey, K.W. McAlister, L.W. Carr, S.L. Pucci, An experimental study of dynamic stall on advanced airfoil sections. Volume 1. Summary of the experiment. Technical report, DTIC Document (1982)

G. Medic, G. Kalitzin, D. You, V.D. Weide, J.J. Alonso, H. Pitsch, Integrated RANS-LES computation of an entire gas turbine jet engine, in *45th AIAA Aerospace Sciences Meeting and Exhibit* (2008). AIAA Paper 2007–1117

E.R. Meinders, Experimental study of heat transfer in turbulent flows over wall-mounted cubes. PhD thesis, Delft University of Technology (1998)

E.R. Meinders, K. Hanjalić, Vortex structure and heat transfer in turbulent flow over a wall-mounted matrix of cubes. Int. J. Heat Fluid Flow **20**(3), 255–267 (1999)

F.R. Menter, Zonal two equation Kappa-Omega turbulence models for aerodynamic flows. AIAA Paper 93-2906 (1993)

D. Micallef, D. Witteck, A. Wiedermann, D. Klub, R. Mailach, Three-dimensional viscous flutter analysis of a turbine cascade in subsonic and transonic flows, in *Proceedings of ASME Turbo Expo 2012* (2012). ASME Paper GT2012-68396

C. Michler, E.H. Van Brummelen, R. De Borst, An interface Newton-Krylov solver for fluid-structure interaction. Int. J. Numer. Methods Fluids **47**(10–11), 1189–1195 (2005)

K. Nakahashi, F. Togashi, Unstructured overset grid method for flow simulation of complex multiple body problems, in *Proceedings of ICAS 2000 Congress* (2000). Paper No. ICAS 0263

B. Ničeno, A.D.T. Dronkers, K. Hanjalić, Turbulent heat transfer from a multi-layered wall-mounted cube matrix: a large eddy simulation. Int. J. Heat Fluid Flow **23**(2), 173–185 (2002)

M.V. Ol, L. Bernal, C.K. Kang, W. Shyy, Shallow and deep dynamic stall for flapping low Reynolds number airfoils. Exp. Fluids **46**(5), 883–901 (2009)

J.H. Page, P. Hield, P.G. Tucker, Inverse design of 3D multistage transonic fans at dual operating points, in *Proceedings of ASME Turbo Expo 2013* (2013). ASME Paper GT2013-95062

M.A. Potsdam, G.P. Guruswamy, A parallel multiblock mesh movement scheme for complex aeroelastic applications. AIAA J. **716**, 2000 (2001)

F. Rahman, J.A. Visser, R.M. Morris, Capturing sudden increase in heat transfer on the suction side of a turbine blade using a Navier-Stokes solver. J. Turbomach. **127**(3), 552–556 (2005)

P. Rautaheimo, T. Siikonen, Flow in a matrix of surface-mounted cubes description of numerical methodology for test case 6.2, in *8th ERCOFTAC/IAHR/COST Workshop on Rened Turbulence Modeling*, Laboratory of Applied Thermodynamics, Espon, Helsinki University of Technology, Helsinki, Finland (1999), pp. 31–36. Report 127

S. Salvadori, G. Riccio, M. Insinna, F. Martelli, Analysis of combustor/vane interaction with decoupled and loosely coupled approaches, in *Proceedings of ASME Turbo Expo 2012* (2012) ASME Paper GT2012-69038

N. Sayma, Personal Communication (2011)

J.A. Sethian, Fast marching methods. SIAM Rev. **41**(2), 199–235 (1999)

M. Shahi, J.B.W. Kok, P.R. Alemela, Simulation of 2-way fluid structure interaction in a 3D model combustor, in *Proceedings of ASME Turbo Expo 2012* (2012). ASME Paper GT2012-69681

M. Shevtsov, A. Soupikov, A. Kapustin, Highly parallel fast KD-tree construction for interactive ray tracing of dynamic scenes, in *Computer Graphics Forum*, vol. 26 (Wiley Online Library, New York, 2007), pp. 395–404

W.A. Silva, R.E. Bartels, Development of reduced-order models for aeroelastic analysis and flutter prediction using the CFL3Dv6.0 code. J. Fluids Struct. **19**(6), 729–745 (2004)

P.R. Spalart, Trends in turbulence treatments. American Institute of Aeronautics and Astronautics (2000). AIAA Paper 2000-2306

P.R. Spalart, S.R. Allmaras, A one equation turbulence model for aerodynamic flows. Rech. Aérosp. **1**, 5–21 (1992)

P.R. Spalart, D.R. Bogue, The role of CFD in aerodynamics, off-design. Aeronaut. J. **107**(1072), 323–329 (2003)

D.B. Spalding, Calculation of turbulent heat transfer in cluttered spaces, in *Proceedings of the 10th International Heat Transfer Conference* (Society for Industrial and Applied Mathematics, Brighton, 1994)

C.H. Stephens, A.S. Arena Jr., K.K. Gupta, C.A. Edwards, Application of the transpiration method for aeroservoelastic prediction using CFD. AIAA J. **2071** (1998)

M. Sussman, E.G. Puckett, A coupled level set and volume-of-fluid method for computing 3D and axisymmetric incompressible two-phase flows. J. Comput. Phys. **162**(2), 301–337 (2000)

J. Thomas, E. Dowell, K. Hall, C. Denegri Jr., Modeling limit cycle oscillation behavior of the F-16 fighter using a harmonic balance approach, in *45th AIAA/ASME/ASCE/AHS/ASC Structures, Structural Dynamics and Materials Conference* (2004). AIAA Paper 2004-1696

J.P. Thomas, K.C. Hall, W.S. Clark, Computation of unsteady nonlinear flows in cascades using a harmonic balance technique. AIAA J. **40**(5) (2012)

P.G. Tucker, *Computation of Unsteady Internal Flows: Fundamental Methods with Case Studies* (Kluwer Academic, Norwell, 2001)

P.G. Tucker, Computation of unsteady turbomachinery flows: Part 1 progress and challenges. Prog. Aerosp. Sci. **47**(7), 522–545 (2011a)

P.G. Tucker, Hybrid Hamilton-Jacobi-Poisson wall distance function model. Comput. Fluids **44**(1), 130–142 (2011b)

P.G. Tucker, P.S. Keogh, On the dynamic thermal state in a hydrodynamic bearing with a whirling journal using CFD techniques. J. Tribol. **118**(2), 356–363 (1996)

P.G. Tucker, C.L. Rumsey, R.E. Bartels, R.T. Biedron, Transport equation based wall distance computations aimed at flows with time-dependent geometry. NASA-TM 2003-212680 (2003)

P.G. Tucker, C.L. Rumsey, P.R. Spalart, R.E. Bartels, R.T. Biedron, Computations of wall distances based on differential equations. AIAA J. **43**(3), 539–549 (2005)

N.R. Vadlamani, High fidelity large eddy simulation of turbines: current status and future outlook. PhD thesis, University of Cambridge (2013)

M. Vahdati, A.I. Sayma, C. Bréard, M. Imregun, Computational study of intake duct effects on fan flutter stability. AIAA J. **40**(3), 408–418 (2002)

J.F. van Kampen, Acoustic pressure oscillations induced by confined turbulent premixed natural gas flames. PhD thesis, University of Twente (2006)

J.A. Verdicchio, J.W. Chew, N.J. Hills, *Coupled Fluid/solid Heat Transfer Computation for Turbine Discs*. Proceedings of ASME Turbo. Expo. 2001 (2001-GT) (2001), p. 123. ASME Paper GT2001-0205

S. Vergne, J.M. Auger, F. Périé, A. Jacques, D. Nicolopoulos, Aeroelastic noise, in *Large-Eddy Simulation for Acoustics*, ed. by C. Wagner, T. Huttl, P. Sagaut (2007), pp. 272–293

M.R. Visbal, Numerical investigation of deep dynamic stall of a plunging airfoil. AIAA J. **49**(10), 2152–2170 (2011)

C. Voigt, C. Frey, H. Kersken, Development of a generic surface mapping algorithm for fluid-structure interaction simulations in turbomachinery, in *V European Conference on Computational Fluid Dynamics*, Lisbon, Portugal (2010)

L.B. Wigton, Optimizing CFD codes and algorithms for use on Cray computer, in *Frontiers of Computational Fluid Dynamics* (1998), pp. 1–15

J. Wissink, W. Rodi, DNS of a laminar separation bubble affected by free-stream disturbances. ERCOFTAC Ser. **9**, 213–220 (2004)

H. Xia, P.G. Tucker, Finite volume distance field and its application to medial axis transforms. Int. J. Numer. Methods Eng. **82**(1), 114–134 (2010)

H. Xia, P.G. Tucker, Fast equal and biased distance fields for medial axis transform with meshing in mind. Appl. Math. Model. **35**(12), 5804–5819 (2011)

H. Xia, P.G. Tucker, W.N. Dawes, Level sets for CFD in aerospace engineering. Prog. Aerosp. Sci. **46**(7), 274–283 (2010)

S. Xu, D. Rempfer, J. Lumley, Turbulence over a compliant surface: numerical simulation and analysis. J. Fluid Mech. **478**(1), 11–34 (2003)

G. Yang, D.M. Causon, D.M. Ingram, R. Saunders, P. Batten, A Cartesian cut cell method for compressible flows—Part B: moving body problems. Aeronaut. J. **101**(1002), 57–65 (1997)

S.Y. Yoon, Z. Lin, W. Jiang, P.E. Allaire, Flow-rate observers in the suppression of compressor surge using active magnetic bearings, in *Proceedings of ASME Turbo Expo 2012* (2012). ASME Paper GT2012-70011

M. Zhang, A. Hou, S. Zhou, X. Yang, Analysis on flutter characteristics of transonic compressor blade row by a fluid-structure coupled method, in *Proceedings of ASME Turbo Expo 2012* (2012). ASME Paper GT2012-69439

B. Zhong, P.G. Tucker, k–l based hybrid LES/RANS approach and its application to heat transfer simulation. Int. J. Numer. Methods Fluids **23**(10), 983–1005 (2004)

R.J. Zwaan, Data Set 1, NACA 64A006 Oscillating Flap. Compendium of Unsteady Aerodynamics Measurements, AGARD (1982a), p. 1. Report No. 702

R.J. Zwaan, Data Set 4, Supercritical airfoil oscillatory pitching and oscillatory flap. Compendium of Unsteady Aerodynamics Measurements, AGARD (1982b), pp. 4.1–4.25. Report No. 702

Chapter 8
Future Outlook

8.1 Concluding Remarks

Chapter 1 outlined the tremendous range of unsteady flow source mechanisms and the challenge with regards to grid requirements in capturing these. The brief overview of numerical methods, given in Chap. 2, showed a wide range of schemes to choose from. However, exploring the use of these in later chapters indicated that just small subsets of these find practical use, the most popular being relatively standard schemes. Grid quality is shown, like numerical scheme choice, to have potentially a strong solution impact. Hence, particularly when making LES computations, attention should be paid to the numerical scheme, grid quality and how these two elements interact.

In Chap. 3 turbulence modelling methods were overviewed. For URANS, the occurrence of a spectral gap in many turbomachinery zones was shown to be uncertain. A hierarchy of key elements for industrial LES was proposed. Different types of LES filters were outlined. Their potential strong impact, above that of the LES model formulation itself, on results, was noted. Importantly, for flows without transition, the actual explicit LES model came low down in the proposed LES hierarchy, other key elements being, problem definition, the numerical scheme and near wall treatments. From Chap. 3 it is clear that considerable expertise is needed to safely use hybrid RANS-LES and even LES techniques. Hence, there seems the need for best practice guidelines (see later). The extensive range methods for generating turbulence inflow was outlined. However, many of these have limited applicability to complex engineering systems. For these, the Reynolds numbers are high and geometries complex. The use of idealized synthetic inflow and basic outflow boundary conditions seems excessively simplistic in complex coupled systems like gas turbine engines. These can give rise to multiple inflow boundaries and system coupling can occur.

In Chap. 4, the various unsteady modelling hierarchies in aerospace were reviewed. These ranged from linear harmonic to DNS. Unsteady, reduced order modelling encompassing deterministic stresses and body forces was discussed.

P.G. Tucker, *Unsteady Computational Fluid Dynamics in Aeronautics*, Fluid Mechanics and Its Applications 104, DOI 10.1007/978-94-007-7049-2_8,

Mixed fidelity methods were proposed, where low and high fidelity treatments are combined. It was shown that, in terms of user assimilation, the potential number of modelling and numerical variants poses a serious challenge. This is especially so when reported experiences for the same scheme can differ dramatically.

In Chap. 5, eddy resolving simulations in aerospace were reviewed along with the status of data for validating these. There was shown to be a lack of detailed validation data to explore in depth the performance of LES and thus refine it. This was found to be especially true for turbomachinery. Grids currently used in turbomachinery were shown to clearly be insufficient for the LES model (and numerical schemes) not to be playing an excessively strong role. Most of the practical LES studies reviewed essentially made use of RANS solvers, where robustness featured more greatly than accuracy. Also, few simulations made use of properly constructed, correlated, turbulence inflow. Even where this was attempted, most measurement sets were incomplete and lacked an adequate basis for defining this inflow. Despite the limitations of the excessively dissipative Smagorinsky model, it had wide spread use, especially with dissipative flow solvers. As with RANS, the modelling of transition was identified as a key issue, especially where LES grids tend to be too coarse. For propulsion systems, MILES related approaches, hybrid LES-RANS and more advanced LES models were shown to have an equal frequency of use, none of these approaches being especially avoided. At the higher Reynolds numbers found in airframes hybrid RANS-LES methods dominated.

Chapter 6 gave an overview of computational aeroacoustics. Of the wide range of far field sound prediction techniques available, the Ffowcs Williams–Hawkings (FWH) approach was found to have considerable popularity. As would be expected, the acoustic related problems tended to use more specialized numerical schemes. For acoustics, a roughly equal split between the use of ILES, LES and hybrid RANS-LES methods was found. The use of acoustic subgrid scale modelling is noted as a useful area for future exploration.

In Chap. 7, dealing with coupled simulations, as would be expected there were considerably fewer examples of eddy resolving simulations. Also, detailed validation data appeared scarcer and hence the level of solution validation lower. However, for practical systems, coupled simulations are identified to be of critical importance. Hence, future modelling needs to encompass: multi-fidelity, multi-objective, multi-physics simulations. These are needed to explore tensions between aerodynamic, aeroacoustic, aerothermal, chemical reaction and structural requirements and potential synergies between these elements. Performing large coupled simulations is noted to present a wide range of computational science (and applied mathematical) challenges that need to be dealt with in a parallel environment.

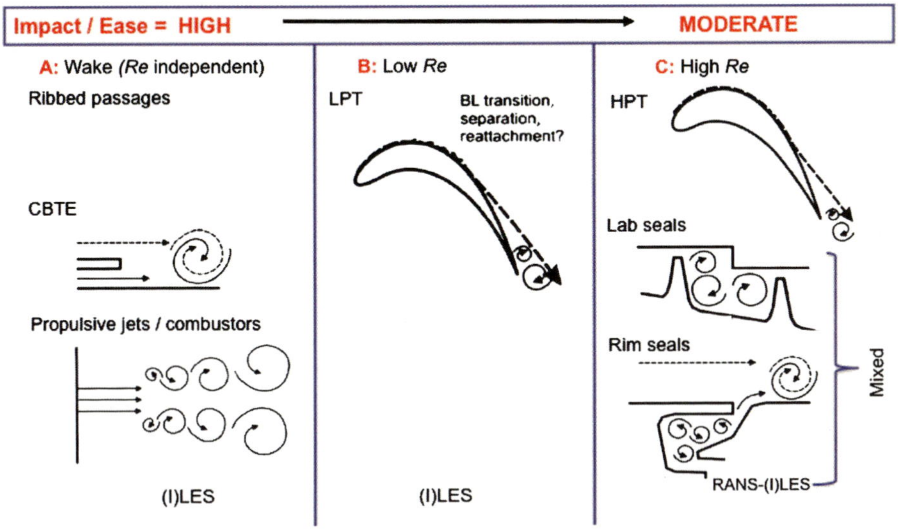

Fig. 8.1 Potential flow taxonomy for gas turbines

8.2 Future Use of Eddy Resolving Simulations

8.2.1 Best Practices

Chapter 5 showed the need for LES best practices. For example, many of the turbomachinery LES studies had under resolved grids. Frequently, they also, made use of inappropriate subgrid scale modelling. The ideal structure for best practices is perhaps unclear. They could be general, as with the ERCOFTAC CFD guidelines or broken down to consider generic building block flow types or be customized to, for example, particular aerospace components. For an engineer strongly engaged in design the latter could perhaps be more useful.

8.2.2 Flow Taxonomies

Looking at the gas turbine, as a specific example, a taxonomy of flow classes could be helpful. Figure 8.1 summarizes a potential one. Note, this is far from exhaustive. However, flows could be broadly classed as follows:

(A) Wake or Reynolds number independent flows;
(B) Low Reynolds number flows;
(C) High Reynolds number flows.

The above are placed in decreasing order of impact relative to ease. For example, the Class (A) flows have limited classical boundary layer content. They have substantial content of large scale eddies playing a critical aerodynamic role. The Class (A) flows, typically involve upstream inflow profiles that are highly inflectional. Hence, they rapidly and naturally generate inflow turbulence. Therefore, this class of flow is forgiving with regards to the inflow and grid. Examples of Class (A) flows, found in gas turbines, would be the cut back trailing edge zones, ribbed passages used to internally cool turbine blades (see Sect. 1.4.1 for details of these zones) and the propulsive jet. The cut back trailing edges contain multiple bluff upstream structures that will naturally generate turbulence. They also have a shear layer between the mainstream gas flow and the flow ejected from the trailing edge. The ribbed passages are specifically designed to generate large-scale structures that are ideal for LES to deal with. Once the propulsive jet nozzle is emerged from, the flow here is also Reynolds number independent. However, clearly the upstream boundary layer is strongly Reynolds number dependent. Hence, it is not possible to form a rigid taxonomy. Nonetheless, largely speaking, the dynamics of the propulsive jet best fits within Class (A). The Low Pressure Turbine (LPT) blade flow is designated as Class (B). It has a relatively modest Reynolds number. This is helpful for LES with regards to grid size. However, on the other hand, the flow is transitional in a complex sense. Hence, the grids needed tend to be quasi-DNS, eradicating uncertainties with regard to subgrid scale model impact. Also, the complex flow physics gives rise to the need for more accurate numerical treatments. Also, the transition process is strongly defined by the inflow turbulence. Hence, high fidelity turbulence inflow boundary conditions are needed for the Class (B) flow. The Class (C) flows are high Reynolds number, immediately giving more severe grid resolution demands. The top right hand frame of Fig. 8.1 is intended to represent a High-Pressure Turbine blade (HPT). These are immediately downstream of the combustor. Hence, they are subjected to hot streaks from the combustor and the complex migration process of these (see Sect. 1.4.1). Therefore, correctly formulated inflow turbulence is of critical importance. Also, real HPT blades have fine surface cooling holes. Again, these increase the modelling and mesh resolution challenges. The far right hand frames, below the HPT blade schematic, represent labyrinth and rim-seal flows. Initially they look as if they might be Class (A) flows. However, they have rotating surfaces, creating a strong classical boundary layer content. Hence, to get the tangential velocity field correct, the flow physics of boundary layers must be relatively well resolved. The classifications outlined largely define the modelling challenges and potential LES benefits relative to resource levels.

8.2.3 Expert Systems and Industrial Simulation Process

The taxonomy above could potentially assist in forming an expert system to safely guide a user, engaged in design, through the simulation process. Figure 8.2 presents a very initial basis for an expert system and process for industrial LES. The left hand column indicates a process chain. The initial stages of this could be potentially guided by the expert system via some graphical user interface. This would result in

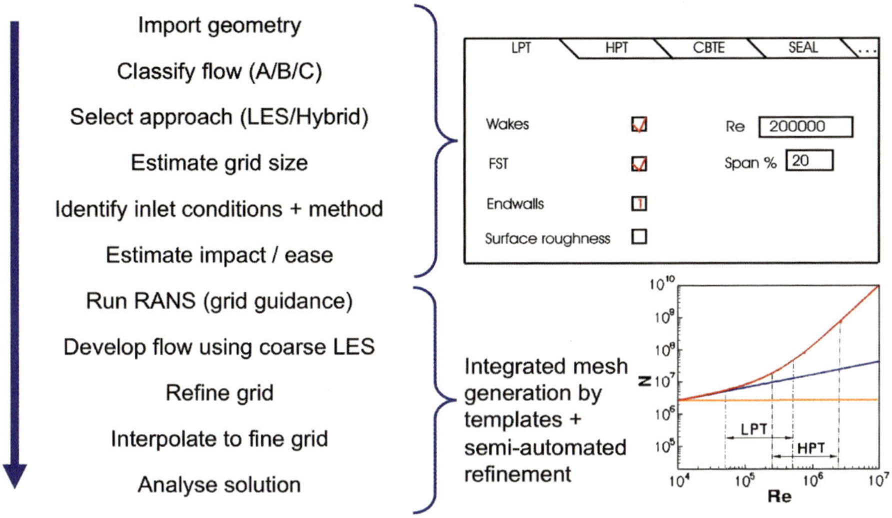

Fig. 8.2 Expert system and process for industrial LES

an estimate of the required grid size and impact relative to ease for the simulation. At this point, the user may well find that relative to the computational resources available to them, LES is impractical. If it is feasible, then an initial RANS simulation could be run. The length and time scales from this could be fed into an idealized energy spectrum. This would give an indication, for a specified mesh, of the fraction of turbulence energy that can be resolved. Based on this information, and heuristic data (see Tables 3.4 and 3.5) absorbing the RANS estimate of wall shear stress, a refined and plausible grid could be generated through some form of semi-automated process. Figure 8.3 shows use of a model energy spectrum, informed by RANS data, to refine an LES grid. Frame (a) shows contours of the fraction of turbulence energy resolved on an initial grid. Frame (b) shows a refined grid with an inserted overset wake block. Frame (c) shows the fraction of resolved turbulence on the refined grid. With overset or unstructured meshes such refinement could be relatively automatic. Indeed the energy spectrum data could perhaps be used to directly drive an unstructured grid generation process.

Next, an initial LES flow field could be developed on a relatively coarse mesh (that say resolves 70% of turbulence energy). Again heuristic data would be needed on how long is needed to evolve the particular class of flow being considered. Hopefully this aspect would be expert system driven via a database.

Judging when a flow has reached a cyclic or fully developed state is a challenge even for URANS and especially LES. A Fourier based procedure applicable to URANS is outlined by Ahmed and Barber (2005). However, as shown by Tyacke (2009) this procedure is impractical for LES. For LES substantial periods can be required to evolve a flow and then to average it. For example, with jets, Shur et al. (2003) report complete LES need time periods around $1000D/U$ where D is the jet diameter and U the jet velocity. For $t > 500D/U$, Shur et al. use the current flow

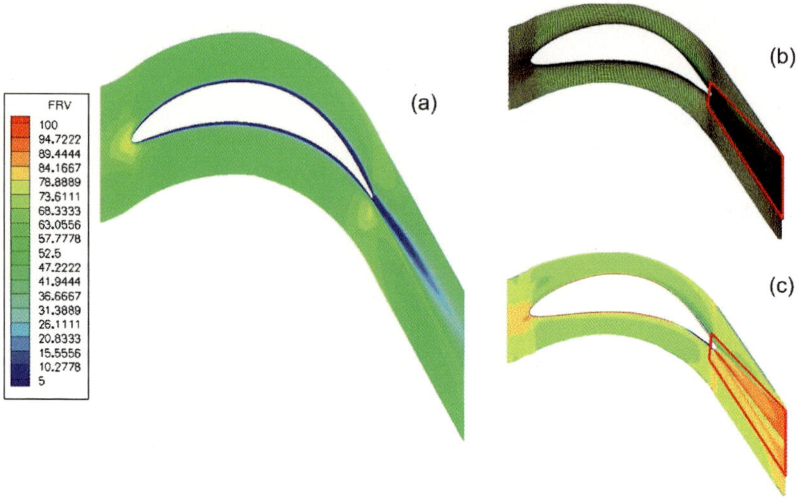

Fig. 8.3 Use of model energy spectrum, based on RANS data, to refine LES grid: (**a**) contours of the fraction of turbulence energy resolved on initial grid; (**b**) refined grid with inserted overset wake block and (**c**) new fraction of resolved turbulence

field to set sponges. For swirling jets, Tucker (2004) uses a total simulation time of $2000D/U$. With low-pressure turbine blade simulations, with no wakes, typically $t \approx 15c/U$ is needed, for complete simulations, where c is chord and U the inlet velocity. Alternatively, for flows with upstream, moving blade wakes, 15 wake passing periods can typically be needed for complete simulations. Hence, evolving flows on coarse meshes seems helpful.

The coarse grid LES would then be interpolated via a fast process such as octree or *k–d* tree based searches. Frequently, with RANS based simulations, the interpolation of an initial coarse grid solution to a fine does not yield much benefit. However, with LES, the fine scales, that are not resolved by the coarse grid have rapid time scales. Hence they quickly respond to changes in the large scales (see Townsend 1980). Certainly, experiences of following this procedure for LES of jets suggests that a considerable speed up can be gained through this grid sequencing process. Again, the final solution analysis process needs some thought with LES. The information available is rich to the point that it could be overwhelming. More sophisticated data analysis techniques such as data mining could be helpful for examining such unsteady flows. Notably, for acoustics the assembly of far field sound spectra can be of importance. Under these circumstances the low frequency component of the spectra, where long run times are needed, can be assembled on the coarse LES grid. The fine grid, interpolated solution, can then be run for a relatively short time to fill in the high frequency part of the spectrum Preliminary tests of this technique for jets has shown some promise.

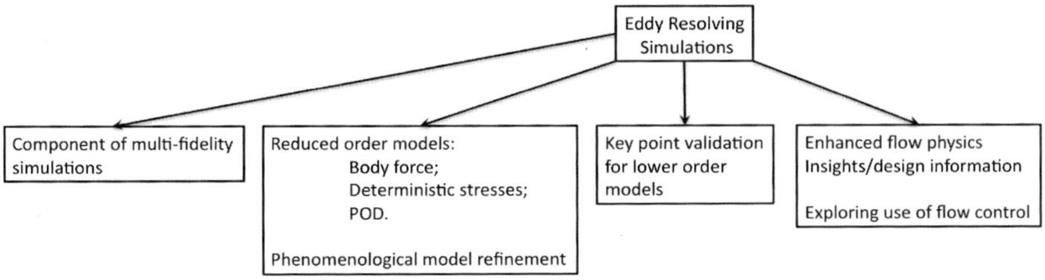

Fig. 8.4 Potential uses of eddy resolving simulations

8.2.4 The Use of LES in Coupled Problems

The increasing use of coupled simulations, to an extent, makes the practical use of LES harder to envision. However, LES can be 'sandwiched' into a larger scale coupled simulation. For example in the Stanford whole engine simulation work, an LES of a compressor is sandwiched between lower fidelity models for the compressor and turbine. Hence, a multi fidelity model is used. The LES can also be used to refine reduced or models. These could be, for example, POD, deterministic stress, or RANS based. Examples have been given of the use of LES in this context in the preceding chapters. Of course the LES can be of direct use for basic high accuracy LES in design, for example exploring altitude influences on LPT aerodynamics (see Medic and Sharma 2012) and understanding basic flow physics. The latter also, of course, has a critical impact on designs. The potential uses of eddy resolving simulations is summarized in Fig. 8.4.

As noted by Jameson (2008), in the past 25 years, computers have become a million times faster. Hence, relative to the time period of many aerospace development projects, there is a rapidly changing landscape in terms of computational power. This is supported by relatively vibrant research into the development of improved numerical methods, parallel performance, LES models, pre and post-processing. Hence, even though there are serious barriers, it seems important for industry to seriously start gearing up for the future use of eddy resolving simulations in design.

References

M.H. Ahmed, T.J. Barber, Fast Fourier transform convergence criterion for numerical simulations of periodic fluid flows. AIAA J. **43**, 1042–1052 (2005)

A. Jameson, Formulation of kinetic energy preserving conservative schemes for gas dynamics and direct numerical simulation of one-dimensional viscous compressible flow in a shock tube using entropy and kinetic energy preserving schemes. Sci. Comput. **34** (2008). doi:10.1007/s10915-007-9172-6

G. Medic, O. Sharma, Large-eddy simulation of flow in a low-pressure turbine cascade, in *Proceedings of ASME Turbo Expo 2012 GT2012*, Copenhagen, Denmark 11–15 June (2012). GT2012-68878

M.L. Shur, P.R. Spalart, M.Kh. Strelets, A.K. Travin, Towards the prediction of noise from jet engines. Int. J. Heat Fluid Flow **24**, 551–561 (2003)

A.A. Townsend, *The Structure of Turbulent Shear Flow* (Cambridge University Press, Cambridge, 1980)
P.G. Tucker, Novel miles computations for jet flows and noise. Int. J. Heat Fluid Flow **25**(4), 625–635 (2004)
J.C. Tyacke, Low Reynolds number heat transfer prediction employing large eddy simulation for electronics geometries. PhD thesis, Swansea University (2009)

Appendix A
Numerical Scheme Performance Studies

A.1 Introduction

To explore the performances of some of the schemes discussed in Chap. 2, examples are considered for acoustic, vortical and Tollmien-Schlichting (T-S) waves. These are propagated in the Fig. A.1 channel. For the T-S, the flow is linearly stable for $Re_c < 5772$ (Orszag 1971). The Reynolds number is based on the half channel height, $h = 1$, and the centre-line velocity. For $Re < Re_c$, instabilities show a downstream decay. Through considering Poiseuille flow, a T-S wave can be obtained from eigensolution of the Orr-Sommerfeld equation. After Navier-Stokes equation linearization and assuming a wave-like solution, the resulting Orr-Sommerfeld equation can be written as:

$$\left[\left(U_0 - \frac{\omega_R}{\hat{\alpha}}\right)\left(\frac{d^2}{dy^2} - \hat{\alpha}^2\right) - \frac{d^2}{dy^2}U_0 + \frac{i}{\hat{\alpha} Re}\left(\frac{d^2}{dy^2} - \hat{\alpha}^2\right)^2\right]\hat{v} = 0 \qquad \text{(A.1)}$$

Here, U_0 is a base flow of the form $U_0(y) = y(2h - y)$. Also, $\hat{\alpha}$ is the complex wave number $\hat{\alpha} = \hat{\alpha}_R + i\hat{\alpha}_I$ and ω_R the real frequency. The variables $\hat{\alpha}_R$ and ω_R are related by the expression $\omega_R/\hat{\alpha}_R =$ constant. After obtaining $\hat{v}$ from Eq. (A.1), the remaining velocity component can be easily calculated via the continuity equation. For the Re considered, the analytic disturbance decays as $e^{-\hat{\alpha}_I t}$ with a rate $\hat{\alpha}_I = -0.0106$ and a wave number $\hat{\alpha}_R = 1.155676$ (Orszag 1971). For the inviscid vortical and cut-on acoustic waves $\hat{\alpha}_I = 0$, and the analytical solutions for these can be directly inferred from the Euler and wave equations. Schemes with dispersion error will affect $\hat{\alpha}_R$. Those with false diffusion affect the decay rate constant $\hat{\alpha}_I$. The performances of numerical schemes are contrasted with these analytic $\hat{\alpha}_I$ and $\hat{\alpha}_R$ values.

Four codes (A-D) with substantially differing numerical algorithms are tested. Full details of the codes can be found in the following references: Jefferson-Loveday (2008); Tucker (2001); Moinier (1999) and Xia (2005), respectively. The key things are that codes (A) and (B) are incompressible structured flow solvers. They are finite difference based with schemes of various orders that can be either centered

P.G. Tucker, *Unsteady Computational Fluid Dynamics in Aeronautics*,
Fluid Mechanics and Its Applications 104, DOI 10.1007/978-94-007-7049-2,

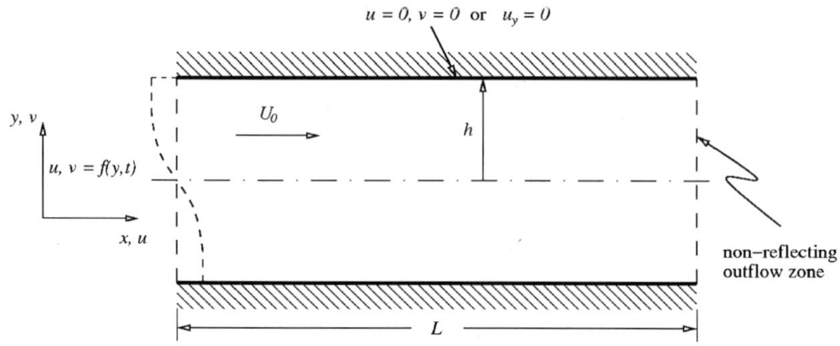

Fig. A.1 Schematic of geometry considered

or upwind biased. Code (A) uses overset curvilinear grids. Code (B) is a strictly Cartesian staggered grid program. Codes (C) and (D) are unstructured compressible flow solvers. The former has a cell vertex basis and the latter cell centered. Both the compressible flow solvers have matrix smoothing, being Roe (1981) scheme based.

A.2 Boundary and Initial Conditions

A.2.1 T-S Wave

For the Code (C) T-S solution, a fixed base flow of the form $U_0(y) = y(2h - y)$ is used and perturbations solved in a linearized manner around this. For the other solvers this field simply corresponds to an initial guess to the solution. For the T-S wave, at the channel inlet (see Fig. A.1), a two-dimensional wave is superimposed on a laminar parabolic profile:

$$u_i'(x = 0, y, t) = A_{TS}\Re\left[\hat{u}_i(y)e^{-i\omega_R t}\right] \tag{A.2}$$

A_{TS} is the amplitude of the inflow perturbations ($A_{TS} < 1$) and $\hat{u}_i$ is the complex velocity vector. This is calculated from the spatial eigenfunctions of the Orr-Sommerfeld equation corresponding to the real frequency ω_R. $\Re$ represents the real part of the complex number. The amplitude of the inflow disturbance $A_{TS} = 10^{-3}$ is defined as the root-mean-square magnitude of the *T-S* wave u component.

For the cut-on acoustic and vortical waves, two-dimensional acoustic and vortical waves are superimposed on uniform mean inviscid flows, the inflow conditions being analytically prescribed using the expressions given below.

A.2.2 Cut-On Acoustic Wave Inlet Boundary Condition

At inflow, the wave pressure fluctuation amplitude can be expressed as

$$p' \sim \cos(w_y y) e^{i\omega t - i w_x x - i w_z z} \tag{A.3}$$

For the current two-dimensional test $w_z = 0$, $w_y h = n\pi$ and $n = 1$. For the axial wave component we have

$$w_x = \frac{-wM + \sqrt{w^2 - (1 - M^2) w_y^2}}{1 - M^2} \tag{A.4}$$

where $w = \omega/c$ and $M = u/c$ where $u = 1$ is the uniform axial velocity and c the speed of sound. The axial velocity fluctuation is given by

$$u' = \frac{w_x}{\rho(\omega - u w_x)} p' \tag{A.5}$$

and the vertical by

$$v' = \frac{-1}{\rho(\omega - u w_x)} \frac{\partial p}{\partial y} = \frac{-k_y \sin(w_y y)}{i\rho(\omega - u w_x)} \tag{A.6}$$

Also, the z component of perturbation velocity is zero and $\rho' = p'/c^2$.

A.2.3 Vortical Wave Inlet Boundary Condition

For the vortical wave, at inflow the pressure distribution is such that

$$p' = 0 \tag{A.7}$$

For the current two-dimensional test $w_z = 0$, $w_y h = n\pi$ and $n = 1$. For the axial wave component we have

$$w_x = w \tag{A.8}$$

where $w = \omega/u$ and where $u = 100$ is the uniform axial velocity. The axial velocity fluctuation is given by

$$u' = \frac{-i w_y}{w_x} \cos(w_y y) e^{-i\omega t} \tag{A.9}$$

and the vertical by

$$v' = \sin(w_y y) e^{-i\omega t} \tag{A.10}$$

Also, the z component of velocity is zero.

A.2.4 Solid Wall Boundary Conditions

For the T-S wave the usual no-slip and impermeability conditions are applied at the solid walls. For the inviscid vortical and acoustic cases just the impermeability conditions are applied. For Code (C), to prevent outflow reflections, characteristic based non-reflecting boundary conditions are used (Giles 1990). Generally, to prevent wave reflection, for the final 40 % of the T-S wave domain, in the streamwise direction, damping is also found necessary and added in a gradual fashion through raising diffusion/upwinding—as discussed in Sect. 2.21. With Code (B), the convective outlet boundary condition of Pauley et al. (1990), discussed in Sect. 2.21, is used.

A.3 General Solution Details

A.3.1 Mesh and Time Step

Simulations generally use hexahedral cell grids with 259 nodes in the streamwise 'x' direction and 67 in the cross-stream 'y' direction. However, for the acoustic wave, to get the same number of points per wavelength, as for the T-S, a doubling of the number of grid cells in the streamwise direction is necessary. This doubled grid, with an additional doubling in the y direction is also briefly tested for interest on the T-S case. However, unless otherwise stated, solutions use the 259 × 67 grid. If a downstream grid expansion is used, to prevent wave reflections, additional nodes are added to the extended domain. The generally modest grids used allow differences between the various spatial schemes to be observed. Grids are refined at solid boundaries using a hyperbolic tangent function. The numerical domain has a height of 2 units in the vertical y direction. The domain in the x direction is uniform extending at least 51 channel heights ($X = 51$). When triangular cells are tested, the nodal locations are, where possible, broadly, the same as for the hexahedral grids the triangulations being formed about these nodes. For all simulations, a dimensionless time-step of $\Delta t^* < 0.05$ is used where $\Delta t^* = \Delta t U_o / 2h$.

A.3.2 Solution Error Estimation

To estimate, in the different codes used here, the $\hat{\alpha}_R$ error, in terms of an approximate kinetic energy deficit, the following is used

$$Error = \left(\frac{v'_7 - v'_A}{v'_{\max}} \right)^2 \tag{A.11}$$

where v'_7 is the amplitude of the 7th wave peak along the channel, v'_A the analytical 7th wave amplitude and $v'_{\max}$ the upstream inlet amplitude. Since the error trait

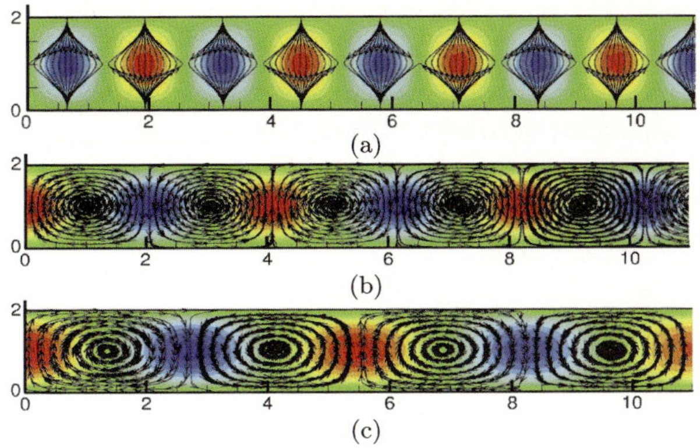

Fig. A.2 Disturbance streamlines for different wave types explored: (**a**) cut-on acoustic, (**b**) vortical and (**c**) T-S

is consistent, the $\hat{\alpha}_R$ values are useful in showing trends. The current approach is more convenient, when dealing with a range of solvers, than performing complete flow variable integrations through time and across the channel. Clearly $\hat{\alpha}_I$ is more easily accurately evaluated and is taken as an average of the wave lengths along the channel. The wavelengths are calculated using the points where $v' = 0$.

A.4 Discussion Results

Plots of predicted disturbance streamlines for the acoustic, vortical and T-S waves, having very different flow physics, are shown in Fig. A.2. The inviscid acoustic and vortical waves, given in frames (a) and (b), respectively are found to be not especially numerically challenging (see later). Hence, most discussion focuses on the T-S wave shown in Frame (c). For this, modelling the delicate balance between convection and diffusion processes is critical and hence the case most challenging. For the less physically complex inviscid waves, where pertinent, comment is made.

First we will consider the influence of spatial and temporal scheme order. This will be carried out for Code (A) with its explicit smoother.

A.4.1 Temporal Scheme Performance Studies

First, we will consider the performances of temporal schemes. Table A.1 tabulates estimated errors in $\hat{\alpha}_R$ and $\hat{\alpha}_I$ for different temporal schemes. It is clear from Table A.1 that the Crank-Nicolson (CN) scheme has greatest temporal accuracy, even being more accurate than the 5th order backwards difference (BD) Gear scheme with its substantial storage demands. It is well known that the stability of BD schemes comes through increased numerical dissipation. Hence, this is the reason why for high quality LES the CN scheme is generally preferred.

Table A.1 Influence of order and nature of temporal differencing schemes

Order	Scheme	$\hat{\alpha}_R$ Error (%)	$\hat{\alpha}_I$ Error (%)
1	BD	41.0	0.059
2	BD	7.2	0.462
5	BD	3.5	0.289
2	CN	2.5	0.279

Table A.2 Influence of order and nature of spatial differencing schemes

Order	Scheme	$\hat{\alpha}_R$ Error (%)	$\hat{\alpha}_I$ Error (%)
2	CD	1.41	0.89
4	CD	0.19	0.21
1	UP	48.16	–
2	UP	9.61	0.58
3	UP	6.30	0.02
4	UP	0.41	0.26
5	UP	0.26	0.24
Hybrid	CD/UP	0.29	0.23
5 (fine grid)	UP	0.02	0.16

A.4.2 Influence of Spatial Scheme's Order

The effects of the spatial discretization scheme order and nature on $\hat{\alpha}_R$ and $\hat{\alpha}_I$ are summarized in Table A.2. Also, Fig. A.3 plots the variation of v' (vertical velocity fluctuation) with X along the channel centre line. In this figure the $N_{UP} = 1$, 2, and 5, and $N_{CD} = 2$ and 4 results are presented. To avoid graph overcrowding the $N_{UP} = 3$ and 4 results are not included. Like Table A.2, Fig. A.3 shows that the choice of spatial differencing scheme has a strong solution influence. Clearly, and as would be expected, the first order upwind scheme result is very poor. The $N_{UP} = 2$ prediction is much better but by $X \approx 45$ there is a clear amplitude error. At $X = 45$ the $N_{CD} = 4$ has the lowest amplitude error relative to the UP schemes. Away from the channel inlet the $N_{CD} = 2$ has clear amplitude and phase errors. These traits are reflected in Table A.2. The hybrid result entry, in this table, uses the $N_{UP} = 5$ scheme in the streamwise direction and 4th order CD (blended with 25 % 5th order UP) in the cross stream. This scheme (reminiscent of that used by Mahesh et al. 1997) is intended for when there are shocks—the high order upwind being applied/directed, as much as possible, across the shock. For complex shock features this could require careful grid design. The final Table A.2 entry is for the fine grid. Note that there is no Table A.2 entry for $\hat{\alpha}_I$ with $N_{UP} = 1$. The lack of flow structure for this scheme prevented this parameter being defined.

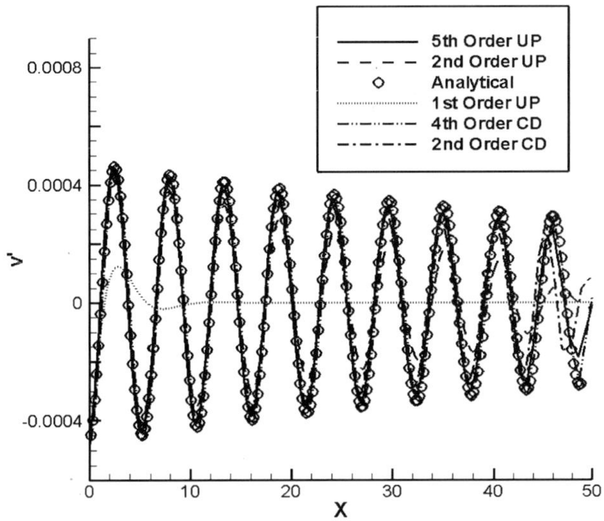

Fig. A.3 Plot of v' against X for different orders and nature of spatial differencing schemes

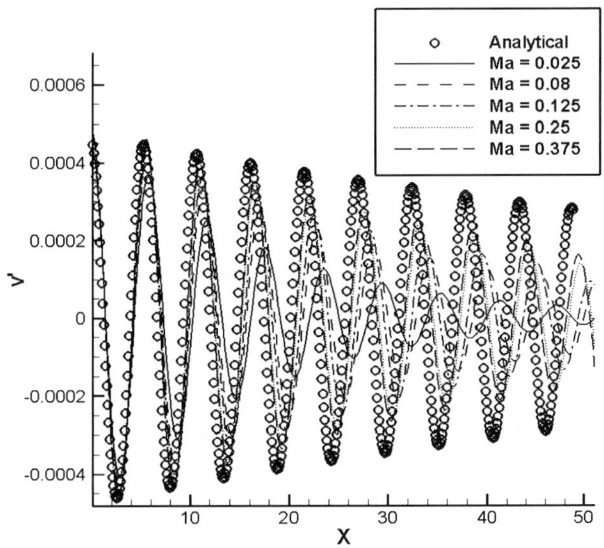

Fig. A.4 Variation of v' against X for different M with $\varepsilon_1 = 0.5$

A.4.3 Mach Number Influences with Classical Compressible Flow Solver

Next we consider performance for a classical high-speed compressible flow solver where there are clear Mach number influences. Figure A.4 plots v' against X for different M. The sudden amplitude decay at low M clearly reflects the well know fact that at low M, schemes related to, for example, flux difference splitting (Roe 1981) appear to become too diffusive. This is because a large disparity in Eigen

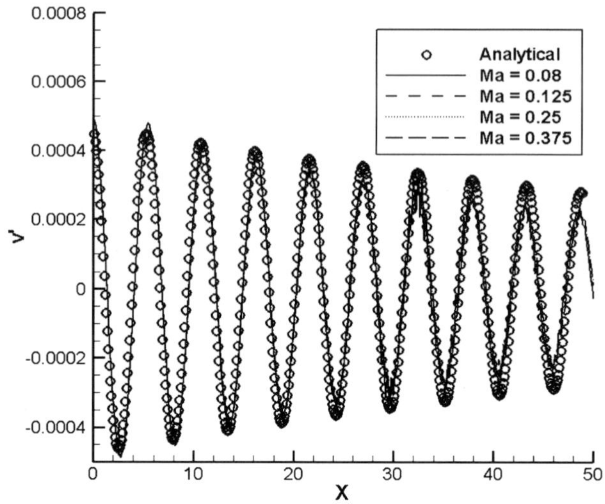

Fig. A.5 Variation of v' against X for different M with $\varepsilon_1 = 0$

values occurs. This results in high levels of diffusion in the equation that furnishes the pressure field.

Figure A.5 shows the variation of v' against X for $M = 0.08$, 0.125, 0.25 and 0.375 with $\varepsilon_1 \to 0$. Setting the latter parameter to zero makes the spatial discretization pure 2nd order CD. This eliminates the scaling issues relating to the matrix of Eigen values. Hence, results for all M are now the same. The flow Reynolds number is sufficiently low that viscous forces keep the equation system stable and free from pointwise oscillations.

Figure A.6 illustrates the dependence of decay ($\hat{\alpha}_R$) and phase ($\hat{\alpha}_I$) errors on both ε_1 and M. Frame (a) plots the (energy) decay error against ε_1 for different M. Frame (b) repeats (a) but this time for phase error. Clearly, as would be expected at higher M there is less sensitivity to theε_1 value. Frame (a) shows that the decay error decreases with ε_1. Frame (b) suggests for phase error, a minimum ε_1 value of about 0.1 is needed. Below this value, presumably due to dispersion effects, arising from using predominantly 2nd order central differencing, the phase error grows with decreasing ε_1. Hence, clearly when using the numerical approaches embodied in compressible flows, for low M DNS/LES care is required. It should also be noted that near solid surfaces $M \to 0$ and so the smoothing terms are expected to play an extreme role. However, for the cut-on acoustic wave the strong $M - \varepsilon_1$ link is lost. This is discussed next.

Through plotting v' against X, Fig. A.7 shows the decay of a cut-on acoustic wave in the density based solver. The same number of grid points per wavelength as for the T-S wave studies are used. The full line represents the analytical solution and the dashed the prediction. For this case, despite $M = 0.003$ and $\varepsilon = 0.5$, the physics of this acoustic wave case results in little apparent dissipation. Hence, as noted earlier, this case appears much less challenging than the T-S.

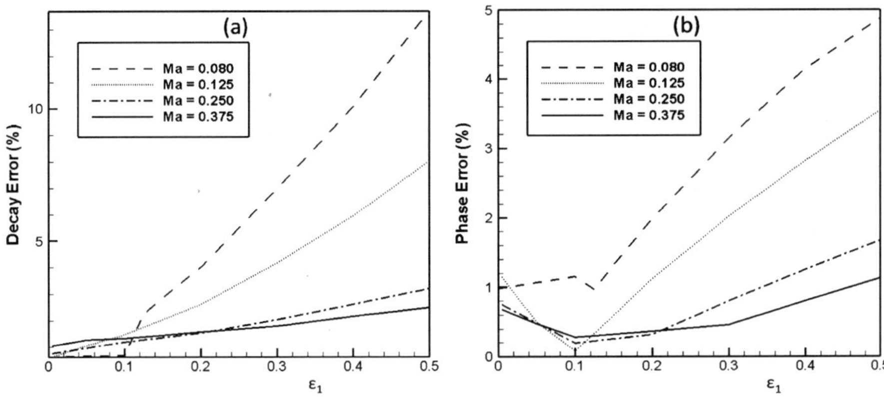

Fig. A.6 Dependence of energy decay ($\hat{\alpha}_R$) and phase errors ($\hat{\alpha}_I$) on ε_1 and M: (**a**) plot of decay error against ε for different M and (**b**) plot of phase error against ε for different M

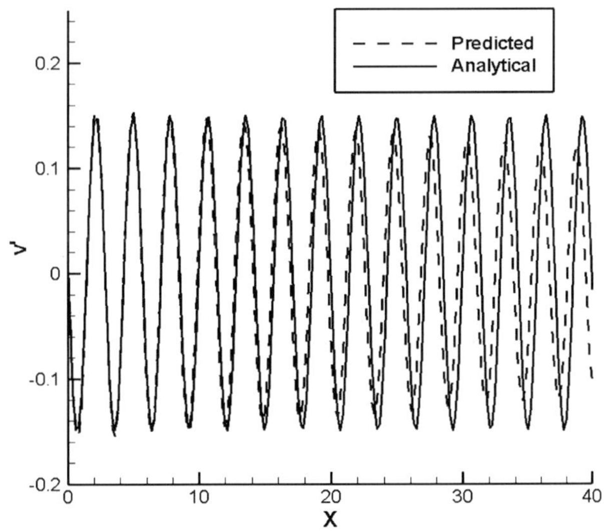

Fig. A.7 Decay of cut-on acoustic wave in density based flow solver for $M = 0.003$ and $\varepsilon_1 = 0.5$

Taking into account both the amplitude and phase errors Table A.3 gives the (best obtainable i.e. stable solutions with the lowest possible value of ϵ_1—where relevant) T-S wave results for the different codes.

A.5 Grids with More Complex Topologies

Clearly, many practical engineering systems exhibit complex geometries. The most realistic options for dealing with such systems is either through use of overset curvilinear grids, unstructured grids or for highly complex problems a combination of

Table A.3 Best obtainable results for the different codes

Code	'α_R' Error (%)	'α_I' Error (%)
A	1.89	1.23
B	0.72	0.27
C	1.17	0.19
D	0.11	0.17

the two. The former is mostly considered in this section with respect to predicting the T-S wave flow. Performance studies for unstructured grids are mostly given in Chap. 2.

A.5.1 Overset Grids

Figure A.8 shows over-set Code (A) results when a distorted grid is embedded in a background rectangular cell grid. Frame (a) shows the grid local to the wavy embedded section. Frame (b) gives the disturbance stream function in the zone of the embedded grid. To allow the grid structure to more easily be seen, the vertical coordinate is stretched by 400 %. This scaling does make the grid distortion in the embedded zone look much bigger than it is in reality. The disturbance streamfunction is well captured. However, in relation to v', the disturbance to the mean flow caused by insertion of the wavy grid is significant, hence $v \neq 0$. Therefore, it proved necessary to subtract the numerically computed mean flow component from the instantaneous to gain the pleasing plot. Hence, it is clear that the overset grid has introduced a numerical disturbance. This in a DNS/LES context would be unhelpful.

To assess the influence of grid metrics terms (associated with curvilinear coordinate transformation) on the above simulation an additional study has been performed. Figure A.9 shows the grid geometry. It consists of a small-distorted block (Block 2) embedded into the larger channel grid (Block 1). The distortion covers the full height of the channel and is geometrically altered so that it is zero at the walls. In contrast to the previous simulation, Block 1 is fully solved behind Block 2 and the main solution in Block 1 is unaffected by Block 2. The aim here is to examine the error of grid distortion relative to the interface error, hence for all cases here the interface points remain in identical positions and aren't point matched. A T-S wave decays/evolves in Block 1 and then interacts with the wavy inner block. The inner block is distorted with a sine wave with maximum amplitudes (A_O) of 0.2, 0.1, 0.05 and 0.0 (no distortion). These values are dimensionless with respect to h. Figure A.10 shows the T-S wave for different distortion amplitudes. Clearly the larger A_O shows the larger error. It can be seen that the error associated with the grid transformation at large distortions is much more significant that of any interface error. Figure A.11 shows the same results but this time with the mean flow component subtracted from the instantaneous field. The resulting plots are far more pleasing.

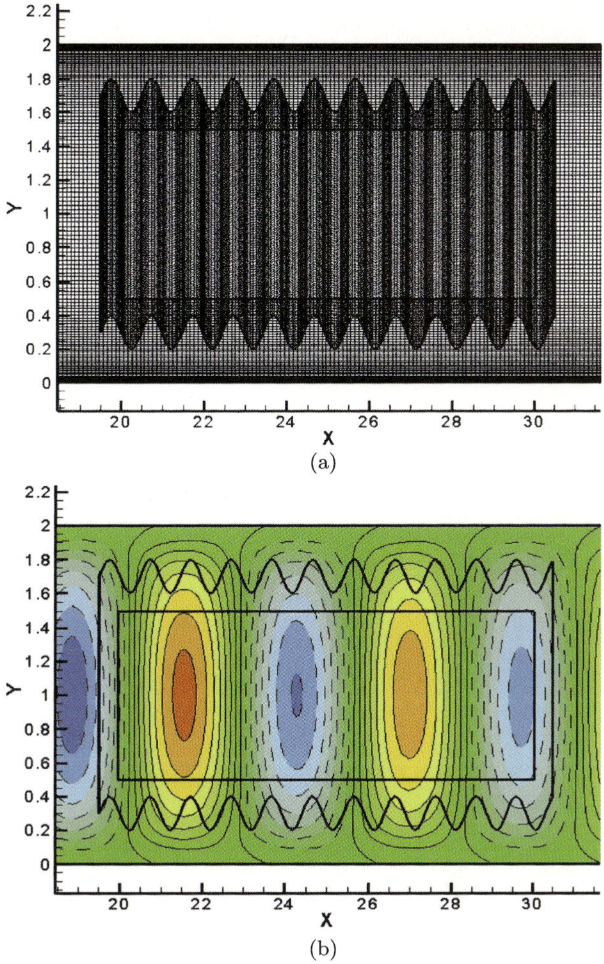

Fig. A.8 Use of overset grid for T-S wave: (**a**) wavy grid overset in centre of channel and (**b**) disturbance stream function in overset grid region

A test is now considered for an undistorted coarse grid sandwiched between two fine grids. The grid is shown in Fig. A.12. Block 2 is half the resolution of blocks 1 and 3 in both the x and y directions.

Figure A.13 shows the resulting wave for the coarse inner block grid as described above. This T-S wave is compared with a fine single block grid of the same resolution as blocks 1 and 3, and a coarse single block grid with the same resolution as Block 2. The wave amplitudes are examined just downstream of Block 2 ($x \approx 30$). As would be expected the fine single block grid gives the better result. This is followed by the Fig. A.12 grid. The single block coarse grid gives the worst result. In the Fig. A.12 embedded grid, the error has increased owing to the coarse section of mesh. However the error here is not as great as that of the single block coarse grid. Therefore, the solution still retains some of the benefits of having the finer mesh upstream in the channel.

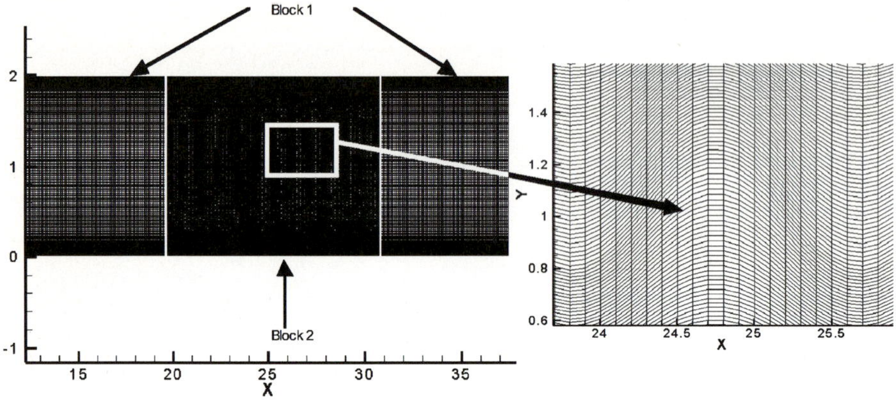

Fig. A.9 Overset grid with $A_O = 0.2$

Fig. A.10 Variation of v' against X for different A_O, using Code (A)

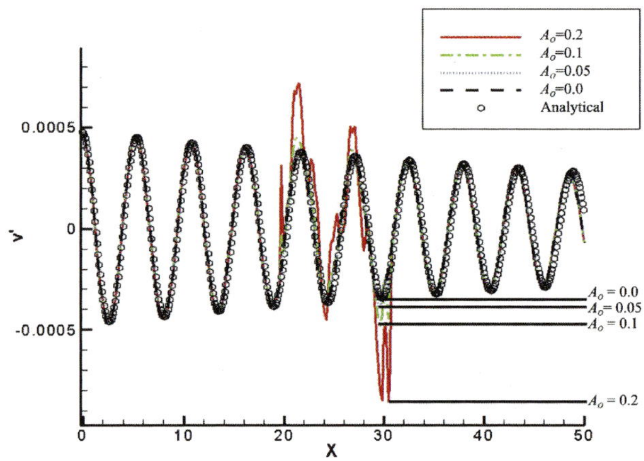

Fig. A.11 Variation of v' against X for different A_O, using Code (A) with mean flow component subtracted from instantaneous

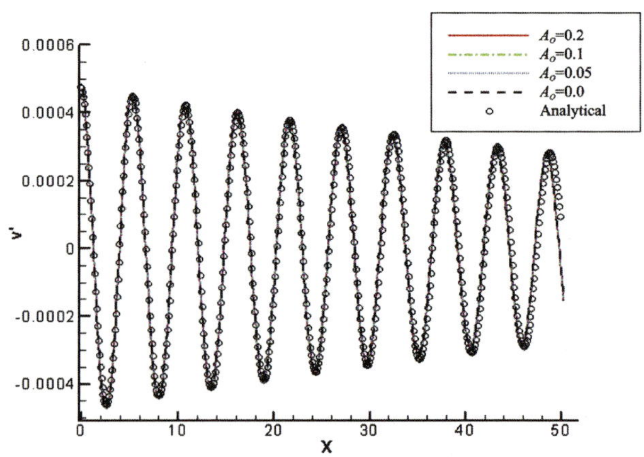

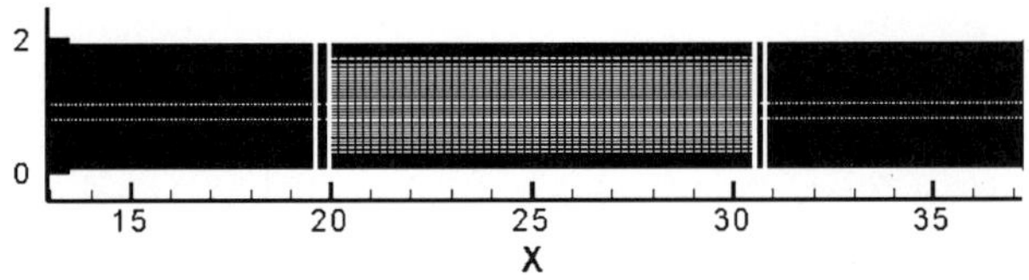

Fig. A.12 Grid with coarse inner block

A.5.2 Unstructured Grids

Figure A.14 gives the impact of different unstructured grid topologies for the T-S waves. Results are also presented at the end of Sect. 2.22. However, the results shown there are for a cell-vertex based unstructured solver—Code (C). The current are for the cell centered Code (D). The results a consistent between the two unstructured codes and the strong impact of grid is clear.

A.6 Conclusions

The propagation of subcritical T-S, inviscid vortical and cut-on acoustic waves was explored. The numerical performances of the very different codes were studied. The unstructured codes did not seem greatly challenged by the inviscid waves. However, for the T-S, where the delicate balance between the convective and diffusive influences is important to characterize, more substantial numerical influence was observed. With respect to numerical order, the 5th order backwards difference time scheme was found less accurate than the 2nd order centered Crank-

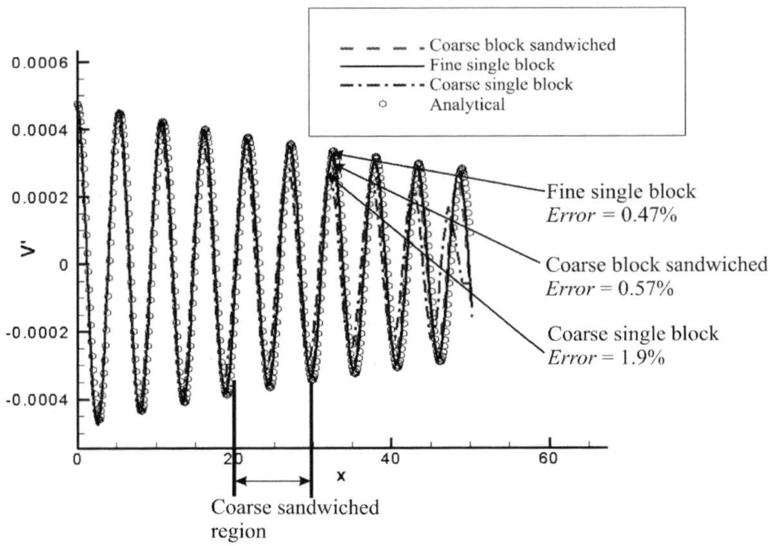

Fig. A.13 Variation of v' against X for different grids, using Code (A)

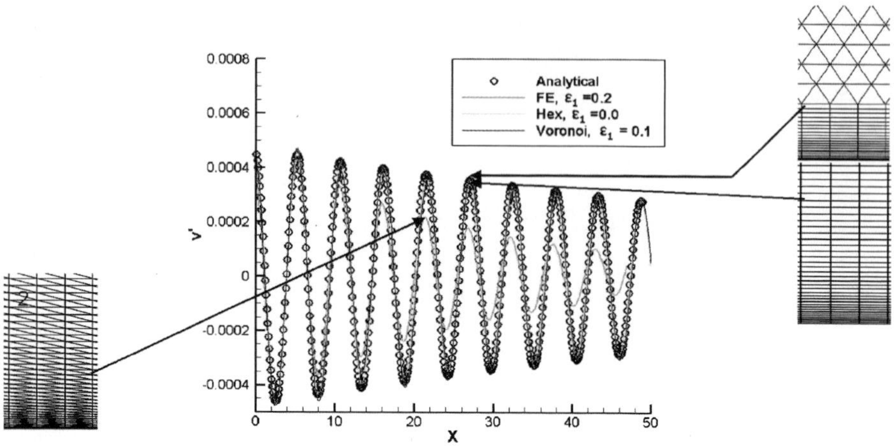

Fig. A.14 Plot of T-S wave decay with streamwise distance for different grid topologies with cell centered unstructured Code (D)

Nicolson scheme. Similarly, in a spatial sense, low order (3rd) UP schemes gave substantial errors, the 4th order centered scheme showing good performance. With shocks in mind, hybridization such as having high order upwind in the streamwise (high Peclet number direction) and centered differencing cross-stream did not seem especially helpful in terms of accuracy. The helpful contribution of the centered scheme appeared much diluted. Results strongly suggested that for accurate T-S wave prediction grids where there is orthogonality between control volume faces and the nodes that straddle them are desirable. This implies the use of hexahedral/rectangular cells or quasi-Voronoi grids. However, the latter are problematic for boundary layer flows where elongated cells can ideally be used to resolve the streak like structures. Overset grids are found to introduce mean flow disturbances comparable to those of the T-S wave motion. However, this effect can also occur with unstructured grids. For DNS/LES simulations this is clearly undesirable.

References

M.B. Giles, Non-reflecting boundary conditions for Euler equation calculations. *AIAA J.* **28**(12), 2050–2058 (1990)

R. Jefferson-Loveday, Wall resolved LES and zonal LES of round jet impingement heat transfer on a flat plate. *Numer. Heat Transf.* **59**(Part B), 190–208 (2008)

K. Mahesh, S.K. Lele, P. Moin, The influence of entropy fluctuations on the interaction of turbulence with a shock wave. *J. Fluid Mech.* **334**, 353–379 (1997)

P. Moinier, Algorithm developments for an unstructured viscous flow solver. PhD thesis, University of Oxford (1999)

S.A. Orszag, Accurate solution of the Orr-Sommerfeld stability equation. *J. Fluid Mech.* **50**, 689–703 (1971)

L.R. Pauley, P. Moin, W.C. Reynolds, The structure of two-dimensional separation. *J. Fluid Mech.* **220**, 397–411 (1990)

P. Roe, Approximate Riemann solvers, parameter vectors and difference schemes. *J. Comput. Phys.* **43**, 357–372 (1981)

P.G. Tucker, *Computation of Unsteady Internal Flows* (Kluwer Academic, Norwell, 2001)

H. Xia, Dynamic grid detach-eddy simulation for synthetic jet flows. PhD thesis, University of Sheffield (2005)

Appendix B
RANS Models

In this section the nature of RANS models is briefly outlined. Appreciation of their traits is critical to understanding the performances of URANS and also hybrid LES-RANS approach. Clearly in both, many of the RANS base model defects will be inherited.

B.1 The Full Reynolds Stress Equations

The Reynolds stress equations can be symbolically expressed as below.

$$\frac{D\overline{u'_i u'_j}}{Dt} = P_{ij} + D_{ij} + R_{ij} - \varepsilon_{ij} \tag{B.1}$$

The left hand side is the convective term. The right hand side terms P_{ij}, D_{ij}, R_{ij} and ε_{ij} correspond to the production, diffusion, redistribution and dissipation terms, respectively. When fully expanded Eq. (B.1) involve hundreds of terms. This production term can be *exactly* expressed as

$$P_{ij} = -\overline{u'_i u'_k}\frac{\partial U_j}{\partial x_k} - \overline{u'_j u'_k}\frac{\partial U_i}{\partial x_k} \tag{B.2}$$

Since, it is exact, it is a useful equation for illustrating physical processes and seeing limitations of simpler turbulence models. Such models are generally based on the Boussinesq approximation

$$-\rho\overline{u'_i u'_j} = \mu_t\left(\frac{\partial U_i}{\partial x_j} - \frac{\partial U_j}{\partial x_k}\right) - \frac{1}{3}\delta_{ij}\overline{u'_k u'_k} \tag{B.3}$$

Where δ_{ij} is the Kronecker delta. Equations (B.3) and (B.1) can be viewed as stress-strain relationships and (like the production terms) it can be useful to contrast these full and reduced models. In this way it can be better appreciated how simpler models tend to fail for challenging flow features including, for example, impingement, separation, high streamline curvature, system rotation and buoyancy.

P.G. Tucker, *Unsteady Computational Fluid Dynamics in Aeronautics*,
Fluid Mechanics and Its Applications 104, DOI 10.1007/978-94-007-7049-2,

Simpler models tend to give:

1. erroneous predicted shear for curved shear layers and adverse pressure gradients;
2. separation suppression on curved surfaces;
3. excessive turbulence in stagnation zones;
4. the wrong behaviour for rotating flows;
5. insensitivity to density gradients;
6. and excessive heat transfer at reattachments points (Leschziner 2000).

They tend to work well for simple 2D shear flows. For such flows the Reynolds stress equations reduce to

$$\frac{D\overline{u'v'}}{Dt} = -\overline{v'v'}\frac{\partial U}{\partial y} + \overline{\frac{p'}{\rho}\left(\frac{\partial u'}{\partial y} + \frac{\partial v'}{\partial x}\right)} - \frac{\partial}{\partial y}\left(\overline{u'v'v'} + \frac{\overline{p'u'}}{\rho}\right) - \varepsilon_{12} \tag{B.4}$$

So taking just the first right hand side term (involving the gradient of a mean and not fluctuating properties) after multiplying through by a timescale $T = Dt$ gives

$$-\rho\overline{u'v'} = -\rho\overline{v'v'}T\frac{\partial U}{\partial y} \tag{B.5}$$

On dimensional grounds $\mu_t \propto \rho \times \text{Length} \times \text{Velocity} \propto \rho\overline{v'v'}T$. Hence, Eq. (B.5) can effectively be stated as

$$-\rho\overline{u'v'}=\mu_t\frac{\partial U}{\partial y} \tag{B.6}$$

This seems in accord with the Boussinesq approximation. Hence, the success of eddy viscosity models for simple shear flows, where shear stress is the only important component.

B.1.1 Compressive/Extensive Strain Problem

Most eddy viscosity based RANS models are calibrated for flows dominated by shear (strains). If compressive or intensive strains dominate serious inaccuracies can arise. From the exact production term for turbulence (Eq. (B.2)) the following kinetic energy production term can be derived for a 2D flow.

$$P_{k,Exact} = -\overline{u'u'}\frac{\partial U}{\partial x} - \overline{v'v'}\frac{\partial V}{\partial y} - \overline{u'v'}S \tag{B.7}$$

Where $S = \partial U/\partial y + \partial V/\partial x$. The substitution of the Boussinesq approximation into Eq. (B.7) gives

$$P_k = \rho\mu_t S^2 \tag{B.8}$$

In Eq. (B.7) the $\partial U/\partial x$ and $\partial V/\partial y$ terms, which arise in the continuity equation, tend to cancel each other. If the former is negative and $\overline{u'u'} > \overline{v'v'}$, $P_{k,Exact}$ will be negative. This can easily happen, but Eq. (B.8) cannot model this. Equation (B.8) quadratic S function means that in stagnation zones turbulence can be massively over-estimated. This problem has given rise to the use of Kato and Launder (1993) and Yap (1987) corrections. Some eddy viscosity models avoid these by modifying C_μ (a key turbulence model constant). Equation (B.8) defects cause the suppression of leading edge separation on blades/aerofoils and excessive predicted heat transfer in this zone. The Yap correction will also prevent the prediction of excessive heat transfer at re-attachment points.

B.1.2 Curvature Problem

The exact production term for a two-dimensional shear flow is

$$P_{12,Exact} = -\overline{v'v'}\frac{\partial U}{\partial y} - \overline{u'u'}\frac{\partial V}{\partial x} \tag{B.9}$$

If there is no streamline curvature the first shear term is greatly dominant. In a flat plate boundary layer the maximum u' is over double that of v'. For a convex curved surface (or concave) as found on aerofoils/blades the $\partial V/\partial x$ term comes more strongly into play, i.e. an extra rate of strain comes into play. This, combined with the dominance of $\overline{u'u'}$ and other factors (Leschziner 2000) makes the last group of terms of some importance. For a convex surface $\partial V/\partial x$ is negative. Hence, shear stress production and shear stress is damped. Indeed experiments show turbulence can be virtually eliminated around convex surfaces (Sloan et al. 1986). A simple linear eddy viscosity model is unable to model this subtle interplay of terms involving Reynolds stress anisotropy. It must be sensitized in some ad hoc fashion using, for example the Richardson correction where essentially

$$Ri = \frac{U_\theta}{R}\frac{\partial U_\theta}{\partial r} \tag{B.10}$$

and R is the surface radius, and U_θ the streamwise velocity parallel to the surface. Hence, if $l = l(1 - C\,Ri)$ since $\mu_t \propto$ Length $\times$ Velocity, on the suction side of a blade, the correction will reduce l and hence μ_t and thus the predicted shear. The 'Richardson Number' correction for curvature are derived based on analogy with body forces such as those occurring when there is rotation or density gradients. Fluid particles when moving along a curved streamline could be viewed as experiencing a body force. With the Richardson analogy the extra strains experienced when a shear layer is curved are accounted for by viewing them as a centrifugal body force. The Richardson number can be expressed in gradient, flux, stress and other forms (Sloan et al. 1986). With the k–ε model (see Sect. B.2.3) the Richardson correction can be incorporated in the source term of the ε equation. However, it could also be used to scale C_μ. This would mean the flow more immediately responds to curvature.

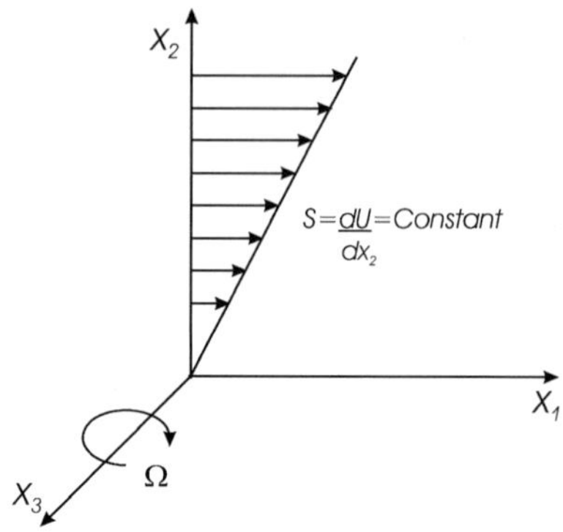

Fig. B.1 Turbulence modification in rotating flows

Other curvature corrections exist involving, for example, modification of the k equation or 'preferential dissipation' modification of the ε equation. The latter is based on the idea that normal stresses promote dissipation more than shear stresses. Also, C_μ can be modified based on contractions of more advanced models such as algebraic stress models. Such corrections frequently improve predictions but can also give worse results.

B.1.3 Body Force Problem

Many flows often involve rotation/swirl. Blade rows rotate and hence any cooling passages inside. In combustors swirl is used to enhance mixing. Aircraft wings have tip vortices. System rotation gives rise to body forces that interact with the turbulence. The nature of the interaction is controlled by mean strain. The radial pressure gradient in a rotating fluid is governed by the equation below.

$$\frac{\partial P}{\partial r} = \frac{\rho U_\theta^2}{r} = \frac{\rho(\Omega + U_\theta')^2}{r} \tag{B.11}$$

where U_θ' is the velocity relative to a system rotating at Ω. If U_θ' radially increases (or more precisely angular momentum $d(rU_\theta)/(rdr) > 0$) pressure increases with radius. Hence if a fluid particle is displaced radially outwards it will experience a radially inwards pressure restoring force. The centrifugal force resists inward movements, the centrifugal force of the inward moving particle being greater than that of displaced particle. Now if U_θ' decreases with radius, a radially displaced particle will have a greater centrifugal force than the fluid it is surrounded by and the local pressure will be insufficient to restore its location. Hence, as shown in Fig. B.1 (relating to a rotating channel), at the outer radius, channel wall, the flow is unstable and at the inner it is stable. To sensitize a turbulence model to this type of behaviour some form of Rossby number is needed of the form

$$Ro = \frac{\Omega}{\partial U/\partial y} \tag{B.12}$$

Or more precisely a gradient based Richardson number (Salhi and Cambon 1997)

$$Ri = Ro(1 + Ro) \quad \text{(B.13)}$$

where

$$Ro = -\frac{2\Omega}{S} \quad \text{(B.14)}$$

For $Ri < 0$ the flow is stable and for $Ri > 0$ unstable. $Ri = 0$ corresponds to $\Omega = 0$ or $2\Omega - S = 0$ (zero absolute vorticity). There are numerous forms of Richardson number along with other forms of correction for the influence of rotation on turbulence. The number of corrections reflects their somewhat patchy performance and anomalous behaviour.

The effects of local flow curvature can have an analogous influence to rotation. Under these circumstances the rotation corrected version of SA-SARC (Spalart and Shur 1997) could be helpful.

When there are substantial density gradients within the flow large scale mixing can be introduced through buoyancy forces. If $\partial T/\partial r < 0$ there will be higher density fluid at the inner radius and lower density at the outer. Then the situation is akin to the classic Rayleigh-Bernard instability, where cold fluid resting on top of hot, with the gravity vector pointing from cold to hot, creates chaotic mixing. For $\partial T/\partial r > 0$ there is stability. Although the full RSM captures the influences of the rotational body forces (hence the physical processes discussed could be elucidated by looking at the terms in this model), the RSM could never account for the consequences of the Rayleigh-Bernard type instabilities noted (Hanjalic 2002). This buoyancy driven instability physics is of key importance for the flow in, for example, aero engine high pressure compressor drums. As noted in Chap. 5, for these, LES type modelling is necessary. Although past efforts have been made to account for the flow in these cavities, through modified RANS models, this approach is notionally flawed.

B.1.4 Flow Acceleration Problem

The flow over aerofoils, within labyrinth type seals, jet nozzles and intakes can undergo substantial acceleration. Acceleration thickens the viscous sublayer. Most popular RANS models are totally insensitive to flow acceleration. Launder and Jones (1969) proposed a provisional linear acceleration parameter against viscous sublayer relationship (based on a sink flow and ensuring C_f remains constant as happens with sink flows). With extreme care, it is possible to use such viscous sublayer scaling in, more advanced turbulence model.

Next the broad nature of different turbulence models is discussed.

B.2 Turbulence Models

B.2.1 Reynolds Stress Model

The Reynolds stress equations form the basis for the Reynolds Stress Model (RSM) for turbulence. The foregoing discussion has to an extent patterned RSM terms to inform us on the performance of simpler models. This might suggest that if an RSM is used it is virtually always expected to give improved agreement. However, this is far from the case and such models contain substantial empiricism. Also, the RSM computational cost is extreme. Like other advanced models, that contain numerous gradient terms, the grid demands (to resolve the gradients) increases. For flows with separation or buoyancy, giving rise to large scale unsteady structures, the RSM, like all other RANS models will fail. Even for attached flows the RSM model performance becomes uncertain near walls. A key problem is modelling dissipation. Near walls the turbulence structures become highly anisotropic and hence so does the dissipation. Also, there are other issues relating to the crude nature of the raw dissipation equation itself which gives rise to many ad hoc potential modifications. The pressure-strain term R_{ij}, in Eq. (B.1) is also a problem. This term controls the redistribution of turbulence energy among the normal stresses. It drives turbulence towards isotropy and reduces shear stresses.

In shear flows, the shear stress is linked to the cross-stream normal stress. Hence, normal stress strongly controls shear. The cross-stream stress is not directly generated in a simple shear flow but arises through R_{ij}. There are a wide range of options for modelling R_{ij} and also the modelling of near wall dissipation. As one might expect, the spawning of a wide range of options suggests a weakness and unresolved issues. These problems have resulted in zonal RANS models, where the RSM is replaced by simpler eddy viscosity models near walls.

B.2.2 Non-linear Eddy Viscosity Models

The extreme computational cost of the RSM but clear weaknesses of simple linear RANS eddy viscosity models, has motivated the quest for some sort of 'half-way-house'. This has given rise ultimately to non-linear eddy viscosity models. Essentially these can all be viewed as supplying extended forms of the Boussinesq approximation with the addition of quadratic (Speziale 1987) and (Gatski and Speziale 1993) and cubic terms (Craft et al. 1996). The model of Speziale (1987) is based on the constitutive relationship for a non-Newtonian fluid (a Rivlin-Erikson fluid). The Gatski and Speziale equation, called the Explicit Algebraic Stress Model (EASM) is derived from an algebraic RSM. The way in which non-linear stress strain relationships can be derived differs greatly but ultimately give rise to expansions involving strain and vorticity tensors. The quadratic terms allow anisotropy to be modelled and cubic the consequences of streamline curvature.

Generally a linear RANS model has a constant value of $C_\mu = 0.09$. The non-linear models can involve variable C_μ formulations based on S and Ω. These will helpfully avoid excessive turbulence prediction at stagnation points. This would give rise to delayed transition (the actual transition model is also important) on the suction side of a turbine blade.

Despite being often marketed as a means of getting RSM type performance at linear eddy viscosity model cost, practice shows this to be far from the case. Certain non-linear model solutions can be nearly as expensive to secure as performing an LES. Also, results can be worse than for linear models, the added non-linearity giving the potential for increased results deviation.

B.2.3 Basic RANS Models

RANS models can range from being highly phenomenological or more theoretical, but even the latter (RSM) contain substantial empirical elements (Spalart 2000).

They can also be classified by the number of differential equations they involve. The mixing length model has no differential equations—and so it is called a 'zero equation model'. The k–ε model involves two substantial differential equations and hence is called a two equation model.

Also the term *realizability* is used. This refers to whether the model is constrained so that it does not break basic physical principles. For example, a realizable model would ensure all the normal Reynolds stresses remain positive along with turbulence kinetic energy.

There are perhaps hundreds of basic RANS models. This, in itself, is indicative of their performance variability and the urgent quest for something better.

B.2.4 Zero Equation Models

Based on dimensional grounds

$$\mu_t \propto \text{Length} \times \text{Velocity} \tag{B.15}$$

With the mixing length essentially we say the length scale $l = \kappa y$ (y is normal wall distance) and

$$\text{Velocity} = l|S| \tag{B.16}$$

The mixing length model is referred to as incomplete. The equation for l takes different forms for different regions of the flow. In the viscous sublayer ($y^+ < 5$)

$$l = \kappa y D \tag{B.17}$$

where D is some form of damping function. The van Driest, is the most well known where $D = 1 - \exp(-y^+/26)$ and y^+ is a form of Reynolds number (It can also be viewed as a dimensionless wall distance). There are numerous forms of D (see, for example, Launder and Priddin 1973). In the outer part of the boundary layer $l = C\delta$ where typically $C = 0.09$. For different free shear flows and wake zones other formulations must be used. For example in a:

mixing layer	$l \approx 0.07 \times$ the layer width;
round jet	$l \approx 0.075 \times$ the jet half width;
plane jet	$l \approx 0.09 \times$ jet half width;
wake	$l \approx 0.16 \times$ the wake half width

and so on. As can be seen, the modeller has numerous potential l choices and these, in themselves, can be subjected to further corrections—for example the Richardson and Rossby number based corrections for curvature and rotation noted earlier.

B.2.5 One Equation Turbulence Models

Zero equation models assume the turbulence is in some form of equilibrium state. They assume that the rate of production and dissipation of turbulence at a particular point are equal. However, this is not always the case. High levels of turbulence can arise at a particular point in space as a result of being generated in another zone and transported by the fluid motion to that point. Hence the modelling of convective processes can be important. All one-equation models essentially solve an equation of the form

$$\frac{D\phi}{Dt} = \Gamma \nabla^2 \phi + S_\phi \tag{B.18}$$

where ϕ is the property transported by the convective and diffusive processes. Γ is a diffusion coefficient and the source term S_ϕ can involve a turbulence production term or, for some models, both production and destruction. The most popular single equation model is the Spalart and Allmaras (1994)—SA model. Essentially with this $\mu_t \propto \phi$. Secundov et al. (2001) and Baldwin and Barth (1990) models also solve a convection-diffusion transport equation essentially for μ_t. Another well known one equation model is the *k*–*l* model (see, for example, Wolfshtein 1969). In this model $\phi = k$ (the turbulence kinetic energy) and l is a mixing length type expression. These one equation models are again relatively forgiving with respect to grid. For this reason the use of the *k*–*l* model is frequently restricted to the near wall region. Hence, it is a component in a zonal model with, say, a two equation model being used away from walls (Liu et al. 2006). Indeed, sometimes zero equation models are used near walls with two equation models being used away from walls. Internal air systems predictions using this strategy can be found in Iacovides and Chew (1993), and Iacovides and Theofanopoulos (1991).

B.2.6 Two Equation Turbulence Models

There are a vast number of two-equation turbulence model variants. Most are based on solving convection-diffusion transport equations for both k and ε (the rate of dissipation of turbulence kinetic energy) the eddy viscosity being computed as

$$\mu_t = \rho C_\mu \frac{k^2}{\varepsilon} \tag{B.19}$$

However, a model by Secundov et al. (2001) solves transport equations for μ_t and l (turbulence length scale). Kim and Chung (2001) solve for μ_t and k. A popular alternative to solving for k and ε is to solve for k and ω (Wilcox 1988) where

$$\mu_t = \rho C_\mu \frac{k}{\omega} \tag{B.20}$$

Hence, the turbulence velocity scale $u = k^{1/2}$ and length scale is proportional to $k^{1/2}/\omega$, ω being a turbulence frequency. Also, $\varepsilon = \omega k$ (Warner et al. 2005) solve equations for k and kl. The k–ω model shows excessive sensitivity to the specified free stream turbulence intensity level. This has given rise to the zonal Menter (1993) model.

B.3 Transition Modelling

Transition is a complex process. It can be influenced by surface roughness, acoustic disturbances and external disturbances, pressure gradient, freestream velocity change, surface curvature, temperature gradients and rotation. The routes to transition are numerous (Coull and Hodson 2011). Some turbulence models can predict the bypass transition process without recourse to special modifications.

The standard procedure with regard to transition is to estimate the transition point using, for example, experimental correlations and then forcing the turbulence model to trip through, for example a crude function multiplying eddy viscosity or more elegant procedures as used with the SA model. Transition zones are characterised by high levels of intermittent unsteadiness. This has given rise to approaches that solve a transport equation for intermittency (the fraction of time that the flow is notionally turbulent). Such approaches are described by Akhter and Ken-Ichi (2007). Even though a transport equation for intermittency is solved the onset of transition location still needs to be specified. For this correlations need to be used (see for example Abu-Ghannam and Shaw 1980 and Arnal 1992). Again there is a wide range of literature and models in this active research area. There do not appear to be currently any methods that can reliably predict transition in the presence of strong pressure gradients and high levels of turbulence, not to mention unsteady wakes. For key, state of the art turbulence transition modelling approaches see, for example, Menter et al. (2006) and Walters and Cokliat (2008).

References

B.J. Abu-Ghannam, R. Shaw, Natural transition of boundary layer the effect of turbulent pressure gradient and flow history. *J. Mech. Eng. Sci.* **22**(5), 213–228 (1980)

M.N. Akhter, F. Ken-Ichi, Development of prediction method of boundary layer bypass transition using intermittency transport equation. *Int. J. Gas Turbine Propuls. Power Syst.* **1**(1), 30–37 (2007)

D. Arnal, Boundary layer transition: prediction, application to drag reduction. Von Karmen Institute of Fluid Dynamics Lecture Series (1992)

B.S. Baldwin, T.J. Barth, A one-equation turbulence transport model for high Reynolds number wall-bounded flows. Technical Report TM 102847, NASA (1990)

J. Coull, H. Hodson, Unsteady boundary-layer transition in low pressure turbines. *J. Fluid Mech.* **681**, 370–410 (2011)

T.J. Craft, B.E. Launder, K. Suga, Development and application of a cubic eddy-viscosity model of turbulence. *Int. J. Heat Fluid Flow* **17**(2), 108–115 (1996)

T.B. Gatski, C.G. Speziale, On explicit algebraic models for complex turbulent flows. *J. Fluid Mech.* **254**, 59–78 (1993)

K. Hanjalic, One-point closure models for buoyancy driven turbulent flows. *Annu. Rev. Fluid Mech.* **34**, 321–347 (2002)

H. Iacovides, J.W. Chew, The computation of convective heat transfer in rotating cavities. *Int. J. Heat Fluid Flow* **14**(2), 146–154 (1993)

H. Iacovides, I.P. Theofanopoulos, Turbulence modelling of axisymmetric flow inside rotating cavities. *Int. J. Heat Fluid Flow* **12**(1), 2–11 (1991)

M. Kato, B.E. Launder, The modeling of turbulent flow around stationary and vibrating square cylinders, in *9th Symposium on Turbulent Shear Flows*, Kyoto, Japan (1993)

S.H. Kim, M.K. Chung, New vt-k model for calculation of wall-bounded turbulent flows. *AIAA J.* **39**(9), 1803–1805 (2001)

B.E. Launder, W.P. Jones, On the prediction of laminarisation. Technical Report No. 1036, Ministry of Technology, Aeronautical Research Council (1969)

B.E. Launder, C.H. Priddin, A comparison of some proposals for the mixing length near a wall. *Int. J. Heat Mass Transf.* **16**, 700–702 (1973)

M.A. Leschziner, Turbulence modelling for separated flows with anisotropy-revolving closures. *Philos. Trans. R. Soc., Math. Phys. Eng. Sci.* **358**(1777), 3247–3277 (2000)

Y. Liu, P. G. Tucker, G. Lo Iacono, Comparison of zonal RANS and LES for a non-isothermal ribbed channel flow. *Int. J. Heat Fluid Flow* **27**(3), 391–401 (2006)

F.R. Menter, Zonal two equation k–ω turbulence models for aerodynamic flows, in *AIAA*. AIAA-93-2906 (1993)

F.R. Menter, R.B. Langtry, S.R. Likki, Y.B. Suzen, P.G. Huang, S. Volker, A correlation-based transition model using local variables—Part I: model formulation. *J. Turbomach.* 128, 413–422 (2006)

A. Salhi, C. Cambon, An analysis of rotating shear flow using linear theory and DNS and LES results. *J. Fluid Mech.* **347**, 171–195 (1997)

N. Secundov, M.K. Strelets, A.K. Travin, Generalization of vt-92 turbulence model for shear-free and stagnation point flows. *J. Fluids Eng.* **123**, 111–115 (2001)

D.G. Sloan, P.J. Smith, L.D. Smoot, Modeling of swirl in turbulent flow systems. *Prog. Energy Combust. Sci.* **12**(3), 163–250 (1986)

P.R. Spalart, Strategies for turbulence modelling and simulations. *Int. J. Heat Fluid Flow* **21**, 252–263 (2000)

P.R. Spalart, M. Shur, On the sensitization of turbulence models to rotation and curvature. *Aerosp. Sci. Technol.* **1**(5), 297–302 (1997)

C.G. Speziale, On non-linear k–l and k–ϵ models of turbulence. *J. Fluid Mech.*, **178**, 459–475 (1987)

D.K. Walters, D. Cokliat, A three-equation eddy-viscosity model for Reynolds-averaged Navier-Stokes simulations of transitional flow. *J. Fluids Eng.* **130**, 121–401 (2008)

J.C. Warner, C.R. Sherwood, H.G. Arango, R.P. Signell, Performance of four turbulence closure models implemented using a generic length scale method. *Ocean Model.* **8**, 88–113 (2005)

D.C. Wilcox, Reassessment of the scale-determining equation for advanced turbulence models. *AIAA J.* **26**(11), 1299–1310 (1988)

K.M. Wolfshtein, The velocity and temperature distribution in one-dimensional flow with turbulence augmentation and pressure gradient. *Int. J. Heat Mass Transf.* **12**, 301–318 (1969)

C.R. Yap, Turbulent heat and momentum transfer in recirculating impinging flows. PhD thesis, University of Manchester, Department of Mechanical Engineering, Faculty of Technology (1987)

Index

P.G. Tucker, *Unsteady Computational Fluid Dynamics in Aeronautics*, Fluid Mechanics and Its Applications 104, DOI 10.1007/978-94-007-7049-2,

Printed by Printforce, the Netherlands